D0081118

ESSENTIALS OF MICRO- AND NANOFLUIDICS

This book introduces students to the basic physical principles needed to analyze fluid flow in micro- and nanosize devices. This is the first book to unify the thermal sciences with electrostatics, electrokinetics, and colloid science; electrochemistry; and molecular biology. The author discusses key concepts and principles, such as the essentials of viscous flows, electrochemistry, heat and mass transfer phenomena, elements of molecular and cell biology, and much more. This textbook presents state-of-the-art analytical and computational approaches to problems in all these areas, especially in electrokinetic flows, and gives examples of the use of these disciplines to design devices used for rapid molecular analysis, biochemical sensing, drug delivery, DNA analysis, the design of an artificial kidney, and other transport phenomena. This textbook includes exercise problems, modern examples of the applications of these sciences, and a solutions manual available to qualified instructors.

A. Terrence Conlisk is Professor of mechanical and aerospace engineering at The Ohio State University. He is an internationally recognized expert in the areas of micro- and nanofluidics, helicopter aerodynamics, and complex flows driven by vortices. He is the author of numerous publications and hundreds of technical presentations and seminars delivered throughout the world. After his PhD thesis (Purdue, 1978) on the prediction of the fluid dynamics and separation of isotopes in a gas centrifuge, he began his work on various aspects of the dynamics of two- and three-dimensional vortices, with a focus on helicopter aerodynamics. Since 1999, he has been involved in modeling ionic and biomolecular transport through micro- and nanochannels for the design of devices used for rapid molecular analysis, sensing, drug delivery, and other applications. Professor Conlisk's wide spectrum of research interests makes him uniquely qualified to write on the thoroughly interdisciplinary fields of micro- and nanofluidics.

ESSENTIALS OF MICRO- AND NANOFLUIDICS

With Applications to the Biological and Chemical Sciences

A. Terrence Conlisk

The Ohio State University

CAMBRIDGE UNIVERSITY PRESS
Cambridge, New York, Melbourne, Madrid, Cape Town,
Singapore, São Paulo, Delhi, Mexico City

Cambridge University Press
32 Avenue of the Americas, New York, NY 10013-2473, USA

www.cambridge.org
Information on this title: www.cambridge.org/9780521881685

First published 2013

Printed in the United States of America

A catalog record for this publication is available from the British Library.

Library of Congress Cataloging in Publication Data

Conlisk, A. Terrence, 1950–
Essentials of micro- and nanofluidics : with applications to the biological and chemical sciences /
A. Terrence Conlisk.
p. ; cm.
Includes bibliographical references and index.
ISBN 978-0-521-88168-5 (hardback)
I. Title.
[DNLM: 1. Thermodynamics. 2. Biomedical Technology. 3. Hydrodynamics.
4. Micro-Electrical-Mechanical Systems. 5. Nanostructures – therapeutic use. 6. Static Electricity.
QU 34]
572′.436–dc23 2011033658

ISBN 978-0-521-88168-5 Hardback

To my mother and father, Ginny and Terry, who first taught me the value of education.

To my brother and sisters, Virginia, Bill, Mary, and Elizabeth for friendship, love and support, and the many good times that so few families experience.

And to my wife, Paulette, and children, Terry and Katie, for their love and understanding; and for putting up with me over these many years. Without you this book would not have been possible.

Contents

Preface *page* xv

1 Introduction and Overview 1

1.1 Micro- and nanofluidics 1
1.2 Some micro- and nanofluidic devices 3
1.3 What is it about the nanoscale? 7
1.4 Nanotechnology 11
1.5 What is a fluid? 13
1.6 Historical perspectives 14
1.6.1 Fluid mechanics 15
1.6.2 Heat and mass transfer 17
1.6.3 Electrokinetic phenomena 19
1.7 The thermal sciences 20
1.8 Electrostatics 23
1.9 Electrolyte solutions 25
1.10 The electrical double layer 26
1.11 Colloidal systems 29
1.12 Molecular biology 32
1.13 The convergence of molecular biology and engineering 34
1.14 Design of micro- and nanofluidic devices 35
1.15 Unit systems 37
1.16 A word about notation 37
1.17 Chapter summary 38

2 Preparatory Concepts 40

2.1 Introduction 40
2.2 Important constitutive laws 41
2.3 Determining transport properties 45
2.3.1 Viscosity 45
2.3.2 Diffusion coefficient 48
2.3.3 Thermal conductivity 52

2.3.4 Electrical permittivity 54
2.3.5 Surface tension and wettability 55
2.4 Classification of fluid flows 59
2.5 Elements of thermodynamics 62
2.6 The nature of frictional losses in channels and pipes 68
2.7 Chapter summary 70

3 The Governing Equations for an Electrically Conducting Fluid 74

3.1 Introduction 74
3.2 The continuum approximation and its limitations 75
3.3 Kinematics 77
3.4 Surface and body forces 83
3.5 The continuity equation 87
3.6 The Navier–Stokes equations 88
3.7 Mass transport 93
3.7.1 Definitions 93
3.7.2 Governing equation 97
3.8 Electrostatics 100
3.9 Energy transport 102
3.10 Two-dimensional, steady, and incompressible flow 106
3.11 Boundary and initial conditions 106
3.11.1 Velocity boundary conditions 107
3.11.2 Mass transfer boundary conditions 113
3.11.3 Electrostatics boundary conditions 114
3.11.4 Temperature boundary conditions 116
3.11.5 Other boundary conditions 117
3.12 Dimensional analysis and similarity 117
3.13 Fluid, electrostatics, and heat and mass transfer analogies 123
3.13.1 Mole fraction and temperature similarity 123
3.13.2 Velocity and electrical potential similarity 125
3.14 Other stress–strain relationships 126
3.15 Mathematical character of partial differential equations 128
3.15.1 Introduction 128
3.15.2 Mathematical classification of second-order partial differential equations 128
3.15.3 Characteristic curves 129
3.15.4 Boundary and initial conditions 130
3.15.5 Classification of the governing equations of micro- and nanofluidics 131
3.16 Well-posed problems 131
3.17 The role of fabrication, experiments, and theory in micro- and nanofluidics 132
3.18 Chapter summary 134

4 The Essentials of Viscous Flow 140

4.1 Introduction 140
4.2 The structure of flow in a pipe or channel 141
4.3 Poiseuille flow in a pipe or channel 143
4.4 The velocity in slip flow 146
4.4.1 Gases 146
4.4.2 Liquids 147
4.5 Flow in a thin film under gravity 148
4.6 The boundary layer on a flat plate 150
4.7 Fully developed suction flows 155
4.8 Developing suction flows 158
4.9 The lubrication approximation 162
4.10 A surface tension–driven flow 166
4.11 Stokes flow past a sphere 169
4.12 Sedimentation of a solid particle 172
4.13 A simple model for blood flow 173
4.14 Chapter summary 174

5 Heat and Mass Transfer Phenomena in Channels and Tubes 180

5.1 Introduction 180
5.2 One-dimensional temperature distributions in channel flow 181
5.3 Thermal and mass transfer entrance regions 184
5.4 The temperature distribution in fully developed tube flow 189
5.5 The Graetz problem for a channel 189
5.6 Mass transfer in thin films 192
5.7 Classical Taylor–Aris dispersion 194
5.8 The stochastic nature of diffusion: Brownian motion 199
5.9 Unsteady mass transport in uncharged membranes 201
5.10 Temperature and concentration boundary layers 205
5.11 Chapter summary 207

6 Introduction to Electrostatics 213

6.1 Introduction 213
6.2 Coulomb's law: The electric field 214
6.3 The electric field due to an isolated large flat plate 216
6.4 Gauss's law 218
6.5 The electric potential 219
6.6 The electric dipole and polar molecules 221
6.7 Poisson's equation 222
6.8 Current and current density 225
6.9 Maxwell's equations 226
6.10 Chapter summary 227

7 Elements of Electrochemistry and the Electrical Double Layer 230

7.1 Introduction 230
7.2 The structure of water and ionic species 231
7.3 Chemical bonds in biology and chemistry 233
7.4 Hydration of ions 234
7.5 Chemical potential 236
7.6 The Gibbs function and chemical equilibrium 240
7.7 Electrochemical potential 243
7.8 Acids, bases, and electrolytes 244
7.9 Site-binding models of the silica surface 246
7.10 Polymer surfaces 249
7.11 Qualitative description of the electrical double layer 251
7.12 Electrolyte and potential distribution in the electrical double layer 253
7.13 Multivalent asymmetric mixtures 259
7.14 The ζ potential and surface charge density: Putting it all together 260
7.14.1 The classical liquid-side view for a symmetric electrolyte 260
7.14.2 The solid-side view and connection to the liquid side 262
7.15 The electrical double layer on a cylinder 265
7.16 The electrical double layer on a sphere 266
7.17 Electrical conductivity in an electrolyte solution 267
7.18 Semi-permeable membranes 270
7.19 The Derjaguin approximation 275
7.20 Chapter summary 278

8 Elements of Molecular and Cell Biology 283

8.1 Introduction 283
8.2 Nucleic acids and polysaccharides 285
8.3 Proteins 287
8.3.1 Protein function 288
8.3.2 Protein structure 289
8.3.3 Some common proteins 292
8.3.4 Few polypeptide chains are useful 295
8.4 Protein binding 295
8.5 Cells 298
8.6 The cell membrane 300
8.7 Membrane transport and ion channels 301
8.8 Chapter summary 304

9 Electrokinetic Phenomena 306

9.1 Introduction 306
9.2 Electro-osmosis 307
9.2.1 The relationship between velocity and potential 307
9.2.2 The Debye–Hückel approximation reviewed 312
9.2.3 Another similarity revealed 312

9.2.4 Asymptotic solution for binary electrolytes of arbitrary valence 313
9.2.5 Walls with different ζ potentials 316
9.2.6 Species velocities in electro-osmotic flow: Electromigration 318
9.2.7 Current and current density in electro-osmotic flow 320
9.2.8 Electro-osmotic flow in an annulus 322
9.2.9 Electro-osmotic flow in nozzles and diffusers 324
9.2.10 Dispersion in electro-osmotic flow 328
9.3 Electrophoresis: Single particles 331
9.3.1 Introduction 331
9.3.2 Electrophoretic mobility 332
9.3.3 Henry's solution 334
9.3.4 The full nonlinear problem 336
9.4 Streaming potential 338
9.5 Sedimentation potential 341
9.6 Joule heating 342
9.7 Chapter summary 344

10 Essential Numerical Methods 348

10.1 Introduction 348
10.2 Types of errors 350
10.3 Taylor series 351
10.4 Zeros of functions 353
10.4.1 Numerical methods 353
10.4.2 Polynomials 358
10.5 Interpolation 359
10.5.1 Linear interpolation 360
10.5.2 The difference table 361
10.5.3 Lagrangian polynomial interpolation 362
10.5.4 Newton interpolation formulas 363
10.5.5 Matlab interpolation functions 365
10.5.6 Cubic spline interpolation 366
10.6 Curve fitting 370
10.7 Numerical differentiation 373
10.7.1 Derivatives from Taylor series 373
10.7.2 A more accurate forward formula for the first derivative 375
10.8 Numerical integration 376
10.8.1 The trapezoidal rule 377
10.8.2 Simpson's rules 380
10.8.3 Matlab integration functions 382
10.8.4 The indefinite integral 382
10.8.5 Other formulas 383
10.8.6 Grid (mesh) size 383
10.8.7 Singularities 384

10.9 Solution of linear systems 386
10.9.1 Solving sets of linear equations in Matlab 389
10.9.2 Iterative solution to linear systems 390
10.9.3 Tridiagonal systems 393
10.9.4 Ill-conditioning and stability 396
10.10 Solution of boundary value problems 398
10.10.1 Introduction 398
10.10.2 Linear equations 399
10.10.3 Nonlinear equations 403
10.10.4 Systems of ordinary differential equations 405
10.10.5 Derivative boundary conditions 407
10.10.6 Convergence tests and Richardson extrapolation 409
10.10.7 Solving boundary value problems with Matlab functions 410
10.11 Solution of initial value problems 411
10.11.1 Introduction 411
10.11.2 Taylor series method 413
10.11.3 Euler methods 414
10.11.4 Runge-Kutta methods 416
10.11.5 Adams–Moulton methods 419
10.11.6 Symplectic integrators 419
10.11.7 Stiff equations and stability 424
10.11.8 Solving initial value problems using Matlab functions 428
10.12 Numerical solution of the PNP system 428
10.13 Partial differential equations 430
10.13.1 Elliptic equations 431
10.13.2 Parabolic equations 432
10.13.3 The Matlab PDE solver 435
10.14 Verification and validation of numerical solutions 435
10.15 Chapter summary 438

11 Molecular Simulations 447

11.1 Introduction 447
11.2 The molecular world 449
11.3 Ensembles 451
11.4 The potentials 451
11.5 Using the Lennard–Jones potential 453
11.6 Molecular models for water 456
11.7 Periodic boundary conditions 457
11.8 The Ewald sum 460
11.9 Numerical issues 463
11.9.1 Time integration 463
11.9.2 Truncation of interactions 464
11.9.3 Boundary conditions 465

11.10 Postprocessing 465
11.11 Nonequilibrium molecular dynamics 467
11.11.1 Introduction 467
11.11.2 Poiseuille flow 468
11.11.3 Electro-osmotic flow 469
11.12 Molecular dynamics packages 471
11.12.1 Introduction 471
11.12.2 What MD/NEMD simulators do 471
11.13 Summary 472

12 Applications 475

12.1 Introduction 475
12.2 DNA transport 476
12.2.1 How does DNA move? 477
12.2.2 Mathematical model 479
12.2.3 Results 481
12.2.4 DNA current 482
12.2.5 Comparison with experiment 483
12.3 Development of an artificial kidney 484
12.3.1 Background 484
12.3.2 The nanopore membrane for filtration 486
12.3.3 Hindered transport 487
12.4 Biochemical sensing 491
12.4.1 Introduction 491
12.4.2 What is a biosensor? 492
12.4.3 Receptor-based classification of biosensors 493
12.4.4 Transducer-based classification of biosensors 494
12.4.5 Evaluation of biosensor performance 495
12.4.6 Nanopores and nanopore membranes for biochemical sensing 496
12.5 Chapter summary 498

Appendix A Matched Asymptotic Expansions 501

A.1 Introduction 501
A.2 Terminology 501
A.3 Asymptotic sequences and expansions 502
A.4 Regular perturbations 503
A.5 Singular perturbations 504

Appendix B Vector Operations in Curvilinear Coordinates 508

B.1 Cylindrical coordinates 508
B.2 Spherical coordinates 508
B.3 Rectangular coordinates 509

Appendix C Web Sites 510

C.1 Fluid dynamics and micro- and nanofluidics 510
C.2 General nanotechnology 511
C.3 Wikipedia 511

Appendix D A Semester Course Syllabus 512

Bibliography 515

Index 533

Preface

The book is meant to be used as a text for an interdisciplinary course in micro- and nanofluidics that includes the study of ionic and biomolecular transport at the advanced undergraduate and beginning graduate levels. The rationale for this book is that most, if not all, problems in the twenty-first century are interdisciplinary in nature, yet no textbooks address the topics required for investigating problems that cut across disciplines in engineering, the physical sciences, and mathematics. The closest approach to this concept is in the several texts that address the thermal sciences at a strictly undergraduate level (Moran *et al.*, 2003). Another set of texts addresses problems in applied mathematics applicable to engineering problems generally at the advanced graduate level (Bird *et al.*, 2002). Still another set of texts under the general area of biophysics links the mathematics and biological sciences, again most often at the advanced graduate level (Murray, 2001, 2003). In contrast, this book aims at the advanced undergraduate and beginning graduate student pool.

A number of other related books are on the market, but all are monographs directed at the senior graduate student (Karniadakis *et al.*, 2005; Masliyah and Bhattacharjee, 2006; Tabeling, 2005; Liou & Fang, 2006; Nguyen & Wereley, 2002; Bruus, 2008; Kirby, 2010; Chang and Yeo, 2010). All these texts emphasize the unique features of transport at the micro- and nanoscale, of which there are many. In contrast, while the reader will be exposed to many of these unique features, it is my contention that transport at the micro- and nanoscale actually unifies all the thermal sciences, fluid dynamics, heat and mass transfer, and thermodynamics; it also, sometimes by necessity, unifies the thermal sciences with electrostatics, electrokinetics, and colloid science; electrochemistry; and molecular biology. This book is the first to show how all these fields are interrelated at the micro- and nanoscale and show how it is essential for a researcher, student, or faculty to acquire an understanding at some level of all these fields. The fundamental concepts within these fields are supported by addressing the continuum and molecular computation techniques that may be employed to solve these problems.

The objective of this book is to introduce students in the physical and mathematical sciences and engineering to the basic physical principles appropriate to analyzing fluid flow in micro- and nanoscale devices. The book will emphasize the fundamental principles involved in the formulation and solution of problems in fluid mechanics and mass transfer for pressure-driven and electrically driven motion of biofluids and electrolyte solutions at the micro- and nanoscale. It will introduce the student to a variety of subject matter spanning the physical sciences, thermal engineering, and applied numerical methods to enable the student to solve problems of an interdisciplinary nature. On completion of the book, the student should be able to extract from a raw physical situation the essential principles from which a useful model for thermal, ionic, and biomolecular transport may be developed.

The primary target audience of this text is the advanced undergraduate and beginning graduate engineering student; however, it is hoped that the book will be accessible to some advanced undergraduate students in physics and the chemical and biological sciences with the appropriate mathematics background.

In writing this book, I have been greatly influenced by the style and format of White (2006), which is aimed at a similar audience and contains exercises at the end of each chapter. White uses canonical problems in the field to illustrate basic fluid dynamic phenomena, and I have followed this style, expanding into heat and mass transfer, electrostatics, electrochemistry, electrokinetics, and molecular biology.

As with any book project, many people have contributed. I am thankful for Dr. David Mott, who read, with a keen eye, an advanced draft of the book, and to Professor Shaurya Prakash, who read a nearly final version of the manuscript. I am very thankful to Professor Susan Olesik, who read an early draft of the electrochemistry chapter, and to Dr. Arfaan Rampersaud, who reviewed the chapter on molecular biology. I am also thankful to Professor Minami Yoda, with whom I have worked over the past seven or so years. We have cut our teeth on micro- and nanofluidics together over that time. I am grateful to my colleagues Professor Shuvo Roy, Dr. Bill Fissell, and Professor Andrew Zydney, who introduced me to the fluid dynamics of the kidney. Professor Sherwin Singer read the molecular simulation chapter and made many suggestions and corrections that have been incorporated. Dr. Harvey Zambrano also helped me greatly with that chapter. I am also grateful to Professors Narayan Aluru and Ron Larson and Dr. Dirk Gillespie for their contributions.

And thanks to those researchers who have contributed the boxed vignettes that are about their work or some aspect of micro- and nanofluidics for which I did not have room. They are acknowledged at the end of the presentation.

I am particularly grateful for all the discussions I have had with faculty at the nanoscale science and engineering center, called the Center for the Affordable Nanoengineering of Polymeric Biomedical Devices (CANPBD).

Thanks go to my present and former students, Prashanth Ramesh, Ankan Kumar, Pradeep Gnanaprakasam, and Devi Pulla, some of whose work appears in the book; to Mike Stubblebine, who helped catalog the figures; and to Professor Subhra Datta and Dr. Lei Chen, who both have done much in the way of research that appears in this book. Subhra wrote first drafts of several sections, and Lei produced many of the figures that appear in the book. Both have read the manuscript, portions more than once. Thanks also to Dan Hoying, an undergraduate physics student who read a nearly final version of the book, and Zhizi Peng and Cong Zhang, who produced several figures; Cong was a significant contributor to the solutions manual; and to Harvey Zambrano who helped me write a section in the molecular dynamics chapter; and to Kevin Disotell who read the final proofs; and to Jim Marcicki who taught me about batteries.

I am grateful to my "Introduction to Micro- and Nanofluidics" class in the Autumn quarter of 2010, who used the book and had many suggestions that were heartily received and implemented. Thanks also go to graduate student Martin Kearney-Fisher, who read the entire manuscript and gave me pages of corrections and suggestions, almost all of which I have incorporated. Martin's suggestions have made the manuscript much better.

Thanks also to my editor, Peter Gordon, who kept me on task and made a number of suggestions on how to write the book, especially the early chapters. He also provided me with additional resources that allowed a more thorough treatment of this rapidly expanding field. And thanks go to Peggy Rote for her diligence in managing the production process.

If I have forgotten to thank someone, I apologize.

The book begins with an introduction and overview of micro- and nanofluidics in Chapter 1, followed by Chapter 2, "Preparatory Concepts." Chapter 2 is meant to unify concepts on two levels: discussing the fundamental roles of transport coefficients in fluid mechanics and heat and mass transfer and the relationship between thermodynamics, the equilibrium science, and heat transfer and fluid mechanics, the nonequilibrium thermal sciences. These two initial chapters are followed by a discussion of the governing equations and boundary conditions associated with micro- and nanofluidics. At the micro and nano levels, several new phenomena come into play:[1]

1. Because of the large surface-to-volume ratio, the characteristics of surfaces play a major role in fluid and mass transport.
2. Classical means of transporting fluids, such as pressure drop, may not be possible.
3. Noncontinuum effects arise when the length scale associated with the fluid transport becomes less than 10 nm.

All these issues are discussed throughout the book, and the second point is the reason that electrokinetic transport methods become important.

The next three chapters cover the fundamentals of viscous flow, heat and mass transport, and electrostatics. While writing this book, I was astonished at how similar these fields are in the way of expressing basic transport phenomena. Many analogies between these three (or four, if mass transfer is considered separately) disciplines are discussed throughout.

[1] We are speaking here primarily of liquid flows. Gas flows are treated extensively by Liou and Fang (2006) and Karniadakis *et al.* (2005).

Following these three chapters are two chapters covering the fundamentals of electrochemistry and molecular and cell biology. These two chapters are meant to reintroduce engineering students to material they may have had in their first-year course work, although parts of each chapter are written at a higher level.

Following these two chapters, electrokinetic phenomena are discussed in Chapter 9. The two most important of these phenomena, electro-osmosis and electrophoresis, are discussed in great detail, and canonical problems of electro-osmosis are described in the spirit of the style of White (2006).

The next chapter covers the basics of numerical methods, from zero finding to the numerical solution of partial differential equations. The primary role of this chapter is to introduce the student to basic numerical methods that can be used to solve problems in micro- and nanofluidics. For some engineering students, this chapter is likely to be a review; however, physics and chemistry students may find this chapter valuable.

Next, the fundamental concepts involved in performing molecular simulations, specifically equilibrium molecular dynamics and nonequilibrium molecular dynamics, are presented along with examples of Poiseuille flow and electro-osmotic flow. It is surprising how different the philosophy and expectations of what is achievable in a simulation on the molecular level are from the continuum perspective. The reader need only compare the presentation of the numerical methods in this chapter with those presented in the previous chapter to see this. Note the differences in how the continuum and molecular results are verified and validated.

The book ends with a chapter devoted exclusively to applications. Applications are too numerous to mention, but I have chosen those applications with which I am familiar. Thus a simple model for DNA transport is presented, along with a section on biochemical sensing and the fluid mechanics and mass transfer involved in the design of a renal assist device.

Each chapter is followed by a set of exercises that range from simple calculations, such as determining the Debye length for a given set of parameters, to finding the solution of viscous flow through an annulus to the calculation of the numerical solution of the Poisson equation to completely open-ended exercises that require a written report. The exercises after Chapter 12 are all open ended. In these open-ended exercises, other applications not included in the book are introduced such as the use of nanoparticles to treat cancer. These exercises have been designed to make maximum use of the Web and emphasize the development of the technical writing skill of the student. A short introduction to technical writing is available from the author, on request.

Several appendices, giving a short introduction to the method of matched asymptotic expansions, the governing equations in cylindrical and spherical coordinates, a list of interesting and useful Web sites, and a prospective syllabus, are also included. Writing this book has been a tremendous learning experience, and I have bought several chemistry and biology dictionaries. I have also acquired more biology and chemistry textbooks in six years than I have in my entire life

(I have Cambridge University Press to thank for some of these). While all this about learning from books is true, I cannot tell you how many times I have been to Wikipedia or used the other Web sites that appear in Appendix B.

Much of the material in the book is gleaned from research papers, from those that are very old and classical to those published very recently. I have tried to be judicious in my choice of references, and to those whom I have overlooked, I apologize. I would be happy to be informed of the omission of a major paper that would contribute to a future manuscript. This has been quite a task, and I and my students, and several faculty, have read parts of the book, as noted earlier. Nevertheless, errors are inevitable, and I would be grateful if I could be informed of any errors that do appear in the book. For these errors, I take full responsibility.

The emphasis in this book has been the interdisciplinary nature of micro- and nanofluidics. This book is me speaking about what I think is important to know in micro- and nanofluidics. Thus I take responsibility for those many topics that are left out. For this, I do not apologize but merely say that tough choices were made. I hope that this book will be read by students with diverse backgrounds and that they will benefit from what I hope is a lucid presentation.

1 Introduction and Overview

1.1 Micro- and nanofluidics

Analyzing and computing fluid flow at small scales is becoming increasingly important because of the emergence of new technologies such as the ability to construct microelectromechanical systems (MEMS). These systems may be used for drug delivery and its control; DNA and protein manipulation and transport; and the desire to manufacture laboratories on a microchip for rapid molecular analysis, requiring the modeling of flows on a length scale approaching molecular dimensions. On these small scales, new flow features appear that are not seen in macroscale flows.

Because of the large surface-to-volume ratio in nanochannels, surface properties become enormously important. Because the pressure drop $\Delta p \sim 1/h^3$, it is prohibitively large for a nanoscale channel. Thus fluid, biomaterials such as proteins, and other colloidal particles are most often transported electrokinetically, and the art of designing micro- and nanodevices requires a significant amount of knowledge of fluid flow and mass transfer (biofluids are multicomponent mixtures) and often heat transfer, electrokinetics, electrochemistry, and molecular biology. To efficiently manufacture laboratories on a microchip, the analysis and computation of flows on a length scale approaching molecular dimensions, the nanoscale, are required.

The common thread is micro- and nanofluidics. Thus micro- and nanofluidics play the role of unifying the fields of fluid mechanics, heat and mass transfer, electrostatics and electrodynamics, electrochemistry, and molecular biology. In particular, nanofluidics opens the door to uncovering the structure and conformation of biomaterials, such as proteins, through molecular simulation.

The objective of this book is to introduce the reader to micro- and nanofluidics, the basic mechanics of modeling fluid flows and heat and mass transfer, that is, at very small scales. The emphasis is on those systems that have biological and chemical applications. These systems are commonly at the microscale, with length scales $\sim$100 μm or 100×10^{-6} meters, and the 100 times smaller nanoscale, at length scales $\sim$100 nm or 100×10^{-9} meters. In many of these systems, transport is from the microscale to the nanoscale and back to the microscale.

Microfluidics in general consists of three distinct components:

1. Modeling: computational and theoretical
2. Fabrication
3. Experimental methods

The emphasis will be on modeling because of the limitations of experimental methods at the micro- and nanoscale, although experimental methods are also discussed, where appropriate; Bohn (2009) discusses some of the experimental methods used to probe single molecules. Because this is primarily a book about modeling, fabrication methods are not discussed.

Government research programs, such as the Defense Advanced Projects Research Agency (DARPA) and Simulation of Biological Systems (Simbiosys), (which ran from 2002 to 2005), deal with the development of microsystems that can be used to identify many types of molecules through their transport characteristics. The ultimate goal of the program was to develop computer-aided design tools (CAD) for applications to chemical–biological warfare defense, infectious disease monitoring, and drug delivery.

The governing equations of fluid flow on a length scale orders of magnitude greater than a molecular diameter are well known to be the Navier–Stokes equations, which are a statement of Newton's law for a fluid. Along with conservation of mass and appropriate boundary and initial conditions in the case of unsteady flow, these equations form a well-posed problem from which, for an incompressible flow (constant density), the velocity field and pressure may be obtained.

Applications in the biomedical field involve, for the most part, internal flows in micro- and nanochannels and tubes. The fluids are generally electrolyte mixtures, with, perhaps, a biomolecular component, usually some protein (say, albumin). Thus mass transfer occurs, and because many biomolecules (e.g., most proteins) are charged, there is an electric field as well. The determination of the identity and rates of transport of ionic and biomolecular species is one of the purposes of many micro–nanoscale devices.

Because micro- and nanofluidics usually involves the transport of charged species, its study requires a multidisciplinary approach. Thus the study of fluid flows at the microscale and nanoscale most often requires expertise in electrochemistry, surface chemistry, electrostatics and electrokinetics, molecular biology, heat and mass transfer, and macro-scale fluid mechanics. To design devices having micro- and nanoscale features requires a team approach involving chemists, biologists, medical practitioners, engineers, and systems analysts. As one might guess, each of these disciplines speaks a somewhat different language, and it is only with some effort that these technical language barriers can be overcome.

There are a number of textbooks on various aspects of electrokinetic phenomena, including the fluid mechanics of electrokinetics; for more details, see Chang and Yeo (2010), Masliyah and Bhattacharjee (2006), Karniadakis *et al.* (2005),

Bruus (2008), Kirby (2010), Tabeling (2005) and Li (2004). All the aforementioned books are essentially monographs written primarily for advanced graduate students beginning their research careers.

It is with this interdisciplinary view of micro- and nanofluidics that our journey begins. We start in this chapter by presenting some examples of micro- and nanofluidic devices, followed by a working definition and a bit of the history of nanotechnology. Then a broad discussion of the fields of fluid mechanics and heat and mass transfer is presented. This is followed by a discussion of electrostatics and the character of electrolytes (charged fluid mixtures). Micro- and nanofluidics often deals with mixtures of liquids and solid particles. If the particles are less than a micron (10^{-6} m) but larger than about 1 nm, the mixture is termed a *colloidal mixture*. Finally, we introduce some of the basic concepts of molecular biology. One of the major objectives of this book is to show how micro- and nanofluidics unifies the fields of the thermal sciences, electrochemical systems, and molecular biology. The unifying features of these fields are described next, followed by a discussion of a typical design procedure. The chapter concludes with two short sections on unit systems and notation, the latter being a very important section. In short, all notation is local.

1.2 Some micro- and nanofluidic devices

In this section, we qualitatively describe several micro- and nanofluidic devices and their applications. An interesting application of microtechnology is small drug delivery devices. These devices can deliver very small and precise doses of medicines quickly and efficiently. Some devices may also be used as biomolecular separators because different species (often electrically charged) and biomolecules travel at different speeds in these channels, due primarily to their differences in size, charge, and shape. Devices of this sort perform analyses faster and are more efficient in a wide variety of applications, including water quality, medical diagnostics, and applications associated with national security such as sensing chemical and biological toxins.

As you reflect on the art of designing micro- and nanodevices, think of the fact that this activity requires a significant understanding of fluid flow and mass transfer and, often, heat transfer and electrokinetics. Mass transfer is especially relevant to biofluids because they are multicomponent mixtures.

A specific example of this sort is an *electro-osmotic pump*, or sometimes *nanopump*, which is used to induce transport of charged molecules. Such a device is depicted in Figure 1.1(a). This biomedical device is useful for the delivery of various types of proteins, such as albumin and immunoglobulin, both of which have many biomedical uses (Peters, 1996). Nine channels are shown in the depiction, but an actual device may employ 20,000–40,000 channels. In general, in biomolecular transport and analysis systems, it is essential to have lots of little channels to distinguish the very small molecules that are being analyzed.

The device depicted in Figure 1.1(a) is often called a *synthetic nanopore membrane*. In biology and chemistry, the term *membrane* is used to describe a thin sheet of porous material that can be either natural (the outer skin of a cell is a membrane) or synthetic.

What is Lab-on-a-chip?

The ability to fabricate devices on the micro and nanoscale has led to the development of devices that can identify different molecules, separate these molecules, manipulate them, and transport these molecules. What had once required a laboratory and large samples can now be done at very small scales.

The generic name given to these types of devices is "Lab-on-a-chip (LOC)," which refers to a type of processing. The key feature of LOC is that the various steps in a diagnostic procedure "the laboratory" are integrated on to one small devices "the chip." These devices have also been called micro total analysis systems (μ TAS) and the two terms are most often taken to be equivalent. Most LOC applications have a biochemical component to them.

Such systems work because in a large or small sample the following general principles regarding the molecules under consideration here apply: Molecules can be identified by properties, such as how they appear under the action of a laser. Molecules of different size and electric charge characteristics move at different speeds allowing them to be separated based on a simple criteria. Molecules may be modified by inducing chemical reactions with other molecules. And finally molecules can be moved from one place to another by the bulk motion of the fluid.

Lab-on-a-chip devices are being developed for bacteria screening, cancer detection, and unicellular exploration.

Such LOC systems have several advantages that make them attractive in biomedical applications:

1. Lower equipment costs and power requirements because of the small length scales,
2. reduced separation and reaction times again because of the small length scales,
3. LOC typically require nano- to picoliter (vs. mL for comparable macroscale analyses) volumes of analyte and reagents, reducing chemical costs and biochemical hazard and waste disposal problems;
4. integrating sophisticated chemistry procedures within a single system, LOC can be used by nonspecialists to perform complex analyses.

The devices from within the chemistry and biochemistry industries described earlier were created using MEMS technology. The broad range of micro- and nanodevices of this sort are commonly referred to as a *lab on a chip*: a device that incorporates all the steps of chemical–biochemical analysis to perform a given measurement. Sometimes these systems are called *micro total analysis systems*, or μTAS.

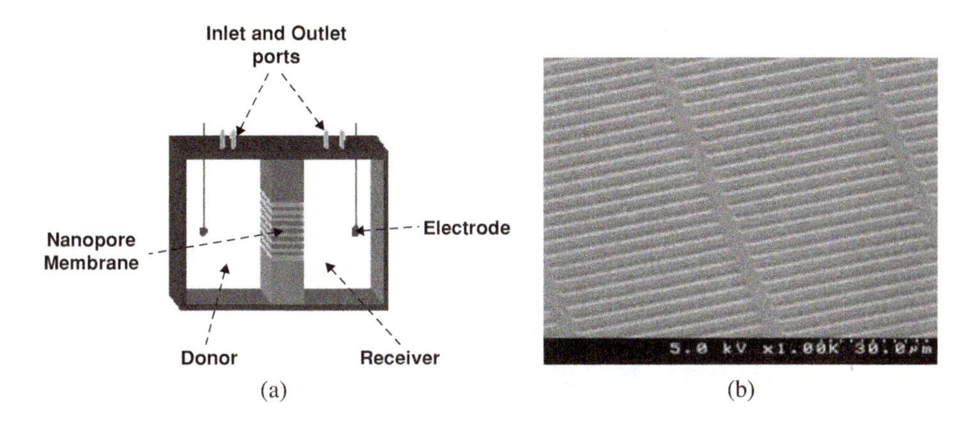

Figure 1.1 (a) Drawing of a nanopump containing a nanopore membrane. (b) Scanning electron microscope (SEM) image of a synthetic nanopore membrane. From Conlisk *et al.* (2009).

As an example of a lab on a chip, Sandia National Laboratory has developed a fully integrated chemical analysis system that has the ability to determine constituents of gas and liquid samples within 1 min (Figure 1.2) through integration of sample collection, separation, and detection steps. It is made up of a compact power source, lasers and photodiodes, a microprocessor, and micro-sized injection and separation channels. In contrast, much of the current chemistry–biochemistry analysis labs consist of equipment as large as a microwave oven for each of the analysis steps.

The use of lab-on-a-chip devices in a chemical–biochemical analysis lab reduces equipment costs but can also reduce the cost of other resources. The advantages of a lab-on-a-chip device are comparable to the advantages of miniaturization seen in the computer industry. For example, when reducing the size of a computer chip, the distance electrons need to travel is much shorter, reducing the processing time. This same principle applies to microfluidic devices: reducing the size of a channel reduces the distance molecules need to travel, therefore

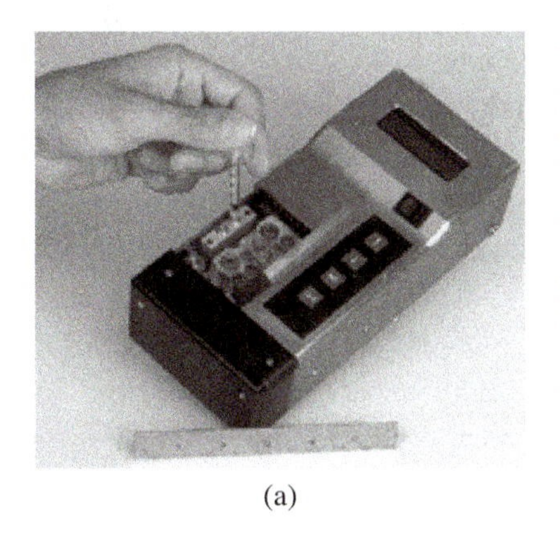

(a)

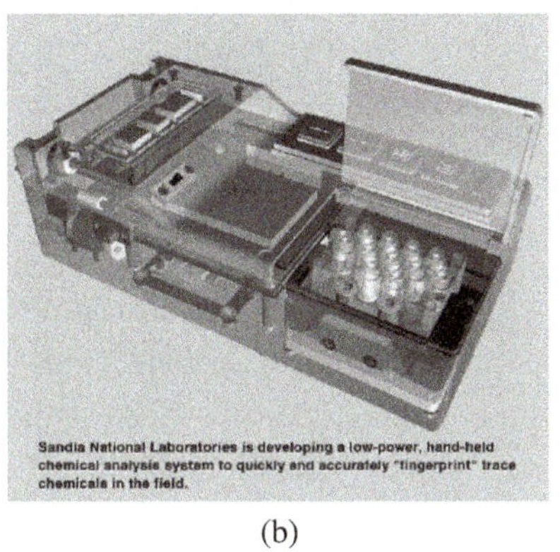

(b)

Figure 1.2 Sandia National Laboratory fully integrated chemical analysis system that can be held in one hand. (a) Final product. (b) Cutaway image of the device, showing many of the components.

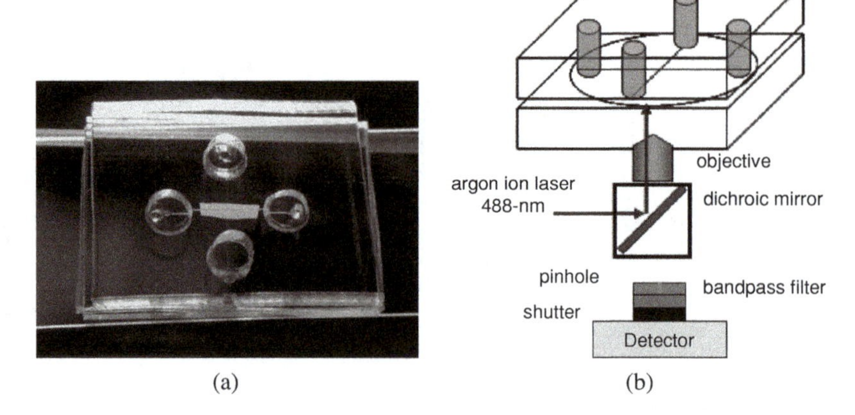

Figure 1.3 (a) Chip used for attomole level chemical detection courtesy of Professor Paul Bohn. (b) The detection system from Kuo *et al.* (2003).

reducing processing time. These devices are highly portable and also require much less *reagent*, or sample, the chemical compound used to detect and identify an analyte. For a standard analysis experiment, microliters or larger of reagents are used for each experiment; however, with lab-on-a-chip devices, only nanoliter or picoliter volumes may be required for each experiment. Think of the savings in chemical use and the environmental benefits of reduced chemical and biotoxic waste.

A number of security applications are associated with these devices. Nanocapillary array membranes of approximately circular cross section (Kemery *et al.*, 1998) are being used to detect biological warfare agents in concentrations at the attomole (10^{-18} mole) level. Such a system is depicted in Figure 1.3. Channels on the order of 10–100 nm are employed to manipulate these biochemicals, which must be handled in very low concentrations. A molecule is identified by a laser-induced fluorescence signal, and the device can identify molecular size and charge based on its transfer characteristics through the channel. A similar device can be employed with an enzyme sample (Gong *et al.*, 2008).

The nanocapillary array membranes (NCAMs) used in the 3-D configuration shown in Figure 1.3 function in the same manner in which individual transistors function in integrated circuits: by controlling the temporal and spatial delivery of ultralow-volume fluid packets. Achieving an all-electronic fluidic switching network with no moving parts is an enabling development for 3-D integrated microfluidic devices. In nearly all cases of multidimensional chemical analysis, the chemical sample needs to be processed through multiple sequential chemical unit operations. These might include separation of a desired component from a raw mixture, subsequent chemical processing (derivatization) to visualize the compound, and placing it in the right spatial location for detection and/or further characterization. In simple 2-D (planar) structures, the management of chip real estate soon becomes a design challenge. Going into the third dimension makes it possible to save real estate and achieve highly compact and efficient designs. Furthermore, there is the possibility of using the individual nanopores to carry out enhanced chemical reactions by taking advantage of the small distances to improve the efficiency of chemical turnover. In this sense, the NCAMs are more than just simple fluidic switching elements; they can be thought of as attoliter-scale chemical reactors with on-demand delivery (or generation) of reagents and removal or collection of products.

Ion Channels

The basic units of all living organisms are cells. In order to keep cells functioning properly there must be a continuous flux of ions in and out of the cell and its components. The cell and many of its components are surrounded by a plasma membrane which provides selective transfer of ions through *ion channels*. The ion channels are embedded in the cell membrane, are usually negatively charged and are about 10 angstroms in diameter. The polarity of the plasma membrane makes it challenging for molecules to move in and out of cells and its components. Thus ions (Ca^{2+}, Cl^-, K^+, Na^+, H^+, Mg^{2+}, HCO^{3-}, $PO_4{}^{2-}$) must selectively move through the membrane via protein channels electrokinetically through a combination of electro-osmosis and electromigration. Both of these fluid dynamic phenomena are discussed in Chapter 9 and ion channels are described in more detail in Chapter 8.

Very recently, on April 26, 2009, new ion channels that govern the function of the inner ear were found in a very surprising place! (http://medicalnewstoday.com/articles/147506.php)

In contrast to the synthetic devices described earlier, natural nanochannels exist in cells for the purpose of providing nutrients and discarding waste; in that sense, they function as natural pumping systems. That is, these natural nanochannels, called *ion channels*, act as electro-osmotic pumps that contain perhaps thousands of nanopores – a natural nanopore membrane. Peter Agre of Johns Hopkins and Roderick MacKinnon of Rockefeller University won the 2003 Nobel Prize in Chemistry for their work on understanding how natural ion channels work.

1.3 What is it about the nanoscale?

Why is everything different at the nanoscale? Or is anything different at all? As the typical length scale of the channel approaches the microscale level and beyond to the nanoscale level (Table 1.1), conventional means of moving fluids, such as with a pressure gradient, become ever more difficult, and the character of the surfaces bounding a fluid becomes ever more important. Consider the channel depicted in Figure 1.4(a). In micro- and nanofluidics, the surface-to-volume ratio is very large, making the nature of the surface (e.g., its charge, roughness, and

Table 1.1. SI units of length measurement

Factor	Prefix	Symbol
10^9	giga	G
10^6	mega	M
10^3	kilo	k
10	deka	da
10^{-1}	deci	d
10^{-2}	centi	c
10^{-3}	milli	m
10^{-6}	micro	μ
10^{-9}	nano	n
10^{-10}	angstrom	A
10^{-12}	pico	p
10^{-15}	femto	f
10^{-18}	atto	a

whether it is hydrophobic or hydrophilic) very important. The surface-to-volume ratio for a channel having dimensions $(L, h, W) = (1 \text{ m}, 1 \text{ m}, 1 \text{ m})$ is

$$\frac{S}{V} = 2\left(\frac{1}{W} + \frac{1}{h} + \frac{1}{L}\right) = 6 \text{ m}^{-1} \tag{1.1}$$

On the other hand, for a channel having dimensions $(L, h, W) = (3\ \mu\text{m}, 1\ \mu\text{m}, 40\ \mu\text{m})$ which is typical of a class of nanopore membranes, the surface-to-volume

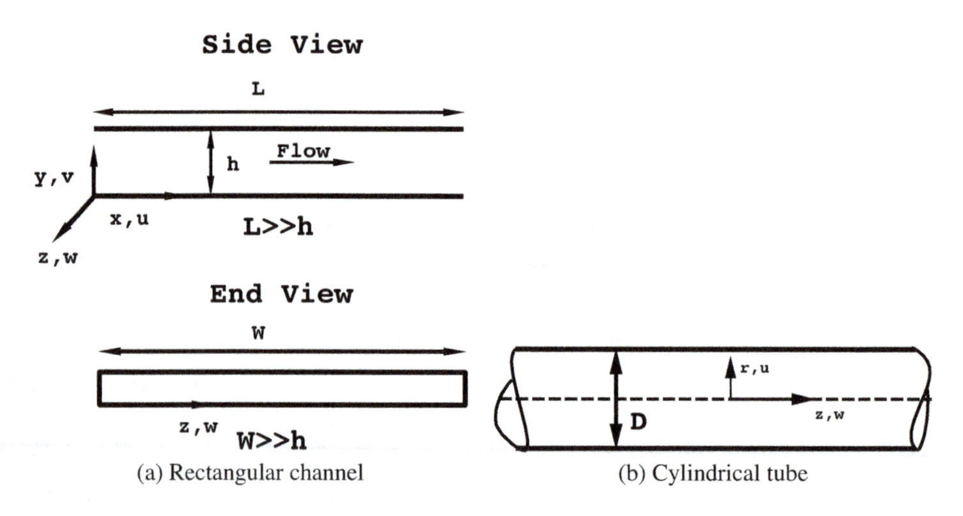

(a) Rectangular channel (b) Cylindrical tube

Figure 1.4 (a) Geometry of a typical channel. In applications, $h \ll W, L$, where W is the width of the channel and L is its length in the primary flow direction. Variables u, v, w are the fluid velocities in the x, y, z directions. (b) Sketch of a cylindrical pore having velocities (u, v, w) in the (r, θ, z) directions.

ratio is

$$\frac{S}{V} \sim 2 \times 10^6 \text{ m}^{-1} \tag{1.2}$$

For a 20 nm channel, $(L, h, W) = (3\ \mu\text{m}, 20\ \text{nm}, 40\ \mu\text{m})$, the surface-to-volume ratio is even higher:

$$\frac{S}{V} \sim \frac{2}{h} \sim 40 \times 10^9 \text{ m}^{-1} \tag{1.3}$$

Because of the large surface-to-volume ratio, a surface roughness, for example, of 5 nm in a 1 μm channel, is negligible, whereas in a 10 nm channel, that same roughness can have a profound effect on the flow. The same situation occurs for a cylindrical tube (Figure 1.4(b)). In this case,

$$\frac{S}{V} = \frac{1}{R} = \frac{2}{D} \tag{1.4}$$

where R is the radius of the tube.

Several comments can be made about fluid flows in channels under 1 μm in minimum dimension:

1. Surface properties of a channel or tube, such as electrical surface charge density and roughness, become very important because of the large surface-to-volume ratio.
2. Significant increases in flow rate may be attained if the surfaces of the channel are hydrophobic (water hating); that is, significant fluid slip may occur at the wall. This occurence is termed *induced slip* or *apparent slip*.
3. The continuum approximation may break down, especially for gas flows.
4. Pressure-driven flow is only viable at very low flow rates, on the order of nL/min or 10^{-9} L/min, in nanoconstrained channels because of the very large pressure drops required otherwise, on the order of atmospheres.
5. Molecular diffusion, which is very slow at the macroscale, is fast at the micro- and nanoscale, the time scale being $t \sim L^2/D_{AB}$.

From the second comment, it is thus seen that for liquids, whether there is slip or no slip at the wall can be a function of surface chemistry, whereas in gases, slip is entirely controlled by the magnitude of the Knudsen number, the ratio of the mean free path to the characteristic length scale. However, it should be mentioned that liquid flows remain in continuum even for channels whose smallest dimensions approach 10 nm.

As mentioned, it becomes increasingly difficult to pump liquids by pressure in nanoscale channels. To see this, let us compare the pressure drop for *electro-osmotic flow* with the corresponding pressure drop for *pressure-driven flow* or *Poiseuille flow*. The *volume flow rate* in electro-osmotic flow may be estimated by

$$Q_e = CU_0hW \tag{1.5}$$

where U_0 is the electro-osmotic velocity scale and is independent of h (as we will see), C is a constant that depends on the concentration of the electrolyte, and W is

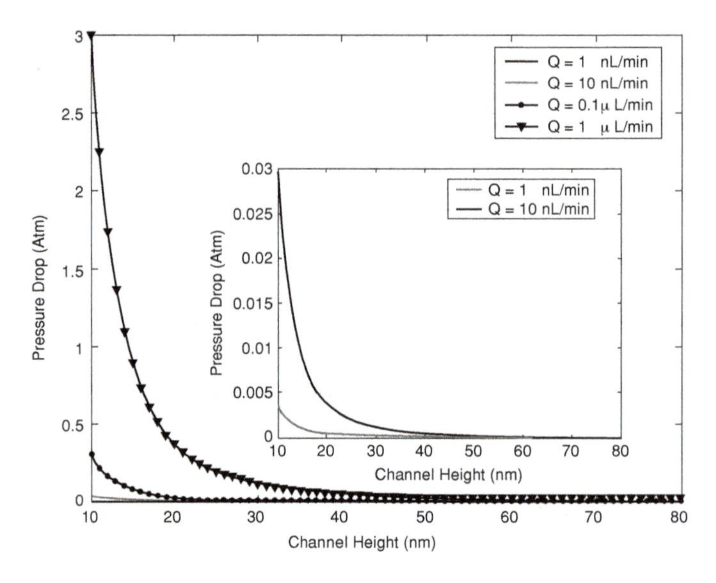

Figure 1.5 Pressure drop as a function of channel height to achieve a volume flow rate of $Q = 10^{-6}$ L/min and several values of the flow rate on the μL scale typical of many existing systems. Here L = liter; 1000 L = 1 m^3. The applied electrical potential for $Q = 10^{-6}$ L/min is very small.

the width of the channel. The velocity scale U_0 turns out to be directly proportional to the imposed electric field, and thus the flow rate is proportional to h. Conversely, for Poiseuille flow, the volume flow rate in a parallel plate channel is given by

$$Q_p = \frac{Wh^3}{12\mu L}\Delta p \tag{1.6}$$

where Δp is the pressure drop. This means that pressure-driven flow requires large pressure drops, as depicted in Figure 1.5; note that at a channel height of 10 nm, 3 atm of pressure drop is required to drive a flow of $Q = 10^{-6}$ L/min, which is a characteristic flow rate in drug delivery applications. However, if the flow rate is $Q = 10^{-9}$ L/min, the pressure drop is not nearly as large. Three atmospheres is a large pressure drop in a liquid, and clearly a relatively large pump would be required to provide this pressure drop. This is a major consideration when designing a nanopore membrane for the applications discussed previously.

As has been seen, it is often not feasible to transport fluids in nanochannels using an imposed pressure drop; electro-osmosis and electrophoresis are often used for transporting both charged and uncharged species and biomolecules. That is, electrokinetic phenomena play a crucial role in micro- and nanofluidics. Moreover, it is important to note that at present, velocity, temperature, and concentration profiles across a channel cannot be measured in channels having at least one dimension under about 1 μm (Sadr *et al.*, 2006; Breuer, 2005). Thus, to understand the physics of flows at those scales, modeling is not only necessary but also essential in describing the important features of the flow within a microdevice having nanoscale features such as a nanopore membrane.

1.4 Nanotechnology

I have heard as many definitions of *nanotechnology* as there are researchers. I will take a stab at an engineering definition: nanotechnology, or simply nanotech, is the study of phenomena on the atomic and molecular scale; that is, phenomena that occur on length scales less than about 100 nm. The *tech* part implies an application. Nanotechnology has been used to create solid materials with novel properties, for the development of medical devices, for rapid molecular analysis, and, as will be seen, in many other disciplines. For example, though the focus of this book is on biomedical applications in the biological and chemical sciences, other applications of micro- and nanotechnology include defense applications of chemical and biochemical sensing, desalination, water purification, and the design of batteries and fuel cells.

Much of what you have learned about fluid mechanics involves flows in pipes on the order of several inches in diameter (heating, ventilating, and air-conditioning), on the scale of meters (propulsion), on the scale of tens of meters (aerodynamics), or on the scale of hundreds or even thousands of meters (weather prediction) to millions and billions of meters flowing ever outward toward the edges of the universe: the astrophysics length scale.

On a smaller scale – but no less exciting and mysterious – this book is all about fluid flows, heat and mass transfer, and electrokinetics at very small scales. Microfluidics is generally viewed as the study of flows whose primary length scales are below about 100 μm = 10^{-4} m (Table 1.1). Nanofluidics refers to flows at length scales below about 100 nm. In recent years, manufacturing techniques have been developed and used to build entire micromachines, including thermal actuators, microvalves, micro- and nanopumps, gears, cantilevers, and other microdevices (Gad-el Hak, 2001). These devices have been used for biomedical applications such as drug delivery, chemical and biochemical sensors, micromixers and microseparators, rapid molecular analyzers, development of an artificial kidney, and DNA analysis and transport, to name only a few applications. Wouldn't it be a major advance in chemotherapy if a drug were so specific to cancer cells that healthy cells and tissues would be left alone? Today, most, if not all, chemotherapy drugs kill both cancerous and healthy cells.

The number of devices having nanoscale features has exploded in the last five years, and the number of companies that nanotechnology has spawned has increased greatly. A Web search on the term "nanotechnology" yields thousands of hits and nearly a dozen suggested new search terms.

The nanoscale is the scale of biology and chemistry. A glance at Figure 1.6 reveals that deoxyribonucleic acid (DNA) and the protein adenosine triphosphate (ATP) have at least one of their dimensions at the nanoscale. The diameters of many important proteins in a globular conformation are within the range of 2–10 nm. Drug delivery systems often transport carrier proteins through a series of channels in parallel that allow uniform delivery of the drug. Lipid spheres,

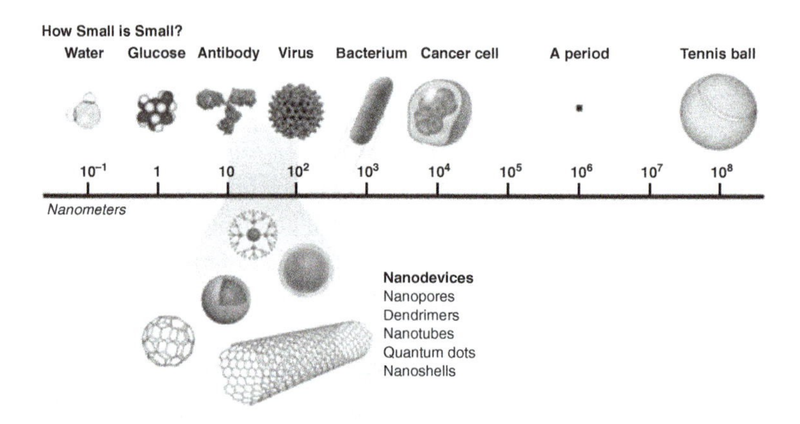

Figure 1.6 The scales of things, courtesy of the National Institute of General Medical Sciences of the National Institutes of Health.

termed *liposomes*, are already on the market; these 100 nm diameter spheres carry anticancer drugs to a target tumor (Malsch, 2005). Biochemical sensing applications involve the transport of an analyte to a target, where it is then identified by an electronic transduction process. In these applications, it is clear that fluid dynamics is the underlying pillar of each process.

The birth of nanotechnology and micro- and nanofluidics is often attributed to one man. Nobel Laureate Richard Feynman was a physicist known for his studies of quantum electrodynamics, the superfluidity of supercooled liquid helium, and high-energy physics. He won the Nobel Prize for his work on quantum electrodynamics in 1965 and was known in his later years for popularizing physics through both books and lectures until his death in 1988.

Many date the drive for miniaturization to the vision Feynman presented in a paper titled "There's Plenty of Room at the Bottom: An Invitation to Enter a New Field of Physics" (Feynman, 1961), presented at the American Physical Society annual meeting on December 29, 1959. In that paper, he asked the question, Why cannot we write the entire 24 volumes of the Encyclopaedia Britanica on the head of a pin? This would require that each of the many thousands of pages of this popular printed encyclopedia be reduced in size by 1/25,000. Feynman's vision in 1959, before the advent of hand calculators, much less laptops and iPads, foreshadowed the revolution in miniaturization that has exploded during the last 20 years and is increasing in scope and speed daily.

Feynman actually issued two challenges, each with a \$1000 prize. While William McLellan built a 1/64th-scale electric motor (Gribbin, 1997) within a year, technically meeting the challenge, it took until 1985 for Tom Newman to reduce a book page to 1/25,000 its size and, in the process, greatly advancing the science.

In a startling prediction foreshadowing the primary objective of drug delivery in the twenty-first century, Feynman revealed that

> a friend of mine (Albert R. Hibbs) suggests a very interesting possibility for relatively small machines. He says that, although it is a very wild idea, it would be interesting in surgery if you could swallow the surgeon. You put the mechanical surgeon inside the blood vessel and it goes into the heart and "looks" around. (Of course the information has to be fed out.) It finds out which valve is the faulty one and takes a little knife and slices it out. Other machines might be permanently incorporated in the body to assist some inadequately functioning organ. Feynman (1961)

This is a stunning prediction in 1959 that is coming true right before our eyes. Indeed, artificial organs are now common, and the design of an artificial kidney is discussed in Chapter 12.

Feynman also suggested the explosion of interest in science and technology at the length scale of biology, the nanoscale. Since 1989, the Foresight Conference on Advanced Nanotechnology has been held annually. It is to Feynman that we owe thanks for the explosion of the fields of micro- and nanotechnology, a branch of which is micro- and nanofluidics.

Nanotechnology and, in particular, micro- and nanofluidics require a multidisciplinary approach. As mentioned earlier, the study of fluid flows at micro- and nanoscales requires expertise in electrochemistry, but real expertise requires an understanding of surface chemistry, electrostatics and electrokinetics, electrochemistry, molecular biology, heat and mass transfer, and macroscale fluid mechanics. To design devices having micro- and nanoscale features requires a team approach involving chemists, biologists, medical researchers and practitioners, engineers, and systems analysts. As one might guess, each of these disciplines speaks a different language, and it is sometimes only with great effort that these technical language barriers are overcome.

1.5 What is a fluid?

Most students reading this book have had a standard undergraduate fluid mechanics class. In this section, we review a few of those topics to refresh your memory and to make sure we are using the same terms and notation you used previously. You have had the experience of diving into a lake or swimming pool and watching as the water parts to make way for your body. This experience is a reflection that unlike a solid, a fluid is a material that continuously deforms under a shear stress, or force per unit area, no matter how small the shear stress may be (Figure 1.7). In contrast, a solid will break once the shear stress rises above a certain level. There are two types of fluids: liquids and gases.

Figure 1.7 The impact of a red-colored drop on a pool of water. A fluid always deforms continuously when impacted by a liquid or solid. Image supplied by Wim van Hoeve, Tim Segers, Hans Kroes, Detlef Lohse, Michel Versluis, Physics of Fluids group, University of Twente, Netherlands.

In a solid, the individual molecules are packed tightly together and will not move unless a large enough shear stress is applied. When a large enough shear stress is applied to a solid, each molecule is displaced in a uniform way. Conversely, in a liquid, the molecules are farther apart so that under an applied stress, different groups of molecules may respond differently.

A gas will always expand to fill a container. In a gas, the molecules are much farther apart than in a liquid. Conversely, a liquid will occupy a finite volume based on the amount of mass present. Thus, if the container volume is larger than the volume of the liquid, a *free surface* will separate the liquid from the gas, which will fill the empty space in the container. A gas flowing at high speed (i.e., near or above the speed of sound) behaves much differently from a liquid flow. However, at low speeds (much less than the speed of sound), roughly at speeds less than 100 mph, gas flows are quantitatively similar to liquid flows.

Much of classical fluid dynamics deals with gases such as air, hydrogen, oxygen, helium, and steam and with liquids such as water, oil, gasoline, and alcohol. The properties of these fluids are well known, having often been measured by a number of different research groups and published as tables in multiple appendices in fluid mechanics, heat transfer, and thermodynamics textbooks. In this text, because many of the most interesting applications are biomedical, we will also deal with *liquid mixtures* such as sodium chloride and water and phosphate buffered saline (PBS), a biofluid that is similar in composition to serum in the blood. Such fluids are called *aqueous solutions* and may be considered to have a constant mass per unit volume, or *density*, as is the case for an *incompressible liquid*. In addition, these liquid mixtures contain charged components, or *ions*, which are atoms that have lost or gained an electron. Solutions containing charged species are called *electrolytes* or *electrolyte solutions*. Electrolyte solutions respond to electric fields and provide the physical mechanism for the occurrence of electrokinetic phenomena. For the case of a *dilute* mixture, or low concentrations of the *solute* (i.e., the sodium chloride), these mixtures have properties near to those of water.

1.6 Historical perspectives

The history of the thermal sciences, defined as the fields of fluid mechanics, thermodynamics, and heat and mass transfer, is fascinating because all the governing

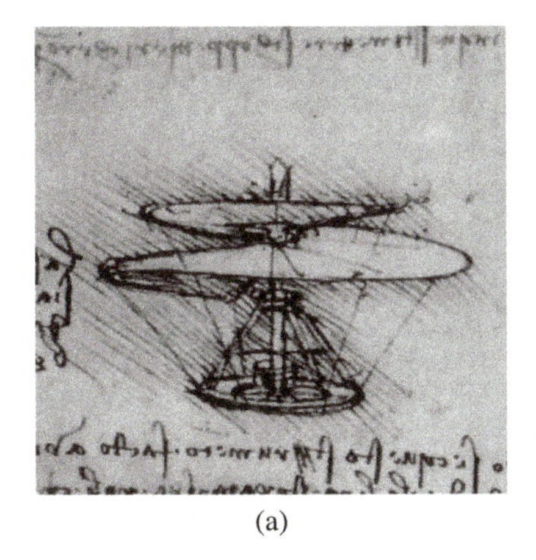

(a)

(b)

Figure 1.8 (a) Da Vinci's drawing of a helicopter. (b) Self-portrait of Da Vinci from Brookhaven National Laboratory.

equations and boundary conditions, except for one, the no-slip condition, were established within a 50 year period from 1800 to 1850. Moreover, James Clerk Maxwell was working on the governing equations of electrostatics in the 1860s. Thus, by around 1870, all the equations and boundary conditions associated with the thermal sciences and electromagnetics were established, except for the no-slip condition. It is for this reason that the history of fluid mechanics is particularly interesting, and we begin there.[1]

1.6.1 Fluid mechanics

Fluid mechanics is one of the component parts of the thermal sciences, a field that includes thermodynamics and heat transfer. As will be seen later in this text, there are many similarities between fluid mechanics and heat and mass transfer and even between the thermal sciences and electrostatics. This is especially true of the boundary conditions that accompany the governing equations derived in rather general form in Chapter 3. Indeed, it will be demonstrated that under rather general conditions, the electrical potential and the fluid velocity are equivalent.

The formal study of fluid mechanics, defined as the study of both stagnant and flowing liquids and gases, may have begun with Leonardo Da Vinci around 1500, when he derived the equation of conservation of mass. He was the epitome of the Renaissance man, a painter, sculptor, botanist, engineer, and scientist, among other talents. His papers include sketches of free jets, vortex shedding behind bluff bodies, and the first outline of a helicopter (Figure 1.8).

In 1687, Isaac Newton recognized the influence of viscosity when he said, "The resistance which arises from the lack of lubricity in the parts of the fluid – other things being equal – is proportional to the velocity by which the parts of the fluid are being separated from each other" (as quoted in White, 2006, p. 2).

[1]Much of the discussion here is based on the book by Tokaty (1971), an excellent reference for the history of fluid mechanics in general.

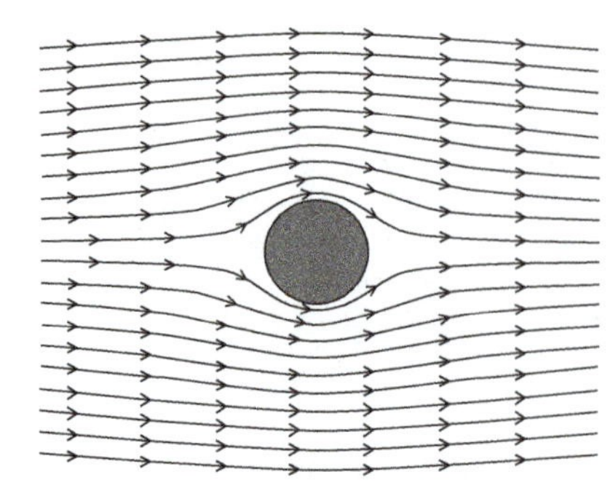

Figure 1.9 Streamlines for Stokes flow past a sphere.

In today's terminology, we would say that stress is proportional to rate of strain; for this reason, we call such fluids *Newtonian*. The shear stress, τ, is defined by[2]

$$\tau = \mu \frac{du}{dy} \tag{1.7}$$

where μ is the *dynamic viscosity*, or simply *viscosity*, and u is the velocity; y is a length. The units of shear stress are N/m^2, or force per area, the same units as pressure. Water, air, oil, and PBS (blood serum) are all Newtonian fluids.

The concept of shear stress originated by Newton would eventually become the central concept in the derivation of the Navier–Stokes equations, the expression of conservation of linear momentum. But first, in the 1660s, Newton had to invent the calculus. The invention of the calculus enabled such applied mathematicians as Daniel Bernoulli and Leonard Euler to approach fluid mechanics problems from a mathematical perspective, resulting in the derivation of the Bernoulli equation in 1738 (appearing in Bernoulli's book *Hydrodynamica*) in a form essentially unchanged today but that neglects the effect of viscosity; the equations governing the flow of an inviscid fluid are called the *Euler equations*.

To incorporate the influence of viscosity, a frictional resistance term was added to Euler's equations by Navier in 1827, by Cauchy in 1828, by Poisson in 1829, by St. Venant in 1843, and lastly, by Stokes in 1845 (White, 2006). Stokes was the first to use the term *coefficient of viscosity*. The resulting equations are now what we call the *Navier–Stokes* equations. The Navier–Stokes equations are similar to the equations of stress in a solid (Dawson, 1976, p. 88), which are termed Navier's equations, and the Navier–Stokes equations are nonlinear, coupled equations encompassing, at a minimum, four dependent variables, three velocity components, and the pressure for an incompressible fluid.

Stokes flow is the name given to very slow flow of a fluid past a body or flow past a very small body. Stokes, in fact, solved the Navier–Stokes equations for flow past a sphere in 1851, thus deriving the drag formula (Figure 1.9)

$$D = 6\pi \mu U_\infty a \tag{1.8}$$

where D is the drag, μ is the viscosity, U_∞ is the velocity far from the sphere, and a is its radius. This formula is among the most famous in all fluid mechanics and is the basis for the derivation of an estimate for the species diffusion coefficient in a liquid mixture.

Even before the Navier–Stokes equations were derived, there was interest in flows within pipes and channels (Tokaty, 1971). It appears as if a Frenchman

[2] Actually, there are six distinct components of stress, as we will see in Chapter 3.

named Chezy wrote the first equation governing flow in a channel around 1813, but it was not until Jean Louis Marie Poiseuille performed experiments in 1838 that the relationship for the volume flow rate, Q (in the notation of Poiseuille),

$$Q \propto \frac{pd^4}{L} \tag{1.9}$$

was established (Tokaty, 1971). In this equation, p denotes the pressure drop, d is the diameter of the tube, and L is its length. Poiseuille was a physiologist who was interested in the laws governing the flow of blood. Subsequently, this relationship was also established by Jacobson in 1860 and by Hagen in 1869 in the form we now call the Hagen–Poiseuille formula. Why Jacobson's name was not added before Hagen's name is unclear. Today, flow in a pipe or channel is called *Poiseuille flow*.

In the same paper by Hagen in 1869, he suggested that in watching the flow of dyed water in a cylinder, the fluid flows in concentric cylinders, which move at different speeds, depending on their distance from the wall. This type of flow has been termed *laminar*. Hagen's experimental result and the hypothesis of the no-slip condition (Goldstein, 1965b; Bernoulli, 1738) allowed Poiseuille to prove theoretically that the velocity across the pipe was parabolic.

Throughout this period, the appropriate boundary conditions to impose on the the aforementioned dependent variables in the Navier–Stokes equations had not been fully identified though the no-slip condition had been suggested. Charles Augustine Coulomb, around 1800, and later Stokes, around 1845, suggested that the fluid should "stick" to a solid boundary, which we now term the *no-slip condition*. It took the pioneering work of Ludwig Prandtl in 1904 to verify the no-slip condition in his discovery of the *boundary layer*, that thin layer near a solid surface in an external flow where the fluid velocity decreases rapidly to zero.[3] The notion of a viscous boundary layer is that viscosity is an important physical effect only in a very thin layer near a solid surface; away from the immediate vicinity of the surface, viscosity has no effect on the flow.

1.6.2 Heat and mass transfer

Heat transfer is the study of energy transport due to a temperature difference (Incropera & Dewitt, 1990). While the basic notions of heat and temperature had been identified in the 1600s, it was not until 1803 that an applied mathematician named Joseph Fourier postulated that heat flows according to a temperature gradient and that at steady state, heat flows from hot to cold, thereby establishing Fourier's law (Figure 1.10):

$$Q_x = -kA\frac{\partial T}{\partial x} \tag{1.10}$$

[3] Actually, the term *boundary layer* has been generalized to include any region where the fluid velocity or any other dependent variable (concentration) rapidly changes.

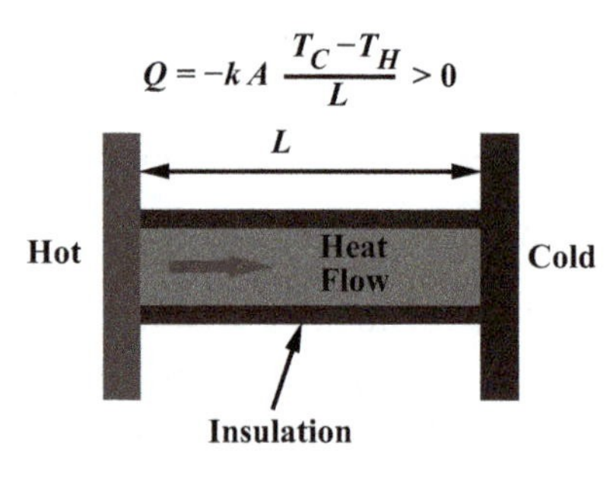

Figure 1.10 Sketch illustrating Fourier's law of heat transfer. In simple cases, the partial derivative is constant and so may be approximated by the finite difference shown.

where Q_x is the heat transfer rate in the x direction in watts (W), A is the area and k is a proportionality constant called the *thermal conductivity*, which has units of W/m° K. This postulate was developed in the course of the development of his famous Fourier series method of approximating functions. However, it was not until the 1850s that the first and second laws of thermodynamics were developed by Rudolf Clausius and William Thomson (Lord Kelvin).[4]

The modern history of mass transfer, the transfer of mass of a species due to a concentration difference of the species (mass transfer requires that the fluid be a mixture), appears to have begun with Thomas Graham, who was interested in the diffusion of gases around 1830 (Cussler, 1997). Graham conducted experiments that took place at constant pressure, a major result being the demonstration that diffusion in liquids is orders of magnitude slower than diffusion in gases. Adolf Fick, who was working in research in the general area of physiology, came across Graham's experiments and postulated that "the diffusion of dissolved material... is left completely to the influence of molecular forces basic to the same law... for the spreading of warmth in a conductor and which has already been applied with great success to the spreading of electricity" (Cussler, 1997, p. 14). In other words, Fick's law was to be of the same form as the law that Fourier developed in 1822 for the heat transfer rate through a body. Thus Fick wrote, for the mass transport of a given species in the x direction (Figure 1.11),

$$J_i = -DA\frac{\partial c_i}{\partial x} \tag{1.11}$$

where D is the diffusion coefficient, A is the cross-sectional area, c_i is the concentration, and J_i is the flow rate of species i in moles/sec, for example. Note the similarity to Fourier's law and to the equation for the viscous force at a surface ($F = A\tau$) defining Newton's law. Indeed, the viscosity, thermal conductivity, and thermal diffusivity or diffusion coefficient all play the same role in these relationships and are called *transport properties*.

Fick published his first paper on diffusion in 1855, just after the Navier–Stokes equations were developed, and he was initially unsure of his hypothesis. He understood the difference between thermodynamic equilibrium and a steady

[4] There are other pioneers of heat transfer associated with convection and radiation. For more on this topic, I found the following Web site interesting: http://www.seas.ucla.edu/jht/pioneers/pioneers.html. Another good Web site on the history of science and engineering is the Lienhard engines site: http://www.uh.edu/engines/.

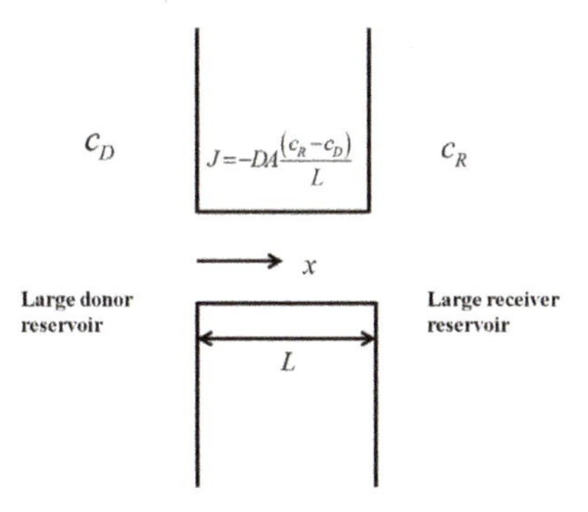

Figure 1.11 Sketch illustrating Fick's law of mass transfer. In simple cases, the partial derivative is constant and so may be approximated by the finite difference shown. The presence of large donor and receiver reservoirs ensures that the concentrations do not depend on time.

state and tried to integrate the unsteady mass transport equation,

$$\frac{\partial c_1}{\partial t} = D\left(\frac{\partial^2 c_1}{\partial x^2} + \frac{1}{A}\frac{\partial A}{\partial x}\frac{\partial c_1}{\partial x}\right) \tag{1.12}$$

but the numerical effort was too much. He even tried to measure the second derivative experimentally and finally succeeded in designing an experimental system consisting of a glass cylinder containing sodium chloride on the bottom and a larger volume of water on the top. After periodically changing the water in the top of the container, he was able to establish a steady concentration gradient (Figure 1.12). There are a great many other pioneers in the history of mass transfer, but Fick and Graham are certainly the leaders in diffusion of mass.

1.6.3 Electrokinetic phenomena

Researchers were aware of electrokinetic phenomena as far back as 1808, when F. F. Reuss discovered that an electric field can induce flow through a tube (Burgeen & Nakache, 1964), which, we have seen, is called *electro-osmotic flow*. After about 50 more years of research, G. Wiedmann deduced that the volume flow rate through a tube was proportional to the applied current. In 1859, Quinke discovered the phenomenon of *streaming potential*.

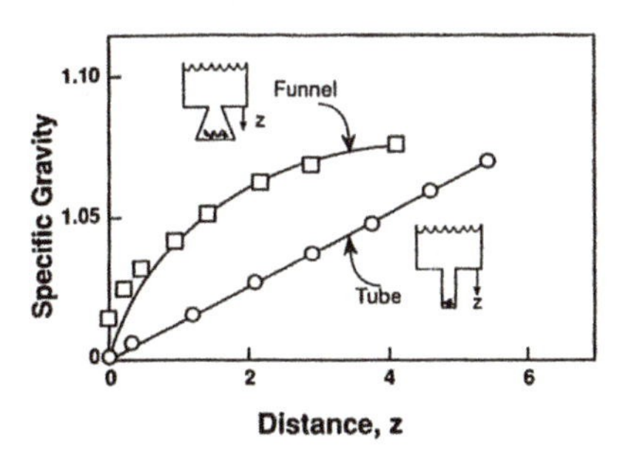

Figure 1.12 Fick's experimental data; the data for the tube indicates steady state mass transfer, and the experimental comparison is very good. This was his third attempt to validate his own law! From Cussler (1997, p. 17). The result for the funnel shows transient behavior.

James Clerk Maxwell laid the foundations of the field of electrostatics with his famous paper that presented what are now called *Maxwell's equations* in 1864. Prior to that, in 1856, he published a paper titled "On Faraday's Lines of Force" (Maxwell, 1847), in which he suggested that "the present state of electrical science seems particularly unfavourable to spaculation" (p. 27). Even in that paper, he recognized the link between the electrical properties of a fluid and electrostatics, discussing, in a section titled "Theory of the Motion of an Incompressible Fluid," a version of the continuity equation and what turns out to be the Navier–Stokes equations!

At about the same time Wiedemann and Quinke were making their discoveries experimentally, James Clerk Maxwell was in the process of deriving the equations governing electromagnetic phenomena. Although he was not truly the originator of the individual equations of Coulomb, Ampere, and Faraday, he was the first to derive them independently in a unified set (Constant, 1958). His equations include an important modification to Ampere's law, the law relating the magnetic field to the current density; the collective set of equations is known as the Maxwell's equations.

In this presentation of the history of the thermal sciences and electrokinetics, it is difficult to know when to stop. In the late 1800s, Helmholtz was working on the *electrical double layer*, and as mentioned earlier, in 1904, Prandtl discovered the boundary layer, thereby suggesting the last boundary conditions required in the theoretical study of the thermal sciences and electrokinetic phenomena. What is clear is that from around 1800 to 1870, the theoretical study of the thermal sciences and electrokinetic phenomena exploded and set the stage for the major discoveries of the late nineteenth and early twentieth centuries.

1.7 The thermal sciences

The design of air-conditioning systems, propulsion and power systems, internal combustion engines, and many other systems involves not only fluid mechanics but also thermodynamics, heat transfer, and mass transfer. Together these disciplines form the core of the subject known as the *thermal sciences*. Fluid mechanics is the primary subject of this text, and in this section, the essence of thermodynamics and heat and mass transfer is briefly described. Often heat and mass transfer are lumped together because the disciplines use similar methods of analysis.

Thermodynamics is the study of

- Energy transformation
- Energy transmission
- Energy conversion

It forms the basis for an engineer's initial analysis of the generation of power in power plants and the operation of pumps, turbines, compressors, and fans (Figure 1.13). The science of thermodynamics determines the direction in which processes occur. Thermodynamics is an equilibrium science in the sense that it can predict the final temperature of a system but not the rate at which a system achieves a final uniform temperature. The dependent variables in a thermodynamic analysis

Figure 1.13 Picture of the Navajo Power Plant in Arizona, courtesy of the United States Geological Survey. Thermodynamics is used as an engineer's initial analysis to determine the operating characteristics of a power plant.

are uniform in space, with no gradients being permitted (Figure 1.14).

As you learned in your undergraduate course work, thermodynamics is characterized by two laws: the first law of thermodynamics, which is an equilibrium expression of conservation of energy, and the second law of thermodynamics, which determines the direction in which a process will occur. Both these equations are algebraic equations typically defining the relationship between work and heat transfer and entropy changes as a function of temperature and pressure.

Josiah Willard Gibbs

Josiah Willard Gibbs (1839–1903) was a Renaissance man being at once an engineer, theoretical physicist, chemist and mathematician. Most students meet Gibbs's work early in their studies as the father of Vector Analysis. The analysis of fluid mixtures is emphasized in this book and the importance of Gibbs's work cannot be underestimated. Gibbs made major advances in our understanding of the thermodynamics of fluid mixtures. In mid-career he published his famous paper Gibbs (1961) "On the Equilibrium of Heterogenous Substances," a paper that is viewed as a significant advance in the theory of physical chemistry. We know Gibbs in thermodynamics for his development of the Gibbs phase rule, Gibbs free energy, the Gibbs-Duhem equation and for the introduction of the chemical potential. The recently published book by Kjelstrup & Bedeaux (2008) is an extension of Gibbs' work valid at equilibrium to non-equilibrium systems.

The rate at which an energy transfer process takes place can be predicted by the discipline called *heat transfer*. Heat transfer describes energy flow as a result of a temperature difference. The combustion process in a gas turbine involves very large temperature gradients and many chemical reactions, and so heat and mass transfer effects must be considered, along with fluid dynamics, in that problem. Heat transfer occurs in a direction of decreasing temperature and is a nonequilibrium phenomenon.

Heat transfer in the absence of a phase change can occur in three different modes:

- Conduction
- Convection
- Radiation

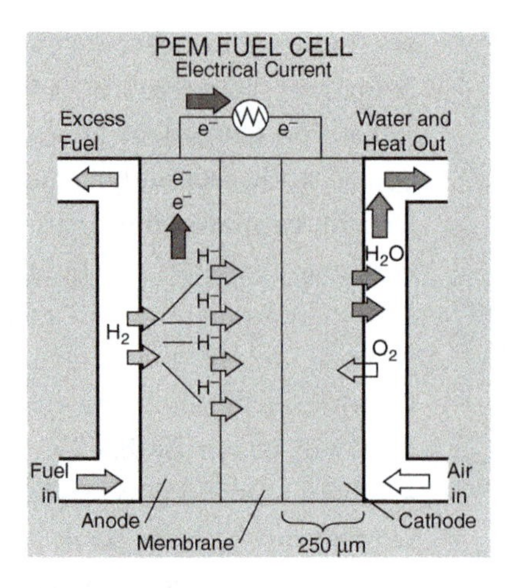

Figure 1.14 A thermodynamic analysis is used as an engineer's initial attempt to determine the operating characteristics of a polymer electrolyte membrane (PEM) fuel cell. Image courtesy of the Savannah River National Laboratory of the Department of Energy.

Conduction is associated with the random molecular motion of the molecules making up the material of interest and is described mathematically by Fourier's law. *Convection* is the transfer of energy between a solid surface and a moving fluid: *forced convection* is the heat transfer within a fluid driven by external means, and *free convection* is the heat transfer through a fluid due to a temperature difference. *Radiation* refers to the heat transfer due to electromagnetic waves moving to or from surfaces. Heat transfer rates are governed by the nonequilibrium form of conservation of energy in which temperature gradients in all three spatial dimensions can occur.

Analogous to heat transfer, *mass transfer* describes the flow of a mass of a given species as a result of a driving mechanism; the type of diffusion that is analogous to Fourier's law of heat transfer is Fick's law, whereby mass transfer occurs as a result of a concentration difference. As with heat transfer, mass of a given species will flow from a higher to a lower concentration according to Fick's law, and we speak of *convective* and *diffusion mass transfer*. When we speak of mass transfer, it is understood that the fluid in question is a mixture or a combination of molecules of distinctly different types, in contrast to a *pure fluid*, which consists of an ensemble of like molecules (Figure 1.15). Water, for example, is a pure fluid, and salt water (i.e., salt fully dissolved in water) is a mixture. Thus we can speak of a *flux* of salt (i.e., mass/area/time) and a flux of water. The total flux of the mixture is obtained by a suitable averaging of the flux of water and the salt (Figure 1.16).

Figure 1.15 Mass transfer occurs in the burning of rocket fuel. In general, the fuel may be a liquid or a solid, and oxygen or air induces combustion. The products of the combustion shown are a mixture of the burned fuel and air.

The three most common modes of mass transfer are

- Diffusion
- Convection
- Chemical reaction

Convection mass transfer is entirely analogous to convection heat transfer. In addition to mass transfer due to a concentration difference, as just described, diffusion mass

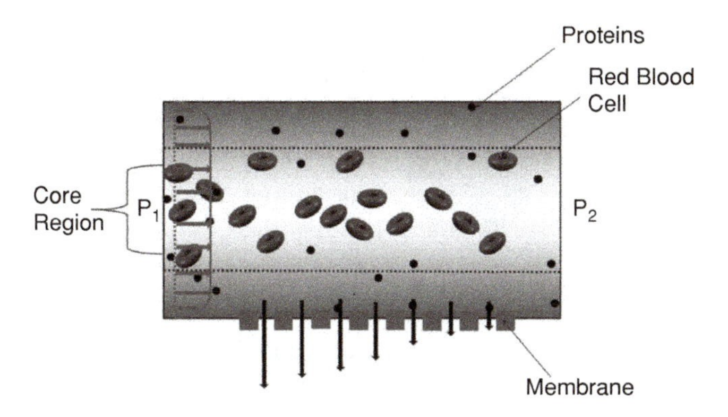

Figure 1.16 Mass transfer takes place in the kidney. Salt and other small molecules are filtered out of blood as urine under the combined action of diffusion and convection.

transfer may also occur as a result of a pressure difference or *pressure diffusion*. *Thermal diffusion* occurs as a result of a difference in temperature. These last two modes of mass transport are usually not important in the applications discussed here.

One final mode of mass transfer, which is nominally termed diffusion, is mass transport due to an electrical potential difference. This mode of mass transfer is the driving mechanism behind electro-osmotic flow.

It is important to note that the term *mass transfer* is not what is meant by flow of a pure fluid such as water in a pipe or channel, even though mass transfer is taking place. Mass transfer in the present context means that a given species within a fluid mixture is transferred relative to bulk fluid motion.

1.8 Electrostatics

To use an electric field to move a fluid, the fluid must be electrically conducting. The electrically charged components of such a fluid are called *electrolytes*; an example of an electrolyte is blood serum in aqueous solution, whose main components are dissociated sodium chloride and water. The electrolyte can be transported through a channel using an electric field, provided that the walls of the channel are electrically charged; charged solid surfaces are discussed in the next section.

That materials could be electrically charged was known to the Greeks as far back as 600 B.C. For example, they found that amber rubbed with wool could attract light objects. Thus amber could acquire an electric charge. The term *electric* is derived from the Greek word *elektron*, which means "amber" (Burgeen & Nakache, 1964).

There are two types of electric charge, negative and positive charge, with the fundamental properties that like charges repel and unlike charges attract. All material is made up of subatomic particles, and these fundamental particles are the negatively charged electron, the positively charged proton, and the neutral neutron.

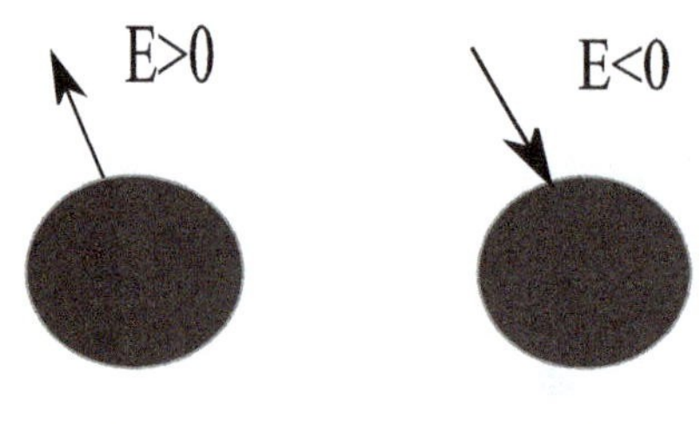

Figure 1.17 Direction of the electric field from (left) positively and (right) negatively charged point spheres.

An atom is a collection of these subatomic particles, with protons and neutrons generally arranged in a sphere called the *nucleus*, around which are the electrons; this atom is electrically neutral in that it has no net charge. The radius of the nucleus of an atom is about 10^{-14} m (Condon & Morse, 1929). Atoms may gain or lose electrons, and the process of doing this is called *ionization*. Loss of an electron means that the atom is positively charged, and the resulting atom is called a *cation*. A gain of an electron means that the atom is negatively charged, and the resulting atom is called an *anion*. Both cations and anions are called *ions*.

Electrically charged materials are generally grouped into two classes: *conductors* and insulators, or *dielectrics*. Metals are, in general, electrical conductors, with electrons in their outer shell free to move, the *free charge*, and thus able to pass electrical current freely; most other materials are electrical insulators that do not pass electrical current in nearly the amount as conductors and in general do not contain free charge. Electrolyte solutions are an exception to the rule that dielectrics do not pass current. On the contrary, electrolyte solutions contain dissociated ions that can move freely in solution; thus electrolyte solutions are *ionic conductors*.

The essence of electrostatics is embodied in *Coulomb's law*. Charles Augustin de Coulomb (1736–1806) showed experimentally that the force acting between two charged particles is

$$F = \frac{ee'}{4\pi\epsilon_0 r^2} \tag{1.13}$$

where ϵ_0 is the *electrical permittivity* of a vacuum, $\epsilon_0 = 8.85 \times 10^{-12}$ C^2/N m^2, r is the distance between the two charges, and e and e' are the charges of the particles. For one proton, $e = 1.602 \times 10^{-19}$ C, where C is coulombs. Note that this law is of the same form as Newton's law of gravitation,

$$F = \frac{Gmm'}{r^2} \tag{1.14}$$

where $G = 6.67300 \times 10^{-11}$ m^3/kg sec^2 is the gravitational constant. The unit of charge is called the coulomb (C), and equation (1.13) is called Coulomb's law.

The *electric field* at any point is defined as the force acting on a single charge e' at that point. The electric field generated by a single particle of charge e is given by

$$E = \frac{F}{e'} = \frac{e}{4\pi\epsilon_0 r^2} \tag{1.15}$$

and the units of electric field are newtons per coulomb, or N/C. Note that if e is positive, the electric field is directed outward in the radial direction (Figure 1.17), and it is directed radially inward if the charge is negative.

If the medium is not a vacuum, then a relative permittivity, or *dielectric constant*, is defined as

$$\epsilon_r = \frac{\epsilon_e}{\epsilon_0} \tag{1.16}$$

The electrical permittivity is a transport property just as is viscosity. Air is a poor conductor of electricity; water is a much better conductor of electricity than air but is still a dielectric.

1.9 Electrolyte solutions

We shall be concerned with the motion, the chemical kinetics, and the colloidal nature of mixtures that involve the study of ions, the solute species, and a solvent, which is usually water. The field of electrochemistry has traditionally comprised two rather separate and distinct subjects (Bockris & Reddy, 1998): the study of ionic solutions and electrodics. Electrodics involves the study of electrodes, usually metallic, which generate an electric field, thus interacting with the ions in solution in the interfacial region between the metallic and solution phases. This portion of electrochemistry is extremely important in understanding the operation of fuel cells. *Ionic solutions* are those mixtures that contain ions or charged species; in this book, most, if not all, the ionic solutions that we consider are aqueous solutions. The charged nature of ionic solutions means that they can be transported using electric fields; thus, although rather distinct subjects in their own right, it is the electrodes which create the electro-osmotic flow of electrolyte solutions discussed in Chapter 9. There are many applications in which electrochemistry is an essential tool, and many of them are described by Bockris and Reddy (1998).

The fluids of interest in this text are aqueous solutions of electrolytes in which water is the solvent and ions are the solute. Sodium and chloride, when dissolved in water, dissociate into Na^+ and Cl^-, and these ions are surrounded by the water molecules, a process known as *hydration*. In electrochemistry, a *buffer* consists of a weak acid and a related salt so that the pH of the solution does not change; sodium chloride, sodium phosphate, and potassium phosphate are the buffers in phosphate buffered saline, a mixture with a composition close to blood serum and which is a common biological fliuid. The composition of PBS is given in Table 1.2.

Table 1.2. Composition of PBS. The pH of standard PBS is 7.4

Symbol	Molarity (M)
Na^+	0.15420
Cl^-	0.14067
K^+	0.00414
$H_2PO_4^-$	0.00147
HPO_4^{2-}	0.00810

Notation alert: In this section, we will employ the notation used most often in the electrochemistry literature, using square brackets to denote activity. This eliminates the need to use long and cumbersome subscripts. *Activity* in chemistry is defined as $a = \gamma c$, where c is concentration in mole/liter and γ is a dimensionless *activity coefficient*. On the other hand, engineers define activity in terms of mole fraction, and activity is thus dimensionless: $a = \gamma X$ (Moran & Shapiro, 2007; Denbigh, 1971).

A *Lewis acid* is a proton (H^+) donor so that in many cases, it is negatively charged, and a *Lewis base* is a proton acceptor and so is positively charged. The pH of a solution is defined as $\text{pH} = \log_{10} 1/[H^+]$ or $[H^+] = 10^{-\text{pH}}$ and measures the acidity of the solution. For pure water, which is electrically neutral $[H^+] = [OH^-] = 1 \times 10^{-7}$ or pH $= 7$. This definition of pH relies on the definition of activity as $[H^+] = \gamma c_{H+}$, which means that there is a dimensional quantity inside the logarithm defining pH, a mathematical difficulty that persists today. This difficulty can be avoided by defining the activity as $a = \gamma c/c_0$ where $c_0 =$ 1 mole/liter. The definition of activity will be clear from the context.

Acids and bases are linked by the general proton transfer reaction

$$\text{HA} + \text{B} \rightleftharpoons \text{A}^- + \text{BH}^+ \tag{1.17}$$

The products of this reaction are termed an acid–base pair. The extent of the dissociation is measured by an equilibrium constant, defined by

$$\text{K} = \frac{[\text{A}^-][\text{BH}^+]}{[\text{B}][\text{HA}]} \tag{1.18}$$

The concept of an equilibrium constant described here is the same as is customarily used in the thermodynamic analysis of combustion, another example of a type of chemical reaction. The limit $K \to 0$ means that little dissociation has occurred. The *valence* of an ion is a whole number and indicates the number of electrons the ion has gained or lost. In aqueous solution, for example, calcium has a valence $z_{Ca} = +2$. The protein albumin has a valence estimated to be $z_{\text{Alb}} = -17$ (Peters, 1996).

Gatorade is an electrolyte mixture that replenishes the electrolytes in the body after intense physical activity. The salts in the blood serum are removed from the body when you sweat.

1.10 The electrical double layer

In dissociated electrolyte mixtures, where a molecule (such as NaCl) has broken up into its ionic components Na^+ and Cl^-, even in the absence of an imposed electric field, an *electrical double layer* (EDL) will be present near the charged surfaces of a channel or tube; this situation is depicted in Figure 1.18. For example,

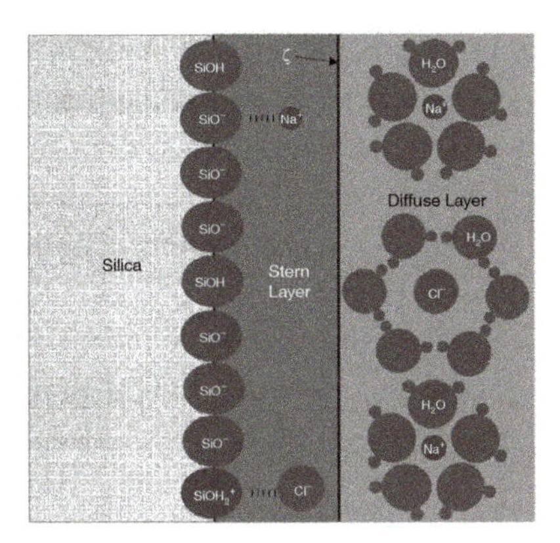

Figure 1.18 Graphic representation of the hydration of ions by water molecules near a charged silica surface. The electric double layer (EDL) consists of a layer of counter ions pinned to the wall, the Stern layer, and a diffuse layer of mobile ions outside that layer. Here the wall is shown as being negatively charged, and the ζ-potential is defined at the Stern plane.

if a wall is negatively charged, an excess of positive charge will accumulate near the wall because opposite charges are attractive. The thickness of the layer near the wall is of the order of the ionic diameter of hydrated counterions, and this layer is called the *Stern layer* (Figure 1.18). Any charged molecule sufficiently far away from the surface will be "screened" from the negatively charged wall by the presence of the positive ions. The layer outside the Stern layer consists of mobile ions and is called the diffuse layer, and the total system, including the Stern layer and the diffuse layer, is called the EDL (Hunter, 1981). The presence of the charged surface and the ions thus creates an electric field.

The nominal length scale associated with the EDL is the *Debye length*, defined by

$$\lambda = \frac{\sqrt{\epsilon_e RT}}{FI^{1/2}} = \frac{1}{\kappa} \tag{1.19}$$

where F is Faraday's constant; ϵ_e is the electrical permittivity of the medium; I is the *ionic strength*, $I = \sum_i z_i^2 c_i$; c_i are the concentrations of the electrolyte constituents at some reference location; R is the universal gas constant; z_i is the valence of species i, and T is the temperature.

Petrus (Peter) Josephus Wilhelmus Debye

Petrus (Peter) Josephus Wilhelmus Debye (1884–1966) received his first degree in Electrical Engineering in 1905 and then a PhD in Physics in 1908 under the tutelage of Arnold Sommerfeld. At Zurich in 1920 as a Professor of Physics Debye along with his assistant Erich Huckel developed a theory of why concentrated electrolytes exhibit behavior different from classical ideal solutions. They attributed the discrepancy to the presence of electrostatic interactions between ions; these results are presented in Debye and Huckel (1923).

Notation alert: In the chemistry literature, the ionic strength is usually defined by $I = 1/2 \sum_i z_i^2 c_i$. A one molar solution (1 M = 1 mole/liter) of dissociated sodium chloride is defined by the concentration of each species being 1 M. With the current definition, the same mixture has the ionic strength equal to 2 M – 1 M sodium and 1 M chloride, for example. The present definition is to avoid a 2 floating around in other equations involving the Debye length. It must be said, however, that the classical chemistry definition works only for monovalent species for which the concentrations are equal. For multivalent ions, the classical definition is extremely confusing. Where absolutely necessary, however, the classical definition will be used.

Internal flows in cylindrical or rectangular channels are the most common form of transport in micro- and nanofluidics. The material used for the walls of these channels is very important because of the large surface-to-volume ratio; silica, when immersed in water, becomes charged and loses a proton, according to the reaction

$$\mathrm{SiOH(solid)} \leftrightarrow \mathrm{SiO^-(solid)} + \mathrm{H^+(aqueous)} \tag{1.20}$$

Thus the silica walls are negatively charged and will attract the positive ions which are in solution, forming a thin layer of positively charged ions pinned at the wall. For a positively charged wall, the roles of the positive and negative charges are reversed.

It should be pointed out that there are many silica surfaces of different chemical composition, and no attempt will be made in this text to make this distinction. The primary difference between silica surfaces is the value of the ζ potential.

The EDL will always be present near a charged wall, even in the absence of a flow. If electrodes are placed upstream and downstream of a rectangular channel, then a separate and distinct electric field is set up, and the ions will be set into motion, dragging the solvent with it and inducing *electro-osmosis*. The device depicted in Figure 1.19 is called an *electro-osmotic pump*. The abundance of cations over anions within the diffuse layer generates the bulk motion of the fluid. In this pump, the anions will be attracted to the positive electrode, which is termed the *anode*. Likewise, the cations will be attracted to the negative electrode, or *cathode*. Thus the cations will move at a speed slightly greater than the bulk electro-osmotic flow, while the cations will move at a speed slightly slower than the bulk electro-osmotic flow. The speed of the ions depends on their valence; the higher the positive valence, the faster the cation speed. The unique feature of electro-osmotic flow is that for a wide range of parameters away from the walls of the channel, the velocity will be constant, unlike for Poiseuille flow, which is parabolic.

The thickness of the EDL is actually an asymptotic property much like the boundary layer thickness in classical external fluid mechanics (White, 2006), which is defined as the location where the velocity reaches 1 percent of its value

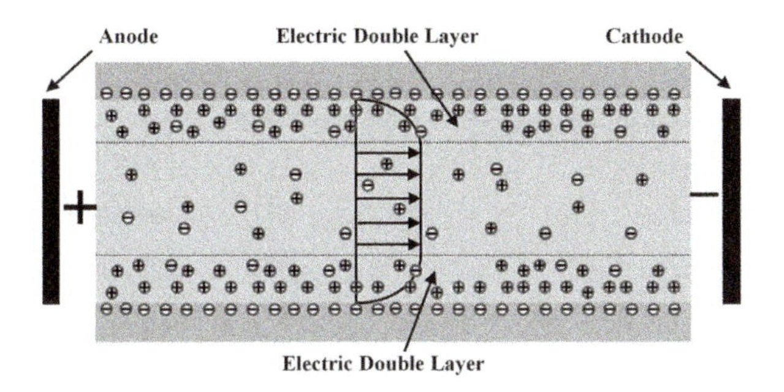

Figure 1.19 Electrodes placed upstream and downstream of a channel having charged walls, generating an electro-osmotic flow.

far from the wall. If we define the dimensionless parameter

$$\epsilon = \frac{\lambda}{h} \tag{1.21}$$

where h is the channel height, then for $\epsilon \ll 1$, the dimensionless thickness of the EDL, defined here as the location where the electrical potential reaches 1% of its value far from the wall, is normally $\delta/h \sim 4\epsilon - 6\epsilon$, depending on the ionic strength. This situation is similar to the formula for the thickness of the Blasius boundary layer on a flat plate, denoted by δ and $\delta/x = 5\ \mathrm{Re}^{-1/2}$, where x measures distance from the leading edge of the plate and Re is the (large) Reynolds number. The Reynolds number is the main dimensionless parameter associated with viscous flow, and for the Blasius boundary layer on a flat plate of length L, it is defined by

$$\mathrm{Re} = \frac{\rho U_\infty L}{\mu} \tag{1.22}$$

where ρ is the fluid density, μ is the viscosity, U_∞ is the velocity far from the plate, and L is its length. If $\epsilon \sim 1$, we say that the EDLs overlap. The ionic concentration and electrical potential distributions within the EDL are discussed extensively in Chapter 7.

1.11 Colloidal systems

The solutions considered by electrochemists often contain embedded particles. The term *colloid science* refers to the science of transporting particles and molecules of a size less than 1 μm and greater than about 10 Å. It should be cautioned that the term *colloid* is very broad, and the intent here is not to define specific terms relating to different colloid systems but to give an overview of the types of colloidal systems that are common in micro- and nanofluidics.

A colloidal dispersion can be a mixture that contains relatively large, soft, or hard polymeric molecules; solid particles; or small molecules such as salt-dissociated ionic species. Most ions have a dimension on the order of angstroms,

Table 1.3. Examples of two-phase colloidal systems. A *gold sol* is a mixture of gold nanoparticles

Continuous	Dispersed	Name	Example
Liquid	gas	foam	beer
Liquid	liquid	emulsion	milk
Liquid	solid	colloidal suspension	gold sol

and the radius of a globular glucose molecule is about 1 nm; the globular protein albumin has a nominal radius of about 3.5 nm. Examples of colloidal systems are given in Table 1.3 (Murrell & Jenkins, 1982). Mixtures containing particles larger than 1 μm have been termed by some researchers as *suspensions* (Raymond, 2007), and modeling of these types of systems is extremely difficult. A *gold sol* is a mixture of gold nanoparticles.

Thus colloidal systems include saltwater mixtures, biofluids such as blood, and systems including proteins and DNA, which is a biopolymer. Blood is a fluid that consists of suspended particles (red and white blood cells) in an aqueous solution of proteins and salts (electrolytes) called *plasma*. Colloidal systems are, more often than not, aqueous systems.

Polymers are large-molecular-weight materials made up of a large number of atoms, linked together in a chain. There may be one or several repeating groups, depending on the specific polymer. Chemists consider a large-molecular-weight polymer to be above approximately $M_w = 500$ g/mol (Murrell & Jenkins, 1982). For example, polystyrene beads, which are often used to measure fluid velocities (Sadr *et al.*, 2004, 2006, 2007), have a molecular weight of $M_w = 10^8$ g/mol. Because polymers are so large, researchers have suggested that solutions containing polymers can never be considered ideal solutions. However, this rather obvious statement is contradicted by evidence that these solutions often behave as if they were ideal (Murrell & Jenkins, 1982).

Polymers are often soft in the sense that they will change shape in solution based on the nature of its constitution. It is not hard to imagine that a polymer may fold in on itself in water if some of its groups are hydrophobic, or "water hating." Polymers can also stretch out linearly, as DNA does when it is exposed to sufficient fluid stress.

Other types of colloid systems are dispersions of soluble and insoluble material in water. The former are called *lyophilic colloids*, and the latter are called *lyophobic colloids*. The dispersed phase (i.e., the particles) can actually aggregate and form larger complexes and, in some cases, must be considered two-phase systems. Lyophilic colloids are commonly treated as a one-phase but perhaps multicomponent mixture. In this case, the *continuous phase* is merely the solvent: water.

Gels are colloidal fluids that contain large macromolecules made up of long polymer chains. Aqueous solutions containing naturally occurring proteins, such as collagen or sugars, are in the gel family. The water molecules interact with one or several of these long chain polymers and occupy the spaces between the polymers. The term *emulsion* is used to denote an aqueous solution containing oils or fat (Table 1.3).

In theory, the most important factors influencing the behavior of a colloidal mixture are particle size, particle shape and rigidity or flexibility, channel and tube surface properties (i.e., surface charge density), particle–particle interactions, and particle–solvent interactions. In practice, however, the most important properties of a dilute electrolyte mixture are the channel and tube surface properties and the particle solvent interaction through hydration (i.e., solvation) (Shaw, 1969) (Figure 1.18). For dilute mixtures typical of biological applications, particle–particle or molecule–molecule interactions should be a second-order effect. Moreover, particle size only becomes important when the ratio of particle size to the channel or tube dimension is $O(0.1)$, although the specific limit may depend on the application.

In colloidal systems, interparticle interactions are usually represented by a potential that incorporates the attractive van der Waals forces and the repulsive electrostatic forces. The van der Waals force represents the effect of the finite size of individual molecules, and the van der Waals equation of state, to be discussed later, represents a correction to the perfect gas law, for which intermolecular forces are not negligible.

If electrostatic forces are negligible, then the van der Waals forces will cause colloidal particles to stick together; this is an indication that the colloidal mixture is unstable. Van der Waals forces are a class of forces between molecules or parts of molecules other than those due to covalent bonds (the sharing of pairs of electrons between molecules) or the explicit electrostatic forces. These forces arise from dipole–dipole interactions, from induced dipole–dipole interactions, and from dispersion forces due to transient dipoles in atoms. For example, water has a permanent *dipole moment* (see Chapter 6 for the definition of *dipole moment*), and so there will be a strong van der Waals force existing between two neighboring water molecules. Van der Waals forces can overwhelm electrostatic forces, even if the particles are charged.

The study of colloidal stability was first described in separate works by Derjaguin and Landau (1941) and Verwey and Overbeek (1948), and for this reason, theoretical work on the stability of colloidal systems is named for these four authors as *DLVO theory*.

Colloidal particles can be used as simulants for DNA and other biomolecules. As will be seen, colloidal particles, such as polystrene beads, can be used to measure velocities near the wall in a microchannel. Negatively charged and fluorescence-labeled polystyrene (PS) beads of size 3–40 nm have been used to examine the transport of particles in micronozzles, as depicted in Figure 1.20 (Wang *et al.*, 2005, 2008). The length of the nozzle is 650 μm, and the left-end and right-end heights are 20 μm and 130 μm, respectively; the nozzle is 40 μm deep, and the walls are negatively charged. The particles are immersed in a 0.1 M NaCl solution, and an 80 V/cm DC electric field was applied across the tapered

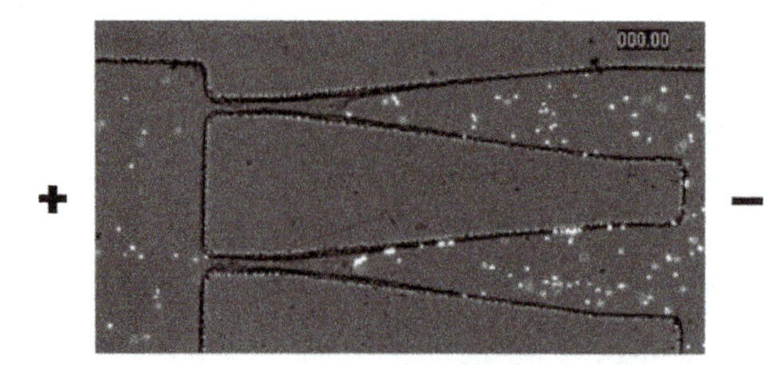

Figure 1.20 Polystyrene beads are sometimes used as biomolecule simulants. The particle donor reservoir is on the right and the particle receiver reservoir is on the left of the diffuser (Wang *et al.*, 2005, 2008). The walls are negatively charged, and the positive electrode is placed in the particle receiver. The flow is from left to right and thus it is a diffuser. The polystyrene beads are shown, and their motion is from right to left, in a direction opposite to the bulk electro-osmotic flow. Picture courtesy of L. J. Lee. Used with permission.

channel, with the positively charged electrode placed in the particle receiver region (left). Since the walls are negatively charged, cations will always be more populous than anions and will move to the negative electrode, dragging the bulk fluid flow from left to right. Thus, in terms of the bulk electro-osmotic velocity, the device is a diffuser. The negatively charged PS beads move from right to left in a direction opposite to the bulk flow, so for these particles, the device is a nozzle. More will be said about such electrokinetic phenomena in Chapter 9.

1.12 Molecular biology

According to Freifelder (1987), the term *molecular biology* was first used by William Astbury in 1945, when he was studying the chemical and physical characteristics of macromolecules. Significant advances were made after that, when it was recognized that the study of biologically simpler entities, such as bacteria and bacteriophages (bacterial viruses), was necessary prior to the study of more complicated entities such as animal cells. This kind of activity led Watson and Crick to make their discovery of the structure of DNA in the 1950s (Crick and Watson, 1953) (Figure 1.21). The sizes of some important biological molecules are depicted in Figure 1.6.

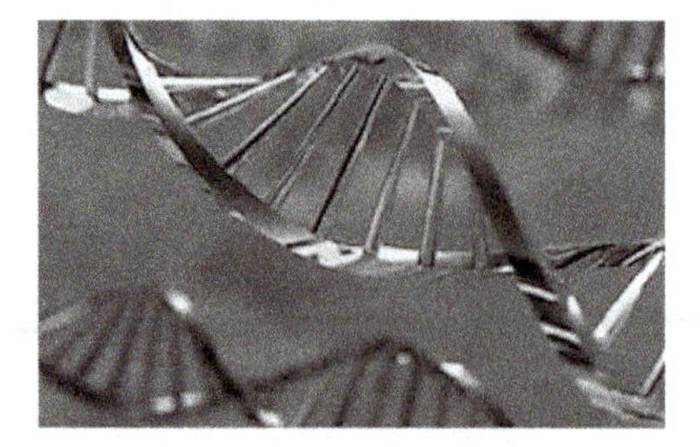

Figure 1.21 Crick and Watson (1953) revealed the structure of DNA in 1945. Sketch courtesy the National Institute of General Medical Sciences of the National Institutes of Health.

Today, more than 50 years later, rapid advances are being made in determining the structure and function of *single* proteins and other biomolecules. Such studies have application in many areas, including rapid molecular analysis, gene therapy, biochemical sensing, and drug delivery. These applications require the development of engineering tools to design entire systems, often specifically for the study of single biological macromolecules. Since the nanoscale is the scale of biology, nanotechnology is being used to create a new field called *molecular medicine* (Roco, 2005). As Roco points out, significant advances have been made in "measurements at the molecular and subcellular levels and in understanding the cell as a highly ordered molecular machine" (Ishijima & Yanagida, 2001, p. 3). Moreover, the molecular origin of disease can be better understood, leading to a much better chance for a cure to such diseases as cancer, diabetes, cystic fibrosis, and neurological disorders, to name only a few.

Cells are populated by *macromolecules*; in vivo, they are very large molecular weight (10^4–10^{12} Da) (1 Da = 1 gram/mol) polymers that inhabit a cell which may contain 10^4–10^5 different types of molecules. Conversely, many of the molecules that inhabit the cell are very small, having molecular weights no more than several hundred Da.

Biomacromolecules or biopolymers are of three different types: proteins, nucleic acids, and polysacchrides (sugars). The three most important properties of these macromolecules from an engineering perspective are their physical size, shape, and charge, if any. For example, most proteins are negatively charged, while glucose is uncharged. As will be seen, this one property of a molecule significantly affects transport mechanisms in nanopore membranes.

The three-dimensional structure of many proteins has been determined experimentally, usually by protein crystallography, and the data have been stored in the Protein Data Bank (PDB) by Brookhaven National Laboratory. The Web site is located at http://www.rcsb.org, and the PDB was created in 1971. The Worldwide Protein Data Bank was created in 2004 to coordinate the worldwide dissemination of protein structures. Ribbon diagrams of protein structure, also known as Richardson diagrams (Richardson, 1973), which are an interpolation of the smooth curve through the polypeptide chain of the protein, can be downloaded in a variety of formats. Over 50,000 structures exist in the Worldwide Protein Data Bank site. A detailed discussion of the structure and function of proteins and other biomolecules appears in section 8.3. One representation of the primary protein structure is depicted in Figure 1.22.

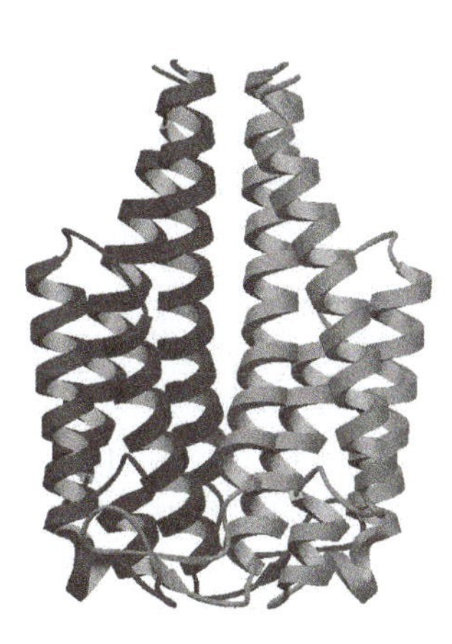

Figure 1.22 Ribbon view of the structure of a typical protein, one type of macromolecule that resides in a cell. Proteins are discussed in detail in Chapter 8. Image courtesy of Brookhaven National Laboratory.

In addition, there has been much work on molecular simulation of proteins. These calculations can identify the conformation and binding characteristics of proteins surrounded by water (Becker & Karplus, 2006; McCammon & Harvey, 1987) and the transport of single- and double-stranded DNA (Cui, 2004). These simulations provide a molecular picture of biomacromolecules and biopolymers not possible with the continuum approach.

Much of the analysis of macromolecules in the balance of the book concerns proteins. The study of the structure and function of a set of proteins, or *proteome*, associated with a given set of genes, or *genome*, is called *proteomics*. Increasingly, micro- and nanofluidic devices are being used to analyze single proteins. For example, a *mass spectrometry* device can be used to measure the charge to mass ratio of various ionic constituents that have been fragmented from specific proteins (Glazer & Nikaido, 2007; Landers, 1994).

1.13 The convergence of molecular biology and engineering

Roco (2005) has pointed out that nanotechnology is beginning to play a revolutionary role in biomedicine. Indeed, all biological systems originate with entities that have spatial dimensions on the nanometer scale. Roco (2005) defines *nanotechnology* as "the ability to measure, design and manipulate on the molecular and supramolecular levels on the scale of about 1 to 100 nm in an effort to understand, create and use material structures, devices, and systems with fundamentally new properties and functions attributable to their small structures" (p. 69). The goal of nanotechnology, he writes, is "to assemble molecules into useful objects hierarchically integrated along several length scales and then, after use, disassemble objects into molecules."

The convergence of molecular biology and engineering may be illustrated by the analogy between natural ion channels and the channels that compose a synthetic nanopore membrane. Natural nanoscale conduits called *ion channels* play a crucial role in the transport of biofluids into and out of cells (Figure 1.23). The basic units of all living organisms are cells. To keep the cells functioning properly, a continuous flux of ions in and out of the cell and the cell components is required. The ion channels play an important role in this process. The walls of the ion channel consist of electrically charged proteins, most often negatively charged, that control the transfer of ions. The polarity of the membrane makes it challenging for some ions and molecules to move in and out of cells and their components. Ion channels are also size selective. Natural ion channels are roughly circular, although the cross-sectional area varies in the primary flow direction.

Advances in micro- and nanofabrication techniques resulting in some of the devices discussed in some of the previous sections have channels with dimensions approaching molecular scales in one dimension. The individual channels in these nanopore membranes may thus be termed *synthetic ion channels*. Moreover, analysis of the natural ion channel is very often performed by the same methods used for the synthetic ion channel (Chen et al., 1995; Gillespie, 1999; Gillespie & Eisenberg, 2001; Hollerbach *et al.*, 2001).

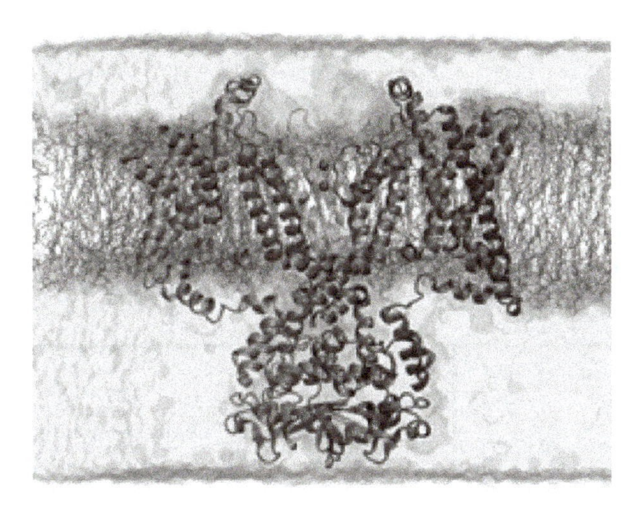

Figure 1.23 Computational simulation of the potassium (K^+) ion channel showing the proteins in a ribbon view that make up the walls of the channel. Ion channels are discussed in Chapter 8. Image courtesy of the National Center for Computational Science.

These so-called nanopumps can be designed for a specific task, whether it be drug delivery, chemical and biochemical sensing and analysis, or molecular separation. The primary design variables are the dimensions of the channels in the array, the materials used for the chip, and the size of the upstream and downstream reservoirs (see Figure 1.1). Often nanopore membranes have been employed for the transport of globular proteins such as albumin and lysozyme and for sugars such as glucose. As mentioned before, these molecules have a typical dimension on the order of $\sim$3 nm, and transport of such proteins has been demonstrated in channels having their smallest dimension on the order of 10 nm (Martin *et al.*, 2005). Having such a small dimension ensures that the amount of the protein delivered through the membrane increases linearly with time – so called zeroth-order release in the drug delivery literature.

The design of these devices thus requires a knowledge of the structure and properties of biomolecules and a good grasp of the basic principles of the thermal sciences, particularly fluid dynamics and mass transfer, and electrostatics and electrodynamics. While the electrostatics–electrodynamics portion of this material is not generally taught within the undergraduate mechanical or chemical engineering curriculum, it is routinely taught within bioengineering graduate programs. Thus the basic principles of molecular biology and thermal engineering must combine to produce efficient and useful design tools.

1.14 Design of micro- and nanofluidic devices

The first step in the design of a micro–nano device may be the development of a theoretical or computational model of the device prior to the fabrication of a model device. Theoretical and/or computational models are generally much less expensive than the actual fabrication of the devices.

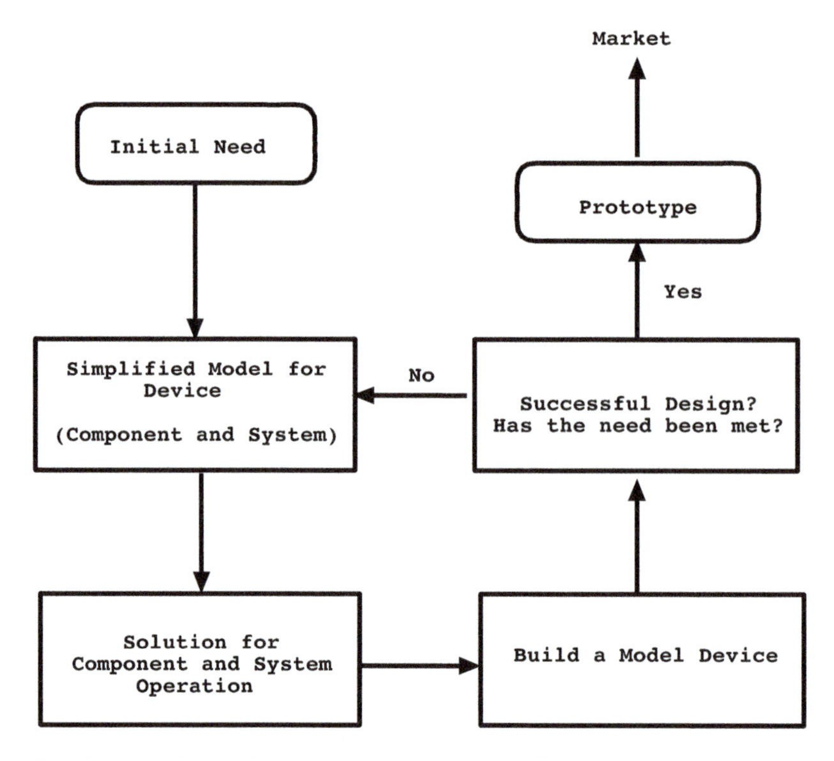

Figure 1.24 Simplified view of the design of a micro–nanofluidic device. The actual process may include a theoretical–computational retro design at the prototype stage.

Developing theoretical and/or computational models of flows at the micro- and nanoscales is especially important because experimental methods are limited to larger scales. To see this, consider the flow of sodium-chloride-water in a square channel of the order of 20 μm on a side. Suppose the walls of the channel are charged and that the motion is driven by an external electric field or electro-osmotic flow. On the wall, there is the EDL, which in general is of the order of 1–100 nm thick. To probe the EDL, spatial resolution on the order of 1/10th of the thickness is required (Sadr *et al.*, 2006). Local electrostatic forces and charge density can be measured using atomic force microscopy (AFM), but this is an intrusive method that disturbs the region of interest. A surface force apparatus (SFA) may also be used, but this method is also intrusive. Both methods have reported spatial resolutions of $O(1\text{–}10\ \text{nm})$ (Sadr *et al.*, 2006).

In general, AFM and SFA techniques cannot be used to probe EDL structure in a flowing fluid. Thus some other nonintrusive technique is required. Sadr *et al.* (2004, 2006) have developed a technique they call nano-particle image velocimetry (nPIV), which uses evanescent waves to measure the average velocity within the first 400 nm next to the wall. Thus velocity, temperature, concentration, and electric potential profiles cannot be measured in channels having dimension(s) less than ~1 μm.

Thus modeling is essential at the micro- and nanoscale. Moreover, using dimensional analysis, as described in Chapter 3, microscale experimental measurements at small ionic strength can often be made equivalent to nanoscale experimental measurements at a higher ionic strength. Thus theoretical and computational results at the nanoscale can validate microscale experiments, and microscale experiments will validate theoretical and computational results at the nanoscale. While experimental measurements at the microscale are difficult, they are an essential tool in the design of microfluidic devices.

It is useful to outline a typical design process in general terms. It is clearly the case that a given design process will vary widely, depending on the application. The development of a microfluidic device and, in particular, a biomedical microfluidic device first requires that a need be identified. For example, there may be a need for an artificial kidney, or a device to treat cancer, or a flow-through biosensor. Such a device could use electro-osmotic pumping, or in the case of the development of an artificial kidney, the natural pressure drop in the body (Fissell, 2006; Fissell & Humes, 2006; Conlisk *et al.*, 2009). Then a simplified model of the device should be developed to identify the physical parameters, such as size, flow rate, and concentrations, required to satisfy the need.

Once the simplified computational model is completed, a model device should be fabricated and experimental results compared with the solution of the simplified model. Once the model device has been validated (see Section 10.14) and the experimental results agree to a specified tolerance, a prototype device, suitable for clinical trials, can be fabricated. Finally, the device will go to market (of course, if the Food and Drug Administration (FDA) agrees). Note that this process involves the triad of theory–computation, fabrication, and experiment, all three of which are essential features in micro- and nanofluidics. A generic design procedure is depicted in Figure 1.24.

1.15 Unit systems

In this book, the System Internationale (SI) unit system will be used exclusively. The units for each major quantity of interest are defined in Table 1.4, where, for example, the unit of force is the newton, N, the unit of mass is the kilogram, kg, and the unit of length and time are the meter, m, and second, sec, respectively. Thus

$$1\ \mathrm{N} = 1\ \mathrm{kg}\frac{\mathrm{m}}{\mathrm{sec}^2}$$

The system of units summarized in Table 1.4 is standard in undergraduate thermal science textbooks.

1.16 A word about notation

A few words on notation in this book are due. In an interdisciplinary book such as this, notation can be a nightmare. Across the disciplines of fluid mechanics,

Table 1.4. SI units used for analysis of an electrically conducting fluid

Quantity	Unit	Symbol
Mass	kilogram	kg
Length	meter	m
Time	second	sec or s
Force	newton $= 1 \frac{\text{kgm}}{\text{s}^2}$	N
Temperature	kelvin	K
Energy	joule	J
Concentration	moles/liter	M
Concentration	kg_i/m^3	
Charge	coulomb	C
Potential	volt $= 1 \frac{\text{Nm}}{\text{C}}$	V

heat and mass transfer, electrostatics, and electrochemistry, the same symbols are often used. In addition, sometimes it proves useful to work in dimensionless form, whereas at other times, it does not. The philosophy I have thus taken is that all notation is local; that is, the meaning of each quantity, where there may be some question of its meaning, is defined in each section or subsection. For example, the internal energy in thermodynamics is denoted by u, which is also most often used for the streamwise fluid velocity.

In the section on electro-osmotic flow in nozzles and diffusers, E_x is used for the dimensionless electric field in the x direction, whereas in the section on Henry's solution for the electrophoretic velocity of a charged particle, the same symbol is used for the dimensional quantity. It simply makes no sense to present Henry's equation in dimensionless form, and it is maddening to continually use a superscript asterisk for dimensional quantities, as is often done.

I considered including a nomenclature so that definitions need not be provided locally. However, this requires the reader to continually flip back and forth to check definitions. Finally, to aid the reader, where appropriate, I have included "notation alerts" to minimize confusion.

1.17 Chapter summary

The purpose of this chapter is to give the reader an overview of the field of nanotechnology, in general, and micro- and nanofluidics, in particular. Principles of micro- and nanofluidics are being used to design devices that can be used for biomedical applications such as drug delivery, chemical and biochemical sensors, micromixers and microseparators, rapid molecular analyzers, development of an artificial kidney, and DNA analysis and transport. During the course of this discussion, it becomes evident that the nanoscale is the scale of biology and chemistry. Most of the concepts introduced in this chapter will reappear later in the book, some, such as the Debye length, many more times.

Micro- and nanofluidics cuts across a number of disciplines. It has elements of fluid mechanics, to be sure, but also mass transport, electromechanics, molecular biology, and electrochemistry. All these fields are introduced in general terms in this chapter. The historical summary clearly indicates that the fields of fluid mechanics, heat and mass transfer, and electrostatics all developed simultaneously in the middle to late nineteenth century.

Many applications to be discussed in this book involve the transport of biomolecules such as proteins. Indeed, the applications cited here require the development of engineering tools to design entire systems, often specifically for the study of single biological macromolecules. A brief introduction to biomolecules (i.e., DNA and proteins) is presented that will lead into the detailed discussion of the structure and function of proteins and other biomolecules that appears in Section 8.3.

One glance at the table of contents and the discussion in this chapter reveals the central theme of this book: micro- and nanofluidics unifies and integrates the fields of fluid mechanics, heat and mass transfer, electrostatics and electrokinetics, and electrochemistry. No other subject can be said to do this.

2 Preparatory Concepts

2.1 Introduction

All of the physical laws of micro- and nanofluidics, whether fluid mechanics, heat and mass transfer, or electrokinetics, involve transport coefficients. The Navier–Stokes equations require viscosity for their application to the viscous flow phenomena so central to micro- and nanofluidics. Each of the other disciplines has its own dependence on transport coefficients, and it is surprising how similar these relationships are. Understanding the similarities and differences in the roles these transport properties play in each discipline forms the basis of this chapter.

Accordingly, in this chapter, we present the basic constitutive laws of fluid mechanics, mass transfer, heat transfer, and electrostatics. These constitutive laws form the basis for the derivation of the governing equations in Chapter 3. We also discuss concepts of surface tension and wettability that are important because of the large surface-to-volume ratio in micro- and nanofluidics. We then discuss how to determine these transport coefficients for use in the solution of problems.

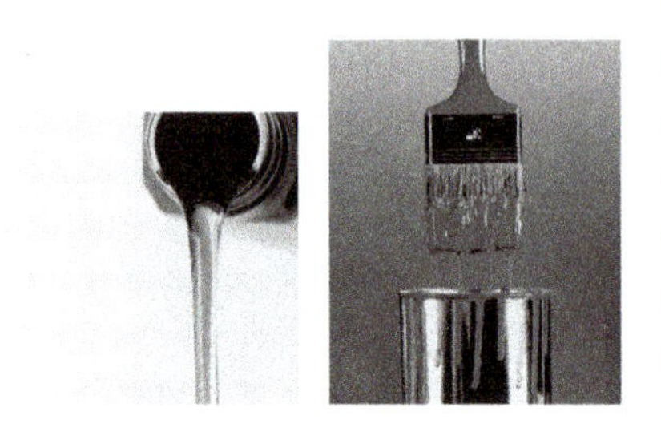

Water, paint and oil have very different values of viscosity.

Having presented these constitutive laws, we discuss the various types of fluid flows, from macroscale down to nanoscale, and the differences among internal and external flow. This section is followed by the presentation of the basic elements of thermodynamics and the interrelationship between fluid mechanics, heat transfer, and thermodynamics for pipe and channel flows, the internal flows that are common in micro- and nanofluidics. The irreversible conversion of mechanical energy to unusable thermal energy creates the pressure drop in a pipe that is directly proportional to the viscosity; inversely proportional to D^4, the diameter of a pipe; and inversely proportional to h^3, the height of a wide channel. It is the last two dependencies that reduce the effectiveness of fluid transport by pressure at the micro- and nanoscale.

This chapter concludes with a summary and set of exercises designed both to refresh the engineer's memory and to introduce potentially new material, specifically in the area of electrostatics.

2.2 Important constitutive laws

All materials, whether solids or fluids, have physical properties that determine the materials' behavior under the influence of an external action. Moreover, there are certain quantities, which may be termed *dependent variables*, that depend on these physical properties and that are important to determine by either experiment or calculation. These variables are identified as follows:

- **Fluid mechanics:** velocity field (u, v, w), pressure p
- **Single-phase heat transfer:** temperature T
- **Mass transfer:** mass density of species A: ρ_A or molar density (concentration) of species A: c_A
- **Electrostatics/dynamics:** electric potential ϕ and/or electric field $\vec{E}$

In identifying these dependent variables, the medium has to be assumed incompressible; if the medium is not incompressible, the density must be added to the dependent variable list. These dependent variables can be measured or calculated as the solution of the partial differential equations that arise in the development of conservation laws such as conservation of mass, momentum, energy, and electric charge. To fully characterize a system, then, a sufficient number of values of the seven unknowns (for a binary mixture) in the preceding list should be measured or calculated. For the case of an N-component mixture, then, there are $6 + N - 1$ unknown dependent variables in the general three-dimensional case.

It should be noted that pressure, density, and temperature are also called *thermodynamic properties* in the sense that the change in these properties from one physical state to another is independent of the process path between the two states. The quantity *entropy*, which determines the direction of a process, is also a thermodynamic property.

The conservation laws mentioned earlier contain parameters that characterize the material of interest, whether it be a solid or a fluid. These material properties are generally grouped into two classes:

1. Transport properties, for example, viscosity, thermal conductivity, diffusion coefficient, electrical permittivity;
2. Thermodynamic properties, for example, density, internal energy, enthalpy, and specific heat.

Transport properties are usually more difficult to determine than thermodynamic properties and are associated with the molecular nature of the material; these properties are generally obtained experimentally, although it should be mentioned that for a gas, the values of these properties may be obtained using the kinetic theory of gases, at least for viscosity, thermal conductivity, and the diffusion coefficient.

Notation alert: In the physical science literature, the viscosity is often denoted by η. In this text, μ will be used for viscosity, according to conventional notation in the engineering community.

The transport properties are defined in terms of the dependent variables in specific ways, and some of these definitions are surprisingly similar in form. This was evident in the historical development of these formulas, as discussed in Section 1.6. In *one dimension*, for which the dependent variables vary only in the direction normal to a wall such as that in a fully developed channel flow or boundary layer, the shear stress for a Newtonian fluid is defined as the derivative of the velocity times the viscosity, or

$$\tau = \mu \frac{du}{dy} \tag{2.1}$$

where τ is the shear stress, u is the velocity, and y is the coordinate normal to the wall. The units of shear stress are force per area, or N/m^2, and so the viscosity has units of N sec/m^2. The derivative of the velocity on the right side of equation (2.1) is called the *rate of strain*. The direction of action of the shear stress is depicted in Figure 2.1. Analogous to the viscosity μ, the *thermal conductivity* k is the proportionality constant in *Fourier's law of heat conduction*:

$$q = -k \frac{dT}{dy} \tag{2.2}$$

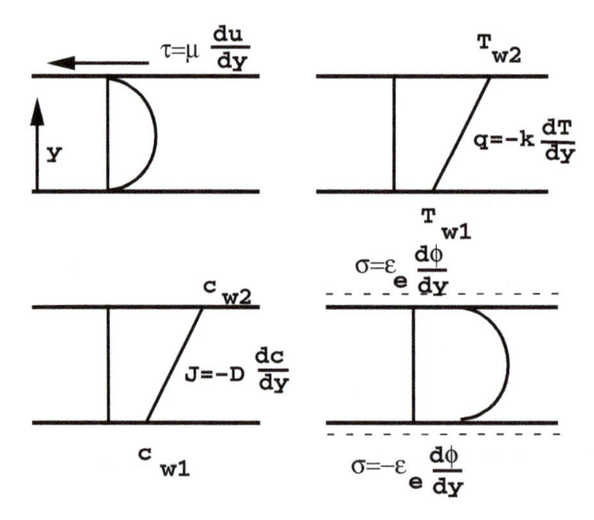

Figure 2.1 Definition of transport properties in one dimension. Note the similarity of the formulas. The surface charge density σ is defined at the charged walls of a channel.

where q is the *heat flux* in W/m^2, where W denotes wat t = 1 Joule/sec, and T is the temperature. The unit of thermal conductivity is W/m°K. As is the case with Newton's law, Fourier's law is an empirical law.

The diffusion coefficient of species A into B is defined by *Fick's law*, and on a molar basis,

$$J_A = -D_{AB}\frac{dc_A}{dy} \tag{2.3}$$

where J_A is the flux of species A, c_A is the local concentration of species A, and D_{AB} is the diffusion coefficient. The units of J_A are mole/m^2 sec if the concentration is given in mole/m^3. Again, as with Fourier's law, Fick's law is an empirical relationship, developed by measuring a known flux as a function of a known concentration gradient. Fick's law on a mass basis is given by

$$j_A = -D_{AB}\frac{d\rho_A}{dy} \tag{2.4}$$

where ρ_A is the *mass density* and the units of j_A are kg/m^2 sec. Generally, engineers prefer mass units, whereas chemists and chemical engineers prefer molar units. The units of the diffusion coefficient are m^2/sec.

Specific to the field of mass transfer, several scales are associated with molar concentration. Typically, chemists work in *molar*, designated by $1M = 1$ mole/liter. The *mole fraction* , $X_A = c_A/c$, is a dimensionless number; here c is the total concentration of the mixture, defined for a *binary mixture* as $c = c_A + c_B$. Often the total concentration remains nearly constant, and in this case, Fick's law can be written as

$$J_A = -cD_{AB}\frac{dX_A}{dy} \tag{2.5}$$

The advantage of working with mass or mole fractions is that both quantities are dimensionless, although equation (2.5) assumes that the total concentration is approximately constant. Similar comments apply to the definition of the mass fraction, $\omega_A = \rho_A/\rho$, where ρ is the total density.

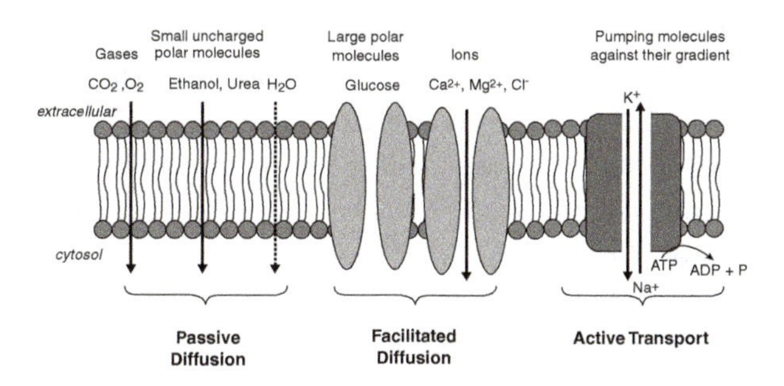

Mass diffusion takes place across the cell membrane, which is porous to gases and small, uncharged polar molecules. A molecule is classified as polar if the locations of the positive and negative charges do not coincide. Water is a polar molecule. From Saltzman (2009).

Table 2.1. Units for the transport properties and their corresponding dependent and independent variables

Property	Units	DQ	Units	DV	Units
μ	$\frac{\text{Nsec}}{\text{m}^2}$	τ	$\frac{\text{N}}{\text{m}^2}$	u	$\frac{\text{m}}{\text{sec}}$
wk	$\frac{\text{W}}{\text{mK}}$	q	$\frac{\text{W}}{\text{m}^2}$	T	K
D_{AB}	$\frac{\text{m}^2}{\text{sec}}$	J_A (mole)	$\frac{\text{mole}}{\text{m}^2\text{sec}}$	C_A	$\frac{\text{mole}}{\text{m}^3}$
D_{AB}	$\frac{\text{m}^2}{\text{sec}}$	j_A (mass)	$\frac{\text{kg}}{\text{m}^2\text{sec}}$	ρ_A	$\frac{\text{kg}}{\text{m}^3}$
ϵ_e	$\frac{\text{C}^2}{\text{Nm}^2}$	σ	$\frac{\text{C}}{\text{m}^2}$	ϕ	volt $= \frac{\text{J}}{\text{Coul}}$

DQ = defined quantity; DV = dependent variable.

Finally, the electrical permittivity is defined in terms of the surface charge density σ and the electrical potential ϕ:

$$\sigma = \epsilon_e \frac{d\phi}{dy} \tag{2.6}$$

where the outward unit normal to the surface is in the positive direction. If the outward unit normal to the surface is in the negative direction, a negative sign is required.

The surface charge density has units of C/m^2 if the potential is in volts. While the preceding equations serve as definitions of the viscosity, thermal conductivity, and diffusion coefficient and are empirical laws, equation (2.6) is obtained from Gauss's law and the fact that the electric field is irrotational: $\nabla \times \vec{E} = 0$. A summary of the one-dimensional relationships complete with units is given in Table 2.1.

It should be noted here that for multidimensional problems, there are differences in the forms of the preceding definitions of the transport properties. For example, because stress is a tensor, there are six distinct components of stress when symmetry is accounted for. Moreover, the heat and mass fluxes are vectors, and the surface charge density is a scalar. All of these transport properties are functions of temperature.

The magnitude of the transport properties for a liquid and a gas indicates the different transport mechanisms of momentum, heat, mass, and ease of transporting electricity. Table 2.2 indicates that the viscosity is much larger in a liquid than in a gas. Thus shear forces are expected to be much larger in a liquid than in a gas for the same velocity gradient. Likewise, liquids conduct energy more efficiently than gases. Conversely, mass transfer is more efficient in gases with

Table 2.2. Approximate magnitudes of the primary transport parameters for liquids relative to gases.

$\frac{\mu_L}{\mu_G}$	$\frac{k_L}{k_G}$	$\frac{\epsilon_{eL}}{\epsilon_{eG}}$	$\frac{D_L}{D_G}$
10 to 100	10 to 100	10 to 100	10^{-4}

a similar concentration difference, and the electric field is inversely proportional to the relative permittivity.

2.3 Determining transport properties

The calculation of transport properties is described in this section. There are several ways to determine these values: first, they can be measured directly or indirectly, as in a parallel plate viscometer for measuring viscosity. For a gas, these transport properties can also be calculated theoretically using the kinetic theory of gases. However, for liquids, there is no such unified theory, and most often, the properties are measured; also, molecular simulations have been used for this purpose. An extensive collection of calculation methods is given by Latini *et al.* (2006).

Theories have also been developed for liquid mixtures; however, for the dilute solutions of interest in this book, the transport properties will be nearly the values appropriate for the solvent, most often water.

The purpose of this section is to present several of the most important formulas used to calculate each of the four transport properties, with the focus being on water. The best source of this kind of information is the open literature in the form of archival papers in which new results for known and novel materials are published. Much in the way of transport properties also appears in the *CRC Handbook of Chemistry and Physics* (Haynes, 2011–2012).

It should be pointed out that all of the methods for empirically predicting transport coefficients are only as accurate as the experimental data used. In particular, all of the schemes used to predict diffusion coefficients in liquids show large fluctuations in percentage error when compared with actual data (Reid *et al.*, 1987).

2.3.1 Viscosity

Viscosity is especially important in internal flows considered in this book, and so it is useful to introduce this transport property here. It is the viscosity that forces fluid to stick to walls of a channel or tube: the no-slip condition. Viscosity of corn syrup, oils, and blood is relatively large, while the viscosity of water is somewhat low, and the viscosity of air is smaller still. Viscosity expresses the effect of molecular motions and interactions on the macroscale. For a perfect gas, an equation to predict viscosity exists (Reid *et al.*, 1987) in terms of particle diameter and molecular weight. The viscosity of air is $\mu \sim 10^{-5}$ N sec/m^2 while the viscosity of water is $\mu \sim 10^{-3}$ N sec/m^2 at room temperature.

Consider the flow between two plates, and suppose that the upper plate is moving at a small speed Δu, as in Figure 2.2, and that the fluid velocity is independent of x; this fluid dynamics problem is called *Couette flow*, and such a device is called a *parallel plate viscometer*. After a small time, an initially

SHEAR STRESS IS RELATED TO VELOCITY

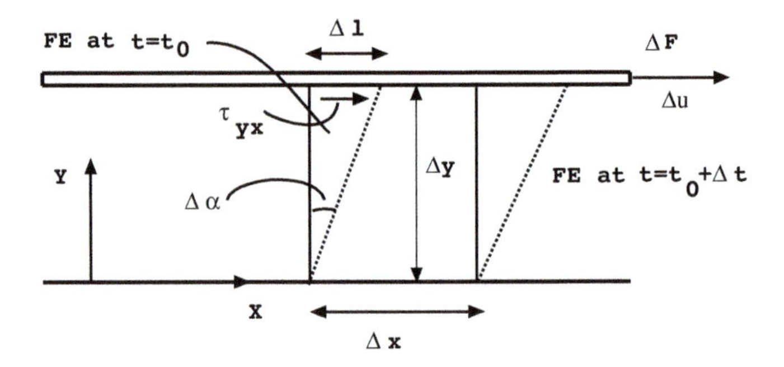

Figure 2.2 Geometry of Couette flow; this device is called a *parallel plate viscometer.* FE = fluid element; the solid line indicates the FE at time t_0.

undeformed fluid cube will undergo a shearing force, as shown in Figure 2.2. The shear stress is the force per unit area on the upper plate and is defined by

$$\tau = \lim_{\Delta A \to 0} \frac{\Delta F}{\Delta A} \tag{2.7}$$

where ΔA is the area over which the force acts. For Newtonian fluids, the shear stress is proportional to the rate of strain, which is

$$\gamma = \lim_{\Delta t \to 0} \frac{\Delta \alpha}{\Delta t} \tag{2.8}$$

To determine γ, note that from the geometry

$$\Delta l = \Delta u \Delta t \tag{2.9}$$

and for small angles $\Delta\alpha$,

$$\Delta y \sin(\Delta \alpha) \sim \Delta y \Delta \alpha = \Delta l \tag{2.10}$$

Substituting for Δl, it is evident that

$$\frac{\Delta \alpha}{\Delta t} = \frac{\Delta u}{\Delta y} \tag{2.11}$$

Thus in the limit as $\Delta y \to 0$ and $\Delta t \to 0$

$$\tau \propto \frac{du}{dy} \tag{2.12}$$

The proportionality constant is the viscosity μ first proposed by Stokes and leads to the definition of the shear stress for a Newtonian fluid discussed previously.

The viscosity can be calculated using the solution for the velocity of Couette flow (Figure 2.3). As will be seen, the solution for the velocity is

$$u(y) = U\frac{y}{h} \tag{2.13}$$

Suppose the velocity of the plate is $U = 0.01$ m/sec, the channel height $h =$ 0.1 cm, the force on the upper plate is measured to be 10^{-6} N, and the width

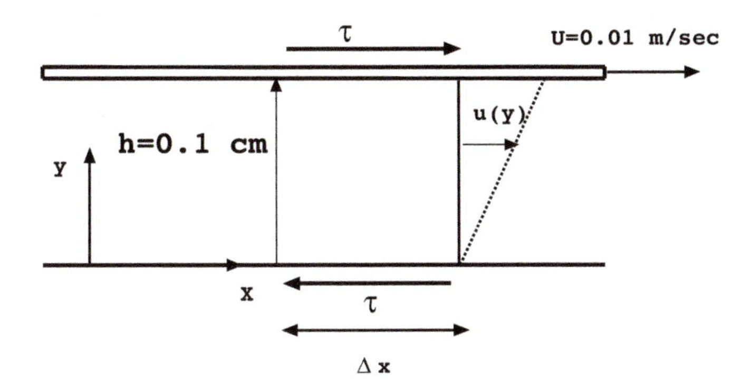

Figure 2.3 Calculation of the viscosity in a parallel plate viscometer.

and length of the cell are $L = W = 10$ cm. To calculate the viscosity given this information, from the definition of the shear stress

$$\mu = \frac{\tau h}{U} \tag{2.14}$$

the shear stress is force per area so that

$$\tau = \frac{10^{-6} N}{0.01 \text{ m} \times 0.01 \text{ m}} = 10^{-2} \frac{\text{N}}{\text{m}^2} \tag{2.15}$$

Thus

$$\mu = \frac{10^{-2} \frac{\text{N}}{\text{m}^2} \times 10^{-3} \text{ m}}{0.01 \frac{\text{m}}{\text{sec}}} = 10^{-3} \frac{\text{N sec}}{\text{m}^2} \tag{2.16}$$

The viscosity is sometimes expressed in Poise and 1 N sec/m^2 = 10 P, where P stands for Poise.

The direction of the stress is important. On the upper wall, the force (stress × area) on the fluid is causing motion of the fluid, and so the stress is in the positive x direction. The bottom plate is preventing motion of the fluid, and so the force on the fluid is in the negative x direction. More will be said about the direction of stress in Chapter 3.

The viscosity of liquid water is a function of temperature, and White (2006) suggests that

$$\ln \frac{\mu}{\mu_0} = a + b \frac{T_0}{T} + c \left(\frac{T_0}{T} \right)^2 \tag{2.17}$$

where $T_0 = 273$ K and $\mu_0 = 0.00179$ N sec/m^2. The values of the constants are (White, 2006) $a = -2.10, b = -4.45$, and $c = 6.55$. According to White (2006), the accuracy of this fit is ±1%. A table of values of the viscosity of water as a function of temperature is presented in Table 2.3. The viscosity of liquids is a very weak function of pressure, which is almost always neglected. Values for the viscosities of several common solvents are depicted on Table 2.4.

Using the values in Table 2.4, we note that the stress exerted on the wall of a channel is about 5 times greater for oil than it is for water and 5000 times greater

Table 2.3. Viscosity of liquid water as a function of temperature

Temperature	Viscosity
T (°C)	μ ($\frac{\text{N sec}}{\text{m}^2}$)
0	1.787×10^{-3}
20	1.0019×10^{-3}
40	0.6530×10^{-3}
60	0.4665×10^{-3}
80	0.3548×10^{-3}
100	0.2821×10^{-3}

From Hardy and Cottingham (1949)

than for water vapor. For example, in a $h = 10$ μm channel for fluid moving at $U = 1$ cm/sec, the shear stress exerted by oil on one wall of the channel is approximately

$$\tau = 4.6 \times 10^{-3} \times \frac{1 \text{ cm/sec}}{10 \times 10^{-6} \text{ m}} = 4.6 \text{ sec}^{-1} \tag{2.18}$$

The electrolyte solutions considered here are dilute, containing at most about 1 M of the electrolyte. Thus, in most of the applications described here, use of the viscosity of water for the viscosity of aqueous solutions does not lead to significant error; this is most often the case in micro- and nanofluidics. Of course, for complex fluids, such as blood, which often exhibits non-Newtonian behavior, this is not the case, and the viscosity will depend, for example, on the concentration of red and white blood cells. More will be said about this when non-Newtonian fluids are discussed in Chapter 3.

2.3.2 Diffusion coefficient

The closest approach to a theory of liquid diffusion coefficient was developed by Stokes in 1850 and Einstein in 1905 (Plawski, 2001). They considered the limiting

Table 2.4. Values of viscosity of selected liquids and water vapor at the indicated temperatures

Fluid	Temperature	Viscosity
	T (K)	μ ($\frac{\text{N sec}}{\text{m}^2}$)
Water (l)	300	0.855×10^{-3}
Water (g)	300	9.09×10^{-6}
Seawater	294	1.06×10^{-3}
Engine oil (unused)	300	4.86×10^{-3}
Methanol	293	0.59×10^{-5}

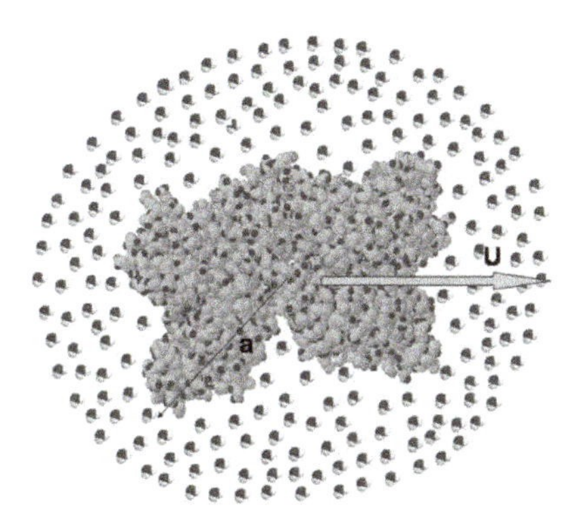

Figure 2.4 Model for calculating the diffusivity of a liquid due to Stokes in 1850 and Einstein in 1905. The solute must be much larger than the solvent molecule. Here a human serum albumin molecule whose radius is about 3.5 nm is depicted surrounded by water molecules of radius about 10 times smaller.

case where a large particle is diffusing through a cloud of smaller particles. This situation is depicted in Figure 2.4. In the Stokes–Einstein picture, a tagged molecule A passes through a sea of smaller particles randomly. They assumed that the force is proportional to the velocity of species A relative to the mass averaged velocity of the mixture:

$$F = f(u_A - u) \tag{2.19}$$

with f being a proportionality constant. There are two cases: in one, the diffusing particle sticks to the fluid molecules, and in the second, it slips. For these cases, the value of f can be shown to be

$$f = 6\pi\mu a \quad \text{no slip (stick)} \tag{2.20}$$

$$f = 4\pi\mu a \quad \text{slip} \tag{2.21}$$

where a is the radius of the diffusing particle. These values of f represent a coefficient in the Stokes drag law multiplying the relative velocity. Einstein defined the force as a gradient in chemical potential; the *chemical potential* will be defined in Chapter 7, but for now, we define the chemical potential in terms of the number density n_A (number of molecules of A per volume) as

$$\mu_A = \mu_A^0 + k_b T \ln n_A \tag{2.22}$$

where $k_b = R/N_A = 1.38 \times 10^{-23}$ J/K is *Boltzmann's constant*, R is the universal gas constant, and N_A is Avagadro's number. The chemical potential has units of energy, and so its gradient will have units of force. Thus Einstein wrote the drag force F as the gradient in chemical potential:

$$\vec{F} = \nabla\mu_A \tag{2.23}$$

so that in one dimension,

$$F = -\frac{d\mu_A}{dx} = \frac{-k_b T}{n_A}\frac{dn_A}{dx} \tag{2.24}$$

In Section 3.7, we will see that

$$u_A - u = -\frac{D_{AB}}{c_A}\frac{dc_A}{dx} = -\frac{D_{AB}}{n_A}\frac{dn_A}{dx} \tag{2.25}$$

where the last equality uses the fact that $n_A = N_A c_A$. Equating the two expressions for the force leads to

$$D_{AB} = \frac{k_b T}{\beta a \mu} \tag{2.26}$$

where $\beta = 6\pi$ for no slip and $\beta = 4\pi$ for slip between molecules. Typically, this equation predicts the diffusion coefficient with an accuracy of 20%.

Albert Einstein and the Diffusion Coefficient

So how did Einstein get his name on the Stokes–Einstein equation? Because he knew a little fluid mechanics and mass transfer! Einstein's PhD Dissertation, "A New Determination of Molecular Dimensions," was published in the same year, 1905, that he published his two papers on the theory of relativity. Einstein was concerned with the determination of the volume of sugar that dissolves in water, so that he could determine if the sugar molecules were hydrated (Stachel, 1998). While we have viewed the Stokes–Einstein equation as determining the diffusion coefficient, Einstein thought of the same equation as determining the molecular size given the diffusion coefficdient! In his dissertation, Einstein's object was to show that "... sizes of molecules of substances dissolved in an undissociated dilute solution can be determined from the internal viscosity of the solution and of the pure solvent, and from the diffusion rate of the solute within the solvent provided that the volume of the solute molecule is large compared to the solvent molecule." And "... such a molecule will behave appproximately like a solid body suspended in the solvent". Stokes drag law applies! Einstein expressed his result in the formula

$$NP = \frac{RT}{6\pi k}\frac{1}{D}$$

in which N is Avagadro's number, P is the radius of the solute, k is the viscosity, and D the diffusion coefficient. This part of his dissertation was published in the paper Einstein (1905b). All five of the papers published in 1905 by Einstein including the two on the theory of relativity are lucidly discussed in the short book by Stachel (1998). The quotes above are taken from that book.

For the general case that lies between perfect slip and perfect no-slip, the diffusion coefficient is given by

$$D_{AB} = \left(\frac{3\mu_B + a\beta_{AB}}{2\mu_B + a\beta_{AB}}\right)\frac{k_b T}{6\pi\mu_B a} \tag{2.27}$$

based on a result for the drag force given by Lamb (1945). Here β_{AB} is the coefficient of sliding friction of the solute molecule.

The Stokes–Einstein equation has been shown to be fairly accurate for describing the diffusion of large spherical particles or molecules in cases where the solvent appears to the diffusing species as a continuum (Bird *et al.*, 2002). The hydrodynamic theory further suggests that the shape of the diffusing species is very important. Tyrrell and Harris (1984) note that the validity of equation (2.26) is also restricted by the fact that the motion of only one diffusing particle is considered, and so the relationship can strictly only apply in the limit of infinite dilution, that is, in the limit of zero solute concentration. Values of diffusion coefficients of several neutral solutes in water are given in Table 2.5.

Table 2.5. Diffusion coefficients at infinite dilution (i.e., very low concentration) for some common fluids in water at $T = 25°C$. From Cussler (1997)

Solute	D_A (cm^2/sec)
Air	2.00×10^{-5}
Nitrogen	1.88×10^{-5}
Oxygen	2.10×10^{-5}
Methanol	0.84×10^{-5}
Ammonia	1.64×10^{-5}

The diffusion coefficient D_{AB} in gases $D_{AB} \sim 10^{-1}$ cm^2/sec and in liquids $D_{AB} \sim 10^{-5}$ cm^2/sec. Biomolecules in aqueous solution have diffusion coefficients that are smaller due to their relatively large size, especially in nanoscale channels; typically, $D_{AB} \sim 10^{-6} - 10^{-7}$ cm^2/sec.

While the Stokes–Einstein equation is a simple and useful tool for estimating diffusion coefficients for very dilute mixtures, more accurate methods have been established. An empirical modification of the Stokes–Einstein equation is the Wilke–Chang correlation (Bird *et al.*, 2002)

$$D_{AB} = \frac{7.4 \times 10^{-8} (\psi_B M_B)^{1/2} T}{\mu_B V_A^{0.6}} \tag{2.28}$$

valid at very low concentration; here B is the *solvent*, the species with the highest concentration. In this equation, ψ_B is an association factor, M_B is the molecular weight, V_A is the partial molar volume, and μ is the viscosity of the solvent in centipoise (1 cP $= 10^{-2}$ P). Wilke and Chang suggest that $\psi_B = 2.6$ for water as the solvent.

Many other correlations have been developed over the years, with widely varying accuracy. Several others are discussed in Bird *et al.* (2002) and Reid *et al.* (1987), and many other diffusion coefficients are given in Reid *et al.* (1987). While these correlations are often useful, it is often easier to use a measured value of liquid diffusivity, and diffusivities for several ions and biomolecules in water are found in Table 2.6. The hydrated radii are usually used in the Stokes–Einstein relation.

Ficoll and dextran are part of a class of hydrophilic polymers that are often used to simulate the action of proteins, specifically in the study of transport in the kidney (Fissell *et al.*, 2007); dextran is a neutral glucose-linked polymer, and Ficolls are sucrose-linked polymers. Both Ficoll and dextran occur in a variety of molecular sizes, and the effective radii for Ficoll and dextran often fall between the Stokes–Einstein value and the hydrated value; the radii can be related to the molecular weight, as described in Section 3.7.

Biofluids are multicomponent mixtures, and the process of predicting diffusion coefficients becomes much more difficult. When all the components of the

Table 2.6. Typical molecular and hydrated radii and diffusion coefficients in bulk water at room temperature (25°C). All diffusion coefficients should be multipled by 10^{-5}

Solute	D_A (cm^2/sec)	Molecular radius (nm)	Hydrated radius (nm)
Ions			
Li^+	1.03*	0.094	0.382
Na^+	1.33*	0.117	0.358
K^+	1.96*	0.149	0.331
Cs^+	2.06*	0.186	0.329
Mg^{2+}	0.71*	0.072	0.428
Ca^{2+}	0.79*	0.100	0.412
Cl^-	2.032*	0.164	0.332
Biomolecules			
Albumin	0.061**	–	7.2
Glucose	0.94***	–	1
IgG	0.04**	–	10
Ficoll	5–10	–	4.81–2.45
Dextran	0.009–0.1	–	27.0–2.36

The asterisk symbol denotes data from Cussler (1997), as calculated from the data of Robinson and Stokes (1959); double asterisks are from Granicka *et al.* (2003), and triple asterisks are from the Web site http://oto.wustl.edu/cochlea/model/diffcoef.htm. Molecular and hydrated radii are from Volkov *et al.* (1997).

mixture are dilute, a good approximation to the diffusivity is to use the binary diffusivity of the given component in the solvent.

As an example, using the Stokes–Einstein relation, with the molecular diameter in Table 2.6, the diffusion coefficient for a sodium ion in water at $T = 25$ C is given by $D_{AB} = 1.86 \times 10^{-5}$ cm^2/sec, a full 40 percent higher than in Table 2.6. However, the molecular radius may not be the appropriate length scale. Using the hydrated radius $a = 0.358$ nm gives $D_{AB} = 0.6 \times 10^{-5}$ cm^2/sec, which is much too low. The average of these two values gives $D_{AB} = 1.23 \times 10^{-5}$ cm^2/sec, which is close to the experimental value. Thus there is some question as to which radius to use, and care should be used in the interpretation of the Stokes–Einstein relation for the diffusion coefficient for ionic species in particular. Note that the influence of valence is incorporated implicitly through the use of the experimental value of the radii. More puzzling is the fact that data from Keilland (1937) suggest that the hydrated radius of sodium be taken as $a = 0.225$ nm, which, when used, gives a result, again, close to the experimental value. Other values for the hydrated and ionic radii are given by Jorgensen (1990).

2.3.3 Thermal conductivity

The *thermal conductivity* is the proportionality constant in Fourier's law and has units of W/m°K. In general, the thermal conductivity of gases is much lower

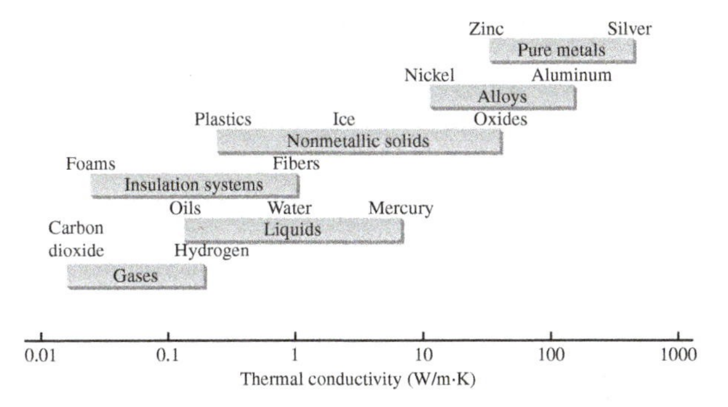

Typical values of the thermal conductivity for a variety of materials. From Turns (2006)

than that for liquids. Solids have the highest conductivity; quartz has a thermal conductivity of 7.6 W/m°K at $T = 20°\mathrm{C}$, while glass has a conductivity of 0.76 W/m°K at $T = 20°\mathrm{C}$. Homogenous materials have a thermal conductivity which independent of direction.[1]

Because of the higher thermal conductivity, solids are better thermal conductors than liquids, which are better conductors than gases. Note that a crystalline structured solid such as quartz has a higher conductivity than glass, whose physical structure is amorphous; that is, it is not densely ordered and behaves more like a liquid. This is the opposite of the diffusion coefficient, which is highest in gases and lowest in solids. Metals have the highest thermal conductivity. As with the diffusion coefficient, there is a well-developed kinetic theory to predict the thermal conductivity of gases. Because of the magnitudes of the solid thermal conductivity, solid surfaces control the amount of heat transfer through a given system.

In general, the thermal conductivity of liquids decreases with temperature, although water is an exception to this behavior, increasing with temperature. Water has a relatively high thermal conductivity compared with other liquids, reaching a maximum of 0.7 W/m°K at $T = 150°\mathrm{C}$. Typical values of the thermal conductivity appear in Table 2.7.

The temperature dependence of many liquids can be described by (Reid *et al.*, 1987)

$$k(T) = A + BT + CT^2 \tag{2.29}$$

where A, B, C are constants. For water, $A = -0.383$ W/m°K, $B = 5.254 \times 10^{-3}$ W/m°K^2, and $C = -6.369 \times 10^{-6}$ W/m°K^3. These constants have been tabulated for many other liquids, and an extensive table is given by Reid *et al.* (1987), along with a discussion of many other methods of estimation. Latini *et al.* (2006) also has a significant amount of information on the calculation of thermal conductivity.

[1] However, fibrous materials such as wood and laminates have different conductivities in different coordinate directions.

Table 2.7. Thermal conductivity of liquid water as a function of temperature

Temperature	Thermal Conductivity
T (°C)	k ($\frac{\text{W}}{\text{mK}}$)
0	0.5619
20	0.5996
40	0.6286
60	0.6507
80	0.6668
100	0.6775

From Kestin (1978).

For restricted ranges of temperature, the thermal conductivity is often represented by

$$k = k_0(1 + bT) \tag{2.30}$$

with T in °C and where k_0 and b are constants.

2.3.4 Electrical permittivity

The electrical permittivity ϵ_r is a transport property like viscosity and, for a vacuum, is defined by

$$c_0^2 \left(10^{-7} \frac{\text{N sec}^{-2}}{\text{C}^2} \right) = \frac{1}{4\pi \epsilon_0} \tag{2.31}$$

where $c_0 = 3 \times 10^8$ m/sec is the speed of light in a vacuum. Thus the permittivity in a vacuum is

$$\epsilon_0 = \frac{1}{36\pi} \times 10^{-9} \frac{\text{C}^2}{\text{Nm}^2} \tag{2.32}$$

If the medium is not a vacuum, a relative permittivity or dielectric constant is defined by

$$\epsilon_r = \frac{\epsilon_e}{\epsilon_0} \tag{2.33}$$

where ϵ_e is the material permittivity. Materials that have a relative permittivity greater than one are called *dielectrics*, and selected permittivity values are given in Table 2.8.

The relative permittivity is a function of temperature, and the permittivity of water decreases with increasing temperature in the form

$$\epsilon_r = Ae^{-\frac{T}{B}} \tag{2.34}$$

Knox and McCormack (1994) suggest that $A = 305.7$ and $B = 219$, and a table of values found in the *Handbook of Chemistry* (Haynes, 2011–2012) appears in

Table 2.8. Relative permittivity or dielectric constant of some materials at room temperature (Israelachvili, 1992)

Material	ϵ_r
Vacuum	1.0
Air	1.000549
Water	78.54
Methanol	33
Salt water	81
Ethylene glycol	40.7
Glass	4–10
Quartz (SiO_2)	4.5
Polystyrene	2.4
Sodium chloride (NaCl)	6.0

High relativity permeability indicates a reduced electric field strength.

Table 2.9. Using the Knox and McCormack (1994) formula, $\epsilon_r = 78.4$, which is consistent with the tabular values.

Another common curve fit for the dielectric constant or relative permittivity of water is the formula developed by Archer and Wang (1990):

$$\epsilon_r = 87.86 - 0.3963T + 7.036 \times 10^{-4}T^2 \tag{2.35}$$

where the temperature is in °C.

2.3.5 Surface tension and wettability

When a liquid interfaces with a gas, say, water with air, a liquid–gas interface is formed, that acts as if it were a stretched membrane and a net pressure change across the surface is present owing to the lack of water molecules at the liquid–gas interface. Consider the simple case of a curved surface depicted in Figure 2.5. The

Table 2.9. Relative permittivity of water as a function of temperature (Haynes, 2011–2012)

Temperature (°C)	ϵ_r
0	87.74
10	83.83
20	80.10
30	76.55
40	73.15

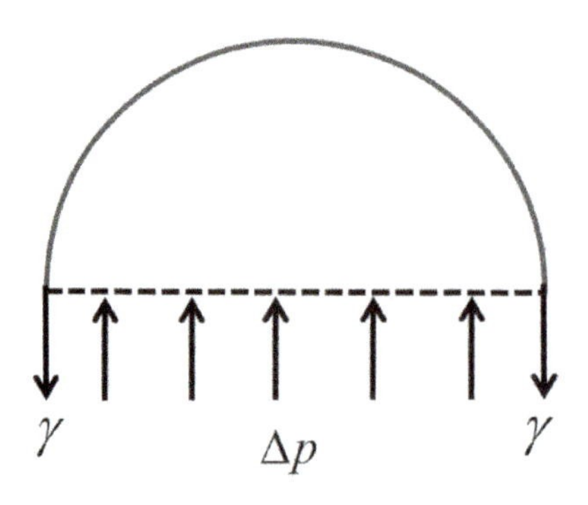

Figure 2.5 Side view of a liquid drop in a gas.

molecules in the liquid are held together by van der Waals forces in a nonpolar liquid and by the short-range hydrogen bonds in a polar liquid like water. These attractive forces are of a different nature from the electrostatic forces that hold the protons, neutrons, and electrons together to form an atom. Thus there is a natural tendency for the surface to contract, and this is done by the surface tension force pulling down on the drop, as shown in Figure 2.5. The pressure inside the drop acts over the drop cross-sectional area πR^2, and the surface tension pulls the drop in the opposite direction around the circumference. Then the net pressure change across the surface is balanced by the tension caused by the significant change in density between the liquid and the gas, and for a spherical drop, this "surface tension" is defined, in the absence of any flow, by the force balance

$$\Delta p = 2\frac{\gamma}{R} \tag{2.36}$$

where γ is the surface tension and has units of N/m and R is the radius of the drop. At $T = 20°\text{C}$, the surface tension coefficient for a clean air–water interface is $\gamma = 0.073$ N/m. For a bubble, a gas inside surrounded by a liquid, the result is

$$\Delta p = 4\frac{\gamma}{R} \tag{2.37}$$

since the liquid bubble surface has two interfaces with the gas.

Finally, for a general curved surface with radii of curvature in each of the two coordinate directions, we have

$$\Delta p = 2\frac{\gamma}{R_1 + R_2} \tag{2.38}$$

These equations relating the pressure to the surface tension are collectively called the *Laplace equations*.

Note that surface tension has units of force per length, and so the free energy associated with a given curved surface is given by $E = \gamma A$, where A is the area of the surface. By the first law of thermodynamics, the energy must also be equal to the work done by the surface tension force $W = E = \gamma A$ (Batchelor, 1967).

A free and curved surface is a surface of minimum energy, and the relationships between the pressure and curvature of the surface shown earlier can be obtained by energy considerations as well. Consider the spherical drop given previously; the total energy of the surface is given by the difference in pressure and the surface tension:

$$dE = -(p_{\text{in}} - p_{\text{out}})dV + \gamma dA = -4\pi R^2(p_{\text{in}} - p_{\text{out}})dR + 8\pi R\gamma dR \tag{2.39}$$

Table 2.10. Surface tension for water as a function of temperature (Karniadakis *et al.*, 2005)

Temperature (°C)	γ $\frac{\text{mN}}{\text{m}}$
10	74.2
20	72.8
30	71.2
40	69.6

The term $mN = 10^{-3} N$.

and the minimum energy is obtained by setting $dE/dR = 0$ so that

$$p_{\text{in}} - p_{\text{out}} = 2\frac{\gamma}{R} \tag{2.40}$$

as obtained earlier.

In the preceding equations, the influence of gravity is neglected. The characteristic length scale associated with gravity is $l_g = \sqrt{\gamma/\rho g}$, which, for water at room temperature, is $l_g = 2.7$ mm. The effect of gravity will then be negligible if the radius of the drop (or bubble) is much less than l_g. The dimensionless number $Bo = D/l_g$ is called the Bond number, where D is the diameter of the bubble, and gravity will be negligible if $Bo \ll 1$.

Surface tension decreases with increasing temperature, and some common values for water are shown in Table 2.10. White (2006) suggests that a good curve fit to these data is

$$\gamma(\text{N/m}) = 0.076 - 0.00017T(^\circ\text{C}) \tag{2.41}$$

The radius of curvature of a sphere is its radius, R. For a general curved surface of the form

$$z = h(x, y) \tag{2.42}$$

there are two principal radii of curvature in each of the coordinate directions; Wehausen and Laitone (1960) provide the detailed expression for the two principal radii of curvature, and for a curve of the form $y = h(x)$, the methods of differential geometry indicate that the radius of curvature is given by

$$R = \frac{\left(1 + \left(\frac{dh}{dx}\right)^2\right)^{3/2}}{\frac{d^2h}{dx^2}} \tag{2.43}$$

In particular, if the slope of the surface is small, $dh/dx \ll 1$, then

$$R = \frac{1}{\frac{d^2h}{dx^2}} \tag{2.44}$$

In differential form, the equations for the pressure difference become

$$\frac{dp}{dx} = \gamma\frac{d}{dx}\left(\frac{1}{R}\right) = \gamma\frac{d^3h}{dx^3} \tag{2.45}$$

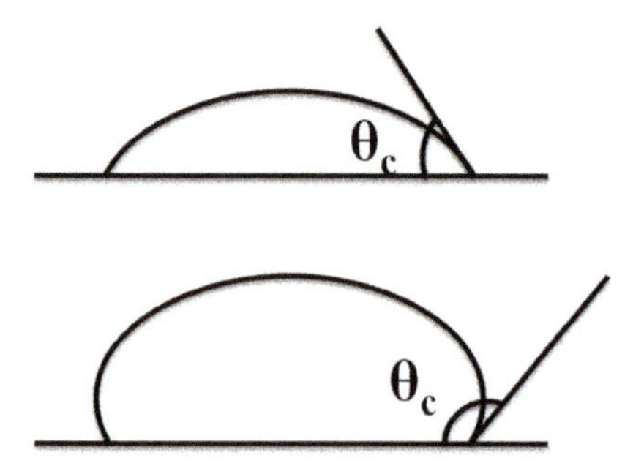

Figure 2.6 Side view of a liquid drop in a gas, indicating the contact angle for a wetting surface (top) and a nonwetting surface (bottom).

There are two dimensionless numbers associated with surface tension, γ. The first is the *Weber number*, defined by $We = \rho U^2 L/\gamma$, and is a ratio between inertial forces and surface tension forces; here U is a characteristic velocity. The other is the *Capillary number*, $Ca = \mu V/\gamma$, and is the ratio between the viscous forces and the surface tension forces.

The concept of surface tension is intimately connected with and determines a fluid's ability to wet a surface. A contact angle is defined by the angle between the liquid and the surface, as depicted in Figure 2.6. If the contact angle $\theta_c < 90°$, the surface is said to be partially wetted; conversely, if the contact angle $\theta_c > 90°$, the surface is said to be nonwetting. If the liquid is water, a partially wetted or fully wetted surface ($\theta_c = 0$) is called *hydrophilic*, while a nonwetting surface is called *hydrophobic*.

Near the contact line, the location where the bubble meets the solid surface, local equilibrium is characterized by three local surface tension coefficients, associated with the contact between the solid and liquid (SL), the solid and gas (SG), and the liquid and gas (LG). A force balance on the contact line in the horizontal direction under equilibrium conditions reveals that (Figure 2.7)

$$\gamma_{\mathrm{SG}} = \gamma_{\mathrm{LS}} + \gamma_{\mathrm{LG}} \cos\theta_C \qquad (2.46)$$

This is called *Young's law* and is a simple first approximation of the macroscopic view of the microscopic region near the contact line. There are several problems with this equation since, assuming that the contact angle can be measured by some means, there are still three unknowns and one equation. Thus at least two of the surface tension coefficients should be measured; in Young's equation, γ_{SL} is often termed the interfacial tension, and values for a number of solvents have been measured and compiled in Israelachvili (1992).

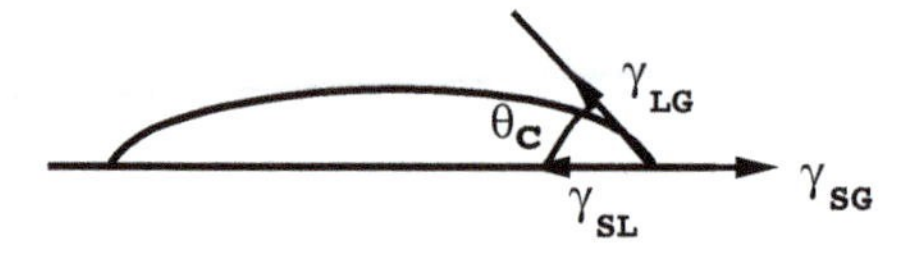

Figure 2.7 Surface tension coefficients for deriving Young's equation.

Unlike the other properties, such as viscosity and thermal conductivity, the surface tension of a given liquid can be altered substantially by the addition of a small amount of surfactant. Surface active agents, or *surfactants*, are long molecules having a hydrophilic head and a hydrophobic tail. They are often added to biofluids to prevent aggregation and may be used to prevent target molecules from adhering to surfaces. In general, surfactants reduce the surface tension of water.

Models for the surface tension of liquid mixtures have been proposed over the years, and a popular formula for binary solutions is given by Shereshefsky (1967), according to

$$\gamma = \gamma_1 - \frac{\Delta\gamma_0 X_2 e^{\frac{\Delta G_s}{RT}}}{1 + X_2\left(e^{\frac{\Delta G_s}{RT}} - 1\right)} \tag{2.47}$$

where γ_1 is the surface tension of the pure solvent, X_2 is the mole fraction of the solute in the bulk, and ΔG_s is the Gibbs free energy change in replacing 1 M of solvent with 1 M of solute in the interfacial region. Values of $\Delta G_s/RT$ for a number of nonelectrolyte mixtures have been compiled by Tahery *et al.* (2005) from work published in the open literature. For example, $\Delta G_s/RT = 1.98$ for a water–methanol mixture at $T = 25°\text{C}$. A molecular model built specifically for polar liquid mixtures has been given by Li and Liu (2001), and for example, the surface tension of a methanol–water mixture for a methanol mole fraction of $X_2 = 0.2$ is calculated to be $\gamma = 40$ mN/m. The surface tension for aqueous electrolyte solutions has been presented by Levin and Flores-Mena (2001) based on the analysis of the Poisson–Boltzman equation of electrostatics; for very low concentrations, the surface tension can be predicted by the Onsager and Samaras limiting law (Onsager & Samaras, 1934). In general, electrolyte solutions exhibit surface tensions higher than pure water.

Surface tension phenomena are essential features in a number of applications, including inkjet printers, electrowetting on dielectric (EWOD), thermocapillary pumping, instabilities of falling films, coating flows, and Marangoni phenomena. The latter refers to the currents set up when the surface tension varies with temperature. Thermocapillary pumping will be discussed later in the book.

2.4 Classification of fluid flows

The field of macroscale fluid mechanics can be classified in a number of ways. As has been discussed, the most important property of fluids is their *viscosity*, which is characteristic of a fluid's ability to stick to walls. A fluid in motion away from the influence of any solid boundaries tends to act as if its viscosity were zero. In this case, we call the flow *inviscid*. Conversely, fluids that move near solid boundaries are flows that are termed *viscous*, based on the fluid's finite viscosity. In general, most flows of interest in micro- and nanofluidics are internal, viscous,

(a)

(b)

Figure 2.8 Examples of macroscale external fluid flows.

incompressible laminar flows. The main dimensionless parameter that describes *viscous flow* is the Reynolds number, which, for an internal flow, is defined by

$$Re = \frac{\rho U d}{\mu} \tag{2.48}$$

where ρ is the fluid density, μ is the fluid viscosity, and U and d are typical velocity and length scales, respectively. For an *internal flow* or a flow bounded on all sides by walls, the length scale D is the diameter for a tube, and for a rectangular channel, its height h or the hydraulic diameter $D_H = 4hW/(2h + 2W)$, where W is its spanwise width. The velocity scale is usually, but not always, taken to be the average velocity at a section, defined by

$$V = \frac{1}{A} \int_A u dA \tag{2.49}$$

where A is the cross-sectional area and u is the velocity. The formal definition of the *volume flow rate* is $Q = UA$ and has units of $Q \sim \mathrm{m}^3/\mathrm{sec}$; the *mass flow rate* is defined by $\dot{m} = \rho Q$. For a very wide channel of height h, $W \gg h$, which is common in micro- and nanofluidics; the average velocity is given by

$$V = \frac{1}{h} \int_0^h u dA \tag{2.50}$$

Fluid mechanics is all around us, from atmospheric flows which involve heat and mass transfer and phase change to isothermal flows of water and air in pipes, the basis of the field of plumbing and air-conditioning, to the aerodynamics of cars and planes, and so on. In Figure 2.8 are two examples of such flows: the flow past a twin rotor Boeing V22 helicopter and the flow field generated by a tornado. A tornado is a region of very rapidly swirling fluid, with varying temperature causing very high winds; such a highly swirling flow is called a *vortex*. The prediction of our daily weather is largely a giant fluid dynamics problem coupled with a variable temperature and mass transfer. There are also electrical effects that occur in a rain storm: lightning.

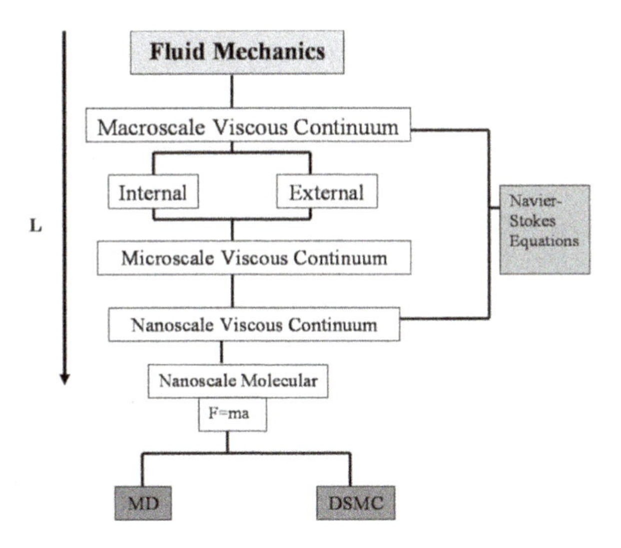

Figure 2.9 Classification of fluid flow problems. Micro- and nanofluidic flows are almost always laminar, viscous, incompressible, internal flows. The symbol L denotes a typical length scale.

In the examples given here, the Reynolds number can be on the order 10^6 or higher and is thus very large. Conversely, in many applications of interest in the biological sciences, such as flows in small capillaries; in small microdevices for analysis of biomolecules; and in sensing and detection of chemical and biological species, the Reynolds number is often orders of magnitude less than 1.

In Figure 2.9 is a classification scheme designed for the flows of liquids from the "small" macroscale down through the nanoscale. As can be seen, as the length scale decreases from the macro- to the "large" nanoscale, the Navier–Stokes equations generally apply for a liquid. However, for a gas, this picture is significantly different, and molecular effects generally appear at the microscale. Molecular effects in aqueous solutions begin to exert themselves at the "small" nanoscale, which is about 20 water diameters or 6 nm (Zhu *et al.*, 2005). Of course, surface effects begin to exert themselves far above this value, and slip flow may occur at hydrophobic walls.

Below the "small" nanoscale regime, molecular simulations, such as molecular dynamics (MD) and nonequilibrium molecular dynamics (NEMD) (i.e., electro-osmotic flow) and, to some extent, Direct Simulation Monte Carlo (DSMC), must be used to describe the flow field. It is interesting to note that DSMC was first conceived for rarefied external gas flows (Bird, 1994) but has begun to be used for selected liquid nanoscale flows.

While the subject of fluid mechanics dominates this text, we will also describe in detail the fundamentals of heat and mass transfer. Together these subjects form two of the three disciplines that are termed the *thermal sciences*, and many problems span all these individual disciplines. Thermodynamics, unlike fluid mechanics and heat and mass transfer, is an equilibrium science and the third rung of the thermal sciences, permeating through fluid mechanics in particular; a brief introduction to thermodynamics is presented next.

2.5 Elements of thermodynamics

Fluid mechanics, heat and mass transfer, and thermodynamics are inextricably linked. Fluid mechanics quantities such as pressure and density (or specific volume $v = 1/\rho$) and the heat transfer property, temperature, are all thermodynamic properties that describe the state of a system. Other thermodynamic properties are internal energy, enthalpy, entropy, and the specific heats c_p and c_v. In this section, a brief introduction to thermodynamics is presented, and in the next section, it is shown how the loss of energy, or irreversibilities, affect flows in pipes and channels.

Pressure is the macroscopic effect of particles within a fluid colliding with themselves. Pressure has units of N/m^2 and is easily measured by using a number of devices (White, 2003; Munson *et al.*, 2005). We usually speak of absolute and gage pressure, which are related by $p_{\text{gage}} = p_{\text{absolute}} - p_{\text{atm}}$. For example, a tire gage measures gage pressure and adjusts the scale to the atmospheric pressure.

A *perfect gas* is one that satisfies the relation

$$p = \rho R T \tag{2.51}$$

where R is the gas constant, p is pressure, and T is the temperature. A perfect gas is one for which intermolecular forces (van der Waals forces) are negligible and the molecules are assumed to have zero mass. The gas constant is gas specific, and for air, $R = 287$ m^2/s^{2}K $= 287$ kJ/kgK. The *universal gas constant* $R_u = R/M_w = 8.314$ kJ/k mole K, where M_w is the molecular weight; for air, $M_w = 28.97$. The density of water is a weak function of temperature, and $\rho \cong 1000$ kg/m^3 at normal room temperatures. For example, at 1 atm of pressure, $p = 1.01 \times 10^5$ N/m^2, and $T = 300°$K, the density of air is $\rho = 1.16$ kg/m^3.

The specific heat is defined as the amount of heat per mass needed to raise the temperature of a material 1°C. The specific heat at constant pressure is defined by

$$c_p = \frac{dh}{dT} \quad \text{at constant pressure} \tag{2.52}$$

where $h = u + p/\rho$ is the enthalpy, and here, u is the internal energy.

Notation alert: In the equations defining the specific heats, h is the *enthalpy*, defined by $h = u + pv$, and u is the *internal energy* and not the fluid velocity.

The specific heat at constant volume is defined by

$$c_v = \frac{du}{dT} \quad \text{at constant volume} \tag{2.53}$$

While the names of these two specific heats imply the opposite, these quantities are defined for processes in which both the temperature and pressure can vary from one value to another. This point is discussed in some detail in most undergraduate

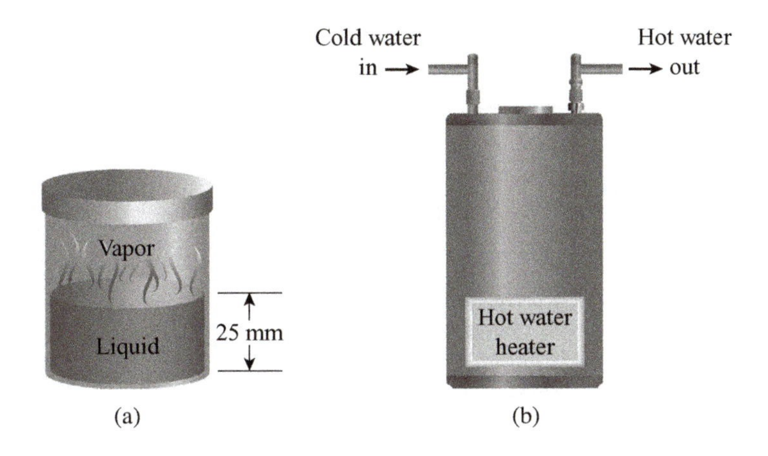

Figure 2.10 Sketch of a closed system (a) and an open system (b).

thermodynamics texts (Moran & Shapiro, 2007). For an incompressible liquid, $c_p = c_v = c$ so that $h = c_p T = c_v T = cT$, and the units of specific heat are kJ/kgK.

Notation alert: In this section, R denotes the gas constant for a specific gas in units of kJ/kgK as opposed to the *universal gas constant* R_u that is in units of kJ/kmole K. Later, however, because in electrokinetic phenomena, the universal gas constant is used exclusively. The context should be clear.

Using the definition of the perfect gas, the specific heats can be related to the gas constant by $R = c_p - c_v$ so that

$$c_p = \frac{kR}{k-1} \quad \text{and} \quad c_v = \frac{R}{k-1} \tag{2.54}$$

where $k = c_p/c_v$ and is constant over a wide range of temperatures.

In thermodynamics, several terms have very specific meanings. A *system* is an arbitrary collection of matter, and the user defines the system of interest. There are two types of systems:

1. Closed system: no mass flow is permitted across the boundaries of a closed system; there *may* be energy flow across the boundaries of a closed system (Figure 2.10a).
2. Open System: mass flow naturally flows across the boundaries of an open system (Figure 2.10b).

In fluid dynamics, open systems are used almost exclusively. A sketch of these types of systems is depicted in Figure 2.10.

Equilibrium is defined as the condition in which the characteristics or properties of a system such as pressure and temperature are not rapidly changing with time. Most often, these properties are independent of time.

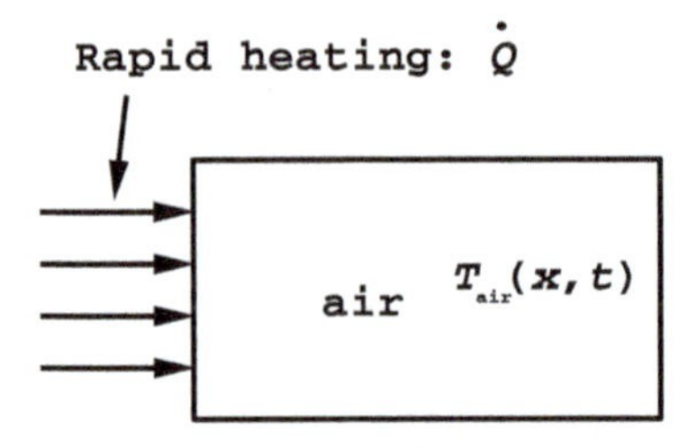

Figure 2.11 Illustration of a nonequilibrium process. Rapid heating causes the entire process to be nonequilibrium, nonquasistatic and thus outside the realm of treatment by thermodynamics.

As an example, suppose there is a gas confined in a closed system. The box is initially at $T = T_0$, and then suddenly, one end is heated, as in Figure 2.11. What is the temperature of the gas in the box for time $t > 0$? The system in this case is not in equilibrium, and thermodynamic methods cannot be used to calculate the spatially varying temperature as a function of time. *Heat transfer* is the discipline that can answer this question. However, in many situations, if heat is added slowly and for a long enough time, the temperature will become independent of time and space, and thermodynamics can predict the final temperature.

Terminology alert: In classical thermodynamics, the term *property* refers to a quantity whose change between two equilibrium states is independent of the process path. Thus enthalpy, internal energy, and entropy are examples of *thermodynamic properties*. Such terminology is normally not used in fluid mechanics or heat and mass transfer.

A *thermodynamic property* is any quantity characteristic of a system, for example, volume, mass, pressure, or temperature. Heat and work are not properties but are processes that act to change the properties of the system. The *state* of a system is the condition of a system described by its properties. For any property, say, enthalpy h, the integral over a process path between states 1 and 2 is

$$\int_1^2 dh = h_2 - h_1 = \Delta h = -\int_2^1 dh \tag{2.55}$$

However, for heat and work, which are not properties,

$$\int_1^2 \delta q = q \quad \int_1^2 \delta w = w \tag{2.56}$$

that is, the process of heat transfer and work being transferred to or from a system depends on the specific way in which it was done.

One might ask the question, How many properties are required to fix the state of a system? The answer is that it depends on whether the fluid is a gas or a liquid and on the number of phases present in the system. For example, for a perfect gas such as air, the pressure is $p = \rho RT$, where ρ is the density, R is the gas constant, and T is the temperature so that two properties in general are required to fix the third. For a liquid in equilibrium with its vapor, that is, in a saturated state, only one property is required.

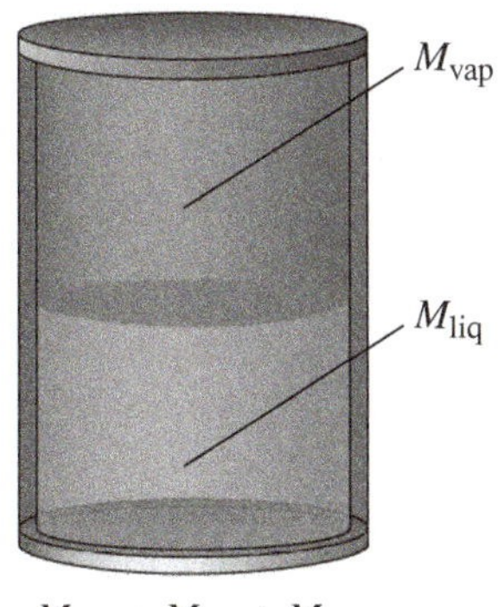

$M_{mix} = M_{vap} + M_{liq}$

One property, such as temperature, fixes the the pressure of the system.

A *process* is the act of changing a system from one equilibrium condition to another. A property has no meaning thermodynamically unless equilibrium is established. When each stage of the process is arbitrarily close to equilibrium, the process is called *quasistatic*. This is the ideal situation, and it is noted that quasistatic processes do not often occur in nature. Nevertheless, calculations are made to approximate the real processes.

The core of thermodynamics are the first law of thermodynamics or *conservation of energy* and the second law of thermodynamics. These two laws are used for a first estimate of the energy output (turbine), the energy input (pump, compressor), and the efficiency of a thermodynamic power or refrigeration cycle.

Consider the steady flow of a fluid through a *control volume* in which mass can cross the boundaries: an open system, as shown in Figure 2.12. The first law of thermodynamics is an expression of the conservation of energy over the control volume and, in the absence of chemical reactions, is given by

$$q - w = \Delta e \tag{2.57}$$

where q is the heat transfer from the system, w is the work output, and Δe is the change in energy in fluid flowing out of the control volume, $\Delta e = e_2 - e_1 = e_o - e_i$. Here the energy is given by

$$e = h + \Delta KE + \Delta PE \tag{2.58}$$

where ΔKE and ΔPE are the change in kinetic energy and potential energy, respectively. The units of all the terms in equation (2.57) are kJ/kg = J/g. The first law of thermodynamics can also be written on a rate basis in the form

$$\dot{Q} - \dot{W} = \dot{m}\Delta e \tag{2.59}$$

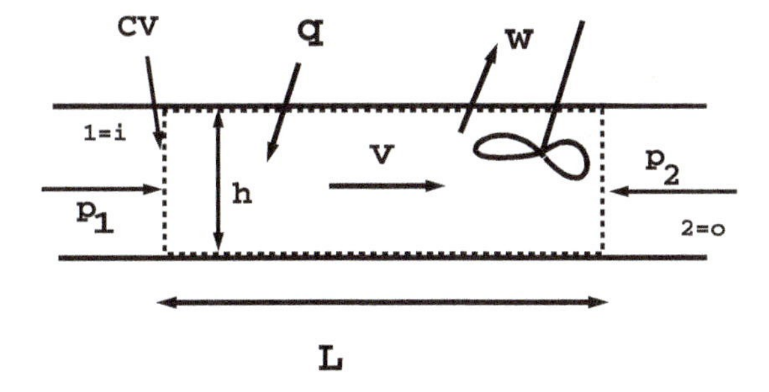

Figure 2.12 Conservation of energy for a control volume with one inlet and one outlet. The directions of positive heat q and work w are shown.

where here

$$\Delta e = h_2 + \frac{V_2^2}{2} + gz_2 - h_1 + \frac{V_1^2}{2} + gz_1 \tag{2.60}$$

In this formulation, work output from a turbine is a positive number.

The second law of thermodynamics is

$$\Delta s \geq \int \frac{\delta q}{T} \tag{2.61}$$

or on a differential basis

$$ds \geq \frac{\delta q}{T} \tag{2.62}$$

where s is the entropy in units of kJ/kg K and δq is used to denote that the heat transfer q is not a property in the thermodynamic sense. In an *ideal* process with no energy losses

$$ds = \frac{\delta q}{T} \tag{2.63}$$

and for no heat losses, $ds = 0$, and the process is *isentropic*. Under this condition, the work output from a turbine, $-\dot{W} = \dot{m}\Delta e$, is a maximum. Equation (2.63) holds for both closed and open systems.

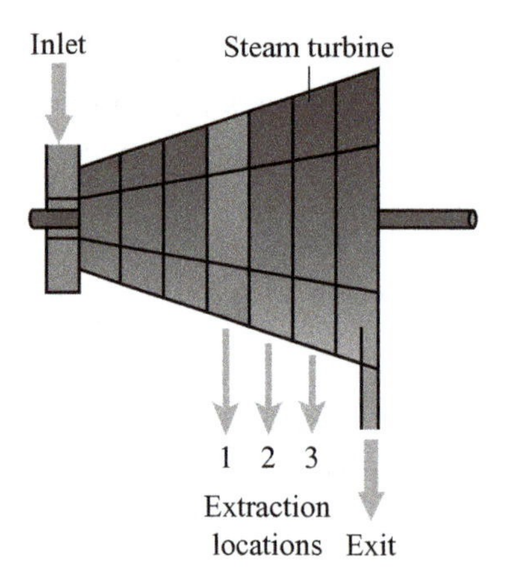

A turbine is a work-producing device and the maximum amount of work output is when the process is isentropic.

Notation alert: In the following analysis of the second law, σ is used to denote *entropy production* and not surface charge density.

Integrating equation (2.62),

$$s_2 - s_1 = \int_1^2 \frac{\delta q}{T} + \sigma \tag{2.64}$$

where σ is called the *entropy production* and $\sigma = 0$ for a *reversible process* and $\sigma > 0$ for an *irreversible process*. A reversible process is one for which there are

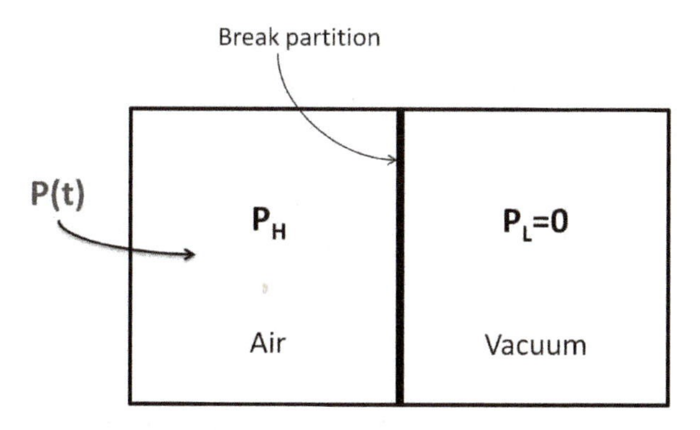

Figure 2.13 The initial condition for the subsequent nonequilibrium process.

no frictional or other losses in the given system; an irreversible process, such as steady viscous flow in a pipe or channel, is one for which there are such losses. A process for which $\sigma < 0$ is impossible.

The magnitude of the entropy change determines the efficiency of a process such as the power production in a typical power plant. An entropy change also determines the direction in which a process occurs and the efficiencies of machines such as pumps, compressors, and fans. For example, it is common knowledge that if a partition separates two compartments of different pressures in a closed system (Figure 2.13), for which no mass can pass its boundaries, and the partition is broken, the final pressure p_f satisfies

$$p_L < p_f < p_H \tag{2.65}$$

The reverse process is never observed to occur spontaneously because energy is required to make it happen. In the same situation as with the pressure, the same is true for heat transfer,

$$T_L < T_f < T_H \tag{2.66}$$

if the system is insulated or $q = 0$. These two common facts are a manifestation of the second law of thermodynamics. In both cases, if both tanks are insulated, the entropy is constant: $\Delta s = 0$.

The specific heats and entropy are related through the Tds equations, which are obtained by combining the first and second laws of thermodynamics. For a closed system, there are no kinetic and potential energy changes, and the first law reads

$$\delta q - \delta w = du \tag{2.67}$$

In a closed system, the only work that can be done is boundary work $\delta w = pdv$, and so combining with the second law

$$Tds = du + pdv = dh - vdp \tag{2.68}$$

where the second equals sign is obtained using the definition of enthalpy. The Tds equations have been derived for a closed system, however, since all of the

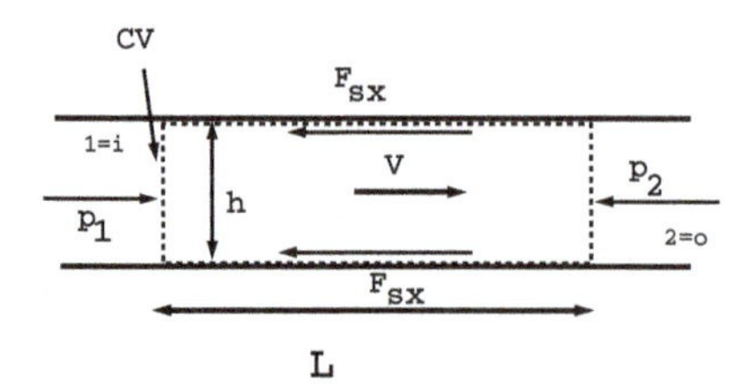

Figure 2.14 Conservation of linear momentum for a control volume with one inlet and one outlet. The number 1 denotes the inlet and 2 the outlet.

variables in equations (2.68) are thermodynamic variables and describe the *state of the system*, they are valid for open systems as well. Equation (2.68) can be integrated for constant specific volume to obtain the entropy change, and

$$\Delta s = c \ln \frac{T_2}{T_1} \tag{2.69}$$

for a liquid undergoing a constant pressure process and where $c_p = c_v = c$ is the specific heat. Thus, in a pipe with $q = 0$, the entropy will rise, resulting in a rise in temperature down the pipe.

Principles of thermodynamics explain the nature of flow in a pipe or channel. Because of viscosity, there is a loss of mechanical energy, which emerges irreversibly as thermal energy due to the rise in entropy. This is discussed next.

2.6 The nature of frictional losses in channels and pipes

Consider the channel of cross-sectional area A depicted in Figure 2.14. Conservation of linear momentum in the streamwise (x) direction over the indicated control volume at steady state leads to

$$F_{sx} + F_{bx} + p_1 A - p_2 A = \dot{m}(V_2 - V_1) \tag{2.70}$$

where F_{sx} and F_{bx} are the surface and body forces acting on the control volume and V_2 and V_1 are the average velocities at sections 1 and 2. Equation (2.70) states that the sum of the forces acting on the control volume is equal to the change in linear momentum over the control volume. In the fully developed region of the channel where the velocity is independent of the streamwise direction (see Chapter 4), $V_2 = V_1$, and we assume $F_{bx} = 0$. In this case,

$$p_1 - p_2 = \frac{F_{sx}}{A} \tag{2.71}$$

The force F_{sx} is due to the viscous stress on the wall, and thus $F_{sx} = 2WL\tau_w$ if $h \ll W \ll L$, where W is the width of the channel, as depicted in Figure 2.14 and τ_w is the shear stress at the wall. Thus

$$p_1 - p_2 = \frac{2\tau_w L}{h} \tag{2.72}$$

Fluid mechanics, heat transfer, and thermodynamics are intimately linked through the coupled nature of the momentum and energy equations. The energy equation, or the first law of thermodynamics, in the absence of any work, results in (for an incompressible fluid)

$$q = u_2 - u_1 + \frac{p_2 - p_1}{\rho} \tag{2.73}$$

where u_2 and u_1 are the internal energies at 1 and 2 and q is the heat transferred across the control volume boundary in kJ/kg. Thus

$$p_1 - p_2 = \rho(u_2 - u_1 - q) = \rho(\Delta u - q) > 0 \tag{2.74}$$

The pressure difference $p_1 - p_2$ on the left hand side of equation (2.74) represents a loss in mechanical energy. Often, the heat transfer $q < 0$ since heat energy is usually lost to the surroundings; moreover, $u_2 - u_1 > 0$. Thus the entire right-hand side of equation (2.74) represents thermal energy, and equation (2.74) is thus a statement that mechanical energy is transformed into unrecoverable thermal energy. This is an example of a thermodynamic irreversibility.

In fluid mechanics, the quantity $p_1 - p_2$ is proportional to a "loss." Thus a *head loss* is defined by

$$p_1 - p_2 = \rho g h_L \tag{2.75}$$

where h_L is the head loss and

$$h_L = f \frac{L}{h} \frac{V^2}{2g} \tag{2.76}$$

where f is called the *friction factor* and g is the gravitational acceleration; for laminar flow, the friction factor in the fully developed region of a wide channel is $f = 24/\mathrm{Re}_h$, where the Reynolds number is based on the channel height h and, for a cylindrical tube, $f = 64/\mathrm{Re}_\mathrm{D}$. Note that as the Reynolds number approaches zero, the friction factor becomes infinite; that is, for Poiseuille flow, resistance to flow in very small channels and tubes is very large.

Note that head loss has units of length and

$$h_L = \frac{u_2 - u_1 - q}{g} > 0 \tag{2.77}$$

Using the Tds equation in the form $Tds = du + pdv$ and the expression defining the entropy production, $\Delta u - q = T\sigma$ for constant temperature. Thus equation (2.77) can be written in the form

$$h_L = \frac{T\sigma}{g} > 0 \tag{2.78}$$

Equation (2.78) is the link between fluid mechanics and thermodynamics for internal viscous flow. The same phenomenon occurs in both pipes and channels of any cross section.

From the definition of internal energy and the definition of the heat transfer at low fluid speeds of interest, here the head loss can also be related to differences in temperature. Thus

$$h_L = \frac{c\Delta T + k\frac{\partial T}{\partial y}|_w}{g} \quad (2.79)$$

where c is the specific heat, ΔT is the temperature difference down the channel, and k is the thermal conductivity, and heat transfer at the wall has been assumed to be by conduction only.

From equation (2.77), it is seen that

$$q = u_2 - u_1 - gh_L \quad (2.80)$$

Here the internal energy $u_2 = cT_2$, where $c = 1$ kJ/kgK is the specific heat, for example. Typically in liquid flows, the temperature rise between two points in a channel or tube in the absence of driven heat transfer to the fluid is very small, perhaps less than 1 K. In very small channels on the order of $h = 1$ μm or less, the heat loss q in kJ/kg is almost all due to the head loss. As an example, for water flowing in a channel of length $L = 10$ μm and height $h = 1$ μm, and a velocity of $0.03\frac{\mathrm{m}}{\mathrm{sec}}$, the head loss $h_L = 0.3$ m and the resulting heat loss is $q = -2.38$ kJ/kg for $T_2 - T_1 = 0.5$K. What will happen if the channel is smaller?

This analysis links in a precise way the three branches of the thermal sciences. Given the friction factor of fluid mechanics, the combination of the heat loss to the surroundings and the increase in internal energy of the fluid in a pipe or channel can be determined. Likewise, the discipline of heat transfer relates the temperature gradient at the wall to the thermodynamic heat loss. It should be mentioned that thermodynamics does not give any information about the internal energy at points *between* 1 and 2 in Figure 2.14.

2.7 Chapter summary

Transport coefficients are fundamental to the development of the governing equations of fluid mechanics and heat and mass transfer and for the conservation of electrical charge for an electrically conducting fluid. Viscosity, thermal conductivity, the diffusion coefficient, and the electrical permittivity are the four transport coefficients that appear in the governing equations of fluid mechanics, heat and mass transfer, and electrostatics. In addition to these fundamental transport coefficients, the concept of surface tension, which is important to some applications in micro- and nanofluidics, is also introduced. The primary purpose of this chapter is to introduce the reader to the process of determining these coefficients from experiment as well as theory. The constitutive laws defining these transport coefficients have been shown to have distinct similarities when the variation of the dependent variables is confined to one dimension.

The fluid mixtures considered in this text are dilute enough that there will be no significant error in assuming that the transport coefficients have the properties of the solvent. The exception is surface tension, which is considerably altered by the addition of a small amount of surfactant.

The form of the constitutive laws in the definition of viscosity, thermal conductivity, and diffusion coefficient were determined empirically from a large number of experiments. Conversely, one interpretation of the electrical permittivity is that it defines the surface charge density.

Fluid mechanics, heat and mass transfer, and thermodynamics are inextricably linked. It is the irreversible conversion of the thermal energy defined by thermodynamics and heat transfer that causes the pressure drop in a pipe. The magnitude of the pressure drop is linearly proportional to the entropy production σ, thereby providing an explicit means of evaluating the entropy production that is not usually possible in a thermodynamic framework.

EXERCISES

2.1 For Poiseuille flow in a channel, the volume flow rate is defined by

$$Q = \int_0^h uW dy$$

where u is the velocity, h is the height of the channel, and W is its width. From methods developed in Chapter 4, it can be shown that

$$Q = \frac{Wh^3 \Delta p}{12\mu L}$$

Calculate the required pressure drop to achieve a volume flow rate of 10^{-6} L/min for a channel of width $W = 100$ μm, length $L = 100$ μm, and for channel heights of $h = 1$ μm, 0.1 μm, and 0.01 μm. Assume the working fluid is water. Repeat for air. Calculate the Reynolds number for each height. Repeat the calculation if the flow rate is 10^{-9} L/min. Put your results in a table with the channel height h as the independent parameter, and compare the results for water and air for each flow rate. Use the channel height h as the length scale in the definition of the Reynolds number. Take the densities to be 1000 and 1.4 kg/m^3 for water and air, respectively, and $.8 \times 10^{-3}$ and 1.8×10^{-5} N sec/m^2 for the viscosities.

2.2 The velocity for Poiseuille flow in a pipe is given by

$$w(r) = -\frac{R^2}{4\mu}\frac{dp}{dx}\left[1 - \left(\frac{r}{R}\right)^2\right]$$

Find the average velocity in the pipe, which is defined as

$$V = \frac{1}{A}\int_0^R 2\pi r w\, dr$$

where R is the pipe radius and $A = \pi R^2$. Show that the flow rate is given by

$$Q = -\frac{\pi R^4}{8\mu}\frac{dp}{dx}$$

Calculate the pressure drop in water required to produce a flow rate of $Q = 1\ \mu\text{L/min}$ through a tube having a radius $R = 10^{-7}$ m using the same values of the length L as in the previous problem.

2.3 Plot the Debye length in water

$$\lambda = \frac{\sqrt{\epsilon_e RT}}{FI^{1/2}}$$

where F is Faraday's constant, ϵ_e is the electrical permittivity of the medium, I is the *ionic strength*, $I = \sum_i z_i^2 c_i$, c_i is the concentrations of the electrolyte constituents at some reference location, R is the universal gas constant, z_i is the valence of species i, and T is the temperature, as a function of ionic strength from $c = c_1 = c_2 = 30 \times 10^{-6}$ M to $T = 1$ M, for a pair of monovalent electrolytes such as sodium chloride in water. For each value of the ionic strength you choose, assuming that the mixture is electrically neutral, calculate the number density of the ionic species if $n_i = c_i Na$, where Na is Avagadro's number. Repeat the calculation for the same values of the concentration for a pair of $+2$, -2 ions.

2.4 A sodium ion has an ionic diameter of about 0.2 nm. Assuming that the ion is a point charge, calculate the electric field and the potential at a distance of $r = 6$ nm from its center in an aqueous system. Repeat the calculation if the solvent is methanol.

2.5 Albumin is a globular protein and so, to a first approximation, it may be assumed to be spherical and to have an ionic diameter of about 7 nm. Estimate its diffusion coefficient in water using the Stokes–Einstein formula.

2.6 Estimate the diffusion coefficient of sodium in water using the Stokes–Einstein relation.

2.7 An equation of state for water relates the pressure to the density by

$$\frac{p}{p_0} = (A+1)\left(\frac{\rho}{\rho_0}\right)^n - A$$

where $A = 3000$, $n = 7$, $p_0 = 1$ atm, and $\rho_0 = 998\ \text{kg/m}^3$. Calculate the pressure required for the density of water to double.

2.8 For thin electric double layers, compared to the smallest channel dimension, the electro-osmotic velocity is constant across the channel and is given by

$$u = CU_0$$

where C is a dimensionless constant dependent on the concentration at some reference location and U_0 is the velocity scale. Find a relationship between U_0 and the pressure drop in Poiseuille flow in a wide channel if the volume flow rate is the same as in the electro-osmotic flow.

2.9 Calculate the heat transfer rate in water if a temperature difference of $\Delta T = 3°\text{C}$ is measured over a distance of $1\ \mu\text{m}$ at $T = 60°\text{C}$.

2.10 Is there a realistic combination of parameters such that the mass transfer rate has the same numerical value as the heat transfer rate if the fluid is water?

2.11 Is there a realistic combination of parameters such that the shear stress has the same numerical value as the heat transfer rate?

2.12 A separation of two species is required based on their valence. Which flow would yield the most efficient separation: EOF or Poiseuille? In considering the EOF, assume that the EDL is thin.

3 The Governing Equations for an Electrically Conducting Fluid

3.1 Introduction

In the first two chapters, the field of micro- and nanofluidics was introduced in a general framework, describing multiple applications as well as the scientific issues associated with fluid flow at small length scales. In Chapter 2, the fundamental transport properties involved in mathematical description of fluid flow, heat and mass transfer, and electrostatics were introduced. In this chapter, we use these concepts to derive, from first principles, the governing equations necessary for analyzing micro- and nanofluidic phenomena. For the purposes of this chapter, it will be assumed that the properties of the fluid medium are constant.

Microfluidic devices are being used for rapid and continuous purification of proteins; a sketch of such a device is shown in Figure 3.1. The device addresses the need for high-throughput purification of very small amounts of proteins and enzymes from the carrier fluid. The term *protein purification* refers to a series of operations meant to isolate a single protein or enzyme in a complicated mixture. Here the microfluidic transport processes involve mass transport of a relatively large number of species with the target molecules present in as little as microgram per liter concentrations. This device can purify a sample in a short period of time and does not require a large amount of sample. The means of developing a model for such a device is discussed in this chapter.

The governing equations of fluid motion on the macroscale are the (incompressible) Navier–Stokes equations. These equations, along with conservation of mass, or the *continuity equation*, enable the calculation of, in the general three-dimensional case, the three velocity components and the pressure. The equations of electrostatics provide information to calculate the *electrical potential* given the distribution of ionic concentrations within the domain of interest. The *energy equation* determines the temperature, and finally, *conservation of species* determines the local concentration of each of the mixture constituents. Together, these equations form a set of $6 + N - 1 = 5 + N$ equations in $5 + N$ unknowns, where N is the number of constituents of the fluid mixture. Note that only $N - 1$ species concentrations are unknown because the sum of the concentrations, the total concentration, is known. The governing equations are presented

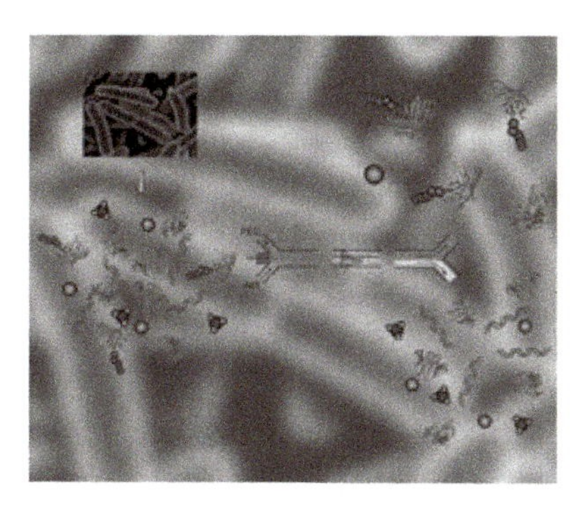

Figure 3.1 The Sandia Lab-on-a-Chip for the high-throughput purification of minute amounts of proteins and enzymes from a carrier fluid (Meagher *et al.*, 2008). Reproduced by permission of The Royal Society of Chemistry.

in Cartesian coordinates, with the relevant operators for cylindrical and spherical coordinates appearing in Appendix B.

After the derivation of each of these governing equations, the appropriate boundary conditions are presented, and then the subject of dimensional analysis is discussed. This issue is an important one, especially for nanofluidic channels for which velocity, concentration, electric potential, and temperature distributions are extremely difficult to measure.

Next, dimensional analysis is used to transform the dimensional governing equations into dimensionless form. This process leads to a discussion of several analogies between heat and mass transfer and velocity and electrical potential. Next, the mathematical character of the governing equations and well-posed problems are discussed, with emphasis on the fact that the character of a partial differential equation determines the numerical method used to solve it. Finally, we discuss the role of fabrication, experiments, and theory in micro- and nanofluidics.

3.2 The continuum approximation and its limitations

In the study of thermal science and the field of electrostatics and electrodynamics, it is tacitly assumed that the fluid may be regarded as a continuum. This means that from a macroscopic viewpoint, the character of the flow, heat and mass transfer, and the electrical properties of the fluid may be described by averaging over some small but finite volume. In practice, the size of this small but finite volume is never specified; however, for the fluid to behave as a continuum, the averaging volume must be much greater than a molecular diameter in a liquid and the mean free path in a gas.

To illustrate the continuum hypothesis, consider the case of interest in this book: a liquid. Consider a small imaginary volume within the domain of flow. The volume is assumed to be so small relative to the smallest macroscopic space dimension that the cube defines a point in the space occupied by the fluid. Consider a region in space as depicted in Figure 3.2, and let us determine the density at the point (x_0, y_0, z_0). In macroscopic terms, the density is defined as

$$\rho = \frac{m}{V} \tag{3.1}$$

but this value may not be the value at (x_0, y_0, z_0). To determine that, we must take the limit

$$\rho = \lim_{V \to V'} \frac{dm}{dV} \tag{3.2}$$

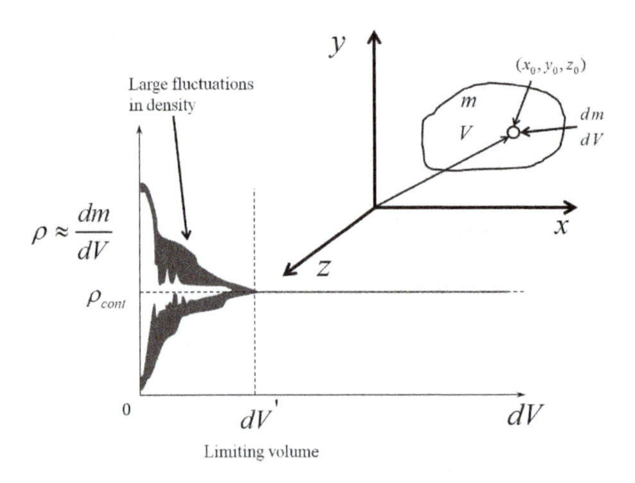

Figure 3.2 Illustration of the applicability of the continuum approach for $V > dV'$.

where V' is some (unspecified) critical volume. As $V \to V'$, the cube may contain only a small number of molecules; that is, the number of individual molecules in the volume is of the same order as the the number of fluid molecules passing into and out of the volume. When this occurs, the continuum assumption breaks down.

The *mean free path* of a gas is the distance a molecule travels between collisions. In air, the mean free path is $\lambda_p \sim 60$ nm. Thus, as shown in Section 3.11, it is expected that the continuum approximation will break down for flow in a channel whose smallest dimension is about 600 nm $= 0.6$ μm $= 10\lambda_p$. In liquids, it makes no sense to speak of a mean free path because the molecules are colliding all the time. For liquids, it makes more sense to speak of a mean molecular spacing. The mean molecular spacing can be taken to be the molecule's diameter, which, for water, is $D \sim 0.3$ nm $= 3$ Å.

If a nominal criterion for the critical volume is taken to be 10 times the molecular diameter, then it is expected that the critical volume for water will be on the order of $V' \sim 10 \times 4/3\pi(\delta^3/8) \sim 140 \times 10^{-30}$ m^3. Thus the continuum approximation for the flow of an aqueous electrolyte (i.e., salt water) is expected to be valid for the smallest channel dimension on the order of 10 nm!

The alternative to continuum approximations of the flow is to treat each molecule individually and solve Newton's law in the form

$$\vec{F} = m\vec{a} \tag{3.3}$$

directly. There are many ways to do this, and these *molecular simulation methods* will be discussed later in this book. Two methods often used when the continuum approximation breaks down are termed *molecular dynamics simulations* (Figure 3.3), often used for liquids, and *Monte Carlo simulations*, most often used for gases. However, it must be said that these simulations are computationally intensive, often taking a month for a single simulation, and many approximations are required even to get to that point. It is for this reason that micro- and nanofluidic

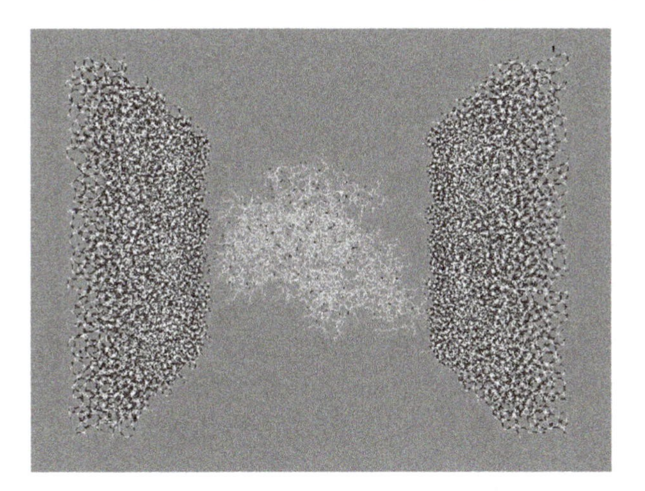

Figure 3.3 Molecular dynamics simulation of an albumin molecule in a channel; the solvent, water, is not shown. The channel walls are also simulated using MD. Without an MD simulation, the *conformation*, or shape and orientation of the albumin molecule, an important protein, cannot be determined. Courtesy of Professor Sherwin Singer. Used with permission.

devices are difficult to design based solely on molecular simulations. Molecular dynamics or some other molecular method is required to determine a molecule's shape and orientation.

3.3 Kinematics

Kinematics is the study of bodies in motion, without reference to the masses or forces that cause the motion. There are two ways of describing the motion of a fluid particle within a given flow field. The motion of a fluid particle could be considered a rigid body, and its trajectory could be followed instantaneously in

Figure 3.4 A Lagrangian description follows each fluid element in time like following the motion of molecules inside a carbon nanotube. Image courtesy of Lawrence Livermore National Laboratory.

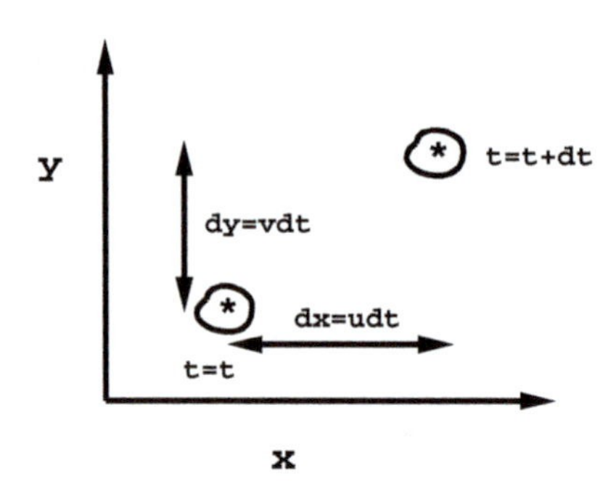

Figure 3.5 Definition of the velocity of a fluid particle.

time. Such a scheme of monitoring the motion of such a particle is called the *Lagrangian description* of fluid motion. Each fluid particle is thus differentiated by its corresponding starting position, and the velocity is described by the function $\vec{V}(x_0, y_0, z_0, t)$, in which the time t is the only variable. The entire flow field consisting of the motion of these fluid particles is represented by determining a number of particle paths originating from different initial positions. This method of describing fluid motion is important for determining residence times of a given analyte in a rapid molecular analysis tool.

Now consider a fluid flow through a microfluidic device such as described in Chapter 1. Instead of tracing a single particle path from an upstream reservoir through a nanofluidic channel (as in Figure 1.1), suppose attention is focused on, say, the space within the microfluidic (or nanofluidic) channels or even in the reservoirs; then the velocity within the given space is a function of position within that space, or $\vec{V} = \vec{V}(x, y, z, t)$. Note that the velocity can vary widely, depending on the position (x, y, z) within the control space of interest and time. If the initial position of such a particle of interest falls within the space, then each value of (x_0, y_0, z_0) corresponds to some value of (x, y, z) at each time t so that

$$\vec{V}(x_0, y_0, z_0, t) = \vec{V}(x, y, z, t) \tag{3.4}$$

and this equation defines the *Eulerian description* of fluid motion.

Any fluid particle has a given fluid velocity, which may vary in time and/or space. Consider a general function $H = H(x, y, z, t)$. The differential of this function is thus

$$dH = \frac{\partial H}{\partial x}dx + \frac{\partial H}{\partial y}dy + \frac{\partial H}{\partial z}dz + \frac{\partial H}{\partial t}dt \tag{3.5}$$

Dividing by dt, then,

$$\frac{dH}{dt} = \frac{\partial H}{\partial x}\frac{dx}{dt} + \frac{\partial H}{\partial y}\frac{dy}{dt} + \frac{\partial H}{\partial z}\frac{dz}{dt} + \frac{\partial H}{\partial t} \tag{3.6}$$

It is recognized that the differentials appearing in equation (3.6) are the local fluid velocities; for example, referring to Figure 3.5, the velocity in the x direction is defined by

$$u(x, y, z) = \frac{dx}{dt} \tag{3.7}$$

and similarly for the other two directions. Thus the *material derivative* is given by

$$\frac{dH}{dt} = \frac{DH}{Dt} = \frac{\partial H}{\partial t} + u\frac{\partial H}{\partial x} + v\frac{\partial H}{\partial y} + w\frac{\partial H}{\partial z} \tag{3.8}$$

The symbol D/Dt for the material derivative is used to distinguish the material derivative from the time derivative of a particle trajectory d/dt.

Applying equation (3.8) to the Eulerian velocity vector $\vec{V}(x, y, z, t) = (u(x, y, z, t), v(x, y, z, t), w(x, y, z, t))$, it is found, for example, that

$$\frac{Du}{Dt} = \frac{\partial u}{\partial t} + u\frac{\partial u}{\partial x} + v\frac{\partial u}{\partial y} + w\frac{\partial u}{\partial z} \tag{3.9}$$

The first term in equation (3.9) is called the *local acceleration* and is nonzero in unsteady or transient situations. The second term is the *convective acceleration*. If the velocity is independent of time, the flow is said to be steady. In the Eulerian description of fluid motion, equation (3.9) shows that a fluid particle may accelerate even if the flow is steady. The form of the material derivative for the other two velocity components is easily obtained.

The material derivative for the temperature and concentration of a given species is required for the energy and species mass balances. These two quantities are obtained by taking $H = T$, the temperature, and $H = c_A$, the concentration in equation (3.8), respectively.

In many cases, the velocity field is independent of one of the coordinate directions. For example, a velocity field of the form $\vec{V} = (u(x, y, t), v(x, y, t), 0)$ is a two-dimensional flow, and if steady, the velocity has the form $\vec{V} = (u(x, y), v(x, y), 0)$. Fully developed flow in the x direction in a channel or pipe is one-dimensional; $\vec{V} = (u(y), 0, 0)$ for a channel (see Figure 1.4), and $\vec{V} = (0, 0, w(r))$ in the (r, θ, z) directions for a pipe of circular cross section. The material derivatives in cylindrical and spherical coordinates are presented in Appendix B.

There are several different ways to visualize fluid flows:

- Velocity vectors in the direction of the velocity at a given point
- Contours of constant velocity
- Streamlines
- Pathlines
- Streaklines

The depiction of streamlines at a fixed time is termed a Eulerian description of the flow field since the flow field is considered a function of position (Figure 3.6). Streamlines are defined at each fixed time by the equation

$$d\vec{r} \times \vec{u} = 0 \tag{3.10}$$

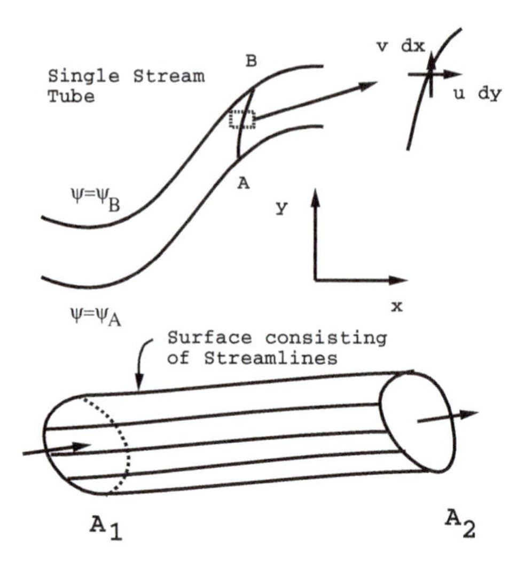

Figure 3.6 Definition of a streamline, a line parallel to the velocity vector.

where $\times$ denotes the cross-product, and this definition is equivalent to

$$\frac{dx}{u} = \frac{dy}{v} = \frac{dz}{w} \tag{3.11}$$

in Cartesian coordinates. By their definition, streamlines are lines parallel to the instantaneous local velocity field.

Another way to define streamlines valid in two-dimensions is to define a *stream function* ψ. Consider the continuity equation in Cartesian coordinates, which, in two dimensions, is

$$\frac{\partial u}{\partial x} + \frac{\partial v}{\partial y} = 0 \tag{3.12}$$

where u and v are the velocities in the x and y directions, respectively. We define the stream function as that function which satisfies

$$u = \frac{\partial \psi}{\partial y} \quad \text{and} \quad v = -\frac{\partial \psi}{\partial x} \tag{3.13}$$

In this case, subject to sufficient restrictions on the mathematical continuity and differentiability of the stream function, which are satisfied in any reasonable physical situation, the continuity equation, or conservation of mass equation, is identically satisfied. It should be mentioned that a stream function cannot be defined in three dimensions. Note also that if the vorticity vector vanishes, $\nabla \times \vec{V} = 0$, then it is easy to show that

$$\nabla^2 \psi = 0 \tag{3.14}$$

The lines $\psi = \text{constant}$ are then the streamlines.

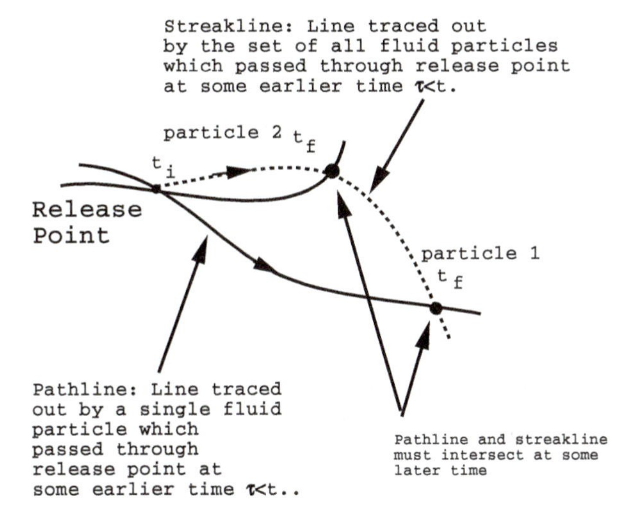

Figure 3.7 Illustration of the difference between pathlines and streaklines.

A *pathline* is a line traced out in time by a given fluid particle. The set of equations satisfied by pathlines are

$$\frac{dx_i}{dt} = u_i(x_i, t) \tag{3.15}$$

subject to initial conditions

$$x_i = x_{i,0} \quad \text{at } t = 0 \tag{3.16}$$

for $i = 1, 2, 3$. These are ordinary differential equations in the time domain for a given particle and describe the Lagrangian representation of fluid motion.

A *streakline* is a line traced out by a neutrally buoyant marker that is continuously injected at a fixed point in space. Suppose the injection point is given by $(x_{1,0}, x_{2,0}, x_{3,0})$. A particle at a given point (x_1, x_2, x_3) at time t must have passed through $(x_{1,0}, x_{2,0}, x_{3,0})$ at some earlier time $\tau < t$. The governing equations to determine streaklines are the same as for pathlines, except that the initial conditions are

$$x_i = x_{i,0} \quad \text{at } t = \tau \tag{3.17}$$

Streaklines are important because it is streaklines that experimentalists visualize when dye or hydrogen bubbles are used to track fluid particles. For steady flow, pathlines, streamlines, and streaklines are all equivalent. However, streaklines are very difficult to calculate because they trace the time history of a potentially large set of fluid particles, all of which left the same point at different times. In contrast, pathlines trace the time history of only a single fluid particle. The difference between streaklines and pathlines is illustrated in Figure 3.7, which shows two particles that left the release point at different times. Their pathlines must intersect the streaklines at two points at a given final time t_f because the pathline of a single particle and one point on the streakline coincide at the initial time at the point of release.

Streaklines in History

When Osborne Reynolds was doing his experiments in 1883 to identify that there are two types of flow, laminar and turbulent, he used streaklines to delineate the two regimes. At low Reynolds number, dye injected into a tube passed neatly in a line down the center of the tube. Turbulent flow was identified by the intense breakup of the dye streak a certain distance downstream of the release point. The history of fluid mechanics was changed forever.

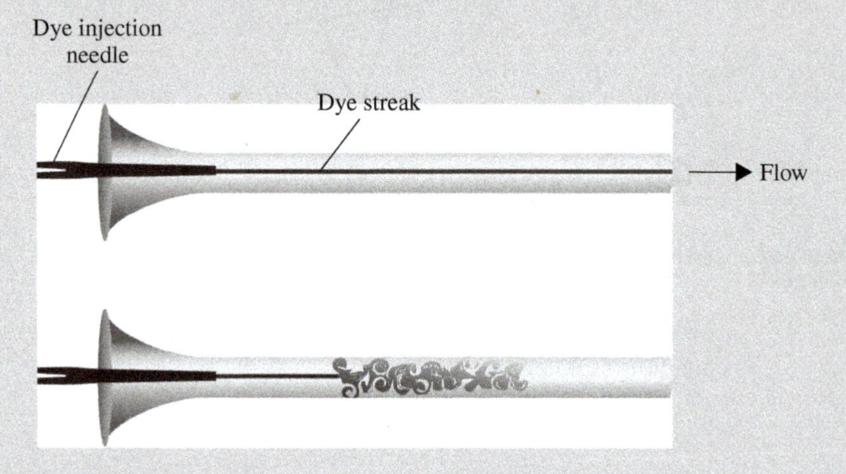

Osborne Reynolds's experiment used streaklines to observe the difference between laminar and turbulent flow. The eddies downstream on the bottom figure indicate that the flow is turbulent. See Goldstein (1965a).

As an example, consider the flow field defined by

$$u = ay, \quad v = -a(x - bt) \tag{3.18}$$

with a and b constant. This is an unsteady two-dimensional flow that satisfies conservation of mass equation (3.12). The equation for the streamlines is

$$\frac{dy}{dx} = \frac{v}{u} \tag{3.19}$$

Separating variables and integrating, it is easily shown that the streamlines are circles with their center at $y = 0$, $x = bt$, defined by

$$y^2 + \frac{(x - bt)^2}{2} = \text{const} \tag{3.20}$$

The pathlines are obtained by solving the set of equations

$$\frac{dx}{dt} = ay \tag{3.21}$$

$$\frac{dy}{dt} = -a(x - bt) \tag{3.22}$$

The solution is obtained by standard techniques as

$$x(t) = A\cos(at) + B\sin(at) + bt \tag{3.23}$$

$$y(t) = B\cos(at) - A\sin(at) + b/a \tag{3.24}$$

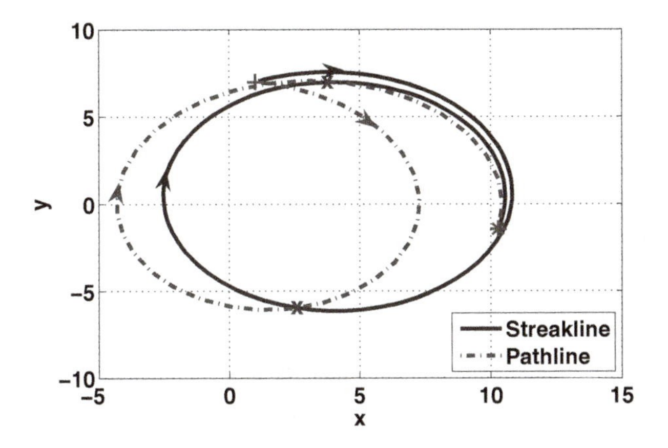

Figure 3.8 Results for the streaklines and pathlines, showing three intersection points. The calculation was carried out from $t = 0$ to $t = 4$, and the results are plotted at $t = 4$. The plus denotes the starting point and the asterisk denotes the end point. The cross denotes the crossing points.

To obtain a particular pathline, initial conditions need to be specified, and if, for example, $y = 7$ at $t = 0$, then the solution is

$$y(t) = (7 - b/a)\cos(at) - \sin(at) + b/a \tag{3.25}$$

Likewise, the streaklines are obtained in the same way as the pathlines, except that the initial condition is expressed as

$$x = 1, \quad y = 7 \quad \text{at } t = \tau \tag{3.26}$$

To determine the constants A and B, we need to solve the system of equations

$$1 = A\cos(a\tau) + B\sin(a\tau) + b\tau \tag{3.27}$$

$$7 = -A\sin(a\tau) + B\cos(a\tau) + b/a \tag{3.28}$$

The results for A and B are as follows:

$$A = (1 - b\tau)\cos(a\tau) - (7 - b/a)\sin(a\tau) \tag{3.29}$$

$$B = (7 - b/a)\cos(a\tau) + (1 - b\tau)\sin(a\tau) \tag{3.30}$$

Substitution of the results for A and B into the expressions for $x(t;\tau)$ and $y(t;\tau)$ and evaluation at $t = 4$ yields the solutions. The results are depicted in Figure 3.8.

3.4 Surface and body forces

The Navier–Stokes equations are the expression of Newton's law $\vec{F} = m\vec{a}$ for a fluid. Thus the forces on a given fluid particle must be evaluated. Instead of working with forces directly, it is customary to work with stress or force per unit area.

Two types of stress may act on a fluid particle: surface stress and body stress. Only those stresses that act in a given direction will cause motion in that direction. Surface forces act directly on the surface of a fluid volume. Body forces, such as

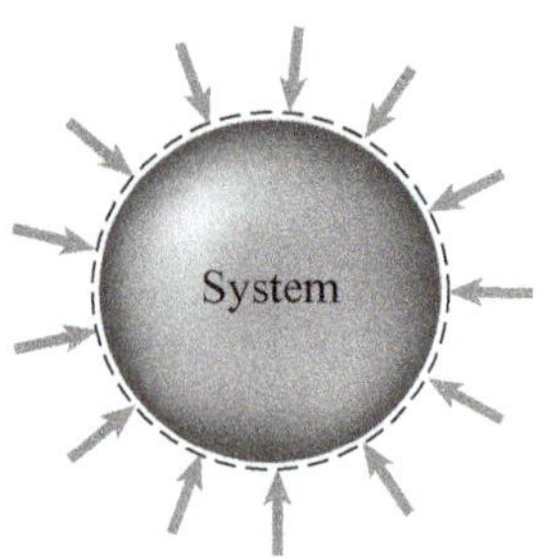

Pressure is a normal stress.

gravity and an electric field, act on the fluid as a whole. The surface stresses in Figure 3.9 have two subscripts corresponding to the surface on which they act and the direction they are pointing. For example, a positive x surface is defined as a surface whose outward unit normal is pointing in the positive x direction. A stress is defined to be positive if it is on a "positive" surface pointing in the positive coordinate direction or on a "negative" surface pointing in the negative coordinate direction. All the stresses in Figure 3.9 have positive magnitudes as drawn.

Wind induces a shear stress on water causing waves to form.

A cube of fluid moving in space may undergo four distinct types of changes:

1. Translation
2. Rotation
3. Extensional strain
4. Angular or shear strain

Translation and rotation cannot generate a stress; the extensional and angular strains are related to gradients of the velocity field and the pressure.

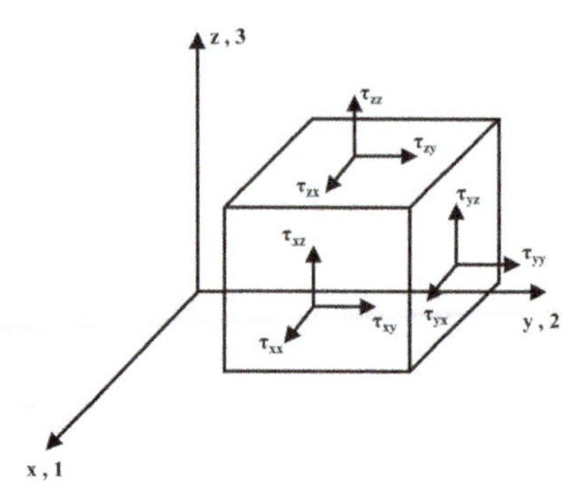

Figure 3.9 Stresses on a fluid cube. Stress is defined as force per unit area. Adapted from Turns (2006).

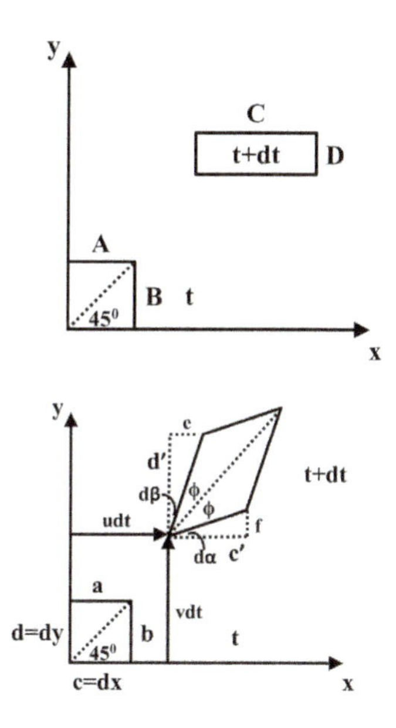

Figure 3.10 Two-dimensional view of (top) extensional and (bottom) shear strain on a fluid cube. Adapted from Currie (2003).

Consider the the extensional and angular strains that may occur to a fluid cube as depicted in Figure 3.10; *strain* is defined as the change in a given length in the fluid cube, and the rate of strain is the quantity of interest.

For the extensional strain, suppose the length $A = dx$ and the length $B = dy$. The notation σ_{xx} denotes the rate of the change in length of a fluid element in the x direction, dx due to flow over its original length, and has units of $\sec^{-1}$. Thus the fractional change in length in the x direction defines the extensional strain rate, and

$$\sigma_{xx} dt = \frac{dx + \frac{\partial u}{\partial x} dx dt - dx}{dx} = \frac{\partial u}{\partial x} dt \tag{3.31}$$

and in a similar fashion,

$$\sigma_{yy} = \frac{\partial v}{\partial y} \tag{3.32}$$

$$\sigma_{zz} = \frac{\partial w}{\partial z} \tag{3.33}$$

Now consider the angular strain rate. The key to determining this quantity is to evaluate the angles $d\alpha$ and $d\beta$ in Figure 3.10 at time $t + \Delta t$ that at time t were zero. Note that if $d\alpha > d\beta$, then the fluid cube has undergone a counterclockwise rotation, whereas if the opposite is the case, a clockwise rotation is obtained, and $e = d' \tan d\beta$ and $f = c' \tan d\alpha$. It is now necessary to determine the lengths $c', d', e,$ and f.

The length of the original face d at $t = t$ is $d = dy$; thus it has changed as a result of the change in the v velocity so that the length d' at $t = t + dt$ is $d' = dy + (\partial v/\partial y) dydt$ using the Taylor series approximation for small dt. Similarly, the length c has changed as a result of the change in the u velocity;

thus $c' = dx + (\partial u/\partial x)\, dxdt$. The length e was originally zero and has become nonzero as a result of the change in u velocity with y. Thus $e = (\partial u/\partial y)\, dydt$, and finally, for the same reason, $f = (\partial v/\partial x)\, dxdt$.

Note that in Figure 3.10, the rotation of the fluid element is about the z axis, the third axis into the page. If counterclockwise rotation is considered positive, let Ω_z denote the average rotation rate of the fluid cube. The angle $d\beta$ indicates a clockwise rotation, and the angle $d\alpha$ indicates a counterclockwise rotation. Then the average rotation rate associated with the fluid cube is

$$d\Omega_z = \frac{1}{2}(d\alpha - d\beta) \tag{3.34}$$

Note that

$$\frac{d\alpha}{dt} = \lim_{dt\to 0} \tan^{-1} \frac{f}{c'} = \frac{\partial v}{\partial x} \tag{3.35}$$

$$\frac{d\beta}{dt} = \lim_{dt\to 0} \tan^{-1} \frac{e}{d'} = \frac{\partial u}{\partial y} \tag{3.36}$$

so that

$$\frac{d\Omega_z}{dt} = \frac{1}{2}\left(\frac{\partial v}{\partial x} - \frac{\partial u}{\partial y}\right) \tag{3.37}$$

A similar analysis shows that the rotation rates about the x and y axes are given by

$$\frac{d\Omega_x}{dt} = \frac{1}{2}\left(\frac{\partial w}{\partial y} - \frac{\partial v}{\partial z}\right) \tag{3.38}$$

$$\frac{d\Omega_y}{dt} = \frac{1}{2}\left(\frac{\partial u}{\partial z} - \frac{\partial w}{\partial x}\right) \tag{3.39}$$

Twice the rotation rate of a fluid particle is an important quantity, and

$$\vec{\omega} = 2\frac{d\vec{\Omega}}{dt} = \nabla \times \vec{V} \tag{3.40}$$

is called the *vorticity*. It is the vorticity that determines the character of the velocity field; note that vorticity is a conserved quantity

$$\nabla \bullet \vec{\omega} = 0 \tag{3.41}$$

If $\vec{\omega} = 0$, the flow is said to be irrotational.

It remains to define the shear strain rate; in keeping with convention (White, 2006; Currie, 2003), the shear strain rate in the xy plane is defined as the average decrease in the angle that is initially 90°, or

$$\sigma_{xy} = \frac{1}{2}\left(\frac{d\alpha}{dt} + \frac{d\beta}{dt}\right) = \frac{1}{2}\left(\frac{\partial v}{\partial x} + \frac{\partial u}{\partial y}\right) \tag{3.42}$$

In similar fashion, it is found that

$$\sigma_{yz} = \frac{1}{2}\left(\frac{\partial w}{\partial y} + \frac{\partial v}{\partial z}\right) \tag{3.43}$$

$$\sigma_{xz} = \frac{1}{2}\left(\frac{\partial u}{\partial z} + \frac{\partial w}{\partial x}\right) \tag{3.44}$$

The shear strain rate is a tensor; note that it is symmetric, $\sigma_{ij} = \sigma_{ji}$, so that there are six distinct values. It is the relationship between stress and rate of strain, termed a *constitutive* relation, that is specific to a fluid.

The pressure is a normal stress and will contribute to the extensional strain rate depicted in Figure 3.10. Thus, if the fluid is Newtonian, the stress tensor is given by

$$\tau_{ij} = -p\delta_{ij} + 2\mu\sigma_{ij} = -p\delta_{ij} + \mu\left(\frac{\partial u_i}{\partial x_j} + \frac{\partial u_j}{\partial x_i}\right) \tag{3.45}$$

for $i = 1, 2, 3$ in tensor notation, where $u_i = (u, v, w)$. The subject of *rheology* is concerned with determining the stress–strain rate relationship for a given fluid or class of fluids. Here δ_{ij} is the Dirac delta function:

$$\delta_{ij} = 0 \quad i \neq j \tag{3.46}$$

$$\delta_{ij} = 1 \quad i = j \tag{3.47}$$

For example, in conventional Cartesian notation for a Newtonian fluid,

$$\tau_{xy} = 2\mu\sigma_{xy} = \mu\left(\frac{d\alpha}{dt} + \frac{d\beta}{dt}\right) = \mu\left(\frac{\partial v}{\partial x} + \frac{\partial u}{\partial y}\right) \tag{3.48}$$

$$\tau_{xx} = -p + \mu\frac{\partial u}{\partial x} \tag{3.49}$$

It is now possible to determine the governing equations for an electrically conducting fluid. A Newtonian fluid having constant properties is assumed.

3.5 The continuity equation

In the sections to follow, the governing equations will be developed in Cartesian coordinates; typically, channels in a nanopore membrane are long and wide, as in Figure 3.11, so the governing equations may be considerably simplified. In your first fluid dynamics course, you studied the integral form of the mass conservation

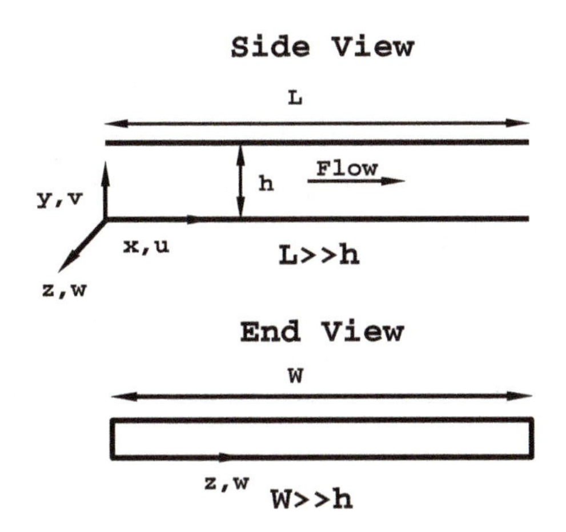

Figure 3.11 Geometry of a typical channel; systems of such channels make up a nanopore membrane.

equation, which is

$$\frac{\partial}{\partial t}\int_{\mathcal{V}} \rho d\mathcal{V} + \int_{S} \rho \vec{V} \bullet d\vec{A} = 0 \tag{3.50}$$

where $d\mathcal{V}$ denotes that the integral is over a volume and $d\vec{A}$ denotes that the integral is over an area. The first term is the time rate of change in mass within the control volume, and the second term is the net mass passing out of the control volume. Using the divergence theorem, which is defined for any vector $\vec{H}$ as

$$\int \nabla \bullet \vec{H} d\mathcal{V} = \int \vec{H} \bullet d\vec{A} \tag{3.51}$$

then

$$\frac{\partial}{\partial t}\int \rho d\mathcal{V} + \int_{V} \nabla \bullet \rho \vec{V} d\mathcal{V} = 0 \tag{3.52}$$

Combining the terms after exchanging the time derivative and the integral, it is noted that the only way for the integral to be zero is for the integrand to be zero, or

$$\frac{\partial \rho}{\partial t} + \nabla \bullet \rho \vec{V} = 0 \tag{3.53}$$

This is the *differential* expression of conservation of mass. Expanding the second term of equation (3.53), we note that equation (3.53) can be written as

$$\frac{D\rho}{Dt} + \rho \nabla \bullet \vec{V} = 0 \tag{3.54}$$

For incompressible flow, defined as

$$\frac{D\rho}{Dt} = 0 \tag{3.55}$$

it is seen that

$$\nabla \bullet \vec{V} = \frac{\partial u}{\partial x} + \frac{\partial v}{\partial y} + \frac{\partial w}{\partial z} = 0 \tag{3.56}$$

This is a single partial differential equation with unknowns (u, v, w). Note that the simpler but more restrictive definition of incompressible flow $\rho =$ constant also satisfies equation (3.55).

3.6 The Navier–Stokes equations

The Navier–Stokes equations are a statement of conservation of linear momentum, or equivalently, Newton's law:

$$\vec{F} = m\vec{a} \tag{3.57}$$

where $\vec{F}$ is the sum of the forces exerted on a fluid element, m is its mass, and $\vec{a}$ is its acceleration. In fluid mechanics, it is customary to write this equation in terms of force per unit volume so that

$$\frac{\vec{F}}{\mathcal{V}} = \rho \vec{a} \tag{3.58}$$

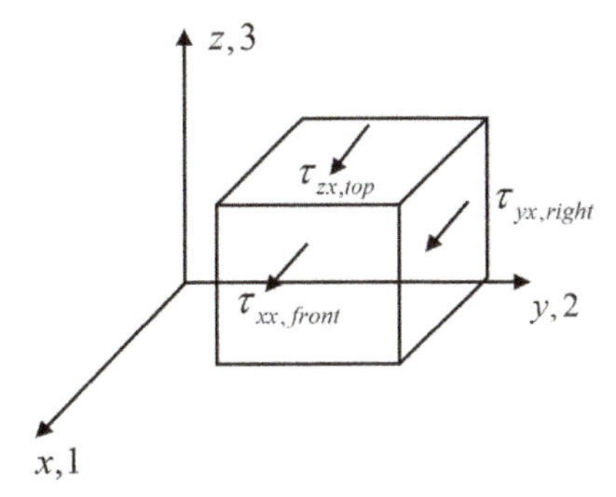

Figure 3.12 Differential element showing pressure forces and viscous forces acting on the element.

In integral form, the incompressible momentum equation takes the form

$$\frac{\vec{F}}{\mathcal{V}} = \frac{\partial}{\partial t}\int_{\mathcal{V}} \rho \vec{V} d\mathcal{V} + \int_{S} \rho \vec{V}\vec{V} \bullet d\vec{A} \tag{3.59}$$

The right-hand side of this equation is equivalent to the material derivative for an infinitesimal control volume, and in the general three-dimensional case, the momentum equation takes the form

$$\rho \frac{D\vec{V}}{Dt} = \vec{F} = \vec{F}_{\text{surface}} + \vec{F}_{\text{body}} \tag{3.60}$$

The surface forces correspond to both pressure and viscous forces, and the body forces correspond to either gravity or electrostatic forces in the case of an electrically conducting fluid such as an electrolyte solution.

Refering to Figure 3.12, consider the net surface force on the fluid cube. In the x direction on the front, right, and top faces, the surface force is

$$dF_x = \tau_{xx,\text{front}} dydz + \tau_{yx,\text{right}} dxdz + \tau_{zx,\text{top}} dxdy \tag{3.61}$$

For the back, left, and bottom faces, a Taylor series expansion can be used; for example, on the back face, the normal stress in the x direction is given by

$$\tau_{xx,\text{back}} = \tau_{xx,\text{front}} - \frac{\partial \tau_{xx}}{\partial x} dx \tag{3.62}$$

The net surface force in the x direction is given by

$$F_{x,\text{net}} = \tau_{x,\text{front}} - \tau_{x,\text{back}} \tag{3.63}$$

so that the net surface force in the x direction is given by

$$F_{x,\text{net}} = \frac{\partial \tau_{xx}}{\partial x} dxdydz + \frac{\partial \tau_{yx}}{\partial y} dydxdz + \frac{\partial \tau_{zx}}{\partial z} dzdxdy \tag{3.64}$$

In a stagnant fluid, this net force is zero, assuming that there are no body forces in the x direction. Dividing through by the volume, since the stress is a symmetric tensor, the net force per unit volume is given by

$$f_{x,\text{net}} = \frac{\partial \tau_{xx}}{\partial x} + \frac{\partial \tau_{xy}}{\partial y} + \frac{\partial \tau_{xz}}{\partial z} \tag{3.65}$$

The procedure is similar for f_y and f_z. Note that

$$f_{x,\text{net}} = \frac{\partial}{\partial x}(\tau_{xx}) + \frac{\partial}{\partial y}(\tau_{yx}) + \frac{\partial}{\partial z}(\tau_{zx}) \tag{3.66}$$

so that $f_{x,\text{net}}$ is the divergence of the first column of the stress tensor. Thus, from the x component f_x, the net surface force per unit volume can be written in the form

$$\vec{f}_{\text{surface,net}} = \nabla \bullet \tau_{ij} \tag{3.67}$$

where ij refers to the triad $(x, y, z) = (1, 2, 3)$.

At this point, it is useful to discuss indicial or tensor notation, which is beneficial when the energy equation is derived. The stress τ_{ij} is a first-order tensor that can be defined by

$$\begin{vmatrix} \tau_{11} & \tau_{12} & \tau_{13} \\ \tau_{21} & \tau_{22} & \tau_{23} \\ \tau_{31} & \tau_{32} & \tau_{33} \end{vmatrix} \tag{3.68}$$

In indicial notation, $(u, v, w) = (u_1, u_2, u_3)$, and the continuity equation is written as

$$\frac{\partial u_i}{\partial x_i} = 0 \tag{3.69}$$

with the occurrence of a repeated index defined as a sum from $i = 1 - 3$. The vector velocity field is denoted simply by u_i in indicial or tensor notation. The advantage of this notation is that equations may be written in tensor notation much more compactly.

Thus Newton's law is given by

$$\rho \frac{D\vec{V}}{Dt} = \vec{b} + \nabla \bullet \tau_{ij} \tag{3.70}$$

where $\vec{b}$ is the body force per unit volume. Note that fluid will experience a convective acceleration if the cross-sectional area contracts and a convective deceleration if the area widens (Figure 3.13).

If the velocity vanishes, the resulting field of fluid mechanics is called *hydrostatics*; in this case, there are no shear stresses, and the normal stress is given by $\tau_{ii} = -p$.

The last step in the derivation of the Navier–Stokes equations is to specify the stress–strain relationship, which may be expressed as

$$\tau_{ij} = \text{function}(\sigma_{ij}) \tag{3.71}$$

for each distinct component. Stokes, back in the mid-nineteenth century, postulated three conditions that a deformation law must satisfy (Tokaty, 1971):

- The fluid is a continuum.
- The fluid is isotropic.
- The relationship must approach the hydrostatic case as the strain rates approach zero.

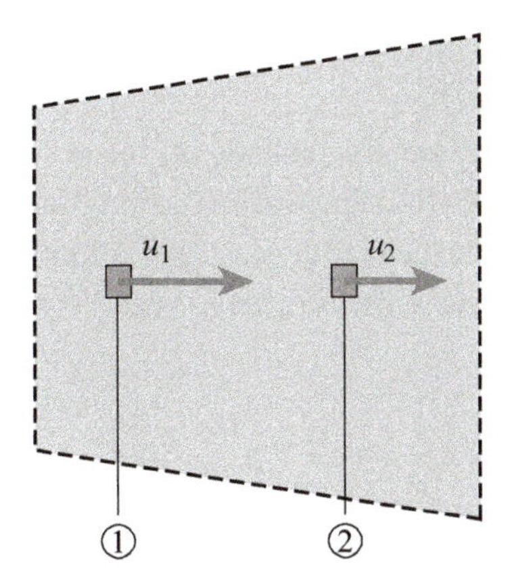

Figure 3.13 In moving through a channel for which the area widens from point 1 to point 2, a fluid element experiences a convective deceleration, $u\,(\partial u/\partial x) < 0$. If the area contracts, a fluid element will experience a convective acceleration, and if the area is constant, the convective term $u\,(\partial u/\partial x) = 0$.

This is the case for a Newtonian fluid for which

$$\tau_{ij} = \mu\sigma_{ij} - \delta_{ij}p \tag{3.72}$$

In the course of determining this relationship, a second coefficient of viscosity emerges that, theoretically, cannot be proven to be zero; however, this term is multiplied by $\nabla \bullet \vec{V}$, which is zero for an incompressible fluid. Thus the second coefficient of viscosity will not be discussed further.

The Navier–Stokes equations in Cartesian coordinates are obtained by substituting the stress–strain relationship into equation (3.60), and the momentum equation in the x direction is given by

$$\rho\frac{Du}{Dt} = b_x - \frac{\partial p}{\partial x} + \text{viscous terms} \tag{3.73}$$

Using the continuity equation to eliminate the terms involving the velocities v and w,

$$\frac{\partial u}{\partial t} + u\frac{\partial u}{\partial x} + v\frac{\partial u}{\partial y} + w\frac{\partial u}{\partial z} = -\frac{1}{\rho}\frac{\partial p}{\partial x} + b_x + \nu\nabla^2 u \tag{3.74}$$

Similarly, in the other two coordinate directions,

$$\frac{\partial v}{\partial t} + u\frac{\partial v}{\partial x} + v\frac{\partial v}{\partial y} + w\frac{\partial v}{\partial z} = -\frac{1}{\rho}\frac{\partial p}{\partial y} + b_y + \nu\nabla^2 v \tag{3.75}$$

$$\frac{\partial w}{\partial t} + u\frac{\partial w}{\partial x} + v\frac{\partial w}{\partial y} + w\frac{\partial w}{\partial z} = -\frac{1}{\rho}\frac{\partial p}{\partial z} + b_z + \nu\nabla^2 w \tag{3.76}$$

where $\nu = \mu/\rho$ is the kinematic viscosity and ∇^2 is the Laplacian in Cartesian coordinates

$$\nabla^2 = \frac{\partial^2}{\partial x^2} + \frac{\partial^2}{\partial y^2} + \frac{\partial^2}{\partial z^2} \tag{3.77}$$

Along with the continuity equation, these are four equations in four unknowns (u, v, w, p) assuming incompressible flow.

The form of the Navier–Stokes equations depends on the coordinate system used; the preceding derivation assumes the standard Cartesian coordinate system. The equations can be put in what is termed *invariant form*, a form valid for both spherical and cylindrical coordinate systems. Written in vector form, the Navier–Stokes equations are

$$\frac{\partial \vec{V}}{\partial t} + (\vec{V} \bullet \nabla)\vec{V} = \frac{1}{\rho}\vec{B} - \frac{1}{\rho}\nabla p + \nu \nabla^2 \vec{V} \tag{3.78}$$

The vector operators in this equation still depend on the coordinate system used; the vector operator $(\vec{V} \bullet \nabla)\vec{V}$ is different in the three coordinate systems. However, using the vector identities

$$\nabla^2 \vec{V} = \nabla(\nabla \bullet \vec{V}) - \nabla \times \nabla \times \vec{V} \tag{3.79}$$

$$(\vec{V} \bullet \nabla)\vec{V} = \nabla \left(\frac{|\vec{V}|^2}{2} \right) - \vec{V} \times \nabla \times \vec{V} \tag{3.80}$$

the Navier–Stokes equations become

$$\frac{\partial \vec{V}}{\partial t} + \nabla \left(\frac{|\vec{V}|^2}{2} \right) - \vec{V} \times \nabla \times \vec{V} = \frac{1}{\rho}\vec{B} - \nabla p - \nu \nabla \times \nabla \times \vec{V} \tag{3.81}$$

for an incompressible fluid. All the vector operations required for this invariant form appear in Appendix B.

Several comments need to be made about these equations. First, they are highly nonlinear; this is manifest in terms auch as

$$u\frac{\partial w}{\partial x} \tag{3.82}$$

in which one unknown multiplies another. Second, they are partial differential equations and can be either elliptic, parabolic, or hyperbolic. Sometimes they can be of different character in different parts of the flow field, as is the case when compressibility of the fluid is important. Third, they are second-order equations because of the second derivatives that appear on the right-hand sides of the equations. Thus, in each coordinate direction, two boundary conditions are required to uniquely solve the equations; one initial condition is required in the case of unsteady flow.

Finally, they are a system of coupled equations. The nature of the fluid velocity in the x direction depends on the velocities in all the other directions. Clearly this makes the solution of these equations difficult.

The Navier–Stokes equations are, in general, difficult to solve in their full form. Flows at small scales are characterized by small velocities and are laminar; that is, the fluid flows in regular layers with little or no mixing between layers. In this case, the nonlinear terms, the convective acceleration terms in the Navier–Stokes equations, can often be neglected. This is the case in micro- and nanofluidics.

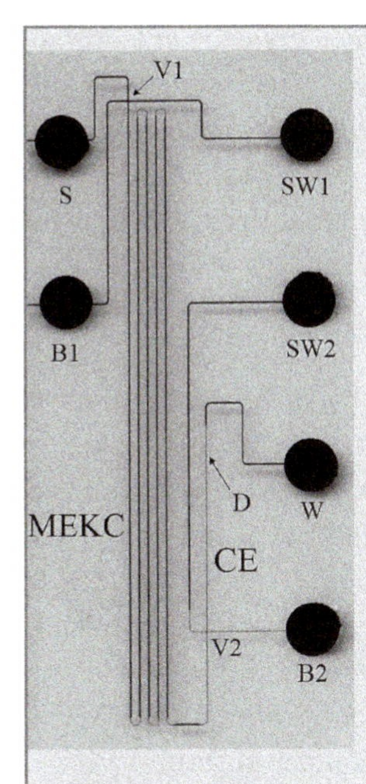

Depicted is a microfluidic chip for performing comprehensive two-dimensional separations of proteins and peptides based on electrophoretic mobility. These devices are fabricated in glass substrates using photolithographic patterning and wet chemical etching methods. The resultant trench features are closed with a glass coverplate with 2-mm diameter vials that act as fluid reservoirs. The channel features have widths ranging from 20 – 75 μm and depths of 10 μm. The first and second dimension channels are ≈20 and 1.5-cm long respectively. The serpentine structure includes asymmetrically tapered turns to control geometrical band broadening. Mixtures of peptides or proteins are introduced at the sample reservoir, S. Buffer reservoir, B1, contains a buffer appropriate for performing micellar electrokinetic chromatography (MEKC), and reservoir B2 contains a buffer for performing capillary electrophoretic (CE) separations. Reservoirs SW1, SW2, and W are waste reservoirs for collecting waste fluids and application of control voltages. Electric potentials are applied to all six reservoirs in a time-dependent fashion to transport materials throughout the structure. The four-way intersections, V1 and V2, act as electrokinetic valves, which are actuated by modulating the voltages applied to reservoirs B1 and B2. The sample mixture is injected through valve V1 with chromatographic separation unfolding in the serpentine channel. Bands eluting from the chromatographic separation channel are sampled with valve V2 and injected into the CE channel. Electrophoretically separated components are detected at position D. The CE injections are performed at 1 Hz, allowing multiple sampling of the chromatographic bands.

Contributed by Professor J. Michael Ramsey.

Reference: Ramsey, J. D., Jacobson, S. C., Culbertson, C. T., Ramsey, J. M. *Analytical Chemistry* 2003, 75, 3758–3764.

3.7 Mass transport

As noted previously, there are several ways to describe mass transport phenomena, depending on whether mass or molar units are used. Thus some definitions are required.

3.7.1 Definitions

Flows in micro- and nanofluidic geometries inevitably require transport of electrolyte mixtures in which mass transport is a crucial feature. A mixture consists of two or more distinct chemical species, and the amount of a given species can be represented by its mass density, ρ_i which has units of kg/m^3, or its molar concentration, c_i, which has units of moles/m^3. The molarity of a mixture is defined as 1 M = 1 mole/L (read 1 molar = 1 mole/liter) and is the most common scale used by chemists.

The mass density and the molar concentration are related by the species *molar mass*:[1]

$$\rho_i = M_i c_i \tag{3.83}$$

The units of M_i are kg/kmole, which, in biochemistry, is called a dalton after the pioneering chemist John Dalton, so that 1 g/mole = 1 kg/kmole = 1 Da.[2] The molar mass is the amount of mass in one mole; the molar mass of a molecule is the sum of the atomic weights of the individual atoms making up the molecule.

For example, the molar mass of water (H_2O) is $1 \times 2 + 1 \times 16 = 18$ Da because the molar mass of hydrogen is $M_H = 1$ Da; the molar mass of oxygen is $M_O = 16$. Often the unit Da is omitted, and it is said that the molar mass of naturally occurring elements ranges from $M = 1$ to 238 and that the molar mass of small and simple chemical compounds ranges from $M \sim 10$ to 1000. The common protein albumin has a molar mass of $M = 66{,}000$. Large polymers can have a molar mass $M \cong 4 \times 10^6$. Selected values of molar mass for a number of compounds are given in Table 3.1.

The molar mass is directly related to the size of the molecule. Venturoli and Rippe (2005) suggest that for globular proteins, the molecular density is about 1.33 g/cm^3 = 1330 kg/m^3, about 30 percent higher than water. On this basis, the molecular weight for a protein with this density, modeled as a hard sphere, is given by

$$M = \frac{4\pi a^3}{3} \rho N_A \tag{3.84}$$

Solving for the molecular radius a,

$$a = \left(\frac{3M}{4\pi \rho N_A} \right)^{0.333} = AM^B \tag{3.85}$$

Results for A and B for selected molecules are given in Table 3.2.

The mixture density is then defined as the sum of the constituent densities, or

$$\rho = \sum_i \rho_i \tag{3.86}$$

Similarly, the total molar density or molar concentration is given by

$$c = \sum_i c_i \tag{3.87}$$

[1] The older term that is often used instead of molar mass is *molecular weight*, which is a dimensionless quantity measured in atomic mass units (1/12 the mass of carbon, C_{12}). The two terms are often used interchangeably. Here we will use the term *molar mass*.

[2] Actually, the dalton is, strictly speaking, a unit of mass: 1 Da = $1.660538782 \times 10^{-27}$ kg.

Table 3.1. Molar masses of some atoms and molecules

Atom/compound	Symbol	Molar mass (Da)
Hydrogen	H	1
Oxygen	O	16
Carbon	C	12
Sodium	Na	23
Chlorine	Cl	35
Water	H_2O	18
Sodium chloride	NaCl	57
Albumin	–	66,000
Glucose	$C_6H_{12}O_6$	180
DNA (ds)	–	660 (single base pair)
RNA	–	300–500
Polyethylene glycol	PEG	1000–35,000
Polymethyl methacrylate	PMMA	80,000–170,000
Ficoll	–	10^4–10^5
Dextran	–	10–10^6

These numbers are rounded; e.g., the molar mass of hydrogen is M = 1.00794. The chemical formulas of the molecules are not shown because they are too long. The length of a single double-stranded (ds) DNA base pair is about 0.34 nm. Ficoll is a sucrose-based polymer and is sometimes used to investigate the sieving function of the kidney. Dextran is a glucose-based polymer and is also sometimes used for kidney sieving analysis.

The mixture composition can also be expressed in terms of mass fraction ω_i,

$$\omega_i = \frac{\rho_i}{\rho} \tag{3.88}$$

or mole fraction,

$$X_i = \frac{c_i}{c} \tag{3.89}$$

Table 3.2. Molar mass and size relationships as compiled by Venturoli and Rippe (2005)

Molecule	A	B
Hard sphere	0.67	0.333
Hydrated hard sphere	0.74	0.333
Globular proteins	0.483	0.386
Ficoll	0.421	0.427
Monodisperse dextran	0.488	0.437
Polydisperse dextran	0.33	0.463

The *mean molar mass* of a mixture is the weighted sum of the individual molar masses, or

$$M = \sum_i X_i M_i \tag{3.90}$$

Note that the mole fraction and mass fractions sum to 1 by definition and are dimensionless quantities. For a mixture of ideal gases, the mole fractions are related to the partial pressure by

$$X_i = \frac{p_i}{p} \tag{3.91}$$

where p is the total pressure. This equation is obtained by assuming that each species exerts its own partial pressure at the mixture temperature

$$c_i = \frac{p_i}{R_{ui} T} \tag{3.92}$$

where R_u is the universal gas constant in kJ/kmole K.

For a binary mixture of species A and B, the mass fraction is related to mole fraction by

$$X_A = \frac{\frac{\omega_A}{M_A}}{\frac{\omega_A}{M_A} + \frac{\omega_B}{M_B}} \tag{3.93}$$

Note that depending on the value of the molar mass, the mole fraction can be much less than the mass fraction. The mass fraction in terms of mole fraction can be obtained by inverting the preceding equation and

$$\omega_A = \frac{X_A M_A}{X_A M_A + X_B M_B} \tag{3.94}$$

Another mass transfer scale that is often used is the *molality*, defined as the number of moles of solute per kilogram of solvent: $m_i = n_i / M_0 n_0$, where 0 denotes the solvent. The ratio

$$\frac{m_i}{X_i} = \frac{1}{M_0 X_0} \tag{3.95}$$

For dilute solutions, $X_i \approx n_i / n_0$ so that in this limit, the molality $m_i \approx X_i / M_0$.

Diffusion mass transport takes place when there is a concentration gradient, just as heat transfer takes place when there is a temperature gradient. Just as heat is transferred from a higher to a lower temperature, diffusion takes place from a higher to a lower concentration. Because gradients of concentration are required for diffusion to take place, diffusion mass transport is a nonequilibrium and irreversible process. Note that there are many similarities between heat and mass transfer, and recall that the two fields developed simultaneously (Figure 3.14). Thus there are many common solutions of the same form, and these analogies will be explored later in this chapter.

Figure 3.14 Prediction of the weather requires the computation of fluid flow and heat and mass transfer of an air-water vapor mixture, in its simplest form. Image courtesy of the National Oceanic and Atmospheric Administration.

An Artificial Kidney

The function of the kidney is to filter out small ions such as sodium and chloride from the blood serum, while retaining larger proteins such as albumin. End Stage Renal Disease (ESRD) is the term given to a dysfunctional kidney that does not retain enough albumin (over 99%). A nanopore membrane can be used as a Renal Assist Device (RAD) to partially supplement the native kidney. A sketch of such a device is depicted on Figure 12.7. Each pore in the membrane is about $h \sim 10nm$ with the flow field generated by the natural pressure drop across the native kidney. Thus the base flow is a Poiseuille flow which must be supplemented by a mass transfer analysis for the diffusion and convection of albumin through the pore. See Chapter 12 for more details.

3.7.2 Governing equation

Let us first consider the case of zero bulk motion. Then mass transport takes place by diffusion only, and the mass flux is given by Fick's law:

$$\vec{j}_A = -D_{AB}\nabla\rho_A \tag{3.96}$$

where $\vec{j}_A$ is a mass flux and has units of kg/sec m^2. If the density of the mixture remains constant, the flux can be expressed in terms of the mass fraction as

$$\vec{j}_A = -\rho D_{AB}\nabla\omega_A \tag{3.97}$$

On a molar basis,

$$\vec{J}_A = -D_{AB}\nabla c_A \tag{3.98}$$

and $\vec{J}_A$ has units of kmole/sec m^2. In terms of the mole fraction,

$$\vec{J}_A = -cD_{AB}\nabla X_A \tag{3.99}$$

if the total concentration remains constant.

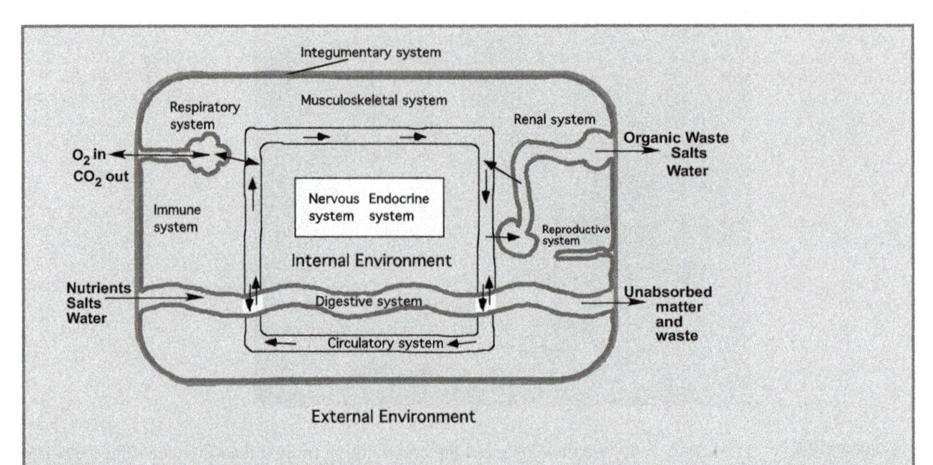

The body is one big mass transfer machine that involves transport of gases through liquids, gases through cells, and transport of liquid water everywhere. The body transports charged liquids, salts in water; this fact means that mass transfer often takes place in the presence of a local electric field. Most proteins are negatively charged and are transported through cell membranes by pressure driven flow and migrate through many regions of the body under a combination of concentration gradients (diffusion) and electric potential gradients; that is, an electric field. A general look at mass transport within the body can be found in Saltzman (2009).

The mass flux of species A relative to a fixed coordinate system is

$$\vec{n}_A = \rho_A \vec{v}_A \tag{3.100}$$

where $\vec{v}_A$ is a species velocity with the corresponding definition for species B:

$$\vec{n}_B = \rho_B \vec{v}_B \tag{3.101}$$

The mass averaged velocity vector in the Navier–Stokes equations $\vec{V}$ of the mixture is then defined by

$$\rho \vec{V} = \vec{n}_A + \vec{n}_B = \rho_A \vec{v}_A + \rho_B \vec{v}_B \tag{3.102}$$

Dividing through by the density ρ,

$$\vec{V} = \omega_A \vec{V}_A + \omega_B \vec{V}_B \tag{3.103}$$

for a binary mixture. The mass flux of species A relative to the mixture mass averaged velocity is

$$\vec{j}_A = \rho_A(\vec{v}_A - \vec{V}) \tag{3.104}$$

and from the definition of $\vec{n}_A$, we have

$$\vec{n}_A = \vec{j}_A + \rho_A \vec{V} \tag{3.105}$$

Substituting from Fick's law, we have

$$\vec{n}_A = -\rho D_{AB} \nabla \omega_A + \rho \omega_A \vec{V} \tag{3.106}$$

Similar manipulations lead to

$$\vec{J}_A = c_A(\vec{v}_A - \vec{V}) \tag{3.107}$$

$$\vec{N}_A = -c D_{AB} \nabla X_A + c X_A \vec{V} \tag{3.108}$$

when the mass transport is described on a molar basis. The preceding equations for the fluxes are appropriate for convection and diffusion mechanisms and do not include an electric component.

As noted previously, an electric field is the most common method of driving the flow of biological fluids in nanoscale channels. For strong electrolytes, a single salt component, such as NaCl, will be entirely dissociated so that nominally, the mixture has three components: undissociated water and positive and negative ions making up the single salt component. In this case, adding the mass transport due to an electric field, the molar flux of species A for a dilute electrically conducting mixture is

$$\vec{N}_A = -D_{AB} \nabla c_A + m_A z_A c_A \vec{E} + c_A \vec{V} \tag{3.109}$$

Here D_{AB} is the diffusion coefficient, R is the universal gas constant, T is the temperature, $z_A m_A$ is called the ionic mobility with $m_A = F D_{AB}/RT$, z_A is the valence, $F = 96{,}500$ Coul/mole is Faraday's constant, and $\vec{E}$ is the electric field. Equation (3.109) is called the Nernst–Planck equation. In this form, the term involving the electric field in the flux equation is called *electrical migration*.

The mass transport equation is then

$$\frac{\partial c_A}{\partial t} + \nabla \bullet \vec{N}_A = 0 \tag{3.110}$$

Expanding the divergence operator, the $\nabla\bullet$ term, the governing equation for the concentration of species A is given by

$$\frac{Dc_A}{Dt} = D_{AB} \nabla^2 c_A + \frac{D_{AB} z_A F}{RT} \left(\frac{\partial c_A E_x}{\partial x} + \frac{\partial c_A E_y}{\partial y} + \frac{\partial c_A E_z}{\partial z} \right) \tag{3.111}$$

where the diffusion coefficient has been assumed constant. This equation is often written in terms of the mole fraction X_A if the total concentration is approximately constant, which is often the case.

Note that the material derivative has appeared in a similar fashion to the momentum equation. By definition of the material derivative, if

$$\frac{Dc_A}{Dt} = 0 \tag{3.112}$$

then in a steady flow field, provided that the concentration field is also steady, lines of constant concentration will coincide with streamlines.

3.8 Electrostatics

At steady state, the electric field $\vec{E}$ must satisfy the reduced set of Maxwell's equations since there is no magnetic field, and in this case,

$$\nabla \times \vec{E} = 0 \tag{3.113}$$

so that the electric field is solenoidal. The electric field must also satisfy

$$\nabla \bullet \epsilon_e \vec{E} = \rho_e \tag{3.114}$$

where ρ_e is the volume charge density. As in the case of potential fluid flow, an electrical potential can be defined as

$$\vec{E} = -\nabla\phi \tag{3.115}$$

and so

$$\nabla \bullet (\epsilon_e \vec{E}) = -\nabla \bullet (\epsilon_e \nabla\phi) = \rho_e \tag{3.116}$$

which is a Poisson equation that determines the electrical potential ϕ. This equation is the differential form of Gauss's law. Here ϵ_e is the permittivity, and the charge density is given by

$$\rho_e = F \sum_i z_i c_i = Fc \sum_i z_i X_i \tag{3.117}$$

where c_i is the molar concentration of species i, X_i is the mole fraction of species i, z_i is its valence, F is Faraday's constant, and c is the total concentration, which usually remains constant. If $\rho_e = 0$ in a given region, that region is said to be *electrically neutral*. For constant permittivity,

$$\epsilon_e \nabla^2 \phi = -Fc \sum_i z_i X_i \tag{3.118}$$

It is important to understand how electric fields can be set up. If the walls of the channel are charged, there is a naturally occuring potential associated with the electrical double layer. This part of the total potential is illustrated in Figure 1.18.

If, in addition, electrodes are placed upstream and downstream of a channel, there is an *external* electric field that causes motion of the fluid, as in Figure 1.19. In most cases of practical interest, the potential associated with the external electric field can be decoupled from the potential associated with the electrical double layer; the volume charge density turns out to be small (see Section 7.18), and then, to leading order,

$$\nabla^2 \phi_E = 0 \tag{3.119}$$

that amounts to neglecting the effect of the electric double layers on the electrodes. Thus, if the external electric field is oriented in the x direction only and the surfaces of the channel are electrically insulating ($\frac{\partial \phi_E}{\partial y} = \frac{\partial \phi_E}{\partial z} = 0$), then the potential associated with the externally imposed electric field satisfies, again to leading order,

$$\frac{d^2\phi_E}{dx^2} = 0 \tag{3.120}$$

and so the external potential is linear with x. In this case, the external electric field is constant, and the total electric field can be written as

$$\vec{E} = E_0\hat{i} - \nabla\phi \tag{3.121}$$

where ϕ is the electrical potential associated with the electrical double layer and E_0 is the imposed electric field.

The Equation of Charge Conservation

The governing equation of mass transport for species A in an electrolyte solution is given by

$$\frac{\partial c_A}{\partial t} = \nabla \bullet (-D_A\nabla c_A + m_A z_A c_A \vec{E} + c_A\vec{V})$$

The current density is given by

$$\vec{J} = F\sum_i z_i\vec{N}_i$$

Multiplying the mass transport equation by $F\sum_{i=1}^{n} z_i$ results in

$$\sum_i z_i F\frac{\partial c_i}{\partial t} = \nabla \bullet \left(-\sum_i FD_i z_i\nabla c_i + \sum_i Fm_i z_i^2 c_i\vec{E} + \sum_i Fc_i z_i\vec{V}\right)$$

But the volume charge density is defined by $\rho_e = F\sum_{i=1}^{n} z_i c_i$ so that

$$\frac{\partial \rho_e}{\partial t} + (\vec{V}\bullet\nabla)\rho_e = \nabla \bullet \left(\sum_i FD_i z_i\nabla c_i + \sigma_e\vec{E}\right)$$

where σ_e is the electrical conductivity and is defined by

$$\sigma_e = F\sum_i m_i z_i^2 c_i$$

and is the principle of *conservation of charge.* Note that this equation is a convective-diffusion equation for the charge density. Note that the higher the mobility, $m_i z_i$, the higher the conductivity. The equation of conservation of charge is a confirmation that charge is conducted through an electrolyte by the motion of ions. If all of the diffusion coefficients are assumed equal and constant, $D_i = D$,

$$\frac{\partial \rho_e}{\partial t} + (\vec{V}\bullet\nabla)\rho_e = D\nabla^2\rho_e + \nabla\bullet(\sigma_e\vec{E})$$

Note that for steady flow,

$$\nabla\bullet\vec{J} = 0$$

The electrical conductivity is normally assumed to be constant, which really means that the concentrations are constant. In this case, and if the fluid velocity is zero, this equation reduces to Ohm's Law $\vec{J} = \sigma_e\vec{E}$.

The current density is a very important quantity for the sensing of molecules in micro- and nanodevices. The dimensional current density on a molar basis is given by (Newman, 1972)

$$\vec{J} = F \sum z_i \vec{N}_i \tag{3.122}$$

where $\vec{J}$ is the dimensional current density. Substituting for the flux of species i in molar form, N_i,

$$\vec{J} = F \sum_i z_i(-cD_{AB}\nabla X_i + cm_i z_i X_i \vec{E} + cX_i \vec{V}) \tag{3.123}$$

Changes in current have been used to sense and analyze biomolecules in nanochannels, and this subject will be considered in Chapters 9 and 12.

It should be noted that strictly speaking, the term *electrostatics* refers to a situation in which the net current density is zero. In the flows of electrolytes described in this text, this is not the case. When charges move, an additional force on the fluid appears in the magnitude $\vec{V} \times \vec{B}$, where $\vec{B}$ is the *magnetic induction*; the magnetic induction arises from motion of the ions in an electro-osmotic flow, for example. (This phenomenon is discussed in Chapter 9). However, in the cases of interest here, to leading order, the effect of magnetic induction can safely be ignored. This point will be discussed in greater detail in Section 6.9.

3.9 Energy transport

The derivation of the thermal energy equation begins with the consideration of the first law of thermodynamics for an open system, which is given by (see Section 2.5)

$$dQ - dW = dE_t \tag{3.124}$$

where Q, W, and E_t are the extensive values of heat transfer, work, and total energy in the system in joules/m^3 or energy per volume, respectively, just as the momentum equation was derived as a balance of forces per volume.[3] Recall that heat and work are dependent on the process path, whereas the energy of the system is independent of the path and is hence a property in the thermodynamic sense. The work term consists of that done by surface forces and

[3] The term *energy equation* is often given to the equation resulting from multiplying a component of the Navier–Stokes equations by the velocity in that component equation. The resulting equation has units of energy per volume, but it is not equivalent to the first law of thermodynamics. Such an equation is called the *mechanical energy equation*.

that done by body forces and any external work fields, such as shaft work in a turbine.

In a turbine, the flow of fluid rotates a shaft, that produces work: shaft work.

The derivation of the energy equation is extremely cumbersome in a Cartesian coordinate system. Thus a mix of indicial and vector notation is used after the manner of the presentation by White (2006). Again, we use a differential control volume, as we did in the derivation of the Navier–Stokes equations.

The total energy per volume in a system is assumed to consist of internal energy, kinetic and potential energy, and the electrical energy present in a conducting fluid. Thus the total energy per unit volume in the system can be written as follows:

$$E_t = \rho\left(e + \frac{1}{2}V^2 + \vec{b} \bullet \vec{r}\right) + \frac{1}{2}\epsilon_e E^2 \tag{3.125}$$

where E is the magnitude of the electric field, $E = |\vec{E}|$, V is the magnitude of the velocity $V = |\vec{V}|$, $\vec{b}$ is the body force per unit volume, e is the internal energy per mass, and $\vec{r}$ is the displacement vector. Note that thermal radiation is not considered.

Notation alert: Here E_t and E are used for total energy and electric field, respectively. The meaning of each quantity should be clear from the context.

To be consistent with the Navier–Stokes equations, the first law should be written on a rate basis using the material derivative

$$\frac{DQ}{Dt} - \frac{DW}{Dt} = \frac{DE_t}{Dt} \tag{3.126}$$

where the units in equation (3.126) are joule/m^{3}sec. Substituting for E_t, it follows that

$$\frac{DE_t}{Dt} = \rho\left(\frac{De}{Dt} + V\frac{DV}{Dt} + \vec{g} \bullet \vec{r}\right) + \epsilon_e E\frac{DE}{Dt} \tag{3.127}$$

Consider the heat transfer term; from Figure 3.15, considering only conduction, for now, and using Fourier's law, the net heat transfer by conduction through the

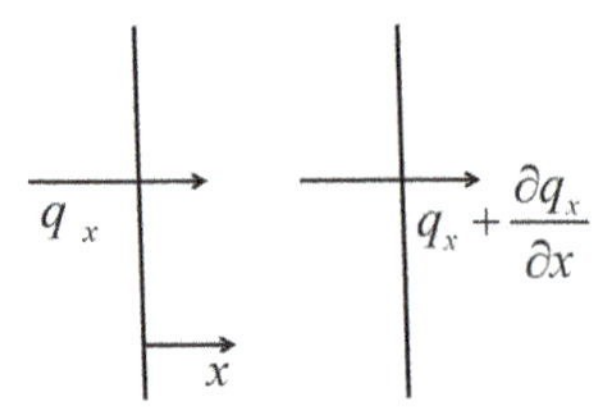

Figure 3.15 Conduction heat transfer in a slab.

slab is given by

$$\left(q_x - \left[q_x + \frac{\partial q_x}{\partial x}\right]\right) dydz = -\frac{\partial q_x}{\partial x} = \frac{\partial}{\partial x}\left(k\frac{\partial T}{\partial x}\right) dydz \tag{3.128}$$

where q_x is the heat flux, that is, the net heat transport per unit area, through the left face of the fluid cube. The other coordinate directions are similar. Thus, after dividing by the volume element,[4]

$$\frac{DQ}{Dt} = -\nabla \bullet \vec{q} = +\nabla \bullet (k\nabla T) + \dot{q} \tag{3.129}$$

where $\dot{q}$ is a volumetric heat generation rate.

The rate of work done by the surface forces is defined as force times velocity. For example, in the x direction,

$$w_{sx} = -(u\tau_{xx} + v\tau_{xy} + w\tau_{xz}) \tag{3.130}$$

Using the same procedure as with the conduction mode of heat transfer,

$$\frac{DW_s}{Dt} = -\nabla \vec{w}_s = \nabla \bullet \vec{V} \bullet \vec{\tau}_{ij} \tag{3.131}$$

The expression $\nabla \bullet \vec{V} \bullet \vec{\tau}_{ij}$ can be expanded as

$$\nabla \bullet \vec{V} \bullet \vec{\tau}_{ij} = \vec{V} \bullet \nabla \bullet \vec{\tau}_{ij} + \tau_{ij}\frac{\partial u_i}{\partial x_j} \tag{3.132}$$

Note from the momentum equation that

$$\nabla \bullet \vec{\tau}_{ij} = \rho\left(\frac{D\vec{V}}{Dt} + \vec{b}\right) \tag{3.133}$$

where $\vec{b}$ is the body force per unit volume vector, and so

$$\vec{V} \bullet \nabla \bullet \vec{\tau}_{ij} = \rho\left(V\frac{DV}{Dt} + \vec{V} \bullet \vec{b}\right) \tag{3.134}$$

The work done by the body forces per unit area is defined by

$$\vec{w}_b = \vec{V} \bullet \vec{b} \tag{3.135}$$

[4] The local time derivative $\partial/\partial t$ appears only in the term DE_t/Dt because the time rate of increase in energy within the volume balances the rest of the terms in equation (3.126).

and we decompose the body force vector as

$$\vec{b} = \vec{g} + \vec{b}_e \tag{3.136}$$

where $\vec{g}$ is the gravitational body force per unit volume and $\vec{b}_e$ is the electrical body force per unit volume. Note that the two terms of the right-hand side of equation (3.134) are the kinetic energy and the energy associated with the body force terms of De/Dt in equation (3.125), and so those two terms cancel with the corresponding terms in the total energy of the system. Thus, using indicial notation, where necessary, to simplify the presentation,

$$\rho \frac{D}{Dt}\left(e + \frac{1}{2}\epsilon_e E^2\right) = \nabla \bullet (k\nabla T) + \tau_{ij}\frac{\partial u_i}{\partial x_j} + \dot{q} \tag{3.137}$$

The units of equation (3.137) are joule/m^{3}sec.

There are many forms of the energy equation; first consider the case where there is no electric field and no source term. Then the energy equation takes the form

$$\rho \frac{De}{Dt} = \nabla \bullet (k\nabla T) + \tau_{ij}\frac{\partial u_i}{\partial x_j} \tag{3.138}$$

The second term can be split into two parts corresponding to the viscous and pressure components of the stress:

$$\tau_{ij}\frac{\partial u_i}{\partial x_j} = \tau'_{ij}\frac{\partial u_i}{\partial x_j} - p\nabla \bullet \vec{V} \tag{3.139}$$

The term

$$\tau'_{ij}\frac{\partial u_i}{\partial x_j} \tag{3.140}$$

is called the *viscous dissipation* and is almost always negligible in flows at low velocities. For incompressible flow, the internal energy $e = c_p T = c_v T = cT$, and thus, for constant specific heat,

$$\rho c \frac{DT}{Dt} = \nabla \bullet (k\nabla T) + \tau'_{ij}\frac{\partial u_i}{\partial x_j} \tag{3.141}$$

which is the most common form of the energy equation. It is the viscous dissipation term that is responsible for the entropy production discussed in the previous chapter.

In addition, for negligible viscous dissipation and for a constant thermal conductivity,

$$\rho c \frac{DT}{Dt} = k\nabla^2 T \tag{3.142}$$

Note that this energy equation is similar in form to the Navier–Stokes equations for the individual velocities in the absence of a pressure gradient, and it turns out that there are analogies that arise in certain cases. These analogies are discussed in Section 3.13.

3.10 Two-dimensional, steady, and incompressible flow

The governing equations for an electrically conducting fluid in the presence of fluid flow and heat and mass transfer are complicated; however, fortunately, they are not often solved in their full form. In the special case of steady, incompressible and two-dimensional flow (the channels in Figure 3.11 are very wide, leading to $\partial/\partial z = 0$) in a fluid with constant properties, which is by far the most common situation, the governing equations become

$$\frac{\partial u}{\partial x} + \frac{\partial v}{\partial y} = 0 \tag{3.143}$$

$$u\frac{\partial u}{\partial x} + v\frac{\partial u}{\partial y} = -\frac{1}{\rho}\frac{\partial p}{\partial x} + b_x + \nu\nabla^2 u \tag{3.144}$$

$$u\frac{\partial v}{\partial x} + v\frac{\partial v}{\partial y} = -\frac{1}{\rho}\frac{\partial p}{\partial y} + b_y + \nu\nabla^2 v \tag{3.145}$$

$$u\frac{\partial c_A}{\partial x} + v\frac{\partial c_A}{\partial y} = D_{AB}\nabla^2 c_A + \frac{D_{AB} z_A F}{RT}\left(\frac{\partial c_A E_x}{\partial x} + \frac{\partial c_A E_y}{\partial y}\right) \tag{3.146}$$

$$\rho c\left(u\frac{\partial T}{\partial x} + v\frac{\partial T}{\partial y}\right) + \epsilon_e E\left(u\frac{\partial E}{\partial x} + v\frac{\partial E}{\partial y}\right) = k\nabla^2 T + \dot{q} \tag{3.147}$$

$$\epsilon_e \nabla^2 \phi = -Fc\sum_i z_i X_i \tag{3.148}$$

where

$$\nabla^2 = \frac{\partial^2}{\partial x^2} + \frac{\partial^2}{\partial y^2} \tag{3.149}$$

Later, the concept of a fully developed regime will be introduced, and in this situation, in the streamwise direction,

$$\frac{\partial}{\partial x} = 0$$

leading to further simplifications. It will be shown that for most relevant cases, the length required to become fully developed is shortest for the electrical potential, being on the order of a Debye length, next shortest for the fluid problem, next shortest for the thermal entry length, and longest for the mass transfer problem. Indeed, mass transfer problems are almost never fully developed.

3.11 Boundary and initial conditions

The governing partial differential equations defined so far are subject to boundary and initial conditions. The nature of these conditions depends on what is known in the particular problem of interest. There are generally three types of canonical conditions, although more complicated combinations do occur. These canonical conditions are as follows:

1. Specified value of an independent variable at the boundary such as the no-slip condition $u = 0$ at a solid boundary

2. Specified gradient or flux at the boundary such as at a horizontal or vertical, constant thickness free surface separating a liquid and a gas or vapor where the shear stress vanishes, $\partial u/\partial n = 0$ where n denotes the direction normal to a free surface, or at the boundary between two media, where the heat flux and electric field normal to the surface are continuous
3. A linear combination of the two types such as in a convection heat transfer problem, $\partial T/\partial y = h(T - T_\infty)$, where h is a heat transfer coefficient.

In the following sections, we discuss each of the fluid dynamics, mass transfer, electrostatics, and heat transfer boundary conditions in some detail.

3.11.1 Velocity boundary conditions

Prandtl, in the early 1900s, performed fundamental experiments demonstrating that in a fluid with finite viscosity, the velocity component parallel to a solid surface will be the velocity of that surface. If the surface is not moving, then the fluid velocity is zero at the surface; that is, the fluid sticks to the surface. Bernoulli (1738) appears to have been the first to recognize this when he noted significant differences between his calculations for the case in which the viscosity $\mu = 0$ and his measurements (Goldstein, 1965b; Lauga *et al.*, 2005). In light of this, Goldstein (1965b) discusses the possible boundary conditions:

- The fluid molecules at the wall are bound to the solid molecules (Coulomb).
- A thin layer of fluid near the wall is bound to the solid (Girard).
- There is a slip at the wall that, in some limit, results in the no-slip condition (Navier).

Ludwig Prandtl was born in 1874 in Bavaria to a family that encouraged his interest in physical phenomena in nature (Anderson, 1982). He received his PhD degree from the University of Munich in 1900 with a doctoral thesis in solid mechanics. Having been appointed as Professor of Mechanics at the Technische Hochschule in Hanover, he continued his new-found interest in fluid mechanics he developed as a designer of a machine that removed material shavings by suction. It was at Hanover that he began developing his famous boundary layer theory, culminating in his famous 1904 paper delivered to the Third Congress of Mathematicians at Heidelberg. Indeed, his most famous contribution has always been the demonstration of the no-slip condition in fluid mechanics. However, he also made significant contributions in other areas of fluid mechanics and heat transfer. During the period from 1905–1910, he became interested in compressible flow and his name appears in compressible flow in the Prandtl-Meyer expansion fan, the flow past a corner whose measure exceeds 90°. After this he focused his attention on low-speed aerodynamics, developing the still-used Prandtl Lifting Line and Lifting Surface theories for calculating lift and induced drag. Prandtl's demonstration of the no-slip condition altered the course of viscous flow research for an entire century.

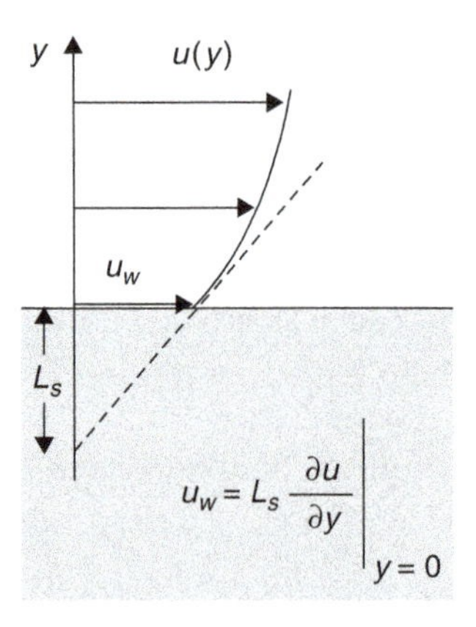

Figure 3.16 Definition of slip length. Sketch by Professor Minami Yoda.

There was some debate as to which of the situations is appropriate, until Prandtl discovered the boundary layer in 1904. His discovery that the boundary layer consists of the regime in which the fluid is brought to relative rest at the wall resolved D'Alembert's paradox that suggested that when $\mu = 0$, there is no drag on a flat plate, which is in conflict with observation. This laid to rest the debate of the previous 100 years over the existence of the no-slip condition until very recently, when the no-slip condition has been questioned in liquid flows in microfluidic devices near both hydrophilic and hydrophobic surfaces, with a consensus building that the no-slip condition may not be appropriate near hydrophobic surfaces.

It should be pointed out that the no-slip condition has been established at the macroscale using empirical means, the result of a large number of experiments begun by Prandtl. It also needs noting that the physical mechanisms for liquid slip are still a matter of some debate, and a detailed discussion of current work is beyond the scope of this text. An excellent review of some recent experimental results showing slip, particularly in liquids, has been presented by Lauga *et al.* (2005), and an excellent discussion of the historical view of liquid slip is given by Karniadakis *et al.* (2005). A short summary of these slip concepts follows.

The amount of fluid slip is quantified by defining a slip length. The slip length in Figure 3.16 can be inferred by indirect techniques, such as flowrate measurement, and also by molecular dynamics simulations and atomic force microscopy.[5] There is evidence to suggest that slip can occur at hydrophobic surfaces, with water molecules being replaced by a gas (e.g., air) layer, inducing a free-surface boundary condition (see later). To distinguish this phenomenon from slip near nonhydrophobic surfaces, this type of slip has sometimes been termed *apparent slip* (Lauga *et al.*, 2005; Bhushan, 2007; Wang *et al.*, 2009). There have been thousands of papers on this phenomenon; see, in particular, Schnell (1956) and Barrat and Bocquet (1999). At hydrophilic surfaces in an aqueous solution, the no-slip condition is generally thought to hold, although this entire field is still a matter for continuing research (Koplik *et al.*, 1989; Honig & Ducker, 2007).

The amount of slip, if it occurs, is also a function of the roughness of the surface (Priezjev & Troian, 2006). If the surface is hydrophilic, roughness decreases slip (Richardson, 1973), whereas for a hydrophobic surface, roughness allows the formation of gas pockets that increase slip (de Gennes, 2002). It is important to note that it has been recognized for a long time that slip can occur in rarefied gases and in non-Newtonian fluids.

[5]The atomic force microscope (AFM) was developed to measure atomic-scale topographical features on a charged or uncharged surface using a highly flexible cantilever beam, the tip of the probe. Forces as small as 1 nN can be measured using an AFM. Detailed information on the operation and applications of the device is given by Bhushan (2007).

Slip Length Increases with Hydrophobicity

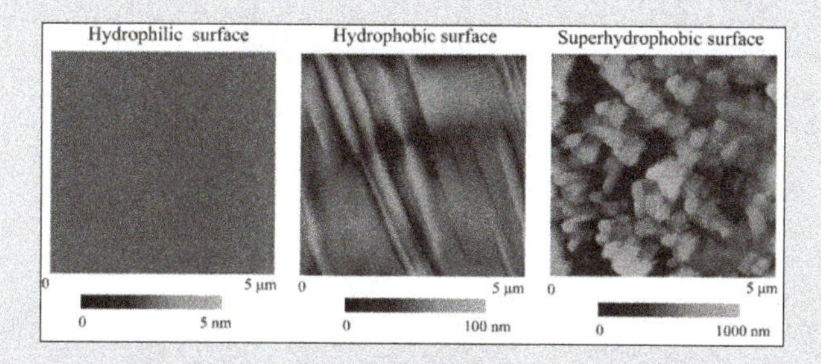

Sketches of hydrophilic and hydrophobic surfaces. Hydrophobicity increases from left to right. Courtesy Bharat Bhushan, used with permission (Wang *et al.*, 2009).

The figure above depicts three surfaces, one a smooth mica surface that is hydrophillic; a hydrophobic surface created by self assembly of alkane n-hexatriacontane and lotus wax. The slip-lengths measured using atomic force microscopy (AFM) are essentially zero for the hydrophilic surface, to 44 nm and 133 nm for the hydrophobic and superhydrophobic surfaces respectively. The rms roughnesses of the three surfaces is 0.2 nm for the mica, and 11 nm and 178 nm for the hydrophobic and superhydrophobic surfaces respectively. Contributed by Professor Bharat Bhushan.

The no-slip condition applies to the vast majority of flows even at the micro- and nanoscale and simply states that the velocity component tangent to the wall equals the wall velocity. The solid wall condition requires the velocity component normal to the wall to be zero. This condition applies to both liquids and gases, provided that the smallest dimension in the physical system of interest is much greater than the mean free path in the case of a gas and a typical molecular diameter in a liquid.

In a channel bounded by two walls at $y = 0, h$, for example, the no-slip condition in three dimensions, is expressed as

$$u = w = 0 \quad y = 0 \quad \text{and} \quad y = h \tag{3.150}$$

whereas the solid wall condition is expressed as $v = 0$ at $y = 0, h$.

Many biomaterials are porous, and for a porous wall at $y = 0$,

$$u, w = 0 \quad v = v_w \quad y = 0 \tag{3.151}$$

The velocity at the wall, v_w, must be calculated from the properties of the porous material.

Whether slip occurs in a gas is dependent on the Knudsen number, and slip will occur if $Kn = \lambda/h = O(10^{-3})$, where λ is the mean free path of the gas. Note that $Kn \sim 1$ if either λ is large, such as in a rarefied gas, or if h, the characteristic dimension, is small, as in a microdevice. The mean free path of air is around $\lambda \sim 60$–70 nm.

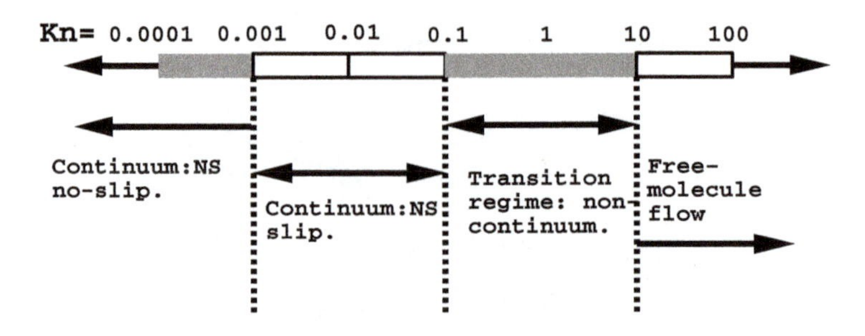

Figure 3.17 The effect of Knudsen number on the nature of the flow.

There is a well-developed theory of the nature of slip flow for gase; the essential features are depicted in Figure 3.17. For $Kn < 0.001$, the flow is essentially of the continuum variety, with the no-slip condition holding. For $0.001 < Kn < 0.1$, the flow is still continuum, but with the slip condition to be discussed here. Next is a transition regime for $0.1 < Kn < 10$, and finally, for $Kn > 10$, free molecular flow occurs. In the transition regime, a molecular simulation must be used; direct simulation Monte Carlo (DSMC) simulations (Bird, 1994) are often used in gases.

The slip condition was apparently first proposed by Navier in 1827, who suggested that the velocity at a solid surface is proportional to the wall shear stress: thus the form of the boundary condition is given by

$$u_{\text{fluid}} = u_{\text{wall}} + L_s \frac{\partial u}{\partial n} \quad \text{at a solid boundary} \tag{3.152}$$

where n is the direction normal to the surface. Here L_s is a slip length; for gases, $L_s = \lambda$ is the mean free path. This equation assumes hard-sphere molecules and perfectly diffuse reflection. For both diffuse and specular reflection,

$$u_{\text{fluid}} = u_{\text{wall}} + \frac{2 - \sigma_v}{\sigma_v} L_s \frac{\partial u}{\partial n} \quad \text{at a solid boundary} \tag{3.153}$$

where σ_v is an accomodation coefficient and is the fraction reflected diffusively. Atomically smooth surfaces are more likely to achieve specular reflection, whereas atomically rough surfaces are more likely to exhibit diffuse reflection, with fluid molecules rebounding at random angles. The Navier slip condition can also be used for liquids, with the slip length taken to be a molecular diameter.

Equation (3.153) is first order in the Knudsen number and is valid for an isothermal flow. This equation can be extended to second order by retaining the next term in the Taylor series expansion for the velocity near the wall and

$$u_{\text{fluid}} = u_{\text{wall}} + \frac{2 - \sigma_v}{\sigma_v} \left(L_s \frac{\partial u}{\partial n} + \frac{1}{2} L_s^2 \frac{\partial^2 u}{\partial n^2} \right) + \cdots \quad \text{at a solid boundary} \tag{3.154}$$

In dimensionless form, this equation can be written

$$u_{\text{fluid}} = u_{\text{wall}} + \frac{2 - \sigma_v}{\sigma_v} \left(Kn \frac{\partial u}{\partial n} + \frac{1}{2} Kn^2 \frac{\partial^2 u}{\partial n^2} \right) \quad \text{at a solid boundary} \tag{3.155}$$

where now the variable n has been scaled on L_s. Karniadakis *et al.* (2005) write this second-order condition in the form

$$u_{\text{fluid}} = u_{\text{wall}} + \frac{2 - \sigma_v}{\sigma_v} \frac{Kn \frac{\partial u}{\partial n}}{1 - B(Kn)Kn} \text{ at a solid boundary} \tag{3.156}$$

where, to match the Taylor series,

$$B = \frac{1}{2} \frac{u''}{u'} \tag{3.157}$$

where the prime denotes differentiation with respect to n. They suggest that B be considered an empirical parameter so as to extend the validity of this slip boundary condition for larger Knudsen numbers. For compressible gases, these slip boundary conditions are considerably more complicated and are coupled to the temperature distribution.

The concept of liquid slip is much less developed than that for gases. For one thing, there is no such thing as a mean free path. Since liquid molecules are always colliding with other molecules a molecular diameter away, an argument can be made that the mean free path is simply a molecular diameter. At higher shear rates in liquids, Thompson and Troian (1997) have suggested, instead of the no-slip condition,

$$u = L_s \frac{\partial u}{\partial n} \tag{3.158}$$

where $\dot{\gamma}_c$ is a slip length that depends on the local shear rate

$$L_s = L_s^0 \left(1 - \dot{\gamma}_c^{-1} \frac{\partial u}{\partial n}\right)^{-1/2} \tag{3.159}$$

where γ_c is a critical rate of strain obtained from experiment. L_s^0 is a scale length, which is taken to be 17 molecular diameters in water by Thompson and Troian (1997). On the basis of these results, the Navier no-slip condition can be viewed as the low strain limit of a more general form that diverges at a critical value of the strain rate.

A *free surface* is the boundary between a liquid and a gas or vapor. In this case, the tangential velocity and the tangential component of shear stress on either side of the free surface must be continuous; in the simple case where the pressure is uniform and there is negligible curvature in the surface,

$$\vec{V}_1 = \vec{V}_2 \quad \text{and} \quad \tau_1 = \tau_2 \tag{3.160}$$

where

$$\tau_1 = \mu_1 \frac{\partial u_1}{\partial n} \tag{3.161}$$

In three dimensions, in the absence of surface tension, the stresses in each of the coordinate directions are continuous across the free surface. It is to be noted that the viscosity of a gas is much smaller than that of a liquid. If the liquid is denoted by 1 and the gas by 2, because viscosity of a gas is much smaller than that of a liquid, usually, $\tau_1 \sim 0$.

Experimental Data on Slip Length

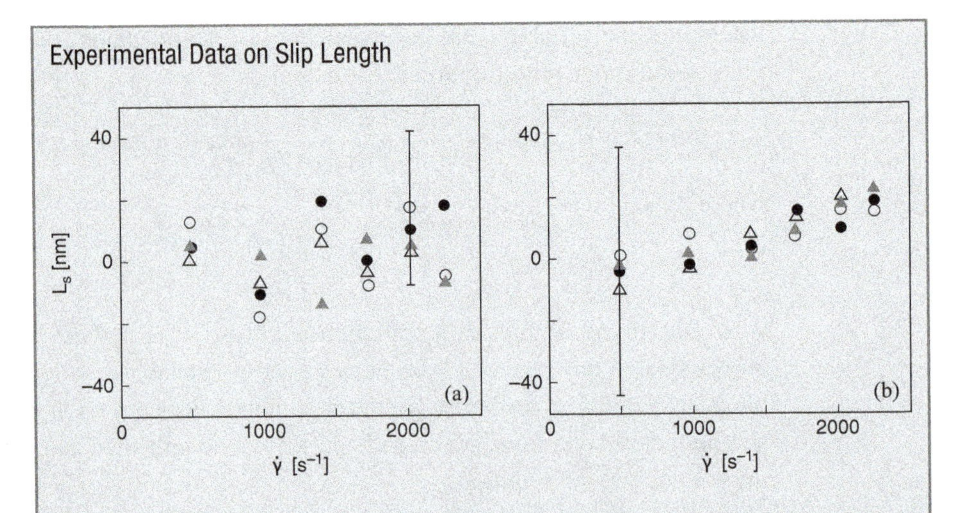

These two figures show the slip length as a function of the shear rate $\dot{\gamma}$ for steady, fully-developed and creeping (Reynolds numbers Re $<$ 0.25) Poiseuille flow of 2 mM (○) and 10 mM (●) ammonium acetate (CH_3COONH_4) and 2 mM (△) and 10 mM (▲) ammonium bicarbonate (NH_4HCO_3) aqueous solutions through 33 μm deep microchannels (Li & Yoda 2010). The slip lengths were estimated using a local method, evanescent wave-based multilayer nano-particle image velocimetry (Li & Yoda 2008), by extrapolating the velocity profile, which is effectively linear within the first 0.5 μm next to the channel wall, from three independent velocity measurements in this near-wall region. The plot on the left (*a*) gives slip lengths for naturally hydrophilic fused-silica channels, while that on the right (*b*) gives slip lengths for identical channels (within fabrication tolerances) made hydrophobic by a self-assembled monolayer of octadecyl trichlorosilane (OTS). In all cases, the results, for microchannels with a cross-sectional aspect ratio exceeding 15, give shear rates within 5% on average of analytical predictions for two-dimensional Poiseuille flow after the data are corrected for the effects of nonuniform tracer distribution. The error bars denote the maximum standard deviations in the slip lengths of (*a*) 25 nm and (*b*) 40 nm, based on the uncertainties in the linear curve-fit and the actual velocity data.

Reference: H. F. Li and M. Yoda (2010) An experimental study of slip considering the effects of nonuniform colloidal tracer distributions. *J. Fluid Mech.*, vol. 662, pp. 269–287, 2010.

Contributed by Professor Minami Yoda

In free surface problems, the interface shape $z = \eta(x, y)$ is an unknown that must be obtained in the course of the solution. In this case, the kinematic condition is that the velocity normal to the free surface must be equal to the fluid velocity at the free surface:

$$w(x, y) = \frac{D\eta}{Dt} = \frac{\partial \eta}{\partial t} + u\frac{\partial \eta}{\partial x} + v\frac{\partial \eta}{\partial y} \tag{3.162}$$

These free surface boundary conditions also hold at a liquid-liquid interface separating two fluids having different properties.

For the case of a curved surface, surface tension effects can be significant. The effect of surface tension is to produce a discontinuity in the normal stress that is proportional to the mean curvature of the surface (Wehausen & Laitone, 1960). The expressions for the stresses are presented earlier in this chapter, and the general conditions for free surfaces in the presence of surface tension are presented by Wehausen and Laitone (1960) and Levich and Krylov (1969). In particular, for two-dimensional flow, it can be shown that the two conditions reduce to

$$[p]\frac{dh}{dx} - \left(2[\mu u_x]\frac{dh}{dx} - [\mu(u_y + v_x)]\right) = \gamma \frac{\frac{d^2h}{dx^2}}{\left(1 + \left(\frac{dh}{dx}\right)^2\right)^{3/2}} \frac{dh}{dx} \tag{3.163}$$

$$[p] + [\mu(u_y + v_x)]\frac{dh}{dx} - 2[\mu v_y] = \gamma \frac{\frac{d^2h}{dx^2}}{\left(1 + \left(\frac{dh}{dx}\right)^2\right)^{3/2}} \tag{3.164}$$

where $[p] = p_1 - p_2$ denotes, for example, the jump in the pressure across the interface and $u_x = \partial u/\partial x$.

3.11.2 Mass transfer boundary conditions

Most often, these boundary conditions fall into the cases of specified concentration or specified flux or a combination of these. Thus, on a molar basis,

$$c_A = c_{A0} \quad \text{at a solid boundary} \tag{3.165}$$

if the specified concentration is known or

$$N_{nA} = N_{A0} \quad \text{at a solid boundary} \tag{3.166}$$

where N_{nA} is the flux in the direction normal to the boundary. At a solid wall, $N_{A0} = 0$.

Finally, a mass transfer convection coefficient, $\bar{h}_m$, can be defined in terms of the gradient of the concentration in situations where flow past a boundary occurs, and

$$-\frac{\partial c_A}{\partial n} = \bar{h}_m(c_A - c_{A\infty}) \text{ at a solid boundary} \tag{3.167}$$

where $c_{A\infty}$ is the concentration far from the wall. The mass transfer convection coefficient is an empirically obtained coefficient, and equation (3.167) serves as a definition of $\bar{h}_m$.

In electro-osmotic flow, the flux of a given species is given by

$$\vec{N}_A = -D_{AB}\nabla c_A + m_A z_A c_A \vec{E} + c_A \vec{V} \text{ at a solid boundary} \tag{3.168}$$

Normal to the wall, the flux is zero, and so the condition is

$$-D_{AB}\frac{\partial c_A}{\partial n} + m_A z_A c_A E_n = 0 \text{ at a solid boundary} \tag{3.169}$$

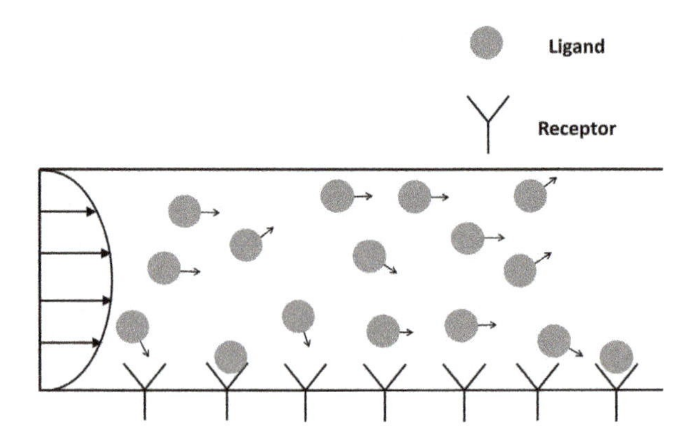

Figure 3.18 Sketch of a biochemical reaction taking place at a surface containing bioreceptors. The *ligand* is the target analyte; ligands are atoms, molecules, ions, or other structures that bind to other structures to form a complex. The terms *ligand* and *receptor* are generic.

where E_n is the electric field component normal to the wall. If a chemical reaction occurs at a surface, there will be a net flux of a species A at the surface. For a homogenous reaction, this boundary condition is

$$\vec{N}_A = K_A c_A \text{ at a solid boundary} \tag{3.170}$$

Note that the rate constant K_A has units of m/sec. The form of the boundary condition depends on the character of the chemical reaction and so depends on the physical problem.

Biochemical reactions occur when biomolecules attach and/or detach from surfaces (Figure 3.18). Typically, these interactions are characterized by two rate constants, which are denoted by K_{on} and K_{off}. The rate constant K_{on} is called the *association rate constant*, associated with adsorption to the surface, and K_{off}, with the *dissociation rate constant* associated with desorption from the surface. There are several forms of these reactions, and one such boundary condition is of the Michaelis–Menten type (Murray, 2001), given by

$$-D_{AB}\frac{\partial c_A}{\partial n} = \frac{\partial c_A}{\partial t} = K_{\text{on}}\, c_A(c_{s0} - c_s) - K_{\text{off}} c_s \tag{3.171}$$

where c_A is the solute concentration on the liquid side at the wall and has units of $M = \text{mole/m}^3$, c_{s0} is the total number of sites available for binding in mole/m^2, and c_s is the surface concentration in mole/m^2. Rhee *et al.* (1989) derives the form of the rate of adsorption–desorption on the right side of equation (3.171).

3.11.3 Electrostatics boundary conditions

In electrostatics, the curl of the electric field vanishes; that is, $\nabla \times \vec{E} = 0$. Following the treatment of Masliyah and Bhattacharjee (2006), applying Stokes's theorem, it follows that

$$\oint \vec{E} \bullet d\vec{s} = 0 \tag{3.172}$$

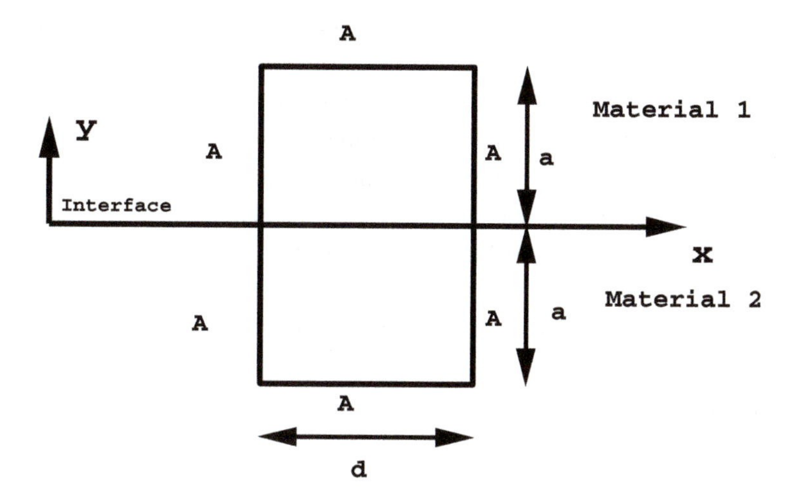

Figure 3.19 An infinitesimal control volume encompassing an interface between two different materials. Here A is the area of the indicated face. Adapted from Masliyah and Bhattacharjee (2006).

where $\oint$ indicates the integral around the boundary of the control volume in Figure 3.19 in a counterclockwise manner. Assuming $d = a$ initially, it follows that

$$E_{x,1} = E_{x,2} \tag{3.173}$$

or the x component of the electric field in each material is the same. This means that the x derivative of the potential is the same, and thus integrating along the interface $\phi_1 = \phi_2 + C$. Letting $a \to 0$ with d fixed, it follows that $C = 0$, and so the potential is continuous at the interface.

For constant electrical permittivity, the electric field satisfies the equation

$$\epsilon_e \nabla \bullet \vec{E} = \rho_e \tag{3.174}$$

Integrating this equation over the control volume of Figure 3.19, and using the preceding condition for the x component of the electric field,

$$\epsilon_{e1} E_{1y} - \epsilon_{e2} E_{2y} = -\epsilon_{e1} \frac{\partial \phi}{\partial n} |_1 + \epsilon_{e2} \frac{\partial \phi}{\partial n} |_2 = \frac{1}{A} \int_{\mathcal{V}} \rho_e d\mathcal{V} \tag{3.175}$$

The term on the right-hand side of this equation has units of Coul/m^2 and is thus a surface charge density. If $\rho_e = 0$, then

$$\epsilon_{e1} \frac{\partial \phi}{\partial n} |_1 = \epsilon_{e2} \frac{\partial \phi}{\partial n} |_2 \tag{3.176}$$

Note that this condition is of the same form as the continuity of heat flux at a solid boundary between two materials or the continuity of heat flux between two materials. The two other components of the electric field that are parallel to the wall must also be continuous. If material 1 is water and material 2 is silica, then $\epsilon_{e1} \gg \epsilon_{e2}$, and approximately,

$$-\epsilon_{e1} \frac{\partial \phi}{\partial n} |_1 = \frac{1}{A} \int_{\mathcal{V}} \rho_e d\mathcal{V} \text{ at a solid boundary} \tag{3.177}$$

where the outward normal to the boundary is assumed to be in the positive direction.

In this formulation, leading to equation (3.175), care must be used in defining the quantity ρ_e because its determination may include surface chemistry reactions on the solid side (say, material 2). Another way to obtain the boundary condition is to integrate the Poisson equation for the electrical potential equation from the Stern plane (Section 1.10) to the bulk, where the potential is uniform, also leading to

$$-\epsilon_{e1}\frac{\partial \phi}{\partial n}\mid_1 = \frac{1}{A}\int_{\mathcal{V}} \rho_e d\mathcal{V} \tag{3.178}$$

For electroneutrality to be preserved,

$$\frac{1}{A}\int_{\mathcal{V}} \rho_e d\mathcal{V} + \sigma = 0 \tag{3.179}$$

leading to the equation

$$-\frac{\partial \phi}{\partial n} = \frac{\sigma}{\epsilon_e} \text{ at a solid boundary} \tag{3.180}$$

where now σ is the surface charge density on the wall. The sign in front of the normal derivative depends on the sign of the outward normal, and in general,

$$\hat{n} \bullet \nabla\phi = \frac{\sigma}{\epsilon_e} \text{ at a solid boundary} \tag{3.181}$$

An alternative to the gradient boundary condition is the specified potential. If the electrical potential at the Stern plane, termed the ζ potential, is known, then

$$\phi = \zeta \text{ at a solid boundary} \tag{3.182}$$

3.11.4 Temperature boundary conditions

These boundary conditions mirror the mass transfer boundary conditions and actually predate their development. Thus

$$T = T_0 \text{ at a solid boundary} \tag{3.183}$$

if the specified temperature is known or

$$-k\frac{\partial T}{\partial n} = q_0 \text{ at a solid boundary} \tag{3.184}$$

At a boundary between two different media,

$$-k_1\frac{\partial T}{\partial n}\mid_1 = -k_2\frac{\partial T}{\partial n}\mid_2 \tag{3.185}$$

that is, the heat flux should be continuous.

Finally, a heat transfer convection coefficient $\bar{h}_T$ can be defined in terms of the gradient of the concentration in situations where flow past a boundary occurs and

$$-k\frac{\partial T}{\partial n} = \bar{h}_T(T - T_\infty) \tag{3.186}$$

where T_∞ is the temperature far from the wall. As with the mass transfer coefficient, the $\bar{h}_T$ is an empirically obtained coefficient. At a free surface, the temperature and heat flux normal to the surface should be continuous.

3.11.5 Other boundary conditions

Other boundary conditions arise in specific situations such as those conditions on the velocity, temperature, and pressure at the inlet and outlet of tubes and channels. Often streamwise gradients of velocity are assumed to vanish at the outlet of a channel or pipe.

When solving the Navier–Stokes equations, boundary conditions on the pressure must be specified. These boundary conditions come from the equations themselves and involve velocity derivatives both normal and tangential to the wall. The bottom line is that the boundary conditions depend on what is known at the ends of the domain for the specific problem of interest.

3.12 Dimensional analysis and similarity

An important question to ask is, Under what conditions are flows of different fluids flowing in two geometrically similar geometries (same shape), with each appropriate length scale of the same ratio, equivalent? Such similar flows are called *dynamically similar*, and the means to determine whether such flows are dynamically similar is called *dimensional analysis*. For two flows of different fluids, at different velocities and different linear dimensions (but with similar geometries, e.g., two spheres), to be dynamically similar, it is reasonable to expect that the ratio of the forces on the two bodies will be fixed at all times.

As you learned in your undergraduate fluid dynamics course, at the macroscale, the most important function of dimensional analysis is to provide a means of designing experiments to be performed on the model scale. The model scale can be an order of magnitude or more smaller than the scale on which information is sought, the *prototype*. In micro- and nanofluidics, of course, it is not necessary to scale down the experiments; indeed, it may be desireable to scale *up* the experiments though in my experience, I have never seen this done! Moreover, there are often difficulties in precisely determining the very dimensions of microfluidic devices (see Section 3.17). The function of dimensional analysis in micro- and nanofluidics is to identify the characteristics of the problem, such as whether it is one-, two-, or three-dimensional, and to identify analogies between fluid flow, heat transfer, mass transfer, and electrostatics. Both these functions lead to an increased understanding of flows at small scales.

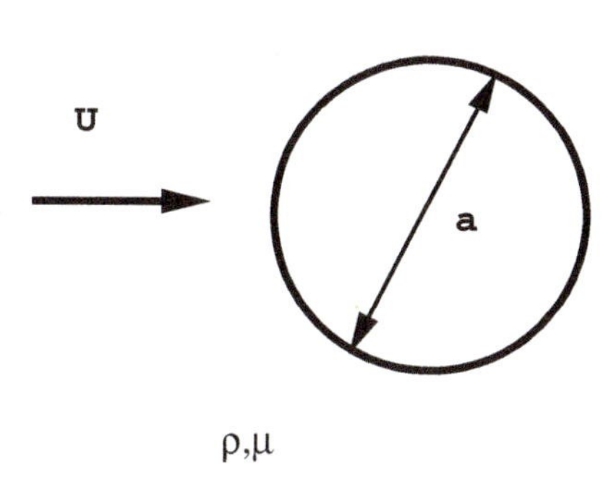

Figure 3.20 Flow past a spherical particle in a fluid with density ρ and viscosity μ.

Consider the case of flow past a smooth, uncharged sphere, as in Figure 3.20. Then the drag on the sphere can be a function of several parameters, namely, $D = D(\rho, \mu, d, V)$, where d is the sphere diameter and U is the velocity far from the sphere.

These five parameters can be reduced to two parameters using dimensional analysis. It is easily shown that the dimensionless drag coefficient on a sphere is defined by

$$c_d = \frac{F}{\rho U^2 d^2} = f(Re) \tag{3.187}$$

where f is a function to be determined and Re is the Reynolds number. Thus flows past two spheres are dynamically similar if the Reynolds numbers are the same in each case. The Reynolds numbers can thus be made equivalent by adjusting the free stream velocity U or the diameter d, or both. Experiments designed for water may be performed in air if the Reynolds number is the same in each case.

Dimensional analysis is the art and science of reducing the number of parameters that constitute the solution to a given problem. Dimensional analysis recognizes that certain physical parameters, including fluid properties and velocities and length scales, appear only in certain groupings. These groupings are *dimensionless parameters*, which are then used to characterize the system. The most important example in viscous flow is the Reynolds number, and for the sphere, $Re = \rho U d/\mu$.

Three procedures may be employed to determine the relevant dimensionless groups, first, by inspection, because certain obvious parameters can be determined by experience such as ratios of length scales and the Reynolds number. The dimensionless parameters can also be determined explicitly, as suggested by the Buckingham pi theorem, which is usually done in undergraduate fluid dynamics courses (Munson *et al.*, 2005). Finally, these parameters may be obtained by writing the governing equations in dimensionless form, and this is the method described here.

The governing equations for the flow of an electrically conducting fluid in the isothermal case indicate that a typical velocity in three dimensions is a function of a number of variables, and in the general case, for steady flow,

$$u = u(x, y, z, U, p, d_1, d_2, d_3, \mu, \rho, D_{AB}, \epsilon_e) \tag{3.188}$$

where d_i are the typical length scales in the three coordinate directions and U is a typical velocity scale. In most cases, a single velocity scale is obvious. This is an overwhelming number of parameters, and dimensional analysis can be employed to reduce significantly the number of parameters. All the parameters on the right-hand side of equation (3.188) are dimensional, and (x, y, z) are variables in the coordinate directions.

Consider first the case of a three dimensional flow in a channel for which the three length scales d_i are all of different magnitudes, as in Figure 3.11. This is the case for the nanopump depicted in Figure 1.1b. In that pump, $(d_1, d_2, d_3) = (L, h, W) = (3\ \mu\text{m}, 4\text{–}50\ \text{nm}, 44\ \mu\text{m})$; that is, each channel in a membrane is nanoconstrained in one dimension, having a (uniform or nearly uniform) pore size ranging from 4 to 50 nm. The number of independent parameters can be reduced substantially by nondimensionalizing the governing equations.

The first step in the process is to define a dimensionless length in each of the coordinate directions; for example, in the streamwise direction of the flow,

$$x^* = \frac{x}{L} \tag{3.189}$$

Substituting into the Navier–Stokes equations with

$$\frac{\partial}{\partial x} = \frac{dx^*}{dx}\frac{\partial}{\partial x^*} \tag{3.190}$$

all the terms involving derivatives in the coordinate directions will be partially nondimensional. Next, a velocity scale must be determined. The velocity scale is normally fairly obvious but is also particular to the problem. Defining a dimensionless velocity,

$$u^* = \frac{u}{U} \tag{3.191}$$

where it is supposed that the value (i.e., a number, $U = 10^{-3}$m/sec) of U is known. (*In what follows, the asterisk will be dropped, and the presence of dimensionless parameters in a given equation will signal the fact that the equation is dimensionless.*)

Typically, the channels in a nanopore membrane are of the shape depicted in Figure 3.11. First, the continuity equation becomes

$$\epsilon_1 \frac{\partial u}{\partial x} + \frac{\partial v}{\partial y} + \epsilon_2 \frac{\partial w}{\partial z} = 0 \tag{3.192}$$

where all three velocities are scaled on the single velocity scale U and $\epsilon_1 = h/L$ and $\epsilon_2 = h/W$. In this convention, h is assumed to be the smallest dimension, and so both ϵ_1 and ϵ_2 will be small. In this situation,

$$\frac{\partial v}{\partial y} = 0$$

and so since $v = 0$ at $y = 0, 1$, according to the solid-wall boundary condition, $v = 0$ everywhere. This leads to the definition of *fully developed flow*, where the streamwise velocity does not depend on the streamwise coordinate: $\partial u/\partial x = 0$ (assuming $w = 0$). This is the case for many micro- and nanoscale membranes used in biological applications.

Using the same procedure in the Navier–Stokes equations, the x momentum equation becomes

$$\frac{\partial u}{\partial t} + Re\left(\epsilon_1 u \frac{\partial u}{\partial x} + v\frac{\partial u}{\partial y} + \epsilon_2 w \frac{\partial u}{\partial z}\right) = -\frac{\partial p}{\partial x} + b_{ndx} + \nabla^2 u \tag{3.193}$$

where b_{ndx} is the dimensionless body force in the x direction. The fluid time scale is $t_f = h^2/\nu$, so the dimensionless time variable for the fluid is $t^* = t/t_f$.

The dimensional body force due to the presence of a constant electric field in the x direction is given by

$$b_x = E_0 \rho_e \tag{3.194}$$

where $\rho_e = Fc\sum_{i=1}^{N} z_i X_i$ is the volume charge density, z_i is the valence of the ionic species i, and F is Faraday's constant. The dimensionless body force emerges from the nondimensionalization process after multiplication of the quantity $h^2/\mu U$ and

$$b_{ndx} = \frac{FcE_0h^2}{\mu U}\sum_{i=1}^{N} z_i X_i = \frac{\beta}{\epsilon^2}\sum_{i=1}^{N} z_i X_i \tag{3.195}$$

where $\beta = c/I$ is the ratio of the total concentration to the ionic strength and is usually large. It is also common to scale the concentrations on the ionic strength, in which case, $\beta = 1$. For electro-osmotic flow, the velocity scale turns out to be $U = \epsilon_e E_0 \phi_0/\mu$, where $\phi_0 = RT/F$.

In the preceding equations, the lengths are nondimensionalized on the triad (L, h, W) (Figure 3.11), and (u, v, w) are the dimensionless velocities in each of the coordinate directions (x, y, z); also

$$\nabla^2 = \epsilon_1^2\frac{\partial^2}{\partial x^2} + \frac{\partial^2}{\partial y^2} + \epsilon_2^2\frac{\partial^2}{\partial z^2} \tag{3.196}$$

Consider first the case in which the body force term $b_{ndx} = 0$ and the flow is thus pressure driven only. Then

$$u = u(x, y, z, U, p, d_1, d_2, d_3, \mu, \rho) \tag{3.197}$$

After nondimensionalizing the problem, the number of parameters in the list has been reduced; in the full three dimensional case, $u = u(x, y, z, p, \epsilon_1, \epsilon_2, \text{Re})$, where now x, y, z are dimensionless; note that there are three fewer parameters than in the dimensional specification.

In the case of fully developed flow in a wide and long channel,

$$\frac{\partial}{\partial x} = \frac{\partial}{\partial z} = 0 \tag{3.198}$$

so that $u = u(y, p, \text{Re})$. However, for fully developed flow, the convective terms in equation (3.193) vanish, and thus $u = u(y, p)$. Actually, it will be shown that $u = u(y, dp/dx)$ and that dp/dx is constant for incompressible flow of a liquid. For a gas, if compressibility effects are important, changes in density will cause the pressure to deviate from the linear behavior of liquids.

The preceding scaling is appropriate for internal flow at low Reynolds numbers. The essential difference between the large and small Reynolds number cases is that for small Reynolds number, the pressure is scaled on the viscosity

$$p = \frac{p^*}{\frac{\mu U}{L}} \tag{3.199}$$

as earlier, whereas in the large Reynolds number case typical of aerodynamic flows, the pressure is scaled as

$$p = \frac{p^*}{\rho U^2} \tag{3.200}$$

Using the same procedure, the mass transfer equation can also be written in dimensionless form, and

$$Sc\frac{\partial X_A}{\partial t} + ReSc\left(\epsilon_1 u\frac{\partial X_A}{\partial x} + v\frac{\partial X_A}{\partial y} + \epsilon_2 w\frac{\partial X_A}{\partial z}\right)$$
$$+ z_A\left(\epsilon_1\frac{\partial X_A E_x}{\partial x} + \frac{\partial X_A E_y}{\partial y} + \epsilon_2\frac{\partial X_A E_z}{\partial z}\right) = \nabla^2 X_A \qquad (3.201)$$

where time is scaled on the fluid time scale. The mass transfer time scale is $t_m = h^2/D_{AB}$, and the fluid time scale is $t_f = h^2/\nu$ so that

$$\frac{t_f}{t_m} = \frac{1}{Sc} \ll 1 \qquad (3.202)$$

for liquids, where $Sc = \nu/D_{AB}$ and, for liquids, $Sc \sim 1000$. This means that on the mass transfer time scale, the velocity is constant, and on the velocity time scale (much smaller), the mole fraction is constant.

The equation for the electric potential may be nondimensionalized in the same way. The dimensionless potential $\phi = \phi^*/\phi_0$, and so

$$\nabla^2\phi = -\frac{Fch^2}{\epsilon_e\phi_0}\sum_i z_i X_i \qquad (3.203)$$

where here ϕ is the dimensionless potential. Taking $\phi_0 = RT/F$, it is noted that

$$\frac{Fch^2}{\epsilon_e\phi_0} = \frac{F^2ch^2}{\epsilon_e RT} = \frac{h^2}{\lambda^2}\frac{c}{I} \qquad (3.204)$$

and thus

$$\epsilon^2\nabla^2\phi = -\beta\sum_i z_i X_i \qquad (3.205)$$

where $\epsilon = \lambda/h$ and λ is the Debye length. Note that there is no time derivative in the potential equation.

Finally, the dimensionless energy equation is given by

$$Pr\frac{\partial\theta}{\partial t} + RePr\left(\epsilon_1 u\frac{\partial\theta}{\partial x} + v\frac{\partial\theta}{\partial y} + \epsilon_2 w\frac{\partial\theta}{\partial z}\right)$$
$$+\Gamma_e\left(\epsilon_1 u\frac{\partial E^2}{\partial x} + v\frac{\partial E^2}{\partial y} + \epsilon_2 w\frac{\partial E^2}{\partial z}\right) = \nabla^2\theta + Ec\Phi + \frac{\dot{q}h^2}{k\Delta T} \qquad (3.206)$$

where θ is a dimensionless temperature $\theta = T - T_0/\Delta T$, where ΔT is a typical temperature difference in the system; Pr is the Prandtl number, $Pr = \mu c_p/k$; and T_0 is a reference temperature. The Eckert number is defined by $Ec = U^2/c_v\Delta T$. The parameter Γ_e is defined by

$$\Gamma_e = \frac{\epsilon_e U E_0^2 h}{k\Delta T}$$

and is usually very small.

The viscous dissipation function Φ is obtained by expanding the second term on the right-hand side of equation (3.137), and

$$\Phi = 2\left(\frac{\partial u}{\partial x}\right)^2 + 2\left(\frac{\partial v}{\partial y}\right)^2 + 2\left(\frac{\partial w}{\partial z}\right)^2 + \left(\frac{\partial v}{\partial x} + \frac{\partial u}{\partial y}\right)^2$$
$$+ \left(\frac{\partial w}{\partial y} + \frac{\partial v}{\partial z}\right)^2 + \left(\frac{\partial u}{\partial z} + \frac{\partial w}{\partial x}\right)^2 \tag{3.207}$$

The energy source term $\dot{q}$ in micro- and nanofluidics is often associated with what is called *Joule heating*. When a fluid with a net electric charge experiences an imposed electric field, there is a rise in temperature due to the resistance caused by electrical current. Joule heating can cause a temperature gradient across the channel and a rise in overall temperature The source term due to Joule heating is given by (Hughes & Gaylord, 1964)

$$\dot{q} = \frac{I^2}{\sigma_e} = \frac{\mid \sigma_e \vec{E} \mid^2}{\sigma_e} = \sigma_e \mid \vec{E} \mid^2 \tag{3.208}$$

where σ_e is the electrical conductivity of the electrolyte, the proportionality constant between the current density and the electric field: $\vec{J} = \sum_i z_i N_i = \sigma_e \vec{E}$. Thus

$$\frac{\dot{q}h^2}{k\Delta T} = \frac{h^2 E_0^2 \sigma_e}{k\Delta T} \mid \vec{E} \mid^2 \tag{3.209}$$

Notation alert: Note that σ_e in this section is the electrical conductivity defined previously and not surface tension or surface charge density.

Equation (3.209) may be interpreted physically as

$$\frac{h^2 E_0^2 \sigma_e}{k\Delta T} \sim \frac{\text{electrical power density}}{\text{thermal power density}}$$

Moreover, the presence of temperature gradients in the system can lead to variations in properties in one or more directions in the preceding equations in which, for simplicity, properties have been assumed constant.

The present section has identified the key dimensionless parameters that arise in the flow of a fluid mixture under electrokinetic effects. These dimensionless parameters have distinct physical meaning as the ratio of key dimensional quantities. Thus the Reynolds number

$$Re = \frac{\rho U_0^2 h^2}{\mu U_0 h} = \frac{\text{inertial force}}{\text{viscous force}}$$

and for small Reynolds numbers, it is thus seen that the flow is viscously dominated.

The Schmidt number

$$Sc = \frac{\frac{\mu}{L^2}}{\frac{\rho D_{AB}}{L^2}} = \frac{\text{viscous diffusion rate}}{\text{mass diffusion rate}}$$

and for heat transfer, the Prandtl number

$$Pr = \frac{\mu c_p}{k} = \frac{\nu}{\alpha} = \frac{\text{viscous diffusion rate}}{\text{thermal diffusion rate}}$$

where

$$\alpha = \frac{k}{\rho c_p}$$

is the thermal diffusivity. The main dimensionless parameter in electrokinetic phenomena is the ratio of two lengths

$$\epsilon = \frac{\lambda}{h}$$

where λ is the Debye length.

3.13 Fluid, electrostatics, and heat and mass transfer analogies

As noted in the previous section, at the micro- and nanoscale, the process of dimensional analysis reveals similarities between heat transfer, mass transfer, velocity, and electrical potential. In this section, three types of similarities between profiles are discussed:

- Mole fraction and temperature similarity in transient one-dimensional problems
- Velocity and electrical potential similarity in electro-osmotic flow
- Velocity and temperature at large Reynolds number

In some cases, such as velocity–electrical potential similarity, the two distributions are identical. Conversely, the distributions may be of the same form and the problems solved by the same techniques, even if the distributions differ in scale due to the differing values of the Reynolds, Prandtl, and Schmidt numbers.

3.13.1 Mole fraction and temperature similarity

Suppose that the energy and mass transfer balances are between convection and diffusion. Suppose also that the pressure gradient vanishes. Then the governing equations in the case of constant properties for two-dimensional flow are given by the following:

$$\frac{\partial u}{\partial t} + Re\left(\epsilon_1 u\frac{\partial u}{\partial x} + v\frac{\partial u}{\partial y}\right) = \nabla^2 u \tag{3.210}$$

$$\frac{\partial v}{\partial t} + Re\left(\epsilon_1 u\frac{\partial v}{\partial x} + v\frac{\partial v}{\partial y}\right) = \nabla^2 v \tag{3.211}$$

$$Sc\frac{\partial X_A}{\partial t} + Re\,Sc\left(\epsilon_1 u\frac{\partial X_A}{\partial x} + v\frac{\partial X_A}{\partial y}\right) = \nabla^2 X_A \tag{3.212}$$

$$Pr\frac{\partial \theta}{\partial t} + Re\,Pr\left(\epsilon_1 u\frac{\partial \theta}{\partial x} + v\frac{\partial \theta}{\partial y}\right) = \nabla^2 \theta \tag{3.213}$$

Note the similarity in all these equations; if the boundary conditions for all the dependent variables are the same, and $Pr = Sc = 1$, then $u = v = \theta = X_A$; that is, all the dependent variables have the same solution.[6] The boundary conditions for internal flow are usually not the same, so these variables will not all be equivalent. Nevertheless, the equations are analogous in that in many situations, these equations can be solved numerically by the same procedure. Note that in liquid flows, $1 < Pr \ll Sc$ so that the time scales for each of the velocities and the mass fraction and temperature are different.

Many heat and mass transfer problems have equivalent solutions, often in the transient fully developed regime. Consider two-dimensional heat and mass transfer in a channel in the case that the velocity field $(u, v) = (0, 0)$; then the governing equations for the temperature and mole fraction are

$$Sc\frac{\partial X_A}{\partial t} = \frac{\partial^2 X_A}{\partial x^2} + \frac{\partial^2 X_A}{\partial y^2} \tag{3.214}$$

$$Pr\frac{\partial \theta}{\partial t} = \frac{\partial^2 \theta}{\partial x^2} + \frac{\partial^2 \theta}{\partial y^2} \tag{3.215}$$

where we have assumed a single length scale $\epsilon_1 = 1$. Suppose that the temperature satisfies $\theta = 1$ at $x = 0$ and $\theta \to 0$ as $x \to \infty$ and that the sidewalls are insulated: $\partial\theta/\partial y = 0$ at $y = 0, 1$. Also suppose that $\theta = 0$ at $t = 0$. Then the heat transfer is one-dimensional in the x direction, and the solution for the dimensionless temperature is

$$\theta(x, t) = \text{erfc}\left(\frac{x\sqrt{Pr}}{2\sqrt{t}}\right) \tag{3.216}$$

Here erfc is the complimentary error function defined by

$$\text{erfc}(x) = \frac{2}{\pi}\int_x^{\infty} e^{-\xi^2} d\xi \tag{3.217}$$

and, for example, the dimensionless temperature may be defined by

$$\theta(x, t) = \frac{T - T_i}{T_0 - T_i} \tag{3.218}$$

where the initial temperature is denoted by T_i and $T = T_0$ at $x = 0$. Similarly, the mass transfer problem having the same boundary and initial conditions has the solution

$$X_A(x, t) = X_{A0}\text{erfc}\left(\frac{x\sqrt{Sc}}{2\sqrt{t}}\right) \tag{3.219}$$

Note that the independent variable in equation (3.219) is

$$\eta = \frac{x\sqrt{Sc}}{2\sqrt{t}} \tag{3.220}$$

[6] They say "never say never," but the author cannot think of a case in which this may be true for liquids since at a minimum, $Sc \sim 1000$, and at a maximum, $Pr \sim 50$–100 (generous).

a specific combination of the dimensionless variables (x, t); η is called a *similarity variable* and indicates the absence of specific length and time scales. Putting this into dimensional form,

$$\eta = \frac{x^*}{2\sqrt{D_{AB}t^*}} \tag{3.221}$$

and so the length and time scales cancel. Solutions as in equations (3.216) and (3.219) are called *similarity solutions.*

For example, the concentration profile in equation (3.219) corresponds to the transport of a drug into a membrane at early times. The species flux is given by

$$J_A(0,t) = -D_A \frac{\partial c_A}{\partial y}\bigg|_{y=0} = c_{A0}\sqrt{\frac{D_A}{\pi t}} \tag{3.222}$$

The total amount of species A or the weight gain per unit area delivered to the half-space $y > 0$ over time is given by

$$w_g(t) = \int_0^t J_A(0,t)dt = 2c_{A0}\sqrt{\frac{D_A t}{\pi}} \tag{3.223}$$

Many drugs are bound to proteins, particularly albumin. Albumin is a charged molecule, and so the problem of the delivery of a charged species to a permeable membrane is an important one.

3.13.2 Velocity and electrical potential similarity

In the second analogy, recall that in the previous section, it was shown that the dimensionless electrical body force in the streamwise momentum equation is given by

$$b_{ndx} = \frac{FcE_0h^2}{\mu U}\sum_{i=1}^{N} z_i X_i = \frac{\beta}{\epsilon^2}\sum_{i=1}^{N} z_i X_i \tag{3.224}$$

Now assume that there is no pressure gradient and that the Reynolds number is very small: Re $\ll$ 1. Then, for incompressible flow, the streamwise momentum equation becomes

$$\nabla^2 u = -\frac{\beta}{\epsilon^2}\sum_{i=1}^{N} z_i X_i \tag{3.225}$$

Comparing this equation with the equation for the electrical potential, which is

$$\nabla^2 \phi = -\frac{\beta}{\epsilon^2}\sum_{i=1}^{N} z_i X_i \tag{3.226}$$

it is seen that the two equations are identical.

Suppose that the velocity satisfies the no-slip condition at the wall and that the electric potential satisfies $\phi(0) = \phi(1) = 0$. Then both the equations and the boundary conditions are identical, and on a dimensionless basis, $u(y) = \phi(y)$. In reality, the potential does not vanish at the wall but satisfies $\phi = \zeta$ at $y = 0, 1$, and thus $u = \phi - \zeta$. The equivalence of the dimensionless velocity and the dimensionless potential is important because then measurements of the velocity are equivalent to measurements of the electric potential. This analogy is also true for a free surface at $y = 1$. Can you see why?

Moreover, it is noted that a Debye length of $\lambda = 1$ nm in a $h = 100$ nm channel gives the same value of $\epsilon = 0.01$ as a $\lambda = 100$ nm Debye length in a $h = 10$ μm channel so that microscale measurements can be validated by nanoscale computations, and conversely, nanoscale computations can be validated by microscale experiments. Note that this similarity analysis does not apply for unsteady flow because there is no time derivative in the potential equation.

3.14 Other stress–strain relationships

Fluids such as water, air, gasoline, and oil and simple mixtures such as blood serum (phosphate buffered saline, or PBS), which is essentially a mixture of water and sodium chloride, are all Newtonian fluids in that the stress varies linearly with rate of strain. Newton's law of viscosity is an empirical law, having been developed by a series of measurements using viscometers and other measurement techniques. *Rheology* is the study of the deformation and flow of matter under an applied stress, and although the Navier–Stokes equations still hold for a non-Newtonian fluid, it is the stress–strain relationship that changes.

Non-Newtonian behavior is exhibited by more complex mixtures such as suspensions, emulsions, ointments, gels, and polymers. As was noted previously, a colloid, for example, is a suspension of particles in a liquid such as water, where the particles range in size from 1 nm to 1 μm. These solutions can be either dilute, ranging in concentration from 1 μM (micromolar) to 1 mM (millimolar), or concentrated, which chemists consider to be greater than 1 M. Conversely, engineers consider mixtures to be concentrated up to the solubility limit, which can be up to a mole fraction $X_A = 0.6$ in lithium bromide, water mixtures. The particle phase is called the dispersed phase, and the liquid is the solvent phase.

The simplest forms of the stress–strain relationships for non-Newtonian fluids are depicted in Figure 3.21. Power law fluids can be either *shear thinning* (pseudoplastic) or *shear thickening* (dilatant). Power law fluids are described by the expression

$$\tau_{yx} = \mu_a \left[\frac{du}{dy}\right]^n \tag{3.227}$$

where n is an experimentally determined parameter and μ_a is termed the *apparent viscosity*. This expression is for the case of fully developed flow in a channel. In the more general case, the full expression for the strain tensor must be used.

There are several objections to the usefulness of the formula for a power law fluid, one being that n may not be constant over the entire range of interest for a given fluid. Also, the units of the apparent viscosity depend on the value of n, and

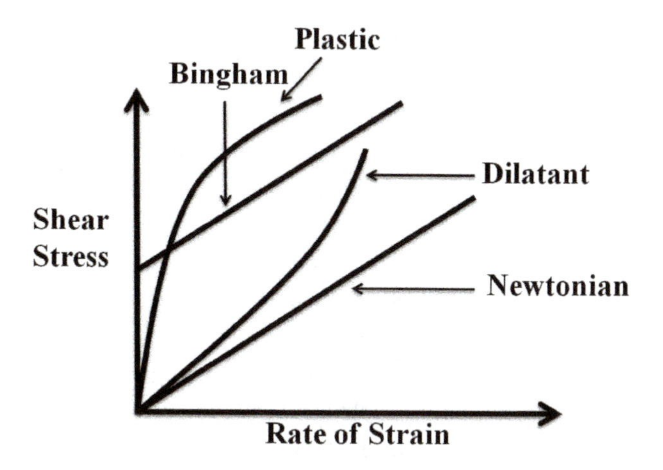

Figure 3.21 Some non-Newtonian viscosity curves. Adapted from White (2006).

the formula cannot hold for a strain rate less than zero. Some examples of power law fluids are blood (in some parameter ranges), milk, some colloidal solutions, and concentrated solutions of sugar and water.

To overcome the last objection, the shear stress is written in a modified form:

$$\tau_{yx} = \mu_a \alpha \left[\frac{du}{dy}\right]^n \tag{3.228}$$

where α is defined by

$$\alpha = \frac{(du/dy)^n}{\mid du/dy \mid^n} \tag{3.229}$$

A better empirical formula that fits most data better than the simple power law is the Carreau equation (Bird *et al.*, 2002):

$$\frac{\mu_a - \mu_\infty}{\mu_0 - \mu_\infty} = (1 + (\lambda_0 \dot{\gamma})^2)^{\frac{n-1}{2}} \tag{3.230}$$

where $\dot{\gamma} = du/dy$ is the strain rate. Values of the zero shear rate viscosity (μ_0) and the infinite shear rate viscosity (μ_∞) and λ for several fluids are given by Bird *et al.* (2002).

Many colloidal systems exhibit what is called a *Bingham type* of behavior, for which there is a limiting stress below which the material acts as if it were a solid. Above this stress, the material flows like a fluid. The expression for the shear stress, in this case, is

$$\tau_{yx} - \tau_B = \mu_a \frac{du}{dy} \quad \text{for } \tau_{yx} > \tau_B \tag{3.231}$$

and $\tau_{yx} = 0$ for $\tau_{yx} < \tau_B$.

The vast majority of the phenomena considered in this text can be analyzed using the assumption of a Newtonian fluid.

3.15 Mathematical character of partial differential equations

3.15.1 Introduction

The governing equations for an electrically conducting fluid are a complicated set of nonlinear partial differential equations (PDEs) that are commonly solved numerically. The mathematical structure of a given PDE is important because it determines the analytical or computational techniques that are best suited to solve the problem. It is worthwhile to note that use of the wrong numerical technique can lead to results that are not only inaccurate but wrong. Moreover, information concerning existence and uniqueness for nonlinear equations is limited. Generally, the engineer, in modeling a physical process, assumes that a solution is possible and develops a model on this basis. If the physical process is properly modeled, there should be a solution to the mathematical idealization of the process. If such a solution cannot be found, either analytically or numerically, then the assumptions of the physical model should be reexamined.

We begin by describing the different types of *linear* PDEs and the differences between each. The nature of each of these equations determines the numerical method used to solve it, if required.

3.15.2 Mathematical classification of second-order partial differential equations

Because the most common forms of PDE encountered in the fields described here are of second order, as exhibited by each of the governing equations discussed in this chapter, we begin with this type of equation in the classification discussion.

There are three types of partial differential equations: elliptic, parabolic, and hyperbolic. In an *elliptic* equation, information, can travel in all directions, and the solution of the equation at one point affects the solution at all other points in the domain. The prototype of an elliptic equation is Laplace's equation. This equation arises in the slow flow of a highly viscous fluid, in conduction-dominated heat transfer problems, and in mass diffusion–dominated problems. The electrical potential is also an elliptic equation.

In a *parabolic* equation, information can travel in only one direction. Thus the solution at a given point is dependent only on what the solution is in a given subset of the domain. The prototype of a parabolic equation is the unsteady heat or mass equation.

The third type of equation is called *hyperbolic*, and the prototype is the wave equation. In a second order wave equation, information is carried in two directions, called *characteristic directions*. This type of equation describes the production and travel of sound. The occurrence of the second order wave equation in heat and mass transfer is not common. First order equations arise in the study of *chromatography*, the process of separating chemical species by selective adsorption of the species on a solid substrate. See Rhee *et al.* (1986, 1989) for details.

Bearing in mind that the Navier–Stokes equations and the Euler equations are highly nonlinear, we begin first with a discussion of the linear case. Some of these ideas can be carried over directly to the nonlinear case, as will be seen. Consider, then, the most general two-dimensional, linear, second-order equation, defined by

$$a\frac{\partial^2\phi}{\partial x^2}+b\frac{\partial^2\phi}{\partial x\partial y}+c\frac{\partial^2\phi}{\partial y^2}+d\frac{\partial\phi}{\partial x}+e\frac{\partial\phi}{\partial y}+f\phi+g=0 \tag{3.232}$$

Then the equation is said to be hyperbolic at a point (x, y) if $b^2 - 4ac > 0$, parabolic if $b^2 - 4ac = 0$, and elliptic if $b^2 - 4ac < 0$. An equation is hyperbolic, parabolic, or elliptic in a domain if it is hyperbolic, parabolic, or elliptic at every point in the domain.

We can define new coordinates (ξ, η) such that equation (3.232) becomes particularly simple. Let

$$\xi = \xi(x, y)$$

$$\eta = \eta(x, y)$$

define a transformation of coordinates such that

$$J = \xi_x\eta_y - \xi_y\eta_x \neq 0 \tag{3.233}$$

where J is called the *Jacobian* of the transformation. The subscripts in this equation denote differentiation with respect to the indicated variable. Then the second-order equation becomes

$$A\phi_{\xi\xi} + 2B\phi_{\xi\eta} + C\phi_{\eta\eta} + \cdots = 0 \tag{3.234}$$

where the coefficients A, B, and C are related to the transformation by

$$A = a\xi_x^2 + 2b\xi_x\xi_y + c\xi_y^2 \tag{3.235}$$

$$B = a\xi_x\eta_x + b\xi_x\eta_y + b\xi_y\eta_x + c\xi_y\eta_y \tag{3.236}$$

$$C = a\eta_x^2 + 2b\eta_x\eta_y + c\eta_y^2 \tag{3.237}$$

The way information is carried by the PDE is embodied in *characteristic curves*. A second-order hyperbolic equation has two characteristic curves: a parabolic equation one characteristic curve, and an elliptic equation no characteristic curves. We discuss the determination of these curves for a hyperbolic equation next.

3.15.3 Characteristic curves

While hyperbolic equations are rare in micro- and nanofluidics, it is customary to indicate how to determine the characteristic curves when the equation is hyperbolic because for a parabolic equation, there is a single characteristic curve that is known (say, time in the unsteady heat equation).[7] Then, with a particular definition of new coordinates ξ and η, we can force $A = C = 0$, and equation (3.234) becomes

$$\phi_{\xi\eta} = 0 \tag{3.238}$$

This is called the *canonical* form of a linear hyperbolic equation. For the wave equation, $A = C = 0$ if we choose (ξ, η) such that they satisfy the first-order equations

$$\xi_x = \lambda_1 \xi_y \tag{3.239}$$

$$\eta_x = \lambda_2 \eta_y \tag{3.240}$$

In order for $A = 0$, we must have

$$a\lambda^2 + 2b\lambda + c = 0 \tag{3.241}$$

This is a quadratic equation that has two distinct roots, say, $\lambda_1(x, y)$ and $\lambda_2(x, y)$. Requiring $C = 0$ yields the same equation, and the two roots $\lambda_1(x, y)$ and $\lambda_2(x, y)$ are called the characteristic curves of the hyperbolic partial differential equation. The calculation of the characteristic curves depends on the precise form of the coefficients of the PDE.

As an example, consider the second-order wave equation in one spatial dimension, given by

$$\phi_{tt} - \phi_{xx} = 0 \tag{3.242}$$

where again, the subscripts denote differentiation. In this case, the characteristic curves are determined by the fact that $a = 1$ and $c = -1$ so that

$$\xi = x + t \tag{3.243}$$

$$\eta = x - t \tag{3.244}$$

These are the directions along which information propagates in a hyperbolic equation. The general form of the solution of the wave equation is thus

$$\phi = F(x + t) + G(x - t) \tag{3.245}$$

which can be obtained directly from the canonical form.

3.15.4 Boundary and initial conditions

In the preceding solution to the wave equation, F and G are arbitrary functions. To determine these functions, initial and boundary conditions must be given.

[7] However, the phenomenon of chemical adsorption discussed in Chapter 7 has been traditionally analyzed as one or more one-dimensional quasilinear equations. See the two-volume set by Rhee *et al.* (1986, 1989).

For the second-order wave equation in a single spatial dimension, two initial conditions and two boundary conditions are required; for the heat equation, one initial condition and two boundary conditions in each direction, and for Laplace's equation, two boundary conditions in each direction, are required.

3.15.5 Classification of the governing equations of micro- and nanofluidics

It is well known that the steady Navier–Stokes equations are elliptic because the viscosity $\nu > 0$; however, the boundary layer equations are parabolic because $a = 0$ if x is considered to be the streamwise direction. The steady energy and mass transfer equations are also elliptic, as is the Poisson equation for the electrical potential. The unsteady flow and their corresponding analogies in heat and mass transfer discussed in Section 3.13 are parabolic. The occurrence of hyperbolic equations in micro- and nanofluidics is somewhat less common but can occur in unsteady mass transfer when diffusion is minimal. Of course, first-order equations in the time domain are by their nature hyperbolic and occur in a variety of problems including chromatography as described by Aris (1956, 1959).

Most of the situations described in this book are elliptic in nature; *ordinary* second-order differential equations of the boundary value type are elliptic by nature.

3.16 Well-posed problems

In engineering and applied science, mathematical models of physical processes must be accurate and robust, characteristics of the solution that are normally termed *stable*. Moreover, though it is common that there may be many different mathematical models for a given physical process, when specific assumptions are made, the solution to the problem should be unique; that is, though different assumptions lead to different mathematical models, the solution to a single mathematical model of a physical process must be unique.

These considerations lead to the definition of a *well-posed problem*. A mathematical model of a physical process is well posed if

1. A solution exists
2. The solution is unique
3. The solution is continuously dependent on the data

The last requirement is especially important. This means that if we change a boundary or initial condition by a small amount, the solution changes by a small amount. Moreover, in engineering problems, this concept should be extended to small changes in material properties such a viscosity of a fluid, thermal conductivity of a solid, and Young's modulus. The concept of continuous dependence on the data is thus related to the stability of the solution. Stability of mathematical models is a subject for an entire book.

3.17 The role of fabrication, experiments, and theory in micro- and nanofluidics

The field of micro- and nanofluidics inherently involves the consideration of length scales on the order of millimeters, the size of reservoirs in drug delivery systems, at the very least, down to the nanometer scale, the size of the pores in a nanopore membrane. Because of the small length scales involved, determining precisely the dimension of the fluid conduits is often difficult. Thus fabrication of microfluidic devices requires precise control at every step of the fabrication process, and even with precise control, fabrication tolerances are much greater, relatively speaking, than in macroscale manufacturing (Gad-el Hak, 2001; Nguyen & Wereley, 2002; Lee, 2006). Moreover, there are inherent limitations to what can be measured at the submicron scale. Consequently, research at the microscale on down requires that the fabricator, experimentalist, and analyst work closely and in tandem in designing micro- and nanofluidic devices.

Most, if not all, experimental methods appropriate for microfluidic applications use tracer particles. Many of the experimental methods used today in the microfluidics area originated in high Reynolds number flows; examples include laser Doppler velocimetry (LDV), molecular tagging velocimetry (MTV, Lempert *et al.*, 1995), and particle image velocimetry (PIV, Adrian & Yao, 1985; Adrian, 1991). These experimental techniques can be distinguished by the type of tracer used and the spatiotemporal characteristics of the measurement: LDV, for example, yields a point measurement of velocity with excellent temporal resolution (up to MHz), whereas PIV, the most commonly used velocimetry method, gives two or three velocity components in a 2-D plane (2D-2C or 3D-3C), while volumetric (3D-3C) PIV is being extended from the macroscale to the microscale (Yoda, 2006). MTV, conversely, typically gives a measurement of up to two of the three velocity components over a one-dimensional line or a two-dimensional plane. An excellent reference book that describes these methods in more detail is the monograph edited by Breuer (2005).

Given the overall dimensions of micro- and nanofluidic channels, <10–$100\ \mu m$, nearly all flows in the micro- and nanofluidic regime are laminar and often fall within the Stokes flow regime. Moreover, to make accurate flow measurements in such small devices, with their limited space, the method should also be nonintrusive. Thus intrusive methods such as atomic force microscopy (AFM), which are useful for probing static surfaces and fluids, are not appropriate for micro- and nanofluidic applications because the fluid will not remain static, especially if the AFM tip is vibrating.

The previously mentioned experimental visualization techniques have a minimum spatial resolution of about $50\ \mu m$ for LDV and $500\ \mu m$ for PIV in macroscale applications (Yoda, 2006). Spatial resolutions as fine as $\sim 2.5\ \mu m$ have been

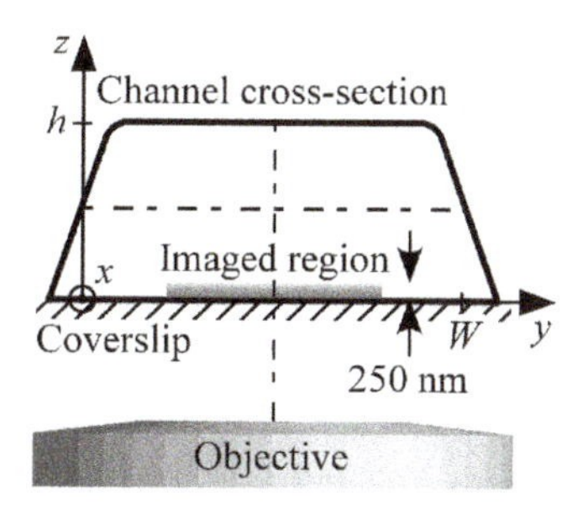

Figure 3.22 Sketch of the field of view in the experimental setup used by Yoda and coworkers (Sadr *et al.*, 2004, 2006).

achieved with LDV and PIV for applications at the microscale (Czarske *et al.*, 2002). However, even these "best case" limits exceed the typical pore dimensions in a nanopore membrane by about 4 orders of magnitude. Thus it is impossible with these techniques to measure velocity, concentration, and temperature profiles in a channel smaller than these resolution limits of a few microns. The conclusion must be that these two velocimetry methods should be interpreted as bulk flow measurement techniques at the nanoscale. The level of resolution of these methods at the micro- and nanoscale means that developing analytical and computational models for micro- and nanoscale flows is not only necessary, but essential.

Despite this limitation, both PIV and MTV are currently used to probe microscale flows in gases and liquids, including electro-osmotic flow (Devasenathipathy & Santiago, 2005; Sadr *et al.*, 2004, 2006). For example, in μPIV, an aqueous solution is often seeded with 200 nm–1 μm diameter fluorescently dyed polystyrene spheres having a density just slightly greater than water ($\rho \sim 1050$ kg/m^3). Illumination by light at a wavelength required to excite the fluorescent tracer particles is viewed by a microscope and then recorded by a digital (CCD = charge-coupled device, CMOS = complementary metal oxide semiconductor) camera.

Typically, the depth of the view (the dimension normal to the field of view) in μPIV is on the order of 1–2 μm, and the velocity field obtained from a 2-D plane is averaged over the *depth of correlation*, which is at least 2 μm. Sadr *et al.* (2004, 2006) and Wereley and Meinhart (2010) have extended the μPIV technique to the submicron scale using evanescent-wave illumination so that the depth of view in their nPIV is ~200 nm. Nano-PIV is, however, limited to accessing the region near the wall, whereas μPIV and MTV are more bulk flow techniques. By comparing pairs of images, all these techniques then yield the average velocity of the fluid within the field of view. A sketch of the field of view used in Yoda's experiments is depicted in Figure 3.22. It should also be mentioned that MTV and μPIV use volume illumination, whereas nPIV uses evanescent wave illumination with an intensity that decays exponentially away from the wall.

In all these methods, the velocity along the optical axis should be negligible for these methods to accurately measure the tracer displacement in the primary flow direction, which must be at a 90° angle to the optical axis. The tracer particles must be carefully tuned to the dimensions of the interrogation window, the field of view, and the minimum spatial resolution of the image (Yoda, 2006). Furthermore, the observation time between the two images must be small enough to permit a reasonably accurate estimate to the velocity of the particles, which is assumed to be the local fluid velocity. It should be noted, however, that Brownian motion, which statistically has a negligible effect on particle motion in the bulk, varies with distance from the wall and may therefore have a nonstochastic influence on particle motion near walls (Sadr *et al.*, 2007; Yoda, 2006).

3.18 Chapter summary

In this chapter, the governing equations for mass momentum and heat and mass transport and for electrostatics have been derived. The equations have been presented in both dimensional and dimensionless form. Dimensional analysis will prove to be a powerful tool for the analysis and solution of the wide range of problems addressed in this text. The main dimensionless parameters associated with these problems have been identified, and their physical meaning has been explained.

In presenting the governing equations and their boundary conditions, it is evident that there are a number of similarities between the fields of heat mass, momentum, and mass transport in mixtures. All equations but the basic equation for electrostatic potential are convective–diffusion equations in some form. Moreover, the boundary conditions, especially those associated with heat and mass transport in mixtures, are similar, if not identical, in form. In fact, it will be seen that in fully developed electro-osmotic flow, it is common for the dimensionless velocity to be equivalent to the electrical potential up to a constant. This situation is described in Chapter 9 and leads to the conclusion that the electrical double layer structure in terms of the electrical potential can be determined by measurement of the velocity within the electrical double layer.

The existence of key dimensionless parameters has been emphasized in this chapter, and dimensional analysis has revealed several analogies: those situations that result in similar, if not identical, solutions to problems in the different fields of fluid mechanics, heat and mass transfer, and electrostatics. This fact inevitably leads to a comment on notation. In a multidisciplinary book of this type, it is impossible to standardize the notation completely in that the same symbol is used for the same quantity throughout. For example, the symbol μ is used for viscosity by mechanical engineers and for electro-osmotic mobility by chemists. In addition, the symbol u may be a dimensional or dimensionless velocity or, as has been seen, internal energy. For this reason, I have tried to make clear the meaning of each symbol as it appears; in addition, I have endeavored, with I think some success, to make each section of the book as self contained as possible. Finally, in this regard, I must emphasize that the appearance of dimensionless parameters in an equation signifies that all quantities in the equation are dimensionless. Conversely, the appearance of properties, such as viscosity and electrical permittivity, in an equation signifies that each quantity in the equation is dimensional, with each term in the equation having the same units. This philosophy will continue throughout the text.

The governing equations have been presented in their full three-dimensional and unsteady form. However, the reader will struggle to find a situation where these equations are solved in their full form. Where appropriate in the following chapters, simplifying assumptions are made to reduce the complexity of the equations while, at the same time, providing a reasonable solution to a given problem. For example, in the design of nanopore membranes, it is often possible to achieve a reasonable solution for determining flow rate using the fully developed form of the equations since the entrance length is very short. Thus Section 3.6 and the sections immediately following may effectively be skipped in many situations.

The fluid has been assumed to be a continuum, and it will be shown later in the text that this is the case as long as the smallest dimension of a channel or membrane is no smaller than ~20 solvent diameters. Molecular simulation methods that must be used for smaller channels are described in Chapter 11. That being said, most nanopore membranes for most applications are bigger than 20 solvent diameters. The exception may be membranes used for desalination, for which the pore size can be on the order of 1–3 nm. Such a process is called *ultrafiltration* (Zeman & Zydney, 1996). In the majority of the remaining chapters of this text, the important problems in fluid mechanics, heat and mass transfer, and electrostatics and dynamics are solved using the these governing equations as a starting point.

EXERCISES

3.1 The velocity field near the core of a tornado is sometimes approximated by the velocity field

$$\vec{u} = -\frac{Q}{2\pi r}\hat{i}_r + \frac{k}{2\pi r}\hat{i}_\theta$$

where (r, θ) are polar coordinates, $r^2 = x^2 + y^2$, and k and Q are constants.

a. Write the velocity components in Cartesian coordinates.

b. Determine the equation describing the streamlines for incompressible flow, and plot several streamlines.

3.2 Sketch the streamlines for the two-dimensional flow

$$u = \alpha x \quad v = -\alpha y$$

with α as a positive constant.

3.3 Consider air to be a binary mixture of 79 percent nitrogen and 21 percent oxygen on a molar basis. Find the mass concentrations.

3.4 Consider a mixture of water and sodium chloride to consist of 95 percent water and 0.05 percent sodium chloride on a molar basis. Find the mass concentrations, and determine the ionic strength of the mixture.

3.5 Starting with the fully three-dimensional and steady Navier–Stokes equations in dimensional form, determine the equations governing fully developed flow in a wide channel. Repeat for a cylindrical tube.

3.6 The velocity distribution in a channel is parabolic, as we know, and the expression for the velocity is

$$u(y) = \frac{1}{2\mu}\frac{dp}{dx}(y^2 - hy)$$

where h is the channel height. Write this equation in dimensionless form, and identify the velocity scale.

3.7 Water flows continuously down a very small crack in cement under the action of gravity and the equation governing flow is

$$\mu\frac{\partial^2 u}{\partial y^2} = -\rho g$$

where μ is the dynamic viscosity, ρ is the density and g is the acceleration due to gravity. There is no pressure gradient. Assume the crack is rectangular and its height h is much smaller than the width W.

a. Using the channel height h as the length scale, and V as the velocity scale, write the equation in non-dimensional form and identify a possible velocity scale V.

b. Using appropriate conditions at the channel walls (which are solid) at $y = 0, h$ determine the velocity distribution in m/sec and the volume flowrate per width of the channel. The channel height h =0.005 cm. What is the maximum velocity?

3.8 The vorticity is defined as the curl of the velocity field

$$\vec{\omega} = \nabla \times \vec{u}$$

Show by direct calculation that the vorticity satisfies

$$\frac{D\omega}{Dt} = \nu\nabla^2\omega$$

Write the equation in dimensionless form. What are the units of vorticity?

3.9 Neglecting internal energy generation and viscous dissipation, show that the two-dimensional, steady state energy equation in a channel for small Γ_e reduces to

$$RePr\left(\epsilon_1 u\frac{\partial\theta}{\partial x} + v\frac{\partial\theta}{\partial y}\right) = \nabla^2\theta$$

where ϵ_1 is as defined in the text. If the temperature boundary conditions are $T = T_0$ at $x = 0, L$ and $t = 0$ at $y = 0$, with

$$-k\frac{\partial T}{\partial y} = q_0 \text{ at } y = h$$

and q_0 = constant, write the boundary conditions in dimensionless form. Discuss the qualitative features of the problem based on the relative magnitude of h and L. Would you expect the solution to depend on x? In what parameter regime? Discuss the choice of the temperature scale.

3.10 Write the corresponding analogue of this problem for the concentration, assuming the walls at $x = 0, 1$ are impermeable and discuss its physical meaning. Would you expect the solution to depend on x? Discuss.

3.11 The governing equation for mass transport in a channel is given by

$$u\frac{\partial c_A}{\partial x} + v\frac{\partial c_A}{\partial y} = D_{AB}\nabla^2 c_A$$

A heterogenous chemical reaction of first order takes place at the surface $y = 0$ so that the boundary condition is

$$-D_{AB}\frac{\partial c_A}{\partial y} = k_r c_A \quad \text{at } y = 0$$

$$\frac{\partial c_A}{\partial y} = 0 \quad \text{at } y = h$$

and at $x = 0$. Write the equation in dimensionless form, and identify the *Damkohler number* as the dimensionless group $Da = k_r h / D_{AB}$. Assume that the velocity scale is the average velocity in a Poiseuille flow. Can the concentration distribution be fully developed?

3.12 In some situations, electrical potential and the velocity are equivalent. Consider fully developed channel flow, and begin with the governing equations and boundary conditions for the potential and the velocity in the form (assuming no pressure drop)

$$\epsilon_e \frac{d^2\phi^*}{dy^{*2}} = \rho_e$$

where $\rho_e = -F\sum_i z_i c_i$ and

$$\mu\frac{d^2u^*}{dy^{*2}} = b_x^*$$

where $b_x^* = \rho_e E_0$. Define dimensionless variables

$$y = \frac{y^*}{h}$$

$$u = \frac{u^*}{U_0}$$

$$\phi = \frac{\phi^*}{\phi_0}$$

and show explicitly that the equations for the potential and the velocity are equivalent for fully developed flow if the velocity scale is taken to be

$$U_0 = \frac{\epsilon_e E_0 \phi_0}{\mu}$$

where E_0 is the applied electric field and $\phi_0 = RT/F$, and where the electrical potential ϕ represents the perturbation to the ζ potential at the wall, that is, the electric potential satisfies the boundary conditions $\phi(0) = \phi(1) = 0$.

3.13 Using the fluid–electrostatics analogy, show that the friction coefficient

$$C_f = -\frac{2\Sigma}{Re}$$

where

$$\Sigma = \frac{\sigma h}{\epsilon_e \phi_0}$$

is a dimensionless parameter.

3.14 Suppose there is an alternating electric field $E^*(t)$ imposed on an electrolyte solution in a wide, two-dimensional channel in the x^* direction. Show that on the concentration time scale, the governing equations are

$$\frac{\partial X_A}{\partial t} - \frac{\partial^2 X_A}{\partial y^2} + \frac{\beta}{\epsilon^2} z_A X_A \sum_{i=1}^{n} z_i X_i - z_A \frac{\partial \phi}{\partial y}\frac{\partial X_A}{\partial y} = 0$$

$$\frac{\partial^2 u}{\partial y^2} + \frac{\beta E_x(t)}{\epsilon^2} \sum z_i X_i = \alpha \frac{\partial u}{\partial t}$$

$$\frac{\partial^2 \phi}{\partial y^2} + \frac{\beta E_x(t)}{\epsilon^2} \sum z_i X_i = 0$$

where $\alpha = 1/\text{Sc}$. For large Schmidt number, the velocity and potential are still approximately equal (as long as the boundary conditions are the same for each) at long times.

3.15 Repeat the previous problem, except scale time on the fluid time scale. What do you conclude about the similarity of the velocity and electric potential?

3.16 According to White (2006), Prandtl, in 1932, proposed the equation

$$\epsilon \frac{d^2 u}{dy^2} + \frac{du}{dy} + u = 0$$

for $\epsilon \ll 1$ as a model for the velocity in a boundary layer. Take the domain of solution to be $y = 0$ to $y = 5$ and the boundary conditions $u(0) = 1$ and $u(5) = 1$. (I have modified the boundary conditions somewhat from what is in White.) Solve this equation for $\epsilon = 1, 0.01, 0.0001$ analytically. What do you find? A numerical solution of this equation would be very difficult for small ϵ. Is there a way to rescale the equation for small ϵ to remove this difficulty?

3.17 The concentration of a pollutant in a fluid is found to be

$$c(x, y, t) = \beta x^2 y e^{-\alpha t}$$

for the flow field given by

$$u = \alpha x \quad v = -\alpha y$$

Calculate Dc/Dt. What do you find? Plot lines of constant concentration at several times.

4 The Essentials of Viscous Flow

4.1 Introduction

As discussed in Chapter 1, many of the current or planned microdevices contain rectangular channels arranged in an array; such channel systems are termed *nanopore membranes*. For example, the nanopore membranes depicted in Figure 1.1 contain up to 30,000 channels, depending on the channel height, which is the nanoscale dimension. Scanning electron microscope (SEM) images of nanopore membranes fabricated by Dr. Shuvo Roy's group at the Cleveland Clinic[1] are depicted in Figure 4.1. You will recognize Figure 4.1(a) as Figure 1.1(b), and is repeated for clarity. The nanopore membrane is fabricated so as to contain channels of a uniform nanoscale height. They fabricated membranes with a variety of pore sizes, ranging from $h = 6.3$ nm to $h = 91$ nm, and there are 11, 500 pores per membrane. Similar membranes are now being used in the development of an artificial kidney (Fissell, 2006). Obviously, it would be difficult to model the flow in such an array from first principles.

There are at least two ways to overcome this problem: (1) treating the membrane as a porous media problem, the membrane being characterized by a permeability and a porosity, and (2) investigating the flow in each channel separately, assuming that the flow behaves similarly in each individual channel. It is the latter approach that is explored here.

In the design of devices employing nanopore membranes, there are questions that must be answered. Consider two limiting cases, in which the membrane is to be used for delivery of a substance and in which the membrane is used for filtration of a solute of specified size; that is, some solutes are allowed to pass through the membrane, whereas others are not, for example,

Design question 1: A device must be designed to deliver a specified amount of a certain drug imbedded in a particle or protein having a nominal size of $a = 50$ nm. What should be the size of the pores in this membrane?

Design question 2: A device needs to filter all particles or molecules below 3 nm nominal size, while retaining all particles or molecules above 3 nm in size.

[1] Dr. Roy is now in the Department of Bioengineering and Therapeutic Sciences at the University of California, San Francisco.

Figure 4.1 Microscopic images of nanopores. (a) Top view of membrane; from Conlisk *et al.* (2009). (b) Top view of a single nanopore. Approximate length scale is shown. There are 11, 500 pores per membrane. Courtesy of Shuvo Roy and Cleveland Clinic. Used with permission.

The particles are charged. Should the membrane be charged? How much? What should be the size of the pores in this membrane?

Nanopore membranes are often designed to carry fluids containing charged species. Electrolyte solutions have been discussed earlier, and in this chapter, a number of internal viscous flow problems relevant to microfluidics and nanofluidics are described. In particular, the reason that electrokinetic phenomena are used so often in micro- and nanofluidics is evident from the calculations described in this chapter (see also Figure 1.5). It is useful to point out that most channels of interest have one very small dimension, perhaps as small as tens of nanometers, whereas the other two dimensions are usually on the micron scale. Thus one-dimensional flow is a good approximation for most membranes.

In this chapter, pressure-driven flow is considered exclusively, and only steady flows are presented. Once the fundamentals of electrostatics and electrochemistry are described, electrokinetically driven flow of an electrolyte solution, which also involves principles of mass transfer and electrostatics, will be combined with the results of this chapter. In the next section, the structure of steady flow in a pipe or channel is described, followed by the presentation of Poiseuille flow in a channel and pipe. This is followed by a discussion of the slip boundary condition and the velocity profile for both gases and liquids. Next a series of other viscous flow problems are discussed, including thin film flow, the Blasius boundary layer, fully developed and developing suction flows, lubrication, flow around a bubble, Stokes flow past a sphere, and a simple model for blood flow.

4.2 The structure of flow in a pipe or channel

The structure of flow in a pipe or a channel is depicted in Figure 4.2 for a pipe. For a channel, replace r with y and D with h or $2h$. Suppose that there is a large reservoir upstream of the entrance to the pipe or channel, as is the case

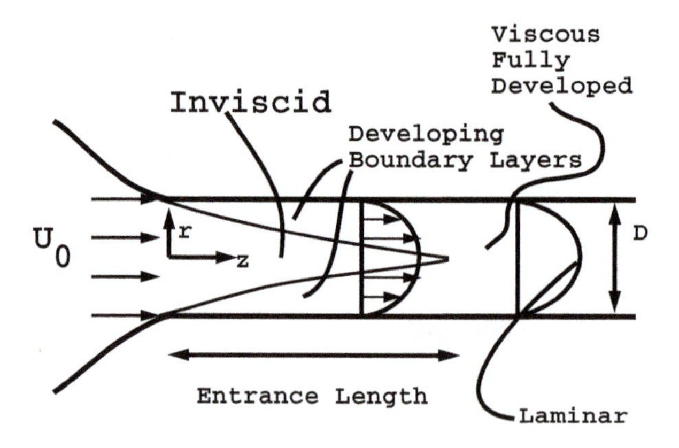

Figure 4.2 Flow through a pipe, showing the entrance region followed by the fully developed region.

in many of the microdevices already described.[2] Then the velocity profile at the inlet will be relatively uniform. To adjust to the presence of the solid walls, the fluid at the wall must assume the wall velocity and the no-slip condition, which, in the case of a stationary wall, is zero. This adjustment region, is the *entrance region*, and its length, the *entrance length*, is a function of the Reynolds number. For a pipe, the entrance length for laminar flow is conventionally taken to be (Langhaar, 1942)

$$\frac{L_e}{D} = 0.06\,Re \tag{4.1}$$

for large Re, in laminar flow. Note that $L_e \to 0$ as $\mathrm{Re} \to 0$. A formula that includes the necessary short entry region as $\mathrm{Re} \to 0$ is given by Chen (1973),

$$\frac{L_e}{D} = \frac{0.6}{1 + 0.035\,Re} + 0.056\,Re \tag{4.2}$$

for a circular pipe, and in the limit $Re \to 0$, $L_e/D = 0.6$. A similar formula due to Atkinson *et al.* (1969) is given by

$$\frac{L_e}{D} = 0.59 + 0.056\,Re \tag{4.3}$$

In the entrance region, the flow resembles a boundary layer because it is unbounded on one side, at least for large Reynolds number.

For a rectangular channel, the hydraulic diameter is given by

$$D_h = \frac{4hW}{2W + 2h} \tag{4.4}$$

which is just 4 times the cross-sectional area divided by the wetted perimeter. For a wide channel, $D_h \sim 2h$, and the preceding formulas hold for channels provided that the hydraulic diameter is used for the length scale.

[2] For the purposes of this discussion, the pipe or channel walls are assumed to be uncharged.

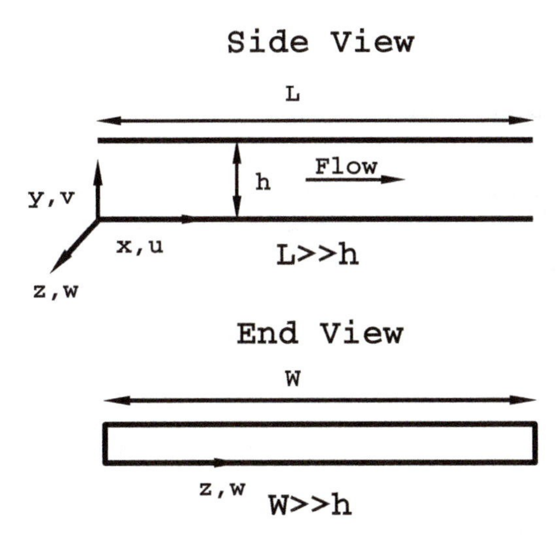

Figure 4.3 Geometry for Poiseuille flow in a channel.

After the boundary layers meet, the velocity no longer depends on x, and a simple parabolic profile for the velocity can be obtained for the case of laminar flow. This region is called the *fully developed regime*, and many microflows discussed in this book can be assumed to be fully developed because at low Reynolds numbers, the entrance region is very short.

4.3 Poiseuille flow in a pipe or channel

Consider the geometry of the channel depicted in Figure 4.3 for a channel so that r is replaced with y. The region of interest is the fully developed zone in a very wide channel so that the velocity does not depend on the streamwise variable x nor on the spanwise variable z because $W \gg h$ (see Figure 4.3). For the purpose of a physical interpretation of the relationship between pressure drop, flow rate, and viscosity, the dimensional form of the equations is used, and incompressible flow is assumed.

The Navier–Stokes equations reduce to

$$\mu \frac{\partial^2 u}{\partial y^2} = \frac{\partial p}{\partial x} - B_x \tag{4.5}$$

$$\frac{\partial p}{\partial y} - B_y = 0 \tag{4.6}$$

Note that the flow does not accelerate because the convective terms vanish identically. The velocity distribution in the channel is one-dimensional, and the flow is said to be hydrostatic in the y direction because a velocity does not appear in the equation ($v = 0$). In this section, the body force in the streamwise direction $B_x = 0$. Later, it will be seen that this body force will be nonzero for electroosmotic flow driven by an external electric field.

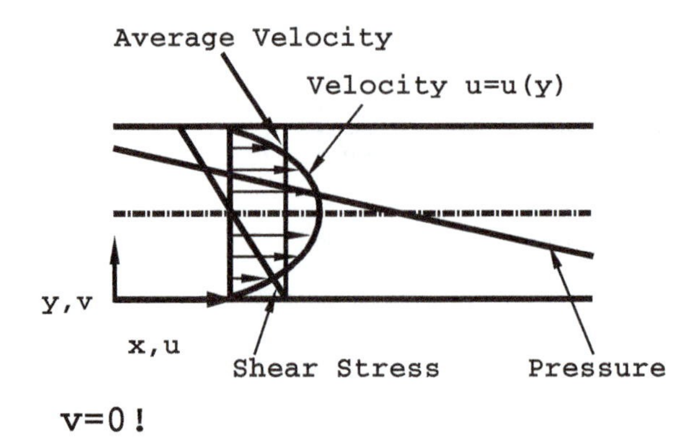

Figure 4.4 Velocity, shear stress, and pressure distribution in fully developed flow in a channel. Velocity and shear stress are plotted as functions of *y* and the pressure as a function of *x*.

If the body force term is conservative, as in the case of gravity, then $f_g = -g\nabla z$, and it can be absorbed into the pressure gradient term; thus, if the conservative body force is neglected, then $p = p(x)$. A function of x cannot be equal to a function of y unless the function is constant. Thus we can easily integrate the governing momentum equation

$$u = \frac{1}{2\mu}\frac{dp}{dx}y^2 + \frac{C_1}{\mu}y + C_2 \tag{4.7}$$

where C_1 and C_2 are constants. Applying the no-slip condition on the two walls $y = 0, h$, we can determine C_1 and $C_2 = 0$ so that

$$u = \frac{h^2}{2\mu}\frac{dp}{dx}\left[\left(\frac{y}{h}\right)^2 - \frac{y}{h}\right] \tag{4.8}$$

and the flow is thus in the direction of decreasing pressure. This *Poiseuille* flow velocity distribution is a simple parabolic distribution originally suggested by Jean L. M. Poiseuille (1799–1869), and thus the shear stress is linear:

$$\tau(y) = \mu\frac{\partial u}{\partial y} = \frac{dp}{dx}\left(y - \frac{1}{2}h\right) \tag{4.9}$$

The velocity shear stress and pressure gradient are depicted in Figure 4.4. Note that the pressure is linear: $p = p(x) = p_0 + (p_L - p_0)\,x/L$ and $dp/dx = (p_L - p_0)/L = -\Delta p/L = \text{constant}$.

Jean L. M. Poiseuille was a practicing physician who realized that "progress in physiology demands a knowledge of the laws of motion of the blood" (p. 87) Tokaty (1971). Thus one of the most famous velocity distributions in fluid mechanics was developed by a medical doctor.

The maximum velocity occurs at the centerline and has the magnitude

$$u_{\max} = -\frac{h^2}{8\mu}\frac{dp}{dx} \tag{4.10}$$

Table 4.1. Results for the calculation of the flow rate through a channel

h (μm)	D_H (μm)	$V(\frac{m}{sec})$	$Q(\frac{m^3}{sec})$	Re
1	2	8.3×10^{-5}	8.3×10^{-14}	1.7×10^{-4}
23	45	0.04	1.0×10^{-9}	2.0
45	86	0.17	7.6×10^{-9}	14.5
78	145	0.51	4.0×10^{-8}	73.0
100	182	0.83	8.3×10^{-8}	151.5

The volume flow rate in m^3/sec is

$$Q = \int_0^h uWdy = -\frac{h^3 W}{12\mu}\frac{dp}{dx} \tag{4.11}$$

and note that flow rate is inversely proportional to viscosity and proportional to h^3. In terms of the pressure drop,

$$\frac{Q}{W} = \frac{h^3 \Delta p}{12\mu L} \tag{4.12}$$

$$\Delta p = \frac{12\mu L Q}{W h^3} \tag{4.13}$$

If the flow rate is held fixed, the required pressure drop increases dramatically as $h \to 0$; the preceding formula for Δp is used to create Figure 1.5. In dimensionless form,

$$\frac{\Delta p}{\rho \frac{U^2}{2}} = \frac{24}{Re}\frac{L}{h} \tag{4.14}$$

where $f = 24/\mathrm{Re}$ is called the friction factor. The average velocity is $U = Q/A$, where $A = Wh$ is the cross-sectional area. As the channel height decreases, the required pressure drop to get a fixed flow rate increases dramatically. The average velocity is $U = Q/A$, where $A = Wh$ is the cross-sectional area. It should be noted that the Reynolds number in this formulation is based on the channel height h. In channels of different shape, a hydraulic diameter should be used.

As a numerical example, suppose the pressure drop $\Delta p = 10^4$ N/m^2 (0.1 atm) for water with the width $W = 10^{-3}$ m and length $L = 0.01$ m. The volume flow rates for various channel heights may easily be calculated; these values are shown in Table 4.1.

The corresponding equations for axisymmetric flow in a tube are (see Appendix B)

$$\frac{1}{r}\frac{\partial r u}{\partial r} + \frac{\partial w}{\partial x} = 0 \tag{4.15}$$

$$\mu \frac{1}{r}\frac{\partial}{\partial r}\left(r \frac{\partial w}{\partial r}\right) = \mu \left(\frac{\partial^2 w}{\partial r^2} + \frac{1}{r}\frac{\partial w}{\partial r}\right) = \frac{\partial p}{\partial x} - B_x \tag{4.16}$$

$$\frac{\partial p}{\partial r} - B_r = 0 \tag{4.17}$$

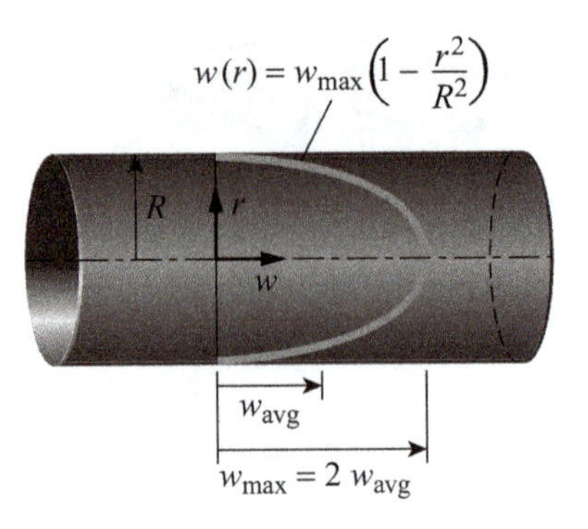

Figure 4.5 Geometry for flow in a cylindrical pipe or tube.

Equation (4.16) may be integrated twice, and apply the boundary condition that $\partial u/\partial r = 0$ at $r = 0$ to ensure that the velocity is finite at $r = 0$ and

$$w(r) = -\frac{R^2}{4\mu}\frac{dp}{dx}\left[1 - \left(\frac{r}{R}\right)^2\right] \tag{4.18}$$

where R is the radius of the tube and the volume flow rate is

$$Q = -\frac{\pi R^4}{8\mu}\frac{dp}{dx} \tag{4.19}$$

and the corresponding pressure drop is

$$\Delta p = \frac{128\mu L Q}{\pi D^4} \tag{4.20}$$

where D is the diameter, and in dimensionless form,

$$\frac{\Delta p}{\rho \frac{U^2}{2}} = \frac{64}{Re}\frac{L}{D}$$

Note that for both the channel and the tube (Figure 4.5), the average velocity scales on the pressure drop as

$$U = C\frac{dp}{dx}. \tag{4.21}$$

In practice, the constant C can be chosen to coincide with the average velocity or the maximum velocity in the channel or tube. For example, if the maximum velocity is chosen, for the channel,

$$C = -\frac{h^2}{4\mu} \tag{4.22}$$

and it is thus seen that $w_{\max} = 2w_{\text{ave}}$.

4.4 The velocity in slip flow

4.4.1 Gases

Fluid slip at a wall arises when the mean free path of the molecules is comparable to the geometrical length scale. When a gas becomes rarified, that is, when the mean free path of the molecules becomes of the same order as a given space dimension, then

$$Kn = \frac{\lambda}{h} = O(1) \tag{4.23}$$

Another way for the Knudsen number to become $O(1)$ is for the channel height h to become small at atmospheric conditions. The mean free path of air is around $\lambda \sim 60$–70 nm, depending on temperature.

Let us consider the case in which the Knudsen number is in the range such that the flow may be defined as continuum flow with slip. Let us also assume that the flow is low speed so that the density remains constant. Then the governing equation

$$\mu \frac{\partial^2 u}{\partial y^2} = \frac{\partial p}{\partial x} - b_x \tag{4.24}$$

but now with the slip condition given by equation (3.153). The solution in the absence of body forces is, for $\sigma_v = 1$,

$$u = \frac{h^2}{2\mu} \frac{dp}{dx} \left[\left(\frac{y}{h} \right)^2 - \frac{y}{h} - \frac{\lambda}{h} \right] \quad \text{for } \frac{\lambda}{h} \ll 1 \tag{4.25}$$

and note that the velocity distribution is merely offset by a fixed distance. Note that to leading order in y, the velocity vanishes at $y = -\lambda$, and this is the definition of *slip length*.

The shear stress is thus the same as for no slip, and

$$\tau(y) = \mu \frac{\partial u}{\partial y} = \frac{dp}{dx} \left(y - \frac{1}{2} h \right) \tag{4.26}$$

Note that the velocity is still a maximum at the centerline and that the pressure drop is linear. The volume flow rate is given by

$$Q = -\frac{W}{12\mu} \frac{dp}{dx} h^3 \left(1 + 6 \frac{\lambda}{h} \right) \tag{4.27}$$

Note here that for continuum approximation to hold, $Kn < 0.1$; above this value, the flow is in the transition regime, where noncontinuum methods must be used, as depicted in Figure 3.17.

4.4.2 Liquids

Using the boundary condition suggested by Thompson and Troian (1997) in equation (3.159), the velocity is

$$u = \frac{h^2}{2\mu} \frac{dp}{dx} \left[\left(\frac{y}{h} \right)^2 - \frac{y}{h} \right] - L_s^0 \left(1 + \dot{\gamma}_c^{-1} \frac{h}{2\mu} \frac{dp}{dx} \right)^{-1/2} \frac{h}{2\mu} \frac{dp}{dx} \tag{4.28}$$

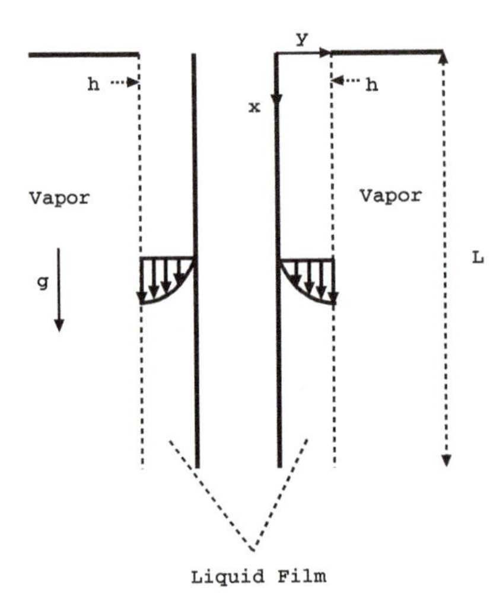

Figure 4.6 A thin liquid film falling under gravity.

The volume flow rate is easily calculated, as we have done before:

$$Q = -\frac{Wh^3}{12\mu}\frac{dp}{dx}\left(1 + 6L_s^0\left(1 + \frac{h^2}{2\mu}\frac{dp}{dx}\dot{\gamma}_c^{-1}\right)^{-1/2}\right) \tag{4.29}$$

Note that as in gases, the flow rate is increased by a constant amount.

The validity of the no-slip condition has been verified at macroscale. In liquids at microscale and below, recent experimental work has shown that slip can occur in liquids as a result of surface chemistry, especially at hydrophobic surfaces, as discussed in Section 3.11. It does need noting that the physical mechanisms for liquid slip are still a matter of some debate; an excellent discussion of the historical view of liquid slip is given by Karniadakis *et al.* (2005).

> **More on Slip!**
>
> According to Karniadakis *et al.* (2005), evidence of fluid slip between solid surfaces and water was found as early as the 1860s. However, other experiments refuted these results and Karniadakis *et al.* (2005) suggest that experiments by Whetham around 1890 led to the conclusion that the no-slip condition holds at hydrophillic surfaces. Note that this is long before Prandtl's experiments that took place around 1904. In fact, Day (1990), in his analysis of the validity of the no-slip condition, does not even mention Prandtl.

4.5 Flow in a thin film under gravity

In the same way we study Poiseuille flow, consider the case of a fluid running down a solid wall under gravity, as depicted in Figure 4.6. If the entire region is at constant pressure, then there will be no pressure gradient in the liquid film. For fully developed flow, the governing equation for the streamwise velocity is

$$\mu \frac{\partial^2 u}{\partial y^2} = -b_x \tag{4.30}$$

$$\frac{\partial p}{\partial y} - b_y = 0 \tag{4.31}$$

with $b_x = \rho g$ and $b_y = 0$. The boundary conditions are

$$u = 0 \ y = 0 \tag{4.32}$$

and at the free surface,

$$\frac{\partial u}{\partial y} = 0 \quad \text{at } y = h \tag{4.33}$$

As noted in the previous chapter, this second boundary condition indicates that the shear stress at the free surface is zero and comes from a stress balance at the interface in the absence of the effects of surface tension. It is an approximate condition valid when, typically, the vapor or gas has a dynamic viscosity much less than that of water and the film thickness is constant, rendering surface tension effects negligible.

A falling liquid film.

The solution for the velocity is obtained in the same way as for Poiseuille flow, and

$$u = \frac{\rho g h^2}{\mu}\left(\frac{y}{h} - \frac{1}{2}\left(\frac{y}{h}\right)^2\right) \tag{4.34}$$

and so the appropriate velocity scale is $U = \rho g h^2/\mu$. In dimensionless form,

$$u = Y - \frac{1}{2}Y^2 \tag{4.35}$$

where $Y = y/h$. The velocity is sketched in Figure 4.6. It is useful to note that if the film is thin compared to the diameter of a cylinder, the same solution holds in cylindrical coordinates.

The flow rate is given by

$$Q = W\int_0^h u\,dy = \frac{\rho g h^3 W}{3\mu} = \frac{1}{3}UhW \tag{4.36}$$

where W is the width of the film into the page. Note that the h^3 dependence seen in Poiseuille flow carries over to this film flow. If the film thickness is $h = 1$ μm, then the velocity scale for water is $U = 10^{-5}$ m/sec.

Care must be used when applying this very simple velocity distribution because it has been assumed that the film thickness h is a constant or nearly so. However, at the micro- or nanoscale, if the film surface is even slightly curved, surface tension may play a role in the fluid dynamics and destroy the one-dimensional nature of the flow profile.

4.6 The boundary layer on a flat plate

The function of a *boundary layer*, an example of an *external flow* that is unbounded on one side, is to bring the velocity to zero on a solid body. Boundary layers occur at high Reynolds numbers, and at first glance, it might not appear appropriate to discuss the Blasius boundary layer in this text on microfluidics, for which the focus is internal flow. However, given Prandtl's role in the demonstration of the validity of the no-slip condition, it is useful to present. Moreover, boundary layer–type behavior, in which the velocity varies rapidly in a very thin region, occurs in electro-osmotic flow for thin electric double layers, to be discussed in detail in Chapter 9. In addition, the entrance flow in a duct at high Reynolds number is essentially an external flow governed approximately by the boundary layer equations, as we have just seen.

Blasius, a student of Prandtl, was the first to solve the problem of laminar flow past a flat plate. The coordinate system is depicted in Figure 4.7. The Navier–Stokes equations in two-dimensional steady flow are given by

$$\frac{\partial u}{\partial x} + \frac{\partial v}{\partial y} = 0 \tag{4.37}$$

$$u\frac{\partial u}{\partial x} + v\frac{\partial u}{\partial y} = -\frac{\partial p}{\partial x} + \frac{1}{Re}\nabla^2 u \tag{4.38}$$

$$u\frac{\partial v}{\partial x} + v\frac{\partial v}{\partial y} = -\frac{\partial p}{\partial y} + \frac{1}{Re}\nabla^2 v \tag{4.39}$$

The pressure is scaled on the dynamic pressure, or

$$p = \frac{p^*}{\rho U_\infty^2} \tag{4.40}$$

and $Re = \rho U_\infty L/\mu$.

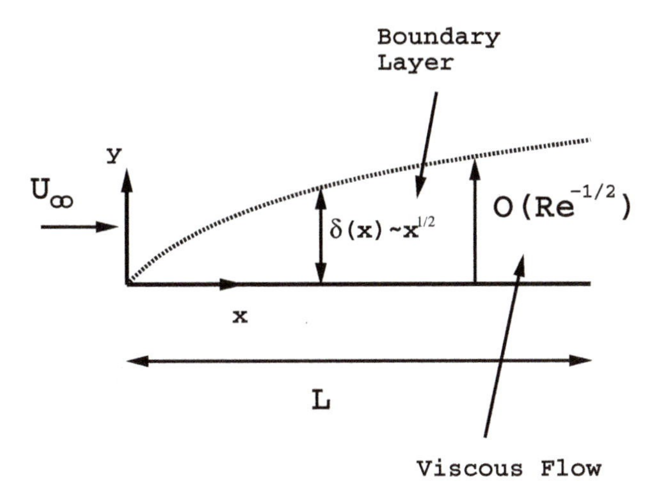

Figure 4.7 Coordinate system for a boundary layer on a flat plate.

The boundary layer is very thin, so let us see if we can simplify the Navier–Stokes equations. From the continuity equation, using dimensional analysis,

$$\frac{\partial u}{\partial x} \sim \frac{U_\infty}{L} \tag{4.41}$$

$$\frac{\partial v}{\partial y} \sim \frac{v}{\delta} \tag{4.42}$$

where δ, the boundary layer thickness, is the scale in the y direction. Thus the v velocity is

$$v \sim U_\infty \frac{\delta}{L} \tag{4.43}$$

and so because $\delta \ll L$, the velocity v is very small. In fact, because of this, the second momentum equation reduces to

$$\frac{\partial p}{\partial y} = 0 \tag{4.44}$$

that is, the boundary layer is so thin that the pressure does not vary across it.

By the same arguments, since $\delta \ll L$,

$$\frac{\partial^2}{\partial x^2} \ll \frac{\partial^2}{\partial y^2} \tag{4.45}$$

and so the streamwise momentum equation becomes

$$u\frac{\partial u}{\partial x} + v\frac{\partial u}{\partial y} = -\frac{\partial p}{\partial x} + \frac{1}{Re}\frac{\partial^2 u}{\partial y^2} \tag{4.46}$$

So now a reduced but still accurate system of equations has been developed, and in dimensionless form (see Section 3.12 for the mechanics of nondimensionalizing the equation) the boundary layer equations are given by

$$\frac{\partial u}{\partial x} + \frac{\partial v}{\partial y} = 0 \tag{4.47}$$

$$u\frac{\partial u}{\partial x} + v\frac{\partial u}{\partial y} = -\frac{\partial p}{\partial x} + \frac{1}{Re}\frac{\partial^2 u}{\partial y^2} \tag{4.48}$$

$$\frac{\partial p}{\partial y} = 0 \tag{4.49}$$

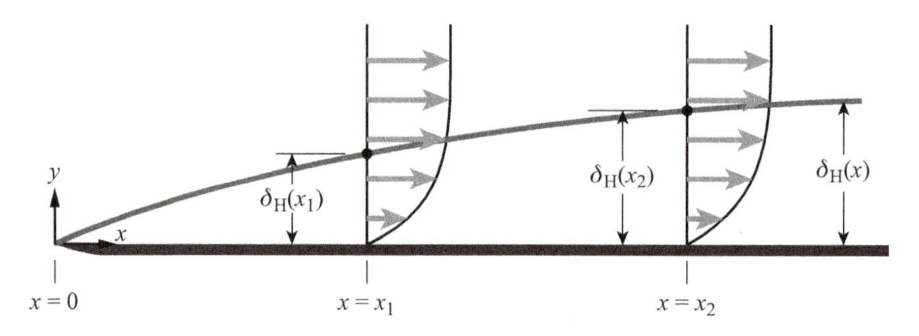

Sketch of the velocity distribution in a boundary layer on a flat plate.

These equations are much simpler than the two-dimensional Navier–Stokes equations. Since the pressure does not change across the boundary layer, its value inside the boundary layer must be the same as its value outside the boundary layer, where the flow is inviscid. Outside the boundary layer, the fluid acts as if it were inviscid, and in dimensional terms, the effective kinematic viscosity $\nu = 0$; in dimensionless terms, this means Re $= \infty$. Of course, the Reynolds number is never really ∞, it is just very large. In this case, the momentum equation reduces to

$$u\frac{\partial u}{\partial x} = -\frac{\partial p}{\partial x} = \frac{dp}{dx} \tag{4.50}$$

Integrating this equation between two points results in

$$p_1 + \rho\frac{U_{\infty,1}^2}{2} = p_2 + \rho\frac{U_{\infty,2}^2}{2} \tag{4.51}$$

for any two points 1, 2. Since U_∞ is constant for flow over a flat plate, $p_1 = p_2$ for any two points, and thus

$$\frac{dp}{dx} = 0 \tag{4.52}$$

This is true for any body that does not have curvature. For example, there will be a streamwise pressure gradient for flow past a sphere or cylinder. The *Blasius boundary layer equations* are the resulting equations:

$$\frac{\partial u}{\partial x} + \frac{\partial v}{\partial y} = 0 \tag{4.53}$$

$$u\frac{\partial u}{\partial x} + v\frac{\partial u}{\partial y} = \frac{1}{Re}\frac{\partial^2 u}{\partial y^2} \tag{4.54}$$

and now there are two equations in two unknowns (u, v) because the pressure has been removed from the problem; in the more general case, the pressure is known from the inviscid flow outside the boundary layer.

These equations are nonlinear, and the Reynolds number is large. Thus we will have a tough time solving these equations numerically because a very small parameter $1/\mathrm{Re}$ multiplies the highest-order term in the equation. This is an example of a *singular perturbation* problem, and a short discussion of this type of problem appears in Appendix A. To remedy this situation, let us scale out the Reynolds number by writing

$$\hat{y} = \frac{y}{Re^{-1/2}} \tag{4.55}$$

$$v = Re^{-1/2}\hat{v} + \cdots . \tag{4.56}$$

Then

$$\frac{\partial u}{\partial x} + \frac{\partial \hat{v}}{\partial \hat{y}} = 0 \tag{4.57}$$

$$u\frac{\partial u}{\partial x} + \hat{v}\frac{\partial u}{\partial \hat{y}} = \frac{\partial^2 u}{\partial \hat{y}^2} \tag{4.58}$$

Now far away from the plate, where $y \sim 1$, say ($y^* = L$, y^* dimensional) $\hat{y} \sim y/\mathrm{Re}^{-1/2} \gg 1$, is very large because the Reynolds number is large. Thus the boundary conditions are

$$u = v = 0 \quad \hat{y} = 0 \tag{4.59}$$

$$u \to 1 \quad \text{as } \hat{y} \to \infty \tag{4.60}$$

The first boundary condition is the no-slip and solid wall condition, and the second condition is a matching condition with the free stream. This last condition means that the *dimensionless velocity u* approaches the appropriate free stream value as $y \sim \delta$, where δ is the boundary layer thickness.

The boundary layer equations can be transformed into an ordinary differential equation that is much easier to solve. This is done through *similarity analysis*. Similarity solutions arise, in general, when there is no physical length scale associated with the problem. A fluid particle inside the boundary layer does not "see" the leading or trailing edge of the flat plate. In this case, define a similarity variable of the form $\eta = \hat{y}/\xi(x)$. Defining the fluid velocities in the form

$$u(\eta) = f' \tag{4.61}$$

and from the continuity equation

$$\hat{v} = -f\xi' + \eta\xi' f' \tag{4.62}$$

and substituting into equation (4.58) leads to

$$f''' + \xi\xi' f f'' = 0 \tag{4.63}$$

The requirement for similarity is that $\xi\xi' = C$, where C is a constant. This constant is arbitrary, and taking $C = 1$ leads to $\eta = \hat{y}/\sqrt{2x}$. This is now

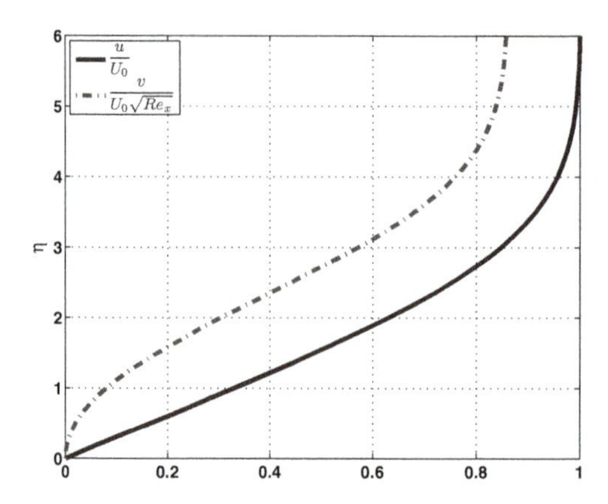

Figure 4.8 Solution for the velocity distribution in a Blasius boundary layer. The v velocity is scaled by the Reynolds number so that it fits on the same plot.

a third-order differential equation that can be solved numerically, subject to the boundary conditions

$$f'(0) = f(0) \quad = 0 \qquad (4.64)$$
$$f'(\infty) = 1$$

and now the original partial differential equation has been transformed into an ordinary differential equation. This makes the numerical solution a lot easier, and it takes mere seconds to solve on a desktop or laptop computer. We will discuss the methods to solve this problem numerically in Chapter 10.

The solutions for the velocities are shown in Figure 4.8. Here the velocity normal to the plate is scaled by the Reynolds number to fit on the same plot. Thus what is plotted is $\hat{v} = -f\xi' + \eta\xi' f'$. Note that both functions approach a constant as the boundary layer edge is approached.

The wall shear stress is calculated as

$$\tau_w(x) = \frac{du}{d\eta}\frac{\partial\eta}{\partial\hat{y}} \quad \text{at } \hat{y} = 0 \qquad (4.65)$$

and using the numerical solution to obtain the value of $f' = du/d\eta$ at the wall, in dimensionless form, the *skin friction coefficient* is given by

$$C_f = \frac{\tau_w}{\frac{1}{2}\rho U_\infty^2} = C_f = 0.664\, Re_x^{-1/2} \qquad (4.66)$$

where $\mathrm{Re}_x = U_\infty x/\nu$. The actual boundary layer thickness is defined as the point where the velocity profile reaches 99 percent of its free stream value, and the result is

$$\delta = 5x^* Re_{x^*}^{-1/2} \qquad (4.67)$$

Note that δ is dimensional; recall that $\mathrm{Re}_{x^*} \gg 1$, except very near the leading edge of the plate.

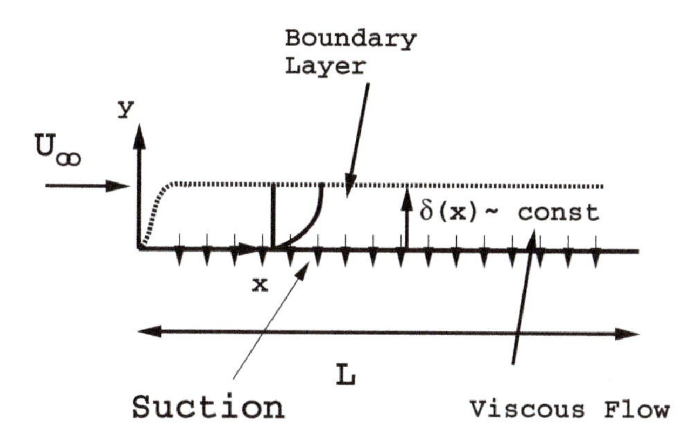

Figure 4.9 Boundary layer over a flat plate with suction.

The velocity u in Figure 4.8 looks almost parabolic, and a simple formula that approximates the numerical solution is

$$u^* = U_\infty(2\eta - \eta^2) \tag{4.68}$$

where now $\eta = y^*/\delta$. The coefficient for the skin friction coefficient has the value 0.730 instead of 0.664, and the coefficient for the boundary layer thickness has the value 5.477 instead of 5 – not bad for such a simple formula.

4.7 Fully developed suction flows

There is also a class of flows with porous walls for which analytical solutions are possible. Let us consider an external flow such as a simple boundary layer flow over a flat plate, as depicted in Figure 4.9. The fluid may be a liquid or a gas, as long as the velocity of the gas flow is much less than the speed of sound.

As with the boundary layer, the streamwise pressure gradient vanishes, and the streamwise momentum equation is assumed to be

$$\mu \frac{d^2u}{dy^2} = \rho v \frac{du}{dy} \tag{4.69}$$

The flow is assumed to be fully developed, which means

$$\frac{\partial u}{\partial x} = 0 \tag{4.70}$$

then $\partial v/\partial y = 0$ so that at most, $v =$ constant. For this balance to occur,

$$\mu \frac{U_\infty}{\delta^2} \sim \frac{\rho v U_\infty}{\delta} \tag{4.71}$$

where δ is the width of the layer. This means that

$$Re_L \frac{v\delta}{U_\infty L} \sim 1 \tag{4.72}$$

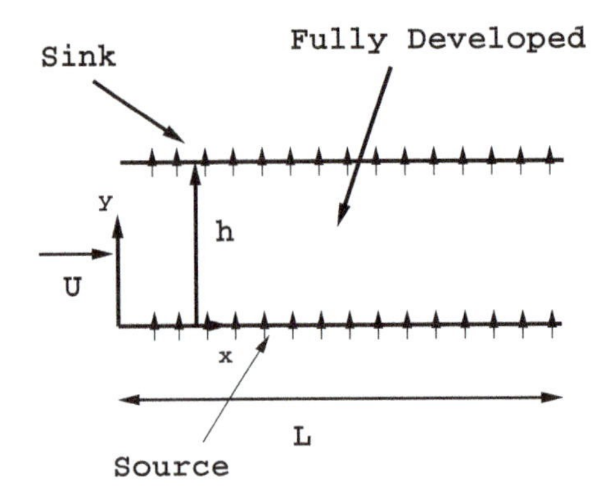

Figure 4.10 A duct flow with constant suction.

In addition, from the continuity equation, the only solution can be

$$v = v_w = \text{constant} \tag{4.73}$$

and thus the equation for the streamwise velocity is

$$\mu \frac{d^2u}{dy^2} = \rho v_w \frac{du}{dy} \tag{4.74}$$

The solution to this equation subject to the no-slip condition at the wall and the matching condition at the edge of the boundary layer is

$$u(y) = U_\infty \left[1 - e^{y v_w/\nu}\right] \tag{4.75}$$

It is clear that v_w must be negative and cannot be positive. The streamwise velocity is sketched in Figure 4.9. Similar to the Blasius boundary layer, the thickness of the layer δ is defined as the location where u reaches 99 percent of the value U_∞, that is, $u = .99U_\infty$, and this means

$$1 - e^{\delta v_w/\nu} = 0.99 \tag{4.76}$$

or

$$\delta = -4.6 \frac{\nu}{v_w} \tag{4.77}$$

For $U_\infty = 0.001$ m/sec in water, where the kinematic viscosity $\nu = \mu/\rho \sim 10^{-6}$ m^2/sec, a value of $v_w = -0.00001$ m/sec corresponds to $\delta \sim 0.05$ m. Conversely, in air, $\nu = 2 \times 10^{-5}$ and $\delta = 1$ m.

Now consider the fully developed flow between two parallel plates: the Poiseuille flow, only now with a source on one wall and a sink on the other, as depicted in Figure 4.10. Again, just as before, and for the same reason,

$$v = v_w = \text{constant} \tag{4.78}$$

and the governing equation for the streamwise velocity is

$$\mu \frac{d^2u}{dy^2} - \frac{dp}{dx} = \rho v_w \frac{du}{dy} \tag{4.79}$$

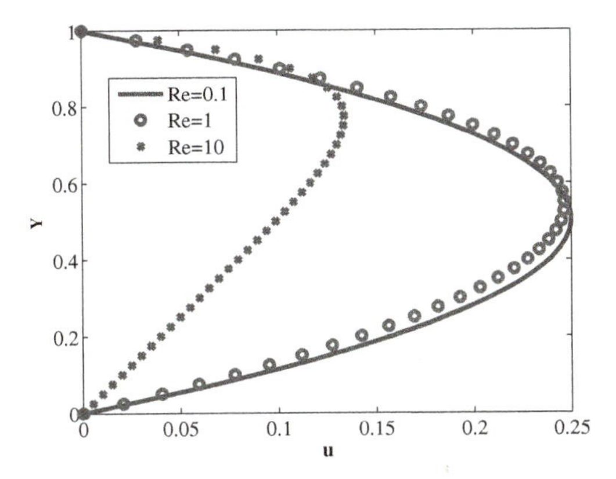

Figure 4.11 Velocity profile for several Reynolds numbers for the porous duct.

where we consider only pressure driven flow.

Recall that the solution for the Poiseuille flow is given by

$$u = \frac{h^2}{2\mu}\frac{dp}{dx}\left[\left(\frac{y}{h}\right)^2 - \frac{y}{h}\right] \tag{4.80}$$

It is clear from this equation that the velocity scale can be taken to be

$$U = -\frac{h^2}{2\mu}\frac{dp}{dx} \tag{4.81}$$

in m/sec if the pressure is given in newtons. The same velocity scale should be used in the present problem to match up the solutions as $v_w \to 0$. In dimensionless form,

$$\frac{d^2u}{dY^2} = -2 + Re\frac{du}{dY} \tag{4.82}$$

where Re $= \rho v_w h/\mu$ is the suction Reynolds number based on the wall velocity and $Y = y/h$. The solution is

$$u = A + Be^{ReY} + \frac{2}{Re}Y \tag{4.83}$$

The constants are found by applying the boundary conditions that $u = 0$ at $Y = 0, 1$ and

$$u = -\frac{2}{Re}\frac{1 - e^{ReY}}{1 - e^{Re}} + \frac{2}{Re}Y \tag{4.84}$$

For small suction Re, the solution approaches the Poiseuille flow, whereas for large Re, the velocity profile is linear over most of the duct and drops rapidly to zero near the upper wall; moreover, the maximum velocity no longer occurs at the centerline. In fact, singular perturbation methods can be used to show that the leading-order term in the boundary layer that forms near $Y = 1$ is given by

$$u_{BL} = \frac{2}{Re}(1 - e^{-\hat{y}}) \tag{4.85}$$

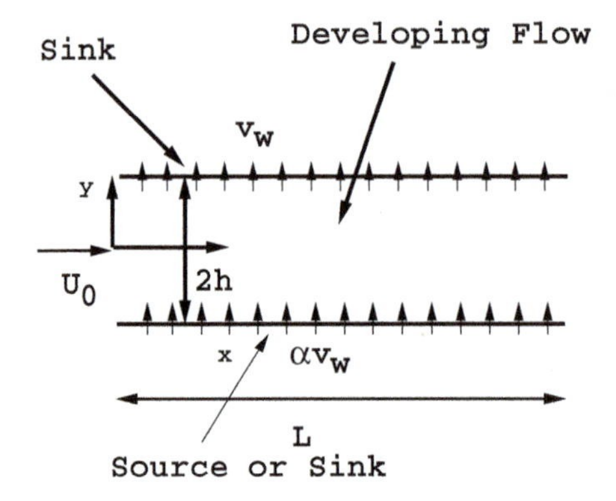

Figure 4.12 A developing suction flow in a duct. The magnitude and direction of suction–injection may be varied. The case $\alpha = -1$ indicates sinks on both walls.

where $\hat{y} = (1 - Y)Re$. This local solution valid near $Y = 1$ matches with the "outer solution" $u_{\text{outer}} = 2Y/Re$, valid in the region away from $Y = 1$. Note that as the suction velocity gets large, the axial velocity must get smaller to achieve mass flow conservation.

The dimensional volume flow rate is given by

$$Q = WhU \int_0^1 udY = WhU_0 \frac{2}{,} Re \left[\frac{1}{2} - \left(\frac{1 - \frac{1}{Re} e^{Re}}{1 - e^{Re}} + \frac{\frac{1}{Re}}{1 - e^{Re}} \right) \right] \tag{4.86}$$

and the average velocity is

$$U_{\text{ave}} = U \int_0^1 udY = U_0 \frac{2}{Re} \left[\frac{1}{2} - \left(\frac{1 - \frac{1}{Re} e^{Re}}{1 - e^{Re}} + \frac{\frac{1}{Re}}{1 - e^{Re}} \right) \right] \tag{4.87}$$

As a numerical example, for a dimensionless pressure gradient $dp/dx = -2$ at a suction Reynolds number $Re = 2$, the dimensional volume flow rate $Q = 1.56 \times 10^{-6}$ m^3/sec for a channel height of 0.1 mm and a channel width of $W = 1$ μm.

4.8 Developing suction flows

There is an interesting class of problems for which the suction–injection velocity is not unidirectional and fully developed, as depicted in Figure 4.12. These problems have been investigated in particular by Berman (1953), White *et al.* (1958), Terrill (1964), Terrill and Thomas (1969), Terrill and Shrestha (1965), and others for both channels and tubes and are termed *developing suction flows*. These types of flows occur in the kidney, as discussed in the box in Chapter 2.

For a channel, a mass balance between $x = 0$ and $x = x$ results in (Figure 4.12)

$$U_0 A_c = (1 - \alpha) v_w A + W \int_{-h}^{h} udy \tag{4.88}$$

where the wall velocity v_w has been assumed constant, $A_c = 2hW$ is the cross-sectional area, and $A = xW$. Here it proves useful to use the half height of the channel for the length scale, and U_0 is the (uniform) velocity at the inlet to the channel. Dividing by W in the two-dimensional case equation (4.88) becomes

$$\int_{-h}^{h} u dy = 2U_0 h - (1-\alpha)v_w x \tag{4.89}$$

so that the flow cannot be fully developed. However, limiting conditions on the magnitude of the suction velocity allow solutions and, as suggested by equation (4.89), define

$$\psi(x, y) = (2hU_0 - (1-\alpha)v_w x)\, f(y) \tag{4.90}$$

In dimensionless form, the stream function equation becomes

$$\psi(x, y) = \left(1 - \frac{1-\alpha}{2}\frac{v_w}{U_0}x\right) f(y) \tag{4.91}$$

where x and y are scaled on h and ψ has been scaled on $2U_0 h$.

The governing equations are the two-dimensional Navier–Stokes equations, and substituting equation (4.90) into the x momentum equation, in dimensional form,

$$\nu f''' - v_w(1-\alpha) f f'' + v_w(1-\alpha) f'^2 = \frac{1}{\rho\,(2hU_0 - (1-\alpha)v_w x)} p_x \tag{4.92}$$

where the differentiation is with respect to y. The dimensionless form of the equation should be (can you show this?)

$$f''' - Re_w \frac{1-\alpha}{2} f f'' + Re_w \frac{1-\alpha}{2} f'^2 = \frac{p_x}{1 - \frac{1-\alpha}{2}\frac{v_w}{U_0}x} = K \tag{4.93}$$

where K is a constant, $p_x = \partial p / \partial x$ is dimensionless, and $Re_w = 2v_w h/\nu$ is the suction–injection Reynolds number and x is scaled on h. The pressure is scaled on the viscosity so that the pressure is scaled by $\mu U_0 / 2h$. The dimensionless axial velocity scaled on U_0 is given by

$$u = \frac{\partial \psi}{\partial y} = \left(1 - \frac{1-\alpha}{2}\frac{v_w}{U_0}x\right) f' \tag{4.94}$$

$$v = -\frac{\partial \psi}{\partial x} = \frac{1-\alpha}{2}\frac{v_w}{U_0} f \tag{4.95}$$

Differentiating equation (4.93) shows that

$$\frac{\partial^2 p}{\partial x \partial y} = 0 \tag{4.96}$$

The boundary conditions are

$$f(1) = \frac{2}{1-\alpha}, \quad f(-1) = \frac{2\alpha}{1-\alpha} \tag{4.97}$$

$$f'(-1) = f'(1) = 0 \tag{4.98}$$

Note that $K = K(\mathrm{Re}_w)$ is unknown.

Consider first the solution to this equation for small suction–injection Reynolds number. For example, if $h = 1\ \mu\text{m}$, $v_w = 0.01$ m/sec, then $Re_w = 0.01$ for water, and the function f can be expanded in a power series, a *regular perturbation expansion*:

$$f = f_0 + Re_w f_1 + Re_w^2 f_2 + \cdots \tag{4.99}$$

Then, differentiating equation (4.93), the leading-order term in the series satisfies

$$f_0'''' = 0 \tag{4.100}$$

The solution is easily shown to be

$$f_0 = \frac{1+\alpha}{1-\alpha} + \frac{3}{2}y - \frac{1}{2}y^3 \tag{4.101}$$

and so the dimensionless stream function is given by

$$\psi_0 = \left(1 - \frac{1-\alpha}{2}\frac{v_w}{U_0}x\right)\left(\frac{1+\alpha}{1-\alpha} + \frac{3}{2}y - \frac{1}{2}y^3\right) \tag{4.102}$$

The corresponding velocities are

$$u_0 = \left(1 - \frac{1-\alpha}{2}\frac{v_w}{U_0}x\right)f' = \left(1 - \frac{1-\alpha}{2}\frac{v_w}{U_0}x\right)\left(\frac{3}{2} - \frac{3}{2}y^2\right) \tag{4.103}$$

$$v_0 = \frac{1-\alpha}{2}\frac{v_w}{U_0}\left(\frac{1+\alpha}{1-\alpha} + \frac{3}{2}y - \frac{1}{2}y^3\right) \tag{4.104}$$

The dimensionless pressure can easily be calculated from both momentum equations. From the x momentum equation,

$$p_x = K(Re_w)\left(1 - \frac{1-\alpha}{2}\frac{v_w}{U_0}x\right) \tag{4.105}$$

and, integrating,

$$p(x, y) = \frac{K}{Re_w}\left(x - \frac{1-\alpha}{2}\frac{v_w}{U_0}\frac{x^2}{2}\right) + p_r(y) \tag{4.106}$$

where

$$p_r(y) = \frac{v_w}{U_0}f' - \left(\frac{v_w}{U_0}\right)^2\frac{f^2}{2} \tag{4.107}$$

For the suction Reynolds number $\text{Re}_w = O(1)$, equation (4.93) is a nonlinear ordinary equation that may be solved numerically; note that the parameter K is unknown and must be determined in the course of the solution. The numerical techniques described in Chapter 10 may be employed.

Results are shown for two cases; in Figure 4.13 are the streamline patterns for two different Reynolds numbers for a sink on the upper wall. Note that the flow is always developing. In Figure 4.14 is the source flow that is symmetric about the centerline. As with both these figures, the wall velocity is assumed known; this, of course, is not true in practice, and v_w must be calculated from an analysis of the region outside the duct.

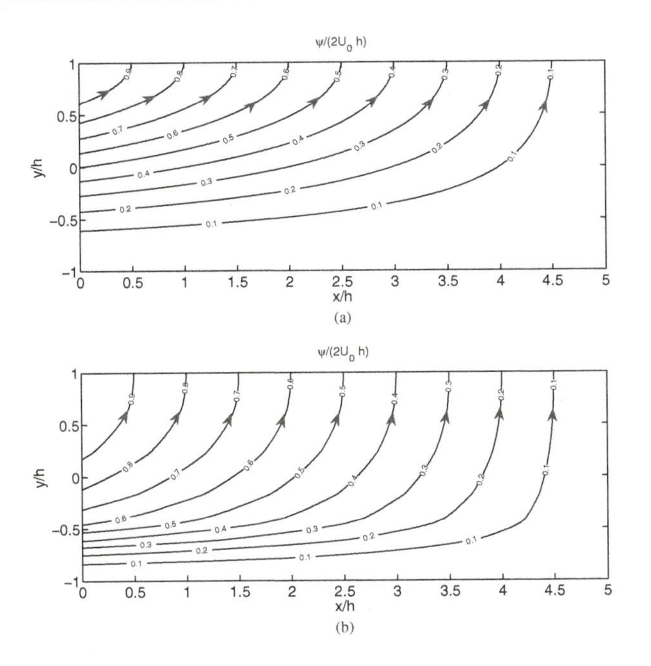

Figure 4.13 Streamlines for $\alpha = 0$ and $v_w/U_0 = 0.4$ for a sink at $y = 1$ and a solid wall at $y = -1$. (a) $Re_w = 0$. (b) $Re_w = 5$.

There are some anomalous results for this class of problems for channels, and these issues are discussed extensively by Drazin and Riley (2006). For example, multiple solutions have been obtained for $Re_w > 12.165$, with one solution exhibiting reduced velocity in the core region of the channel; reversed flow may also occur. There are also multiple solutions for the suction solution in a cylindrical tube (White, 2006), and there are also regimes in Reynolds number where no solutions exist! It is perhaps not surprising that there would be limitations on

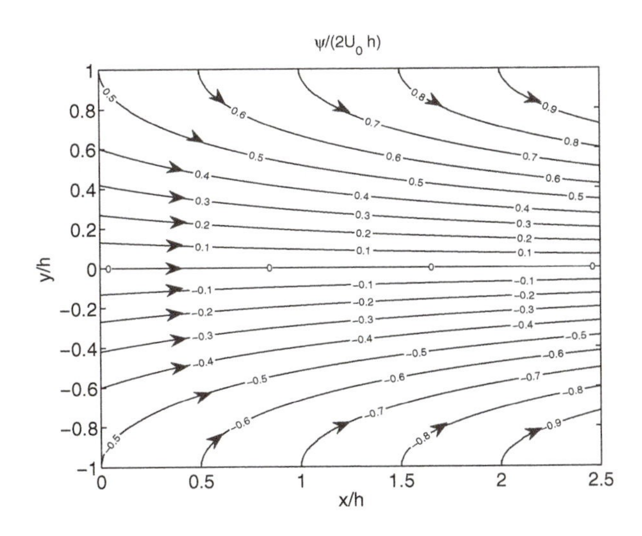

Figure 4.14 Streamlines for sources on both walls $\alpha = -1$ and $v_w/U_0 = -0.2$, $Re_w = -5$.

the source and sink magnitudes, although the precise Reynolds numbers at which the limits occur are not obvious.

The suction velocity at the wall has been assumed to be known, but actually, it must be calculated from the properties of the porous medium. A porous medium is characterized by two primary parameters: porosity and permeability. The porosity is simply the ratio of the void space where fluid can flow to the total volume of the material and is a property of the porous medium itself. The permeability is a measure of how easy it is for fluid to pass through the porous medium. The permeability is defined in one dimension by Darcy's law:

$$Q_p = -\frac{K_0 A}{\mu}\frac{dp}{dx} \tag{4.108}$$

where K_0 is the permeability, and is usually determined empirically, and A is the area. Note that the permeability has units of velocity.

Of course, the analysis of porous media is very important in hydrologic applications and in soil science. In the present context, the nanopore membranes depicted in Figure 4.1 can be considered porous media. The porosity of many membranes of this type is estimated to be about 1% for pores ranging from 10–100 nm in the nanoscale dimension and the ratio $\frac{K_0}{L} \sim \frac{10\,\mathrm{mL}}{\mathrm{min\,mmHg\,m^2}}$ (Fissell and Humes, 2006). The streamwise length of the pores is $L = 5\ \mu\mathrm{m}$, the spanwise width is $W = 45\ \mu\mathrm{m}$, and the overall area of these membranes is typically $A \sim 1\ \mathrm{mm}^2$.

4.9 The lubrication approximation

The classical definition of the term *lubrication* is the art of reducing frictional resistance between two surfaces by placing a liquid there. In classical lubrication problems, the convective terms in the governing equations do not vanish; however, they are small enough that they can be neglected. This is often the case in micro- and nanoflows, whether the variation comes from a slow variation in roughness of the surfaces or slight changes in the surface charge density in electro-osmotic flow.

In recent years, the term *lubrication approximation* has come to mean any flow where the dominant (but not exclusive, as in fully developed flow in a pipe or channel) variation in the velocity or other variable is in the cross-streamwise direction, the y direction. In lubrication problems, the streamwise velocity will vary slowly in x and so a small velocity component must be present in the y direction. In this section, dimensional variables will be considered initially, and then important expressions will be nondimensionalized near the end of the section.

Notation alert: As has been mentioned, an asterisk is often used to denote a dimensional variable in this text. To do that in this section would be particularly awkward, and so several times in this section, we will use the same symbol for both a dimensional variable and a dimensionless variable. The dimensionless versions are indicated by the presence of dimensionless parameters, while the

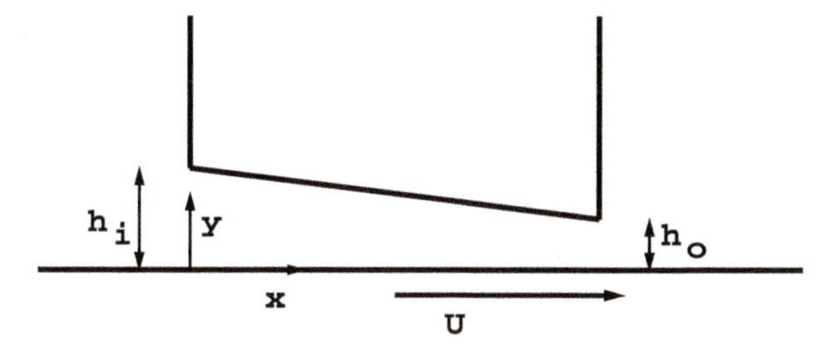

Figure 4.15 A slider bearing to illustrate the lubrication approximation. In recent years, the lubrication approximation has taken on a larger meaning in the sense that it applies to flows having a slight variation in the streamwise direction, no matter the cause.

dimensional form is indicated by the presence of a physical variable such as viscosity.

Consider a typical slider bearing, which is depicted in Figure 4.15. Dimensions may be, for example, for lubricating oil, $h_i = 1$ μm, $L = 1$ cm, $U = 0.25$ m/sec, $\mu = 0.10$ Nsec/m^2, $\rho = 1000$ kg/m^3, where the properties are those of a 10 W oil at $T = 20$°C. The ratio of convective terms to the viscous terms is

$$\frac{\rho u \frac{\partial u}{\partial x}}{\mu \frac{\partial^2 u}{\partial y^2}} \simeq \frac{\rho \frac{UU}{L}}{\mu \frac{U}{h^2}} = \frac{\rho U h}{\mu} \frac{h}{L} = Re_h \frac{h}{L} \tag{4.109}$$

and $Re_1 = Re_h(h/L) \sim 2.5 \times 10^{-5}$. Thus, even though the convective terms are not identically zero, they are far smaller than the viscous terms, and so to leading order, they may be neglected. For example, these same equations apply to the study of small-amplitude periodic roughness in a channel or tube, provided that the roughness wave length is much greater than the amplitude of the roughness.

Thus, if $Re_1 \ll 1$, then the convective terms may be neglected, even if $v \neq 0$ and $u = u(x, y)$. The only restriction is that the v velocity should be small. Indeed, from the continuity equation,

$$v \sim \frac{h}{L} U \ll U$$

since $h/L \ll 1$.

The governing equations are the same as the Poiseuille equations with the boundary conditions

$$u = U, \quad v = 0, \quad y = 0$$

$$u = 0, \quad y = h$$

$$u = v = 0, \quad y = h$$

This is a superposition of Poiseuille and Couette flow except that $h = h(x)$ and the solution for u is

$$u = \frac{1}{2\mu} \frac{dp}{dx} (y^2 - h(x)y) + U \left(1 - \frac{y}{h(x)}\right) \tag{4.110}$$

The difference between this problem and the classical Poiseuille and Couette problems is that $v \neq 0$ since $h = h(x)$ and the pressure gradient is unkown.

To obtain the pressure gradient, note that integration of the continuity equation leads to

$$v = -\int_0^y \frac{\partial u}{\partial x} dy \tag{4.111}$$

so that

$$v(x, h) = -\int_0^h \frac{\partial u}{\partial x} dy$$

To evaluate the integral the Leibnitz rule is used

$$\int_{F_1(x)}^{F_2(x)} \frac{\partial}{\partial x} f(x, y) dy$$
$$= \frac{d}{dx} \int_{F_1(x)}^{F_2(x)} f(x, y) dy + F_1'(x) f(x, F(x)) - F_2' f(x, F_2(x))$$

Here $F_1 = 0$, $F_2 = h(x)$, $f = u(x, y)$. But $F_1' = 0$, $u(x, h(x)) = 0$ so that both the last two terms drop out. Thus since $v(x, h) = 0$ the pressure must satisfy

$$\frac{d}{dx} \int_0^h \left(\frac{1}{2\mu} \frac{dp}{dx} (y^2 - hy) + U \left(1 - \frac{y}{h} \right) \right) dy = 0 \tag{4.112}$$

or

$$\frac{d}{dx} \left(\frac{dp}{dx} h^3 \right) - 6\mu U \frac{dh}{dx} = 0 \tag{4.113}$$

and this is the one-dimensional *Reynolds equation*, named after Osborne Reynolds, the same Reynolds as in the Reynolds number.

For a linear variation of the slider block,

$$h(x) = h_i + (h_o - h_i) \frac{x}{L} \tag{4.114}$$

an analytical solution may be obtained. To show this, one integration of the Reynolds equation yields

$$\frac{dp}{dx} h^3 = 6\mu U (h - h_m) \tag{4.115}$$

where h_m is the value of $h(x)$ at which $dp/dx = 0$, assuming such a point exists. Integrating again,

$$p(x) = 6\mu U \int_0^x \left(\frac{1}{h^2} - \frac{h_m}{h^3} \right) dx \tag{4.116}$$

because $dh/dx = \text{const}$; then

$$p = \frac{6\mu U}{\frac{dh}{dx}} \int_0^x \left(\frac{1}{h^2} - \frac{h_m}{h^3} \right) dh$$
$$= \frac{-6\mu U L}{h_o - h_i} \left(\frac{1}{h} - \frac{h_m}{2h^2} \right) + A \tag{4.117}$$

where A is an integration constant. Now h_m and A are constants that must be determined by applying the boundary conditions; the most common boundary conditions on the pressure in lubrication flows are $p = 0$ at $x = 0, L$ and at $x = 0, h = h_i$. Thus, at $x = 0$,

$$0 = \frac{-6\mu UL}{h_o - h_i}\left(\frac{1}{h_i} - \frac{h_m}{2h_i^2}\right) + A \tag{4.118}$$

and at $x = L$,

$$0 = \frac{-6\mu UL}{h_o - h_i}\left(\frac{1}{h_o} - \frac{h_m}{2h_o^2}\right) + A \tag{4.119}$$

Solving this system of equations leads to

$$p(x) = \frac{6\mu UL}{h^2\left(h_o^2 - h_i^2\right)}(h - h_i)(h - h_o) \tag{4.120}$$

and the maximum pressure is $p_m = p(h_m)$, where

$$h_m = 2\frac{\frac{1}{h_o} - \frac{1}{h_i}}{\frac{1}{h_o^2} - \frac{1}{h_i^2}} = \frac{2h_o h_i}{h_i + h_o}$$

The normalized height expression based on h_i is given by

$$h(x) = 1 + mx \tag{4.121}$$

where here $m = h_o - h_i/h_i$, and the axial coordinate is scaled on L. Thus the dimensionless pressure is given by

$$p(x) = 6\frac{(h-1)(h-h_o)}{h^2\left(h_o^2 - 1\right)} \tag{4.122}$$

where the pressure is scaled as

$$p = \frac{p^*}{\frac{\mu U}{h_i}\frac{L}{h_i}}$$

and p^* is dimensional.

Results for the pressure distribution are shown in Figure 4.16. The pressure increases rapidly as the slope increases in magnitude and the point of maximum

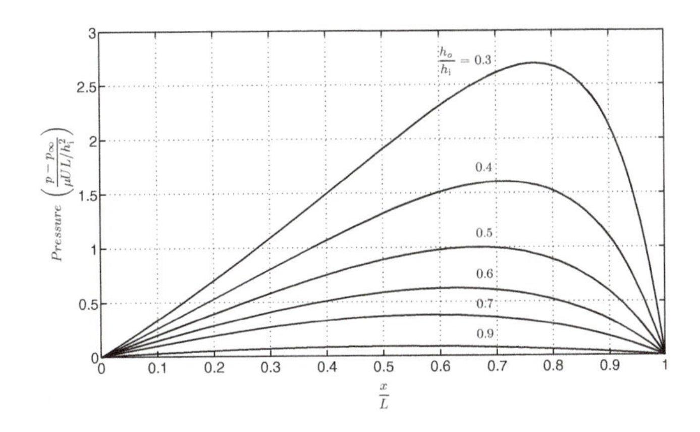

Figure 4.16 Dimensionless pressure as a function of x for a linear variation of the slot. Note that the pressure increases rapidly as the slope increases in magnitude.

pressure moves down the slot. A dimensionless pressure of $p = 2$ corresponds to a dimensional pressure of 5×10^{10} N/m^2 for an oil having a viscosity of 0.1 Nsec/m^2 for the preceding parameters. A large pressure indeed!

The dimensional volume flow rate through the slot is given by

$$Q = \int_0^h u dy = -\frac{1}{12}\frac{dp}{dx}h^3 + \frac{Uh}{2} \tag{4.123}$$

and solving for the pressure gradient and integrating from 0 to L, the volume flow rate is given by

$$Q = \frac{h_i h_o U}{h_i + h_o} \tag{4.124}$$

per width of the bearing. As an example, for the parameters described earlier, the flow rate $Q = 8.3 \times 10^{-7}$ m^3/sec and $h_m/h_i = 2/3$; that is, the maximum in the pressure occurs when the height has decreased by 2/3 and the x location is $x/L = 2/3$. Since the pressure in the gap $p > 0$, there is a net force upward on the surface to support the load.

Note that the solution for the velocity given by equation (4.110) is valid for any variation of $y = h(x)$, as long as the surface variation satisfies the lubrication approximation. Thus the solution is valid for a dimensional variation of

$$h = h_0 + \alpha \sin \omega x \tag{4.125}$$

where $\alpha \ll h_0$ and $\alpha\omega \ll 1$, a formula that corresponds to periodic roughness, as noted earlier.

4.10 A surface tension–driven flow

At small scales, capillary forces can be significant because the pressure force exerted on a bubble is proportional to the inverse of its radius of curvature. Bubbles in microchannels can interfere with the process of filling the channel with liquid. Suppose a bubble is moving steadily in a tube at constant velocity U, small enough that Stokes flow applies. In a reference frame attached to the bubble, the bubble is stationary, and the tube is moving at velocity $-U$ in the opposite direction, as shown in Figure 4.17. Note that the bubble is axisymmetric within the tube and that the gap $(H(x))$ between the bubble and the tube wall is assumed to be very small. The interface can be divided into several regions; the front and rear boundaries of the bubble enclose a region in which the standard lubrication limit can be used. Within this region are two subregimes: one in which the gap between the bubble and the wall is constant and equal to δ and the transition regime from the regions near the bubble ends to the tube.

In the transition regime at the front or rear end, assuming the gap $H \ll a$, Cartesian coordinates can be used, and the governing equation is given by

$$\frac{\partial^2 u}{\partial y^2} = \frac{1}{\mu}\frac{\partial p}{\partial x} \tag{4.126}$$

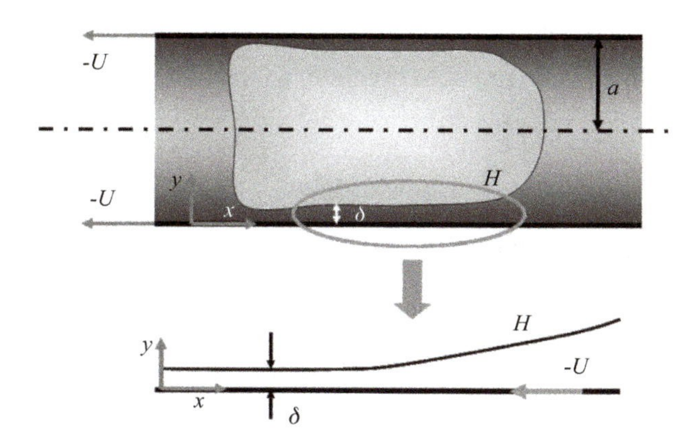

Figure 4.17 Sketch of a bubble in a tube showing the change in coordinate system for which the flow in the gap is steady. The gap between the bubble and the channel walls is H. The coordinate system is fixed to the bubble so that relative to the bubble, the tube is moving at $-U$, where $+U$ is the speed of the moving bubble.

At the channel wall, the velocity boundary condition is $u = -U$ at $y = 0$, and at the gas–liquid interface, the tangential stress at the interface is zero, which gives

$$\frac{\partial u}{\partial y} = 0 \quad \text{at } y = H(x) \tag{4.127}$$

The normal stress condition at the interface is given by the Young–Laplace equation (Bretherton, 1961):

$$p_1 + \frac{\sigma}{a} = -2\mu \left(1 + \frac{H_{xx}}{\left(1 + H_x^2\right)^2}\right) \frac{\partial u}{\partial x} \tag{4.128}$$

where σ is the interfacial tension. Bretherton (1961) shows that the pressure can be approximated by

$$p_1 \approx -\sigma \frac{d^2 H}{dx^2} - \frac{\sigma}{a} \tag{4.129}$$

On the basis of equation (4.126) and the boundary conditions, the solution for the streamwise velocity, assuming that the lubrication approximation holds, is

$$u = \frac{1}{\mu} \frac{dp}{dx} \left(\frac{y^2}{2} - Hy\right) - U \tag{4.130}$$

in a coordinate system traveling with the bubble. Substituting equation (4.128) into equation (4.130), the velocity profile is given by

$$u = -U - \frac{\sigma}{\mu} \frac{d^3 H}{dx^3} \left(\frac{y^2}{2} - Hy\right) \tag{4.131}$$

and so the volume flow rate through through the gap at the rear of the bubble is given by

$$Q = -2\pi a \int_0^{H(x)} u dy = 2\pi a \left(UH - \frac{\sigma H^3}{3\mu} \frac{d^3 H}{dx^3}\right) \tag{4.132}$$

In the region under the bubble, the thickness of the liquid film is a constant $H(x) = \delta$, and there is no pressure gradient, so the flow velocity across the film becomes

$$u = -U \tag{4.133}$$

and the corresponding volume flow rate is

$$Q = -2\pi a \delta U \tag{4.134}$$

From mass conservation, the two flow rates must be equal, and this condition leads to

$$H^3 \frac{d^3 H}{dx^3} - CaH = -Ca\delta \tag{4.135}$$

where $Ca = 3\mu U/\sigma$ is the capillary number and, in this analysis, is assumed small. Defining new dimensionless variables

$$\eta = \frac{H}{\delta} \quad \xi = \frac{x}{\delta}\left(\frac{3\mu U}{\sigma}\right)^{1/3} = Ca^{1/3}\frac{x}{\delta} \tag{4.136}$$

the equation in dimensionless form becomes

$$\eta^3 \frac{d^3\eta}{d\xi^3} = \eta - 1 \tag{4.137}$$

This equation can be solved numerically or by the method of matched asymptotic expansions, as is done by Wilson (1982) for the drag-out problem of film theory.

Note that the thickness of the layer between the bubble and the tube wall δ is unknown and must be calculated along with the solution. Three conditions on the function η are also required. These conditions are discussed in detail by Probstein (1989) based on physical arguments originally given by Landau and Levich (1942). In particular, under the bubble in the lubrication region, $H(x) = \delta =$ constant; near the front and rear ends of the bubble, $\eta \to \infty$, and in this regime, equation (4.137) indicates that $d^3\eta/d\xi^3 = 0$, and so

$$\eta \simeq A\xi^2 + B\xi + C \tag{4.138}$$

where the constants A, B, and C are obtained from the boundary conditions. In particular, from a numerical solution, Bretherton (1961) gives the result

$$\frac{\delta}{a} = 0.643 Ca^{2/3} \tag{4.139}$$

leading to a parabolic profile for the layer between the bubble and the tube:

$$\eta = 0.3215 Ca^{2/3} \frac{x^2}{\delta^2} \tag{4.140}$$

This solution is valid in the transition regime, where both $\eta \gg 1$ and $H/a \ll 1$.

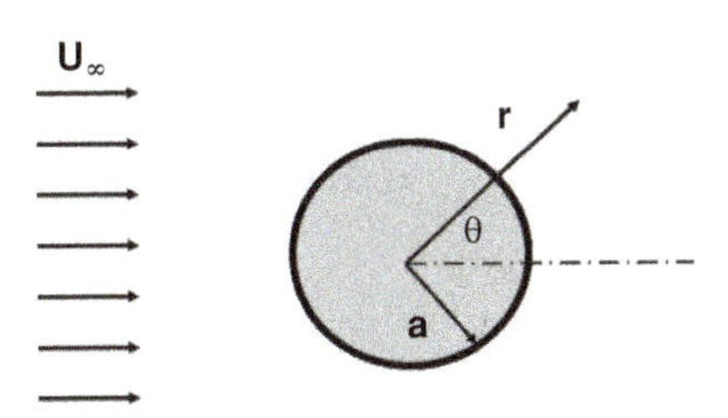

Figure 4.18 Coordinate system for Stokes flow past a sphere.

4.11 Stokes flow past a sphere

Many problems in micro- and nanofluidics involve flow past a sphere as depicted on Figure 4.18. Einstein's derivation of the diffusion coefficient, or, more precisely, the molecular dimension of a solute (Einstein, 1905b), is based on the drag for flow past a sphere derived by Stokes. Assume that the Reynolds number $Re = \rho U_\infty a/\mu \sim 0$ is small, where U is the relative velocity between the sphere and the velocity far from it, and assume that the flow is axisymmetric about the transverse axis and is thus independent of the transverse angle ϕ. The governing equations are thus two-dimensional and are given in dimensionless form by

$$\mathrm{Re}\left(\nabla\left(\frac{|\vec{V}|^2}{2}\right) - \vec{V}\times\nabla\times\vec{V}\right) = -\nabla p - \nu\nabla\times\nabla\times\vec{V} \tag{4.141}$$

For Re = 0, these equations reduce to

$$0 = -\nabla p - \nu\nabla\times\nabla\times\vec{V} \tag{4.142}$$

where here $\vec{V} = (u_r, u_\theta, 0)$. This is the so-called *Stokes flow* past a sphere.

It is simplest to determine the stream function first and then determine the velocity components. The continuity equation in spherical coordinates is given by

$$\frac{1}{r^2}\frac{\partial}{\partial r}(r^2 u_r) + \frac{1}{r\sin\theta}\frac{\partial}{\partial\theta}(u_\theta \sin\theta) = 0 \tag{4.143}$$

and so the stream function can be defined as

$$u_r = \frac{1}{r^2\sin\theta}\frac{\partial\psi}{\partial\theta} \tag{4.144}$$

$$u_\theta = -\frac{1}{r\sin\theta}\frac{\partial\psi}{\partial r} \tag{4.145}$$

Eliminating the pressure by cross-differentiating the momentum equations, the stream function then satisfies

$$\left[\frac{\partial^2}{\partial r^2} + \frac{\sin\theta}{r^2}\frac{\partial}{\partial\theta}\left(\frac{1}{\sin\theta}\frac{\partial}{\partial\theta}\right)\right]^2 \psi = 0 \tag{4.146}$$

The boundary conditions are from no slip,

$$\frac{\partial \psi}{\partial r}(r=1,\theta)=\frac{\partial \psi}{\partial \theta}(r=1,\theta)=0 \tag{4.147}$$

and far from the sphere,

$$\psi \to \frac{1}{2}r^2 \sin^2\theta \tag{4.148}$$

because the velocities far from the sphere are $u_r \sim \cos\theta$ and $u_\theta \sim -\sin\theta$.

As suggested by the far-field boundary condition, let us try a solution of the form

$$\psi(r,\theta)=\left(Ar^4+Br^2+Cr+\frac{D}{r}\right)\sin^2\theta \tag{4.149}$$

The condition $u_r=0$ at $r=a$ requires, in dimensionless form,

$$A+B+C+D=0 \tag{4.150}$$

and for $u_\theta=0$ at $r=a$,

$$4A+2B+C-D=0 \tag{4.151}$$

The far-field condition requires immediately that $B=1/2$ and $A=0$. Thus $C=-3/4$ and $D=1/4$ so that, in dimensionless form,

$$\psi(r,\theta)=\frac{1}{4}\left(2r^2+\frac{1}{r}-3r\right)\sin^2\theta \tag{4.152}$$

The dimensionless velocity components are given by the definition of the stream function as

$$u_r=\frac{1}{2r^2}\left(2r^2+\frac{1}{r}-3r\right)\cos\theta \tag{4.153}$$

$$u_\theta=-\frac{1}{4r}\left(4r-\frac{1}{r^2}-3\right)\sin\theta \tag{4.154}$$

Note that the velocities are everywhere below the free stream values.

The streamlines are shown in Figure 4.19. Note that they are symmetric about $\theta=90°$, and the streamlines are remarkably similar to the potential flow past a

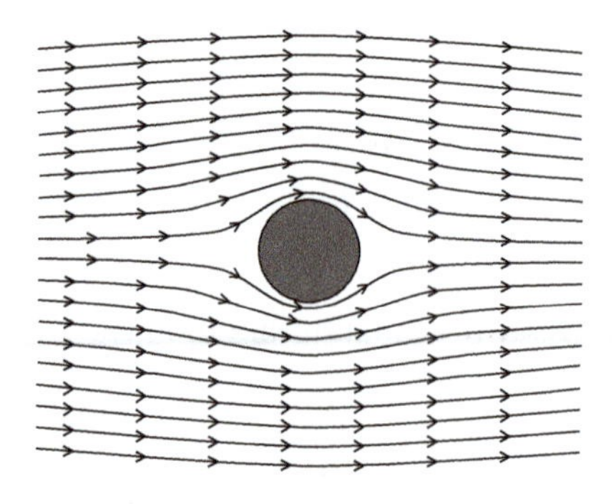

Figure 4.19 Streamlines for Stokes flow past a sphere.

sphere. However, a closer look reveals that the velocities are significantly different (White, 2006). The pressure is given by

$$p = p_\infty - \frac{3}{2}\mu U_\infty a \frac{\cos\theta}{r^2} \tag{4.155}$$

At low Reynolds number, the pressure should scale with the viscosity so that in dimensionless form (as in the developing suction and lubrication flows),

$$\frac{p - p_\infty}{\mu U_\infty / a} = -\frac{3}{2}\frac{\cos\theta}{r^2} \tag{4.156}$$

At low Reynolds number, the drag is dominated by viscous effects, as opposed to form drag, which is dominant at high Reynolds number. The uniform flow tends to push the sphere to the right and is thus positive, and the drag may be obtained by integrating the stress components in the direction of motion. It is the shear stress acting on the radial plane (r) in the positive θ direction that induces the viscous component of the drag, and this component is, in dimensional form,

$$\tau_{r\theta} = \mu r \frac{\partial}{\partial r}\left(\frac{u_\theta}{r}\right) + \frac{\mu}{r}\frac{\partial u_r}{\partial \theta} = -\frac{3\mu U_\infty a^3}{2r^4}\sin\theta \tag{4.157}$$

The pressure also induces a stress, and this component normal to the sphere is

$$\tau_{rr} = -p + 2\mu\frac{\partial u_r}{\partial r} \tag{4.158}$$

The overall drag is calculated from

$$D = -\int_0^\pi \tau_{r\theta}\sin\theta dA - \int_0^\pi p\cos\theta dA \tag{4.159}$$

where $dA = 2\pi a^2 \sin\theta d\theta$ is the area element of integration. Performing the integration, the result is

$$D = 4\pi\mu a U_\infty + 2\pi\mu a U_\infty = 6\pi\mu a U_\infty \tag{4.160}$$

This formula is the basis for the Stokes–Einstein equation used to estimate the diffusion coefficient of a given species in a liquid mixture. If the sphere is moving at speed U_s in a mean streaming motion of speed U_∞, the drag is given by

$$D = 6\pi\mu a(U_\infty - U_s) \tag{4.161}$$

As noted long ago, there is a problem with this solution for small but non-zero Reynolds number. To see this, note that the order of magnitude of a typical convective term is U_∞^2/ar^2 and that the order of magnitude of a typical viscous term is $\sim \nu U_\infty/a^2r^2$ so that these terms are the same order of magnitude when $\mathrm{Re}(r/a) \sim 1$. Thus, far from the sphere, the solution will always break down. This fact was noted by Oseen (1910), who suggested that the governing equations be retained in the form

$$Re\vec{U}_\infty \bullet \nabla\vec{V} = -\nabla p + \nabla^2\vec{V} \tag{4.162}$$

where now $\vec{U}_\infty$ is the velocity field far from the sphere.[3] It can be seen from the preceding discussion that there are two regions: near the sphere, where $r \sim 1$ and the convective terms are negligible, and far from the sphere, where $rRe \sim 1$. The flow far from the sphere requires a correction of $O(Re)$, and the result for the stream function was first given by Oseen (1910); further terms in the asymptotic expansion for small Reynolds number are given by Proudman and Pearson (1957). The Oseen (1910) result for the first order Reynolds number correction is

$$\psi(r,\theta) = \frac{1}{4}\left(2r^2 + \frac{1}{r}\right)\sin^2\theta - \frac{3}{2Re}(1+\cos\theta)\left(1 - e^{-\frac{1}{2}Rer(1-\cos\theta)}\right) \tag{4.163}$$

This leads to a correction to the dimensionless drag coefficient in the form of

$$c_D = \frac{D}{\rho U_\infty^2 a^2} = \frac{6\pi}{Re}\left(1 + \frac{3}{8}Re\right) \tag{4.164}$$

4.12 Sedimentation of a solid particle

Sedimentation is the process by which particles in a solution settle under gravity or another force field such as a centrifugal force. The term *sedimentation* is also used to describe the changes in concentration of a solute caused in part by a gravitational field. In this case, the drag force on the particle must balance the net weight of the particle. For a spherical particle, using Stokes's drag law, the force balance in a stagnant fluid becomes

$$6\pi\mu aU = \frac{4}{3}\pi a^3(\rho_p - \rho) \tag{4.165}$$

so that the sedimentation velocity is given by

$$U = \frac{2}{9}\frac{a^2}{\nu}\left(\frac{\rho_p}{\rho} - 1\right)g \tag{4.166}$$

The particle Reynolds number is thus

$$Re = \frac{\rho Ua}{\mu} = \frac{2}{9}\frac{a^3}{\nu^2}\left(\frac{\rho_p}{\rho} - 1\right)g \tag{4.167}$$

For small particles apart from the very large particle density limit, the particle Reynolds number will be much less than 1.

[3] It is hard to express the Oseen correction equations in the invariant form.

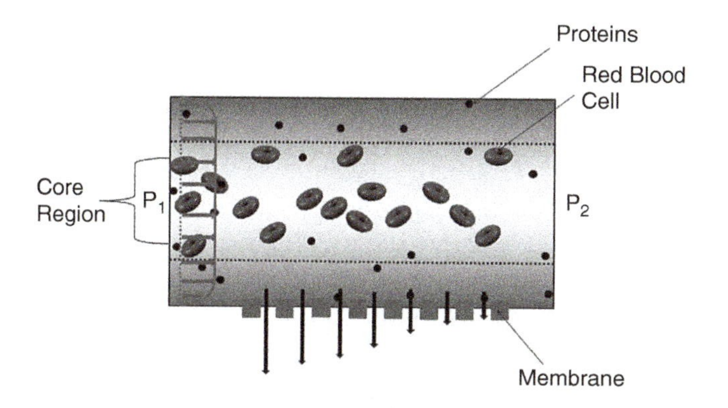

Figure 4.20 Red blood cells and platelets make blood a non-Newtonian fluid in many situations.

Relative to the motion of the particle, the local flow field around a particle is similar to the the flow field past a fixed sphere described in the previous section.

4.13 A simple model for blood flow

Blood is a fluid that contains a number of components, including water, salts, and proteins as well as platelets, leukocytes, and erythrocytes. It is the erythrocytes, or red blood cells, that make blood a non-Newtonian fluid, and the volume fraction of the red blood cells is called the *hematocrit*. The plasma portion of blood (water, salts, and proteins) behaves as a Newtonian fluid (Figure 4.20).

For hematocrit less than 40 percent, blood is usually modeled as a Casson fluid (Fung, 1981) for which the shear stress in one dimension is given by

$$\sqrt{\tau} = \sqrt{\tau_0} + \sqrt{\mu|\dot{\gamma}|} \quad \tau \geq \tau_0 \tag{4.168}$$

and $\dot{\gamma} = \partial w/\partial r = 0$ for $|\ \tau\ | \leq \tau_0$. Here τ_0 is a constant yield stress; the yield stress is very nearly $\tau_0 = 0.5\ \mathrm{N/m^2}$, which is very small (Fung, 1981).

Physiological data in both arteries and veins suggest that the flow is Newtonian near the wall and non-Newtonian in the core of the tube. From a shell balance on an infinitesimal control volume in the fully developed region of the tube,

$$\tau = \left(\sqrt{\mu \mid \frac{\partial w}{\partial r}} \mid + \sqrt{\tau_0}\right)^2 = -\frac{r}{2}\frac{dp}{dx} \tag{4.169}$$

From the definition of the shear stress, the quantity τ_0 can be evaluated where $\partial w/\partial r = 0$ at $r = r_c$, and so

$$\tau_0 = -\frac{r_c}{2}\frac{dp}{dx} \tag{4.170}$$

If r_c is the radius at which the transition to non-Newtonian behavior takes place, then for $\tau > \tau_0$, the Casson equation can be integrated for constant viscosity to give

$$w(r) = -\frac{1}{4\mu}\frac{dp}{dx}\left(R^2 - r^2 - \frac{8}{3}r_c^{1/2}(R^{3/2} - r^{3/2}) + 2r_c(R - r)\right) \quad \text{for } r > r_c \tag{4.171}$$

$$w(r) = w_c = w(r_c) \quad \text{for } r < r_c \tag{4.172}$$

Equation (4.171) can be put in dimensionless form by taking the velocity scale to be $U_0 = -(R^2/4\mu)(dp/dx)$ and

$$w(r) = 1 - r^2 - \frac{8}{3}r_c^{1/2}(1 - r^{3/2}) + 2r_c(1 - r) \tag{4.173}$$

and now r_c is scaled on the radius of the tube R. Note that in dimensionless form,

$$w_c = 1 - r_c^2 - \frac{8}{3}r_c^{1/2}\left(1 - r_c^{3/2}\right) + 2r_c(1 - r_c) \tag{4.174}$$

The velocity profile is depicted in Figure 4.21. The sketch of the velocity profile in Fung (1981) implies that there is a discontinuity in stress at $r = r_c$. However, a quick calculation shows that the stress is zero at $r = r_c$ when approached from either side.

The velocity may be integrated to obtain the volume flow rate:

$$Q = 2\pi\int_0^R rw dr = -\frac{R^4}{8\mu}\frac{dp}{dx}F(\xi) \tag{4.175}$$

where

$$F(\xi) = 1 - \frac{16}{7}\xi^{1/2} + \frac{4}{3}\xi - \frac{1}{21}\xi^4 \tag{4.176}$$

and $\xi = 2\tau_0/R(dp/dx)^{-1}$. Comparing the flow rate result with the Newtonian value

$$Q = -\frac{R^4}{8\mu}\frac{dp}{dx} \tag{4.177}$$

it is seen that the Casson flow rate is always less than its Newtonian counterpart.

4.14 Chapter summary

The objective of the chapter was to introduce the student to a number of viscous flow problems relevant to micro- and nanofluidics. Because micro- and nanofluidics most often involves internal flows, the vast majority of the solutions presented here are for internal flows. In addition, the structure of the flow in channels and tubes, including the concept of an entrance length, is also described.

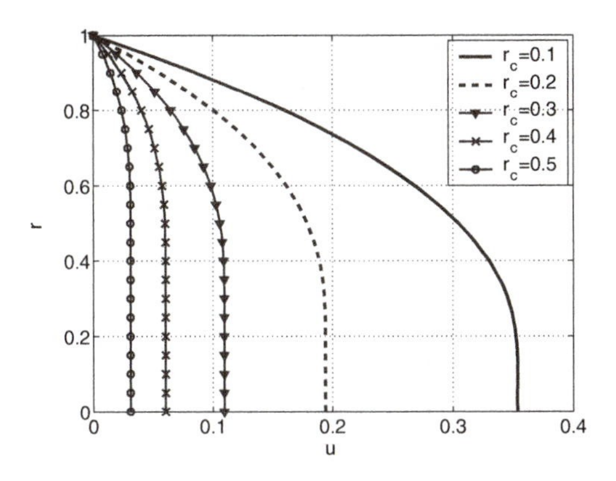

Figure 4.21 Velocity profile for blood flow modeled as a Casson fluid.

In this chapter, the walls of the channel or tube are assumed to be uncharged. Solutions are presented for

- Poiseuille flow in a channel and tube
- Fully developed flow in a thin film
- Slip flow in a channel and tube for both liquids and gases
- Blasius boundary layer on a flat plate
- Fully developed suction flows
- Developing suction flows
- Lubrication approximation in a channel
- Surface tension–driven flow due to a moving bubble
- Stokes flow past a sphere
- Sedimentation
- Casson fluid model of blood flow

Additional viscous flow solutions are presented as exercises. All these flows have the property that the relevant Reynolds number is moderate or small (except for the Blasius boundary layer) so that the flow is laminar, and in many cases, the convective terms can be neglected and the flows are at most two-dimensional. We will revisit some of these solutions later; the presence of Poiseuille flow in a nanopore membrane is an important feature of the sieving efficiency of the artificial kidney discussed in Chapter 12.

The focus in this chapter is on steady flows. However, there are many other applications for which transients are important. Some of these flows are discussed elsewhere (Acheson, 1990; White, 2006) in great detail, and these and other references may be consulted for the details.

EXERCISES

4.1 Water enters a channel of height $h = 1$ μm. Assuming that the boundary layers on the walls can be approximated by a Blasius boundary layer, determine the entrance length for a flow rate of $Q = 1$ mm^3/sec. The width of the channel is $W = 40$ μm. Compare with the given formulas for the entrance length described in Section 4.2.

4.2 Repeat the previous problem for air.

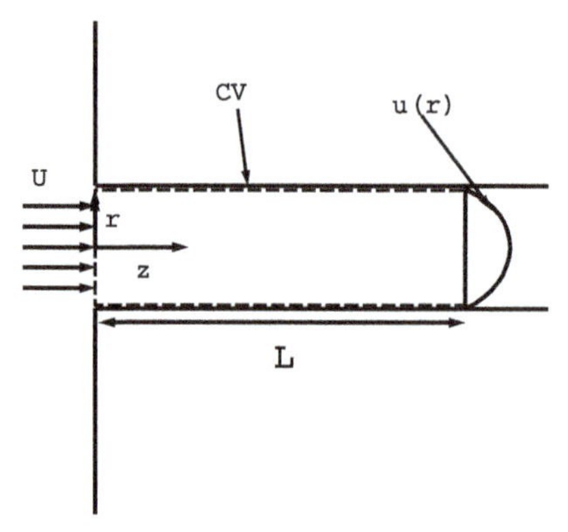

Exercise 4.3

4.3 Using Bernoulli's equation, estimate the pressure drop in the entrance region of a pipe. Using the control volume approach, show that the drag in the inlet region of a pipe is given by

$$D = \pi a^2 \left(p_0 - p_x - \frac{1}{3}\rho U^2 \right)$$

Is the formula different for a rectangular channel?

4.4 An incompressible fluid flows in a channel with a moving upper wall at speed U_w and imposed pressure drop dp/dx. Superpose the Couette and Poiseuille flows and write the velocity in dimensionless form. Calculate the imposed Couette velocity such that the volume flow rate is zero. Plot the streamlines for several values of the dimensionless parameter that governs the motion (i.e., the ratio of the Poiseuille and Couette velocity scales).

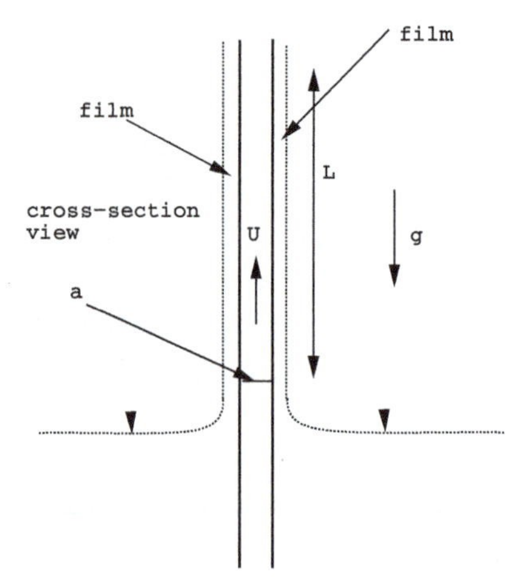

Exercise 4.5

4.5 Consider the viscous flow on a wire that is being extruded from a liquid bath, as in the figure. The diameter of the wire is on the order of the the film thickness so that cylindrical coordinates must be used. Outside the film is air at constant pressure p_{atm}.

a. Neglecting the effects of surface tension, find the solution to the problem and write it in dimensionless form. What is the velocity scale?
b. Plot the velocity distribution across the film in dimensionless form for the ratio of the film thickness h to the wire radius a of 0.3. Take the ratio of the velocity U to the velocity scale found in part a to be 1.
c. Calculate the flow rate and the power per unit length required to pull the wire out of the bath. Use the values $U = 1$ μm/sec, wire radius $a = 1$ μm, film thickness $h = 0.3a$ and the density and viscosity taken to be that of water.

4.6 Consider the steady laminar flow between two coaxial tubes. The flow is maintained by a pressure gradient. Calculate the velocity distribution within the annulus. Calculate the volume flow rate and relate it to the pressure drop. Write the expression for the pressure drop in dimensionless form, thus finding an expression for the head loss parameter.

4.7 Repeat the previous problem for the case in which the inner cylinder is moving at speed U.

4.8 Verify the leading-order expression for the pressure distribution for the developing suction flow at low Reynolds number, which is

$$p(x, y) = \frac{K}{Re_w}\left(x - \frac{1-\alpha}{2}\frac{v_w}{U_0}\frac{x^2}{2}\right) + p_r(y)$$

where

$$p_r(y) = \frac{v_w}{U_0}f' - \left(\frac{v_w}{U_0}\right)^2\frac{f^2}{2}$$

4.9 Numerically solve the developing suction problem for a sink at $y = 1$ and a solid wall at $y = 0$, for $v_w/U_0 = 0.2$, thus validating the results of Figure 4.14. Such a geometry is a simple model for a device to be used for an artificial kidney. Calculate the leading-order term for the velocities and the pressure for Reynolds numbers $Re_w = 0, 1,$ and 3. Plot the streamlines for each case.

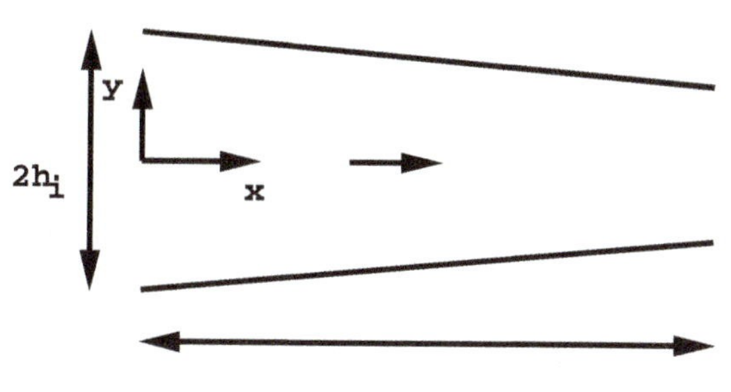

Exercise 4.10

4.10 A pressure-driven flow of water passes through a converging nozzle of length 3 μm. The nozzle is very wide, $W = 40$ μm, and is 50 nm in height at the entrance and 40 nm in height at the exit. Calculate the velocity distribution and volume flow rate under the lubrication approximation. For a pressure drop of 0.01 atm, what is the volume flow rate and the Reynolds number? Plot the axial velocity and the pressure at several locations along the nozzle. How would the results change for a diffuser?

4.11 Repeat the previous problem for a cylindrical nozzle of diameter $D =$ 50 nm.

4.12 Consider the Stokes flow past a circular cylinder. Show that the stream function satisfies

$$\left(\frac{\partial^2}{\partial r^2} + \frac{1}{r}\frac{\partial}{\partial r} + \frac{1}{r^2}\frac{\partial^2}{\partial \theta^2}\right)^2 \psi = 0$$

Along with the no-slip condition on the cylinder surface, which is

$$\frac{\partial \psi}{\partial r}(a, \theta) = \frac{\partial \psi}{\partial \theta}(a, \theta) = 0$$

$$\psi \sim U r \sin\theta \quad \text{as } r \to \infty$$

show that the solution of the form $\psi = Ur \sin\theta$ leads to

$$\psi = \left(Ar^2 + Br\log r + Cr + \frac{D}{r}\right)\sin\theta$$

and hence there is no solution to the problem because no choice of the constants can satisfy all boundary conditions.

Proudman and Pearson (1957) show that using methods of matched asymptotic expansions that near the cylinder,

$$\psi = \frac{\left[\left(r\log r - \frac{1}{2}r + \frac{1}{2r}\right)\right]}{\left(\log\frac{8}{\mathrm{Re}} - \gamma + \frac{1}{2}\right)}$$

where γ is Euler's constant $\gamma \sim 0.58$. Thus the constants B, C, and D can be determined, but the constant A cannot.

4.13 Show that the solution for a moving sphere through a motionless liquid is obtained merely by subtracting the uniform flow term in the Stokes flow past a fixed sphere.

4.14 The slow flow past a spherical bubble of radius a may be modeled by assuming that the boundary condition on the bubble is no tangential stress $(\partial/\partial\theta = 0)$ at the bubble surface. Show that the stream function is given by

$$\psi = \frac{U}{2}(r^2 - ar)\sin^2\theta$$

and that the drag is given by $D = 4\pi\mu U a$.

4.15 In a rapid molecular analysis system, the channels in the membrane making up the system are often rectangular. Calculate the solution for the velocity field of a Casson fluid in fully developed pressure-driven channel flow of height h. At what channel height would you expect the influence of the size of the red and white blood cells to become important? Assume that the channel is very wide, and thus provide the analogue for a channel of blood flow through a tube.

5 Heat and Mass Transfer Phenomena in Channels and Tubes

5.1 Introduction

In the previous chapter, the types of steady viscous flows common to micro- and nanofluidics were presented. Of particular importance is the Poiseuille flow and flow past a sphere. In the present chapter, classical solutions for heat transport and mass transport of uncharged species that may occur in parallel with the Poiseuille flow are described. This chapter will lead into a discussion of electrostatics and electrochemistry, concepts that are so important in micro- and nanofluidics.

Both heat and mass transfer are discussed in this chapter. This is because of the similarity in the forms of the governing equations and boundary conditions and the similar nature of the transport properties: the thermal conductivity and the mass diffusivity or diffusion coefficient. Indeed, heat and mass transfer often occur simultaneously (Figure 5.1). Both heat and mass transfer can be grouped into two modes: conduction heat transfer, diffusion mass transfer, and convective heat and mass transfer. A third mode of heat transfer, radiation heat transfer, is not discussed.

Fortunately, most liquid flows in micro- and nanochannels do not encounter large temperature and concentration gradients so that in general, the fluid properties remain relatively constant. Thus the governing equations can usually be decoupled to some extent, even at Reynolds numbers not normally viewed as being small.

Several basic problems of heat transfer in channels are considered before presenting the formal definition of the fully developed temperature distribution. Unlike the formal definition of fully developed fluid flow, the solution for the fully developed temperature distribution depends on the boundary conditions on the temperature at the walls of a tube or channel. The two cases of constant wall temperature and constant heat flux are considered, and the length of the thermal entrance region can be calculated for both these conditions. It should be noted that because the Schmidt number is so large for liquid flows, mass transfer in channels and tubes is seldom fully developed.

The presentation in this chapter is similar to the presentation of previous chapter. After discussion of the entrance regime, several relevant mass transfer

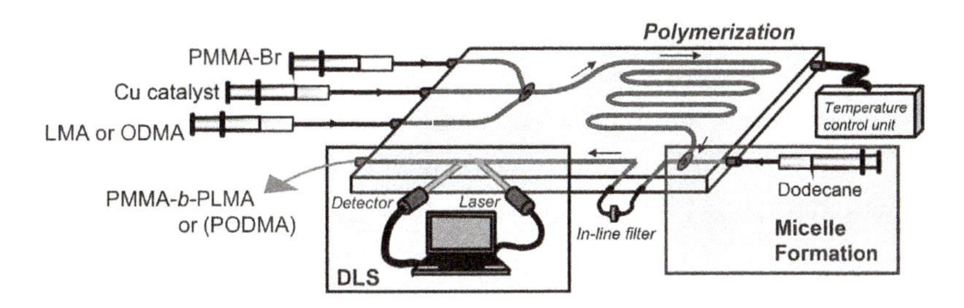

Figure 5.1 The National Institute of Standards and Technology (NIST) Lab-on-a-Chip microreactor for analysis of the influence of polymer additives and the synthesis of copolymer surfactants. Go to www.nist.gov and search polymers for more information.

problems are discussed, including mass transfer in thin films, Taylor–Aris dispersion, unsteady mass transport with drug delivery applications, and a mass transfer view of Brownian motion.

Solutions for steady state thermal and mass transfer boundary layers are presented, illustrating the influence of the relevant dimensionless parameters, the Prandtl and Schmidt numbers. It is apparent from the contents of this chapter that there are a number of analogous problems in the two disciplines of heat and mass transfer, perhaps a result of their parallel historical development, as discussed in Chapter 1.

5.2 One-dimensional temperature distributions in channel flow

In this section, simple one-dimensional temperature distributions in channel flow are described and compared in form to the corresponding velocity distributions. Here it is assumed that the temperature variations are small so that changes in properties with temperature can be neglected. Also a constant wall temperature is assumed, and the temperature $T = T(y)$, where y measures distance from the wall and $0 \leq y \leq h$.

Consider the temperature distribution in a channel containing a liquid flowing at a small Reynolds number where the flow is two-dimensional. The governing equation at steady state for constant properties is given by

$$\rho c_p(\vec{u} \bullet \nabla)T = k\nabla^2 T \tag{5.1}$$

Note that the temperature T and all the other variables are dimensional; the appropriate temperature scale cannot be chosen until the boundary conditions are specified.

In the hydrodynamic fully developed region, the transverse velocity $v = 0$, and at low Reynolds numbers and moderate Prandtl numbers, convective terms are negligible so that

$$\frac{\partial^2 T}{\partial x^2} + \frac{\partial^2 T}{\partial y^2} = 0 \tag{5.2}$$

For example, at physiological temperatures of $T = 37°\text{C}$, the Prandtl number of water is $Pr \sim 4.5$. Thus, for the Reynolds number $Re \sim 10^{-3}$, $Re\,Pr = 0.0045$, and the convective terms in the energy equation will be negligible. If the channel is long so that $\partial/\partial x \ll \partial/\partial y$, then we might expect that the temperature distribution is given by the solution of the one dimensional equation

$$\frac{d^2T}{dy^2} = 0 \tag{5.3}$$

Assuming that the wall temperatures are known, the boundary conditions can be taken to be

$$T = T_{w1} \quad \text{at } y = 0 \tag{5.4}$$

$$T = T_{w2} \quad \text{at } y = h \tag{5.5}$$

Integrating twice and applying the boundary conditions,

$$T(y) = (T_{w2} - T_{w1})\frac{y}{h} + T_{w1} \tag{5.6}$$

This equation sets the temperature scale as $T_{\text{scale}} = \Delta T = T_{w2} - T_{w1}$, and the dimensionless temperature is thus

$$\theta = \frac{T - T_{w1}}{\Delta T} = Y \tag{5.7}$$

where $Y = y/h$. Clearly if $T_{w2} = T_{w1}$, then the temperature is constant across the channel. This temperature distribution is equivalent to the temperature distribution in a solid slab.

Notation alert: In this section, Q refers to overall heat transfer rate in watts and not to volume flow rate.

The heat transfer flux in $\text{W/m}^2\text{K}$ is defined by the equation

$$q = -k\frac{\partial T}{\partial y} = -k\frac{dT}{dy} = \frac{Q}{A} \tag{5.8}$$

in this case, and from this simple linear solution, we find that the total heat transfer rate is

$$Q = -\frac{kA\Delta T}{h} = -\frac{\Delta T}{R} \tag{5.9}$$

where R is the heat transfer resistance. If R is large, the heat transfer rate is small. In microchannel flows, for a channel with $L = 1$ μm in the flow direction and $W = 40$ μm wide, $R \sim 4.4 \times 10^{5}\,°\text{C/m}$ for a channel height of $h = 100$ nm. The negative sign is to ensure that $Q > 0$ as it flows from a higher to a lower temperature. Thus, for $T_{w2} > T_{w1}$, the heat flux is negative (in the negative y direction). This situation is depicted in Figure 5.2. Note also that the heat flux is constant, ensuring an overall energy balance $q_{w1} = q_{w2}$.

The *Nusselt number* in this problem is defined by $Nu = q_w b/k(T_{w1} - T_{w2})$, where here b is an appropriate length scale. Physically, the Nusselt number is the ratio of the convective heat transfer rate to the conduction heat transfer rate,

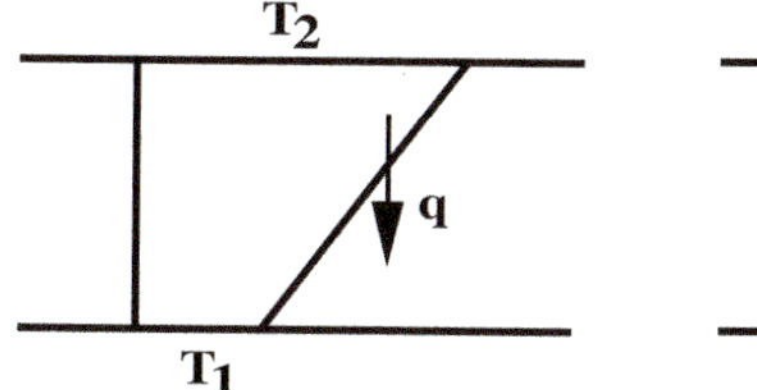

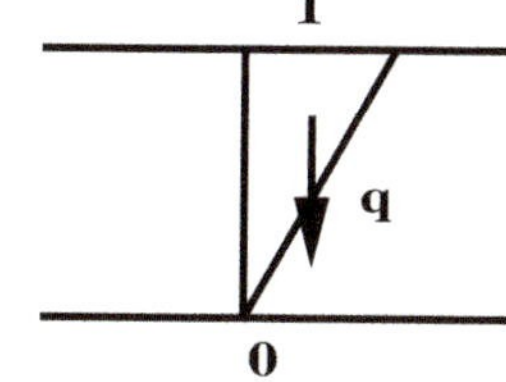

Figure 5.2 Temperature distribution in a rectangular channel due to a temperature difference between the two walls. The figure on the right depicts the dimensionless distribution from equation (5.7).

and this interpretation is similar to that of the Reynolds number being the ratio of the convective forces to viscous forces. In this situation, $b = h$, and using the expression for the heat flux, it is seen that $Nu = 1$.

Now consider the conduction heat transfer through a duct with a uniform source of thermal energy; the equation governing the one-dimensional temperature distribution is

$$\frac{d^2T}{dy^2} + \frac{q_{gen}}{k} = 0 \tag{5.10}$$

where again, a constant thermal conductivity has been assumed. Here q_{gen} has units of W/m^3, and this problem could occur in the flow of an electrolyte solution where q_{gen} arises due to the presence of an electric field, a phenomenon known as *Joule heating*.

For constant temperature boundary conditions, as previously, we can integrate twice to obtain

$$T(y) = (T_{w2} - T_{w1})\frac{y}{h} + T_{w1} - \frac{q_{gen}}{2k}(y^2 - yh) \tag{5.11}$$

Note that for $T_{w2} = T_{w1} = T_w$, the temperature distribution is of the same form as the velocity distribution in fully developed flow in a channel, with the quantity $q_{gen}/2k$ taking the place of $-(1/2\mu)\,(dp/dx)$ and

$$T(y) - T_w = -\frac{q_{gen}}{2k}(y^2 - hy) \tag{5.12}$$

or in dimensionless form,

$$\theta(y) = \frac{T(y) - T_w}{\frac{q_{gen}h^2}{2k}} = Y - Y^2 \tag{5.13}$$

Thus, on a dimensionless basis, $\theta(y) = u(y)$, where $u(y)$ is scaled on the Poiseuille velocity $U_0 = -(h^2/2\mu)\,(dp/dx)$.

In the general case where $T_{w2} - T_{w1} \neq 0$, the solution for the temperature is a superposition of the linear temperature profile and the parabolic distribution due to the heat source, much like the superposition of Poiseuille and Couette flow. For this situation, there are two temperature scales that could be used to form the dimensionless temperature distribution: $T_{scale} = \Delta T = T_{w2} - T_{w1}$ and $T_{scale} = q_{gen}h^2/k$. If both quantities are of the same magnitude, either can be used to form the dimensionless temperature.

Another problem of interest is the balance between conduction and viscous dissipation terms in the energy equation. In this case, the governing equation becomes

$$k\frac{d^2T}{dy^2} = \mu\left(\frac{du}{dy}\right)^2 \tag{5.14}$$

in the fully developed region. If the velocity is given by the Poiseuille flow distribution,

$$u(y) = \frac{1}{2\mu}\frac{dp}{dx}(y^2 - yh) \tag{5.15}$$

then since

$$\frac{du}{dy} = \frac{1}{2\mu}\frac{dp}{dx}(2y - h) \tag{5.16}$$

integrating equation (5.14), the temperature is given by

$$T = T_w + \frac{1}{4\mu k}\left(\frac{dp}{dx}\right)^2\left[\frac{y^4}{3} - \frac{2}{3}y^3h + \frac{h^2y^2}{2} - \frac{h^3}{6}y\right] \tag{5.17}$$

where the wall temperatures are assumed equal and constant: $T_{w2} = T_{w1} = T_w$. Thus the temperature scale can be defined by

$$T_{\text{scale}} = \frac{\left(\frac{dp}{dx}\right)^2 h^4}{4\mu k} \tag{5.18}$$

It is left as an exercise to write equation (5.14) in dimensionless form and to estimate the order of magnitude of the resulting dimensionless parameters such that viscous dissipation is an $O(1)$ effect.

In the present section, the dependence of the temperature on the streamwise coordinate has been neglected in a manner similar to what is the case in the hydrodynamically fully developed region, and several canonical problems have been identified. In the next two sections, the thermal entrance region is formally defined.

5.3 Thermal and mass transfer entrance regions

Just as for fluid flow, there is a thermally developing region where the temperature distribution adjusts to a different boundary condition on the walls of the channel or tube. As with the fluid flow, the temperature initially behaves similarly to an external flow. The expression for the thermal entrance length is of the same form as for the hydrodynamic entrance length, and for a constant tube surface temperature and laminar flow,

$$\frac{L_{eT}}{D} = 0.06\,RePr \tag{5.19}$$

for large Reynolds number in the laminar regime $Re \leq 2300$.

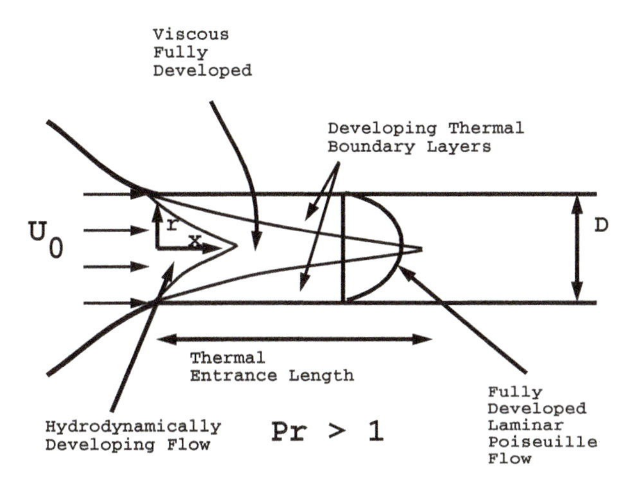

Figure 5.3 Flow through a pipe showing the thermal entrance region for Pr > 1. The picture is similar for a channel with *r* replaced by *y*.

It is interesting to note that some heat transfer texts (Incropera & Dewitt, 1990; Kays and Crawford, 1980) indicate that the constant in the definition of the entrance length is 0.05; this may be a function of the definition of the length of the entrance region as being the length where the temperature approaches to within 95 or 99 percent of the fully developed value. Alternatively, the discrepancy could be due to the interpolation of the individual numerical result (Kays and Crawford, 1980). In any case, the difference is small and irrelevant for applications because usually, only an estimate of the entrance region length is required.

In liquid flows, the Prandtl number $Pr > 1$ and so the thermal entrance length is somewhat longer than the hydrodynamic entrance length (Figure 5.3). For a uniform heat flux, the result is virtually no different than the case of constant surface temperature (Kays & Crawford, 1980).

Similarly, the entrance length for mass transfer in a tube on the macroscopic scale is

$$\frac{L_{em}}{D} = 0.06\,ReSc \tag{5.20}$$

Note that for a Reynolds number of $Re = 10$ and $Sc = 1000$, $L_{em}/D = 600$, which is extremely large, and for this reason, concentration profiles are rarely considered to be fully developed in channels of interest in microfluidics and nanofluidics.

Specifically for microchannels, Lee and Garimella (2006) have performed a number of numerical computations for different channel aspect ratios for a constant heat flux and found that the result for a slit pore is

$$\frac{L_{eT}}{D_H} = 0.057\,RePr \tag{5.21}$$

which agrees with the macroscale computation for the tube at constant surface temperature. Recall that the hydrodynamic entrance length is nonzero for

$Re \to 0$, as described in the previous chapter; this corresponds to the entrance region for the Stokes flow problem in a channel or tube. Based on the analogy between fluid mechanics and heat transfer and, by extension, mass transfer (see equation (5.20)), Lee and Garimella (2006) suggest that the Reynolds number can be replaced by *RePr* for heat transfer and *ReSc* for mass transfer in those formulas.

While the formulas for the thermal entrance length are similar to the hydrodynamic entrance length for both constant temperature and constant heat flux boundary conditions, the formal definition of thermally fully developed flow is somewhat different, depending on the boundary conditions at the wall. The magnitude of the thermal energy rate transported in the axial direction in kJ/sec is typically defined in terms of a mean temperature T_m as

$$Q = \dot{m}cT_m = \rho AUcT_m = \int_A \rho wcT dA \tag{5.22}$$

where U is the average axial velocity and A is the cross-sectional area of the tube. Equation (5.22) defines the area-averaged mean temperature as

$$T_m = \frac{1}{UA}\int_A wT dA \tag{5.23}$$

If T_w denotes the wall temperature, then a dimensionless temperature may be defined by

$$\theta = \frac{T_w - T}{T_w - T_m} \tag{5.24}$$

The flow is said to be *thermally fully developed* if

$$\frac{\partial \theta}{\partial x} = 0 \tag{5.25}$$

This condition is reached if the surface temperature is uniform, as in the previous section, or if the heat transfer rate in the streamwise direction is constant. Note that the simple solutions discussed in the previous section in the channel flow case satisfy these conditions.

For a tube, note that equation (5.25) implies that

$$\frac{\partial \theta}{\partial r} = \text{constant} = -\frac{\frac{\partial T}{\partial r}|_{r=a}}{T_w - T_m} \tag{5.26}$$

which means that the heat flux across the tube is constant. The heat flux normal to the wall can also be expressed in terms of a *heat transfer coefficient* as

$$\dot{q}_r'' = -k\frac{\partial T}{\partial r} = h_T(T_w - T_m) \tag{5.27}$$

and so equation (5.26) implies that the heat transfer coefficient h_T is constant.

For a Prandtl number $Pr > 1$, let us assume that the velocity is fully developed; if the Prandtl number $Pr \sim 1$ or smaller, as is the case for gases, then the flow field and the temperature distribution are fully coupled. For $Pr \gg 1$, the velocity

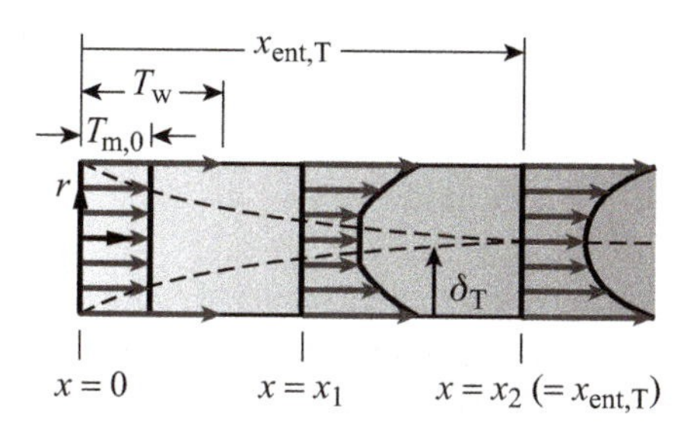

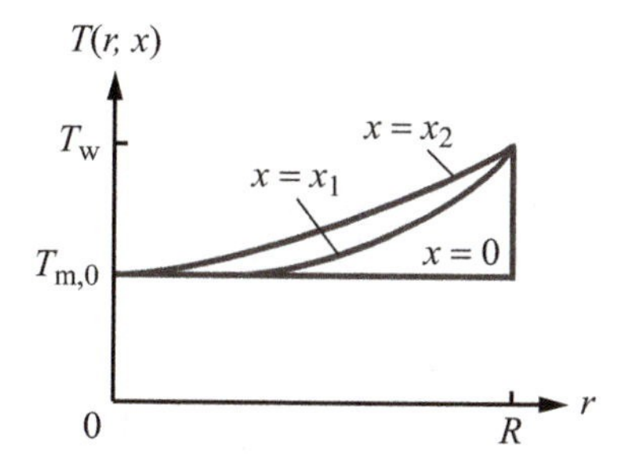

Figure 5.4 Thermal entry region for $T_w > T_m$, showing the temperature distribution in the entrance region of a pipe. At $x = x_2$, the temperature distribution is fully developed. The picture is similar for a channel with r replaced by y.

field becomes fully developed sooner than the temperature, and the governing equation for the temperature in a tube is given by

$$\rho c w \frac{\partial T}{\partial x} \doteq k \left(\frac{\partial^2 T}{\partial r^2} + \frac{1}{r} \frac{\partial T}{\partial r} \right) \tag{5.28}$$

where for a liquid, $c_p = c_v = c$. Using the definition of θ, and since $\partial \theta / \partial x = 0$,

$$\frac{\partial T}{\partial x} = \frac{d T_w}{dx} - \theta \left(\frac{d T_w}{dx} - \frac{d T_m}{dx} \right) \tag{5.29}$$

If $\partial T / \partial x =$ constant, then equation (5.28) can be integrated directly. For constant heat flux, since $h_T =$ constant, then $T_w - T_m =$ constant and

$$\frac{\partial T}{\partial x} = \frac{d T_w}{dx} = \frac{d T_m}{dx} = \text{constant} \tag{5.30}$$

Conversely, if $T_w =$ constant, then

$$\frac{\partial T}{\partial x} = \frac{T_w - T}{T_w - T_m} \frac{d T_m}{dx} \tag{5.31}$$

and the solution to equation (5.28) can be calculated numerically or by iteration, as suggested by Kays and Crawford (1980). The thermal entry region for a tube at constant wall temperature is depicted in Figure 5.4.

Moreover, because of the similarity of the heat and mass transfer equations, similar methods may be used for the mass transfer problem. However, unlike temperature, surface concentrations cannot normally be measured, and so the no-flux condition corresponding to

$$\frac{\partial c_A}{\partial r} = 0 \text{ at the wall} \tag{5.32}$$

is used, leading to a concentration profile independent of r due exclusively to the boundary condition. Moreover, since the Schmidt number is so large in liquids, on the macroscale, as noted above, the concentration distribution is often not fully developed at all. This will be seen later when Taylor dispersion is considered.

There have been some indications in the literature that the various hydrodynamic, thermal, and mass transfer entrance lengths in microscale geometries may be different from those at the macroscale. A careful reading of the literature shows that this is likely not the case. Reasons for the possible discrepancies include the difficulties in measuring anything at such small scales. Because the dimensions of pipes and channels are small, any small error in the determination of the precise dimensions is magnified, leading to results that conflict with the macroscale result. Indeed, Judy *et al.* (2002) suggest that any such differences in the related problem of determining the pressure drop are within the experimental error of the measurement of the dimensions. In the same spirit, Bavier and Ayela (2004) suggest that some experimental methods are poorly adapted to the microscale regime. The conclusion is that these formulas do indeed apply to the microscale.

Jean-Baptiste Joseph Fourier was born 1768 in Auxerre, France. Fourier became obsessed with mathematics and heat at an early age, excelling specifically in his mathematics studies. During his life he was a professor of mathematics at the École Polytechnique for many years. However, it was during his service to Napoleon as Prefect of the Department of Isère in Grenoble, France where he developed his theory of heat conduction by 1807 despite heavy criticisms by contemporaries such as Laplace, Lagrange, and Poisson. This theory was based upon and solved using a series of function expansions now known as Fourier series (Churchill, 2005). Fourier developed his famous mathematics series to describe the heat conduction in solid bodies such as the earth. His studies on heat conduction were combined with his trigonometric series and they were presented together in his famous memoir, *On the Propagation of Heat in Solid Bodies*. Today, the Fourier series has become an essential tool when solving many engineering problems. Unfortunately, his contemporaries mixed reviews and competing theories kept his critical discoveries from being accepted and recognized by the scientific community with the proper respect during his lifetime.

5.4 The temperature distribution in fully developed tube flow

For the constant heat flux boundary condition, the equation governing the temperature in a tube is given by

$$\frac{1}{r}\frac{\partial}{\partial r}\left(r\frac{\partial T}{\partial r}\right) = \frac{2w_{\text{ave}}}{\alpha}\left(1-\frac{r^2}{a^2}\right)\frac{dT_m}{dx} \tag{5.33}$$

Here w_{ave} is the average velocity in the tube and $\alpha = k/\rho c_p$ is the thermal diffusivity.

For constant heat flux, dT_m/dx is constant, and this equation can be integrated twice with respect to the radial coordinate r, with the result

$$T = T_w + \frac{2w_{\text{ave}}}{\alpha}\frac{dT_m}{dx}\left(-\frac{3}{16}a^2 - \frac{r^4}{16a^2} + \frac{r^2}{4}\right) \tag{5.34}$$

The mean temperature is thus given by

$$T_m = T_w - \frac{11}{96}\frac{2w_{\text{ave}}}{\alpha}\frac{dT_m}{dx}a^2 \tag{5.35}$$

and the local heat flux rate is given by

$$q'' = h_T(T_w - T_m) = h_T\frac{11}{96}\frac{2w_{\text{ave}}}{\alpha}\frac{dT_m}{dx}a^2 = \text{constant} \tag{5.36}$$

Applying conservation of thermal energy and neglecting losses over a differential control volume in the tube, the heat flux at the wall is also given by

$$q'' = \frac{\rho c w_{\text{ave}} a}{2}\frac{dT_m}{dx} \tag{5.37}$$

so that the Nusselt number

$$Nu = \frac{h_T D}{k} = \frac{48}{11} \cong 4.364 \tag{5.38}$$

The constant temperature solution is a bit more complex, and Kays and Crawford (1980) suggest the use of successive approximations beginning with the constant heat flux solution. In this case, the Nusselt number in the fully developed zone is $Nu = 3.66$. Sketches of the axial temperature distribution for both the constant temperature and constant heat flux are depicted in Figure 5.5.

5.5 The Graetz problem for a channel

For the *Peclet number* $Pe_T = RePr \gg 1$, the hydrodynamic entrance length is shorter than the thermal entrance length, so the flow can be assumed to be fully developed, as depicted in Figure 5.6. The governing equation and boundary conditions in the thermally developing region for a channel having constant wall temperature in dimensionless form are

$$y(1-y)\frac{\partial \theta}{\partial \xi} = \frac{\partial^2 \theta}{\partial y^2} \tag{5.39}$$

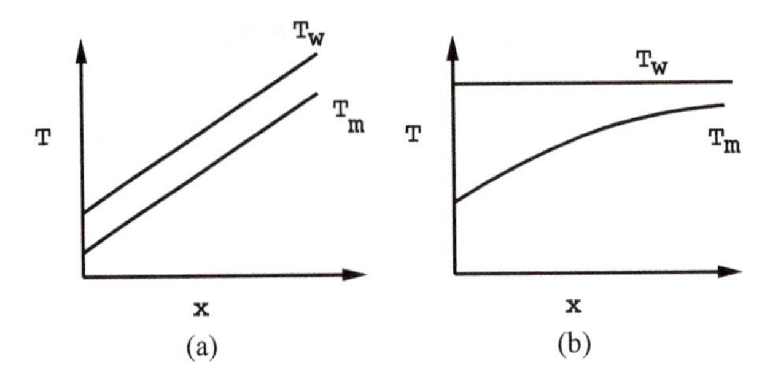

Figure 5.5 Sketch of the temperature in fully developed laminar flow. (a) Constant heat transfer flux. (b) Constant wall temperature.

where $\xi = x/\text{RePr} = x/Pe_T$ subject to

$$\theta(0, y) = 1 \tag{5.40}$$

$$\theta(\xi, 0) = \theta(\xi, 1) = 0 \tag{5.41}$$

The variable $x = x^*/h$, and here θ is the scaled temperature, defined by

$$\theta = \frac{T - T_w}{T_0 - T_w} \tag{5.42}$$

where T_0 is the temperature at the entrance to the channel. This is the *Graetz problem*, and in this section, a solution to this problem is developed.

The solution may be obtained by separation of variables in the form (Brown, 1960)

$$\theta = \sum_{i=1}^{\infty} c_i Y_i e^{-\lambda_i^2 x} \tag{5.43}$$

Substituting into the preceding governing equation, the functions Y_i satisfy

$$Y_i'' + \lambda_i^2 y(1 - y)Y_i = 0 \tag{5.44}$$

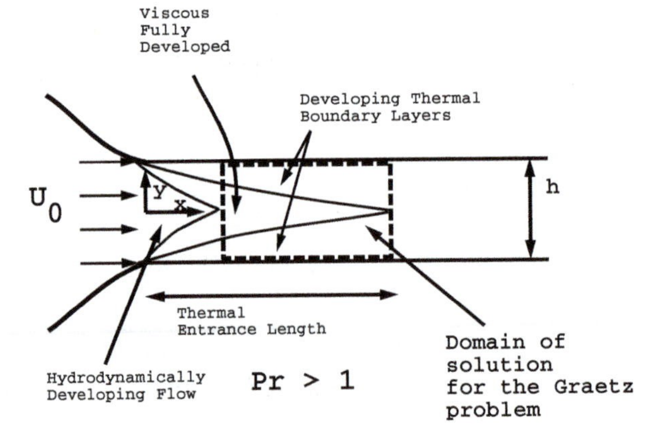

Figure 5.6 The domain of solution for the Graetz problem for $Pr \gg 1$ in a channel.

subject to $Y(0) = Y(1) = 0$. This is what is called a *Sturm–Liouville problem*[1] so that the solution may be written formally in terms of an infinite series expansion of the functions Y_i, and λ_i are the eigenvalues of equation (5.44). These eigenvalues may be calculated numerically, and this has been done for the first 10 eigenvalues by Brown (1960).

The heat flux at the wall is defined by

$$q''_w = -k\frac{\partial T}{\partial y} \tag{5.45}$$

which must be balanced by that convected through the channel at the mean temperature

$$q''_w = h_T(T_m - T_w) \tag{5.46}$$

Equating these two quantities defines the local heat transfer coefficient as

$$h_T = -\frac{k}{h\theta_m}\frac{\partial\theta}{\partial y} \tag{5.47}$$

The Nusselt number is defined by

$$Nu = \frac{hh_T}{k} = \frac{1}{\theta_m}\frac{\partial\theta}{\partial y}|_{y=0} \tag{5.48}$$

Thus all that are needed to determine the Nusselt number are the mean temperature and the temperature gradient at the wall.

Out of this analysis comes the value of the Nusselt number in the fully developed regime, which, for a very wide channel, is

$$Nu = \frac{h_T D_H}{k} = 7.54 \tag{5.49}$$

for constant temperature and $Nu = 8.24$ for the constant heat flux solution. In this case, the hydraulic diameter is $D_H \sim 2h$, and from the numerical solution, the asymptotic Nusselt number is reached when

$$2\xi = 0.12$$

leading to

$$\frac{L_{eT}}{2h} = 0.06\,RePr \tag{5.50}$$

for the entrance length, as stated in Section 5.3. Detailed numerical data for channels of various aspect ratios and the solution to the Graetz problem in a tube are presented by Kays and Crawford (1980).

[1] A Sturm–Liouville problem is an ordinary differential equation with boundary conditions that, in the language of linear algebra, form a complete orthonormal set, and the solution may be written formally in terms of an infinite series expansion of the functions. There are limitations on the form of the coefficients in the equation.

5.6 Mass transfer in thin films

A problem similar to the thermal Graetz problem in a channel considered in the last section is mass transport in a thin film. For simplicity, consider the absorption of, say, oxygen, O_2, into water. Then the governing equation for mass transfer is

$$\rho \left(u\frac{\partial \omega}{\partial x} + v\frac{\partial \omega}{\partial y} \right) = \rho D_{AB} \frac{\partial^2 \omega}{\partial y^2} \tag{5.51}$$

where ω is the mass fraction of the gas in the liquid. The film flow is almost fully developed because the amount of gas absorbed into the film is very small. Indeed, the diffusion coefficient of O_2 in water is only 2.4×10^{-9} m^2/sec. This means that the values of the viscosity and density of the liquid film are not significantly affected so that the velocity is as given in Section 4.5. Because oxygen is being absorbed into the film, the v velocity will be finite but very small (lubrication approximation), and thus, to a very good approximation,

$$u(y)\frac{\partial \omega_A}{\partial x} = D_{AB} \frac{\partial^2 \omega_A}{\partial y^2} \tag{5.52}$$

where

$$u = \frac{\rho g h^2}{\mu} \left(\frac{y}{h} - \frac{1}{2} \left(\frac{y}{h} \right)^2 \right) \tag{5.53}$$

The problem is subject to boundary conditions

$$\frac{\partial \omega_A}{\partial y} = 0 \quad \text{at } y = 0 \tag{5.54}$$

$$\omega_A = 0 \quad \text{at } x = 0 \tag{5.55}$$

$$\omega_A = \omega_{As} \quad \text{at } y = h \tag{5.56}$$

The first boundary condition is appropriate for a solid impermeable wall. This problem is similar, though not identical, to the Graetz problem and has been solved by Pigford.[2]

Because the diffusion coefficient is very small, we expect that the mass transfer is slow so that the diffusion of the gas takes place only near the liquid–gas surface. This means that we can replace the velocity with its value at the interface, and

$$u(h) = U_{\max} = \frac{\rho g h^2}{2\mu} \tag{5.57}$$

approximately, and now y measures distance locally from the vapor–liquid interface. Thus the problem becomes

$$\frac{\rho g h^2}{2\mu}\frac{\partial \omega_A}{\partial x} = U_{\max}\frac{\partial \omega_A}{\partial x} = D_{AB} \frac{\partial^2 \omega_A}{\partial y^2} \tag{5.58}$$

[2] R. L. Pigford, PhD thesis, University of Illinois, 1941.

subject to

$$\frac{\partial \omega_A}{\partial y} = 0 \quad \text{as } y \to \infty \tag{5.59}$$

$$\omega_A = 0 \quad \text{at } x = 0 \tag{5.60}$$

$$\omega_A = \omega_{As} \quad \text{at } y = 0 \tag{5.61}$$

where we have replaced y with $h - y$ and made the approximation that mass transfer takes place only near the surface. The first boundary condition can be replaced by

$$\omega \to 0 \quad \text{as } y \to \infty \tag{5.62}$$

The solution for this problem is given in many texts and can be obtained by using a Laplace transform in x. The result is

$$\frac{\omega_A}{\omega_{As}} = \text{erfc}\left(\frac{y}{\sqrt{4D_{AB}x/U_{\max}}}\right) = \frac{\omega_A}{\omega_{As}} = 1 - \frac{2}{\pi}\int_0^{\eta} e^{-\xi^2} d\xi \tag{5.63}$$

where η is a similarity variable and $\eta = y/\sqrt{4D_{AB}x/U_{\max}}$.

The flux of gas into the film is given by

$$n_{Ay} = -\rho D_{AB}\frac{\partial \omega_A}{\partial y} = -\rho\omega_{A0}\sqrt{D_{AB}U_{\max}/(\pi x)} \tag{5.64}$$

and integrating over a length L, the total amount of mass transferred is

$$M_A = \int_0^W \int_0^L n_{Ay} dz dx = 2\rho W \omega_{A0}\sqrt{D_{AB}U_{\max}L/\pi} \tag{5.65}$$

The result for the velocity and the mass fraction is sketched qualitatively in Figure 5.7.

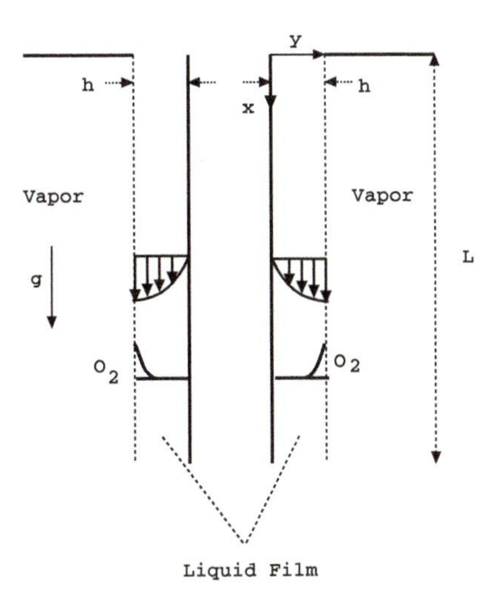

Figure 5.7 A thin liquid film falling under gravity: Velocity and mass fraction of a gas dissolving into a liquid.

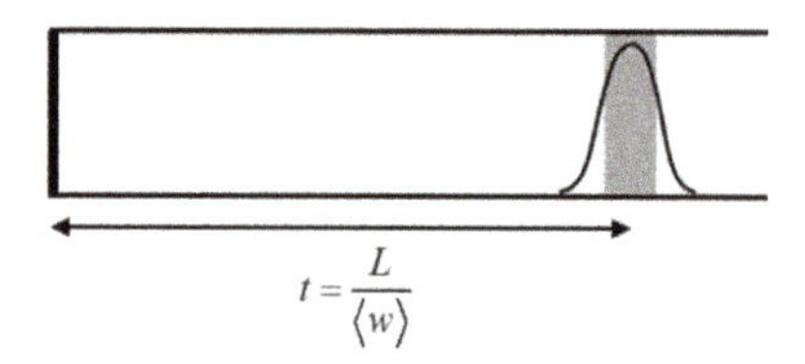

Figure 5.8 Sketch of the broadening of the concentration distribution of a solute as it flows through a channel or tube, Taylor dispersion. The solid line is a Gaussian distribution of solute locally around the point $z = \langle w \rangle t$, provided the dispersion coefficient is used as opposed to the diffusion coefficient.

As an indication of the magnitude of the flux of gas into the film, suppose that oxygen is diffusing into a film of water of height $h = 10^{-4}$ m and length $L = 1$ cm and width $W = 1$ cm. Then

$$U_{\max} = \frac{\rho g h^2}{2\mu} = \frac{1000 \times 9.8 \times 10^{-8}}{2 \times 10^{-3}} \frac{\text{m}}{\text{sec}} = .05 \frac{\text{m}}{\text{sec}} \tag{5.66}$$

Suppose we assume that the amount of O_2 dissolved in the water corresponds to $\omega_s = 0.0013$ at the free surface. Then the mass M absorbed is

$$M = \int_0^W \int_0^L n_{Ay} dz dx = 2\rho W \omega_{As} \sqrt{D_{AB} U_{\max} L/\pi} \tag{5.67}$$

or

$$M = 2\rho W \omega_{As} \sqrt{D_{AB} U_{\max} L/\pi} \tag{5.68}$$

$$= 2 \times 1000 \times 10^{-2} \times 0.0013 \sqrt{2.4 \times 10^{-9} 0.05 \times 0.01/\pi} = 3 \times 10^{-8} \text{kg } O_2 \tag{5.69}$$

a very small number indeed.

5.7 Classical Taylor–Aris dispersion

G. I. Taylor (1886–1975), the "father" of dispersion, was a physicist who made significant discoveries in a variety of areas including applied mathematics, classical physics, and fluid mechanics. His fluid mechanics work was very broad, dealing with applications in meteorology and oceanography in addition to his work in chemical separation science, his theory of dispersion. In oceanography, he is the discoverer of the Taylor column, that column of water in a rotating flow in which the velocity does not vary in a direction parallel to the rotation axis. He was an avid sailor, having, as a boy, designed and built a sailboat that he eventually sailed on the River Thames. George Batchelor, his student, was an eminent fluid mechanician in his own right. See the Cambridge book *The Life and Legacy of G. I. Taylor*, by George Batchelor, published in 1994 for more information.

Notation Alert: In keeping with Taylor's original notation we use z for the axial coordinate instead of x that heat transfer texts use.

In micro- and nanofluidics, it is often desired to follow a solute as it is transported down a tube. A solute's speed can often be used to determine its identity, and different species in a mixture can be separated based on their speed of travel through a tube. In the early 1950s, G.I. Taylor (1953, 1954) considered the transport of a small slug of solute in a tube to investigate the longtime spreading behavior of the solute slug as it moves under a pressure-driven flow in the tube. The solute will spread relative to the local flow speed because of the parabolic variation of the Poiseuille flow, termed *convective dispersion* (Probstein, 1989) and also because of solute radial diffusion and axial convection. Such a process of spreading of the solute is called *dispersion* and is depicted schematically in Figure 5.8.

Taylor (1953, 1954) was interested in the long-time behavior of the solute such that the time scale associated with the convective transport of the solute is much longer than the time scale associated with diffusion or

$$\frac{L}{w_{max}} \gg \frac{a^2}{D_A} \tag{5.70}$$

or, in dimensionless form,

$$\frac{a}{L} Pe = \epsilon Pe \ll 1 \tag{5.71}$$

where $Pe = w_{max} a / D_A$ is the mass transfer Peclet number, a is the radius of the tube, and $w_{\max}$ is the fluid velocity at the centerline of the tube, and D_A is the diffusion coefficient, dropping the solvent subscript, B. Our job here is twofold: to determine the precise distribution of the radially averaged concentration of solute as a function of the streamwise variable and to determine how much the average solute concentration spreads, quantified by the definition of a *dispersion coefficient*.

On the basis of this long–time scale analysis, Taylor began with the convective–diffusion equation for the mass fraction of species A in a fully developed flow, given by

$$\frac{\partial \omega_A}{\partial t} + w(r)\frac{\partial \omega_A}{\partial z} = D_A \frac{1}{r}\frac{\partial}{\partial r}\left(r \frac{\partial \omega_A}{\partial r}\right) \tag{5.72}$$

with the boundary conditions

$$\frac{\partial \omega_A}{\partial r} = 0 \quad \text{at } r = a, 0 \tag{5.73}$$

which expresses that no mass can cross the boundary and that the solution is symmetric about the centerline, respectively. A slug of solute is injected at a position in the channel in fully developed Poiseuille flow at $t = 0$, according to

$$\omega_A = \omega_0 \delta(z - z_0) \tag{5.74}$$

Typically, z_0 is taken to be zero. Taylor assumed that any mass transport is by axial convection and radial diffusion and that the mixture is assumed to be dilute in the species A. The task is to evaluate the axial spreading of the average mass fraction as a function of axial position relative to the average velocity of the flow.

Notation alert: In keeping with Taylor's original notation, $\langle f \rangle$ will denote an average in this section.

In quantifying dispersion by defining a dispersion coefficient, what we will eventually find is that the radially averaged mass fraction of solute will spread according to a Gaussian distribution

$$\langle \omega_A \rangle = \frac{\langle m_{A0} \rangle}{2\pi a^2 \rho \sqrt{\pi K t}} e^{\left(-\frac{(z - \langle w \rangle t)^2}{4Kt} \right)} \tag{5.75}$$

where K is the *dispersion* coefficient. This is the same solution as for the analysis of the diffusion of a δ function source in an unbounded media with the diffusion coefficient replaced by the dispersion coefficient. Skip the math, if you want to, and just plot this as a function of $z - \langle w \rangle t$ for several times.

Taylor wrote the solution in a coordinate system moving with the average fluid velocity $\bar{z} = z - \langle w \rangle t$, where $\langle w \rangle = \mathrm{w}_{\max}/2$, the average velocity in the tube. In this coordinate system, the time derivative in the two coordinate systems are related by

$$\frac{\partial}{\partial t}|_z = \frac{\partial}{\partial t}|_{\bar{z}} - \frac{\partial \bar{z}}{\partial t}_z \frac{\partial}{\partial t}|_{\bar{z}} = \frac{\partial}{\partial t}|_{\bar{z}} - \langle w \rangle \frac{\partial}{\partial t}|_{\bar{z}} \tag{5.76}$$

After substituting for the average axial velocity and neglecting the axial diffusion term, the governing equation becomes

$$\frac{a^2 w_{\max}}{D_A} \left(\frac{1}{2} - \xi^2 \right) \frac{\partial \omega_A}{\partial \bar{z}} = \frac{1}{\xi} \frac{\partial}{\partial \xi} \left(\xi \frac{\partial \omega_A}{\partial \xi} \right) \tag{5.77}$$

where $\xi = r/a$. No time derivative appears in equation (5.77) because the long time solution is quasi-steady in the coordinate system moving with the average velocity $\langle w \rangle$.

The Poiseuille velocity expression is the velocity relative to the average velocity in the tube in dimensionless form, with the axial velocity scaled on the maximum velocity. Taylor argued that at large times, $\partial \omega_A / \partial \bar{z}$ should, to a first approximation, be independent of ξ,[3] an argument similar to the argument used in fully developed heat transfer for constant heat flux. Equation (5.77) can then be integrated twice to obtain

$$\omega_A(\xi, \bar{z}) = \frac{a^2 w_{\max}}{8 D_A} \frac{\partial \omega_A}{\partial \bar{z}} \left(\xi^2 - \frac{1}{2} \xi^4 \right) + \omega_A(0, \bar{z}) \tag{5.78}$$

The average of a quantity is defined in dimensionless form as

$$\langle \omega_A \rangle = \frac{\int_0^1 \omega_A \xi d\xi}{\int_0^1 \xi d\xi} = 2 \int_0^1 \omega_A \xi d\xi \tag{5.79}$$

Taking the average of equation (5.78),

$$\langle \omega_A(\xi, \bar{z}) \rangle = 2 \int_0^1 \omega_A \xi d\xi = \frac{a^2 w_{\max}}{24 \mathrm{D_A}} \frac{\partial \langle \omega_A \rangle}{\partial \bar{z}} + \omega_A(0, \bar{z}) \tag{5.80}$$

[3] An asymptotic analysis of the problem confirms this. See Datta *et al.* (2008).

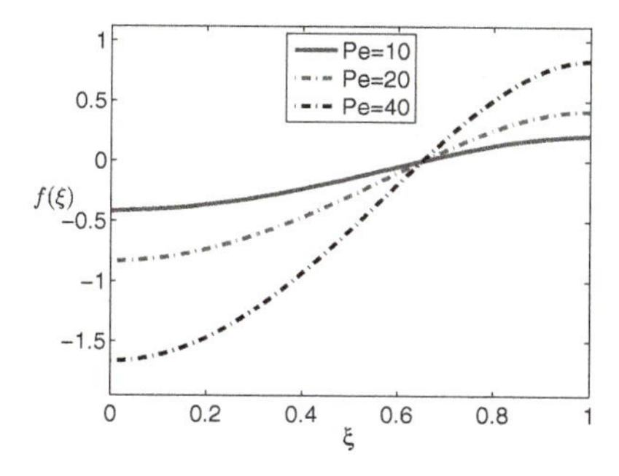

Figure 5.9 Plot of the function $(Pe/8)\,f(\xi)$ from equation (5.81), with $\bar{z}$ scaled by the radius a. The result indicates that the concentration is lower than the average near the center of the tube and higher than the average near the tube boundary as it flows through a channel or tube. Here $Pe = aw_{max}/D_A$. The broadening is called *Taylor dispersion.*

Substituting into equation (5.78), the departure from the average mass fraction is given by

$$\omega_A(\xi, \bar{z}) - \langle \omega_A(\xi, \bar{z}) \rangle = \frac{a^2 w_{\max}}{8 D_A} \frac{\partial \omega_A}{\partial \bar{z}} \left(\xi^2 - \frac{1}{2}\xi^4 - \frac{1}{3} \right) \tag{5.81}$$

Writing the right side of this equation in dimensionless form, we have

$$\omega_A(\xi, \bar{z}) - \langle \omega_A(\xi, \bar{z}) \rangle = \frac{\epsilon Pe}{8} \frac{\partial \omega_A}{\partial \hat{z}} \left(\xi^2 - \frac{1}{2}\xi^4 - \frac{1}{3} \right) \tag{5.82}$$

where $\hat{z} = \bar{z}/L$, $\epsilon = a/L$, and $Pe = w_{\max} a/D_A$. Recalling that from equation (5.71), $\epsilon Pe \ll 1$, so that the amount of dispersion is small.

A plot of

$$f(\xi) = \frac{Pe}{8} \left(\xi^2 - \frac{1}{2}\xi^4 - \frac{1}{3} \right) \tag{5.83}$$

shows that it is negative in the center of the channel and positive at the wall, as shown in Figure 5.9 for $\partial \omega_A / \partial \bar{z} > 0$, so that the mass fraction of the solute is higher near the tube boundary $\xi = 1$. Conversely, for $\partial \omega_A / \partial \bar{z} < 0$, the mass fraction is lower near the wall of the tube. This radial variation in mass fraction originates from the radial variation in w and gives rise to dispersion or axial spreading of the solute, as discussed later. The effect is more pronounced the larger the Peclet number. Thus, near the center at higher velocities, diffusion is outward toward the wall, whereas at the lower velocities near the wall, the diffusion is inward.

To determine the *dispersion coefficient* in the form given by Taylor, we note that the average mass flux of species A through the pulse relative to the average velocity is given by

$$\begin{aligned} \langle n_A \rangle &= \frac{\rho}{\pi a^2} \int_0^a \langle (\omega_A(\xi, \bar{z}) - \langle \omega_A(\xi, \bar{z}) \rangle)(w - \langle w \rangle) \rangle \xi \, d\xi \\ &= \frac{\rho a^2 \langle w \rangle^2}{D_A} \frac{\partial \langle \omega_A \rangle}{\partial \bar{z}} \int_0^1 \left(\xi^2 - \frac{1}{2}\xi^4 - \frac{1}{3} \right) \left(\frac{1}{2} - \xi^2 \right) \xi \, d\xi \end{aligned} \tag{5.84}$$

or

$$\langle n_A \rangle = -\frac{\rho a^2 \langle w \rangle^2}{48 D_A} \frac{\partial \langle \omega_A \rangle}{\partial \bar{z}} = -K\rho \frac{\partial \langle \omega_A \rangle}{\partial \bar{z}} \tag{5.85}$$

Here K is the *dispersion coefficient*

$$K = \frac{a^2 \langle w \rangle^2}{48 D_A} = \frac{D_A}{48} Pe_{\text{ave}}^2 \tag{5.86}$$

The dispersion coefficient measures the deviation of the average solute flux from the product of the averaged mass fraction and averaged velocity. Note that in most practical applications, dispersion should be minimized.

Let us now write a mass balance in terms of the average mass fraction across the tube and across a plane moving with the average fluid velocity gives

$$\rho \frac{\partial \langle \omega_A \rangle}{\partial t} = -\frac{\partial \langle n_A \rangle}{\partial \bar{z}} \tag{5.87}$$

Equations (5.85) and (5.86) can now be used to predict the axial spreading originating from the velocity profile $w(r)$. Substituting for $\langle n_A \rangle$ from equation (5.85),

$$\frac{\partial \langle \omega_A \rangle}{\partial t} = -\frac{\partial \langle n_A \rangle}{\partial \bar{z}} = K \frac{\partial^2 \langle \omega_A \rangle}{\partial \bar{z}^2} \tag{5.88}$$

The solution to this equation for the case where the solute is initially at a point having an averaged mass indicated by $\langle m_{A0} \rangle$ at time $t = 0$ is given by

$$\langle \omega_A \rangle = \frac{\langle m_{A0} \rangle}{2\pi a^2 \rho \sqrt{\pi K t}} e^{\left(-\frac{(z - \langle w \rangle t)^2}{4Kt} \right)} \tag{5.89}$$

Thus the radially averaged mass fraction of species A decays axially according to a Gaussian distribution locally about the point $\bar{z} = \langle w \rangle t$ in the same way that a solute released from a point in free space disperses with the diffusion coefficient here replaced by the dispersion coefficient.

Strictly speaking, the Taylor solution is only valid for large Peclet number and for axial diffusion negligible with respect to the dispersion ($Pe_{\text{ave}} > 7$ approximately); Aris (1956) removed the assumption of large Peclet number and extended Taylor's solution to include axial diffusion. His result for the dispersion coefficient K is

$$K = D_A \left(\frac{Pe_{\text{ave}}^2}{48} + 1 \right) \tag{5.90}$$

Note that for $Pe_{\text{ave}} < \sqrt{48} \simeq 7$, axial diffusion increases axial dispersion, and for $Pe_{\text{ave}} > \sqrt{48}$, axial diffusion decreases axial dispersion. Bird *et al.* (2002) show a figure summarizing the regions of validity of the Taylor and Aris formulas. They suggest that the long time limit corresponds to a dimensionless time of

$$\frac{D_A t}{a^2} \sim 1$$

and so for a diffusion coefficient of $D_A \sim 10^{-9}$ m^2/sec and $a \sim 1$ μm, $t \sim 1$ msec is a small time scale indeed.

It is evident that because of the similarity between the heat and mass transfer equations, there is a similar Taylor–Aris dispersion problem for temperature, with the mass fraction ω_A replaced by a suitably defined average temperature (Leal, 2007).

5.8 The stochastic nature of diffusion: Brownian motion

It was shown that the concentration distribution of a solute undergoing Taylor–Aris dispersion from a point source satisfies a Gaussian distribution. This fact was derived deterministically; however, the Gaussian distribution is most associated with a statistical approach, and this interpretation is discussed here. In this statistical analysis, the diffusion coefficient emerges naturally as the variance of all possible positions that molecules can inhabit.

The molecules in a stagnant gas undergo rapid and very high frequency oscillations arising from the collisions with other molecules; the same is true in a liquid, although the frequency is somewhat different because of the proximity of other molecules. This oscillatory motion occurs simultaneously with any other fluid motion, such as electro-osmosis, and is called *Brownian motion* after the botanist Robert Brown (Daune, 1993), who showed that rapid oscillations of molecules are ubiquitous (Stachel, 1998). Because these oscillations are so rapid, corresponding to about 10^{14} collisions per second for a macromolecule in a liquid (Daune, 1993), the precise trajectory of each molecule cannot be observed.

Robert Brown

Brownian motion is essentially a random process and considers the random motion of n particles in the x direction. Suppose $n(x, t)$ is the number of molecules at the given position x at a given time t. Einstein (1905a, 1905b) first exhibited the stochastic nature of diffusion using a simple approach, as follows, in the second of five papers that changed the course of science in 1905 (Stachel, 1998). (He eventually wrote a book on the subject (Einstein, 1956).) Fixing x at a later time, say, $t + \tau$, the total number of particles is given by the sum over all the possible positions b from which the molecules could move from $x - b$ to x, weighted by a probability function $p(b)$, or

$$n(x, t + \tau) = \int_{-\infty}^{\infty} n(x - a, t)p(b)db \tag{5.91}$$

where the probability function p satisfies the constraints

$$\int_{-\infty}^{\infty} p(b)db = 1 \tag{5.92}$$

$$\int_{-\infty}^{\infty} bp(b)db = 0 \tag{5.93}$$

Expanding the function $n(x, t)$ in a Taylor series in both space and time, assuming small displacements, and then substituting into equation (5.91), it follows that

$$n(x, t) + \tau\frac{\partial n}{\partial t} + \frac{\tau^2}{2}\frac{\partial^2 n}{\partial t^2} = \int_{-\infty}^{\infty} n(x, t)p(b)db - \frac{\partial n}{\partial x}\int_{-\infty}^{\infty} bp(b)db + \frac{\partial^2 n}{\partial x^2}\int_{-\infty}^{\infty}\frac{b^2}{2}p(b)db + \cdots \tag{5.94}$$

The second term on the right-hand side vanishes, and the integral condition on p results in the first term being just $n(x, t)$. Thus, for small τ and small b, to leading order,

$$\frac{\partial n}{\partial t} = \frac{1}{2\tau}\int_{-\infty}^{\infty} b^2 p(b)db\frac{\partial^2 n}{\partial x^2} = D\frac{\partial^2 n}{\partial x^2} \tag{5.95}$$

if the coefficient D is taken to be

$$D = \frac{1}{2\tau}\int_{-\infty}^{\infty} b^2 p(b)db = \frac{\sigma}{2\tau} \tag{5.96}$$

where σ denotes the variance. Equation (5.95) is recognized as the one-dimensional diffusion equation, and D is the translational diffusion coefficient, which is linearly related to the variance of b. Note that because n and the concentration are proportional, equation (5.95) can also be written in terms of the concentration.

The deterministic solution to this equation with a number of molecules n_0 initially concentrated at $x = 0$ is the same as the Gaussian distribution

$$n(x, t) = \frac{n_0}{2\sqrt{\pi Dt}}e^{\frac{-x^2}{4Dt}} \tag{5.97}$$

where $n(x, t)$ is the number of molecules per length and can also be interpreted as the probability that a particle will reach the point x after N steps:

$$P(x, N) = (\pi\sigma^2)^{-1/2}e^{\frac{-x^2}{\sigma^2}} \tag{5.98}$$

where σ is the variance and $\sigma = 2\sqrt{Dt} = \sqrt{N}b$.

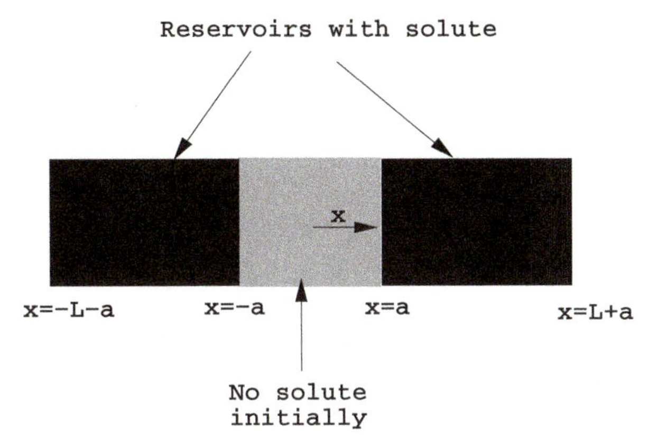

Figure 5.10 Geometry of the problem considered by Wilson. This problem is a simple model for drug delivery to a membrane, here defined as the region $-a \leq x \leq a$.

The same type of analysis is valid for a set of solute particles in a solvent, provided that the particles are large compared to the solvent molecules. In that case, the diffusion coefficient is given by the Stokes–Einstein equation.

The effect of Brownian motion is usually not accounted for in models because the long time effect on the trajectories of molecules and particles integrates out because of the random nature of the process. This is certainly true in the absence of solid surfaces. However, Brownian motion may be a factor when measuring fluid velocities near a wall using neutrally bouyant particles because these effects do not integrate out. The particles can bounce off the walls, which will obviate the random nature of the process.

In the work of Sadr *et al.* (2007), designed to estimate slip length, the authors show experimentally and computationally that Brownian motion can lead to an overestimate of particle velocities near a wall and hence an overestimate for the slip length. Moreover, the effect of Brownian motion may be even more important in nanochannels under $O(100 \text{ nm})$ in the smallest dimension. At such dimensions, the particles are always near a wall, making a suitable Brownian correction essential. Improving the spatial resolution of these velocimetry techniques will require the use of smaller particles on the order of several nanometers in radius, such as quantum dots, in contrast to the $\sim$100 nm particles used by Sadr *et al.* (2007). The accuracy of methods to measure slip lengths at small scales is a matter for future research.

5.9 Unsteady mass transport in uncharged membranes

The delivery of a solute to an uncharged membrane has been analyzed by Wilson (1948), who considered mass transport to a central membrane, as depicted in Figure 5.10. The governing equation is given by

$$\frac{\partial c_A}{\partial t} = D_A \frac{\partial^2 c_A}{\partial x^2} \tag{5.99}$$

where the solute A may be an uncharged biomolecule such as glucose.

The unique feature of Wilson's work is that instead of the concentration, he works with

$$M(x,t) = \int_0^x c_A(x,t)dx \tag{5.100}$$

which is the total amount of the species per unit area of the cross section transported into the membrane from $x = 0$ to $x = x$. As depicted in Figure 5.11, solute is contained in reservoirs surrounding the membrane, with only solvent in the membrane. At time $t = 0$, fluid is allowed to diffuse into the membrane. The membrane is of finite size compared with the upstream and downstream reservoirs so that only a fixed amount of solute can diffuse into the membrane. The concentration of solute in the reservoirs is assumed to be uniform in space but not in time. Again, referring to Figure 5.11, the concentration at $x = \pm a$ will vary with time.

Substituting equation (5.100) into equation (5.99), we find that

$$\frac{\partial M}{\partial t} = D_A \frac{\partial^2 M}{\partial x^2} \tag{5.101}$$

and symmetry of the problem requires

$$\frac{\partial^2 M}{\partial x^2} = 0 \quad \text{at } x = 0 \tag{5.102}$$

The boundary condition at $x = a$ was described by March and Weaver (1928), who analyzed the corresponding heat transfer problem. Suppose initially that the concentration in the reservoir is c_0. Then its concentration at some time t later will be c_0 minus what has been diffused away, or

$$Lc_A(a,t) = Lc_0 - D_A \int_0^t \frac{\partial c_A}{\partial x}|_{x=a} dt \quad \text{at } x = a \tag{5.103}$$

where L is the length scale associated with the reservoirs. Differentiation of this equation with respect to time leads to

$$L\frac{\partial c_A}{\partial t} = -D_A \frac{\partial c_A}{\partial x} \quad \text{at } x = a \tag{5.104}$$

The boundary condition in this form is difficult to implement because it is of mixed type; that is, it contains both a time and space derivative.

Returning to equation (5.103), Wilson argues that the concentration of free solute at $x = \pm a$ must be continuous so that the concentration of solute in the bath is given by $\partial M/\partial x$ at $x = a$. Using the definition of M, and assuming equation (5.101) holds at $x = a$, the condition of the total amount of solute remaining fixed is

$$M + L\frac{\partial M}{\partial x} = Lc_0 \quad \text{at } x = a \tag{5.105}$$

This condition is recognized as a convection-type boundary condition similar to that in heat transfer discussed in Chapter 3. The initial condition is

$$M = 0 \quad \text{for } -a \leq x \leq a \tag{5.106}$$

$$M = c_0 L \text{ in the reservoirs} \tag{5.107}$$

This completes the specification of the problem.

It is, however, useful to remove the constant in the boundary condition at $x = a$. Let $M = Bx + f(x,t)$. Then the problem for f is

$$\frac{\partial f}{\partial t} = D_A \frac{\partial^2 f}{\partial x^2} \tag{5.108}$$

with conditions

$$\frac{\partial^2 f}{\partial x^2} = 0 \quad \text{at } x = 0 \tag{5.109}$$

$$L\frac{\partial f}{\partial x} + f = 0 \quad \text{at } x = a \tag{5.110}$$

and initial condition

$$f(x,0) = -\frac{L c_0 x}{a + L} \quad \text{at } t = 0 \tag{5.111}$$

Here $B = L/a + L\,(c_0)$.

Note that the form of the solution for M is the superposition of the steady state and a transient. The problem is solved using Fourier series methods, and the eigenfunctions given by

$$f_n = e^{-p_n t} \sin \frac{q_n x}{a} \tag{5.112}$$

satisfy the partial differential equation if

$$p_n = \frac{D_A q_n^2}{a^2} \tag{5.113}$$

Thus the general solution is given by

$$f(x,t) = \sum_n A_n e^{-p_n t} \sin \frac{q_n x}{a} \tag{5.114}$$

where the symmetry condition at $x = 0$ rules out the cosine functions. The full solution for the amount of solute in the membrane per unit area of the sheet is given by

$$M(x,t) = \frac{L c_0 x}{a + L} + \sum_n A_n e^{-p_n t} \sin \frac{q_n x}{a} \tag{5.115}$$

The coefficients A_n, p_n, and q_n are obtained by application of the boundary conditions and, for example,

$$\tan q_n = -\frac{L q_n}{a} \tag{5.116}$$

$$A_n = \frac{2aLc_0}{a^2 + aL + q_n^2 L^2} \frac{a}{q_n \cos q_n} \tag{5.117}$$

The values of q_n must be nonzero and positive.

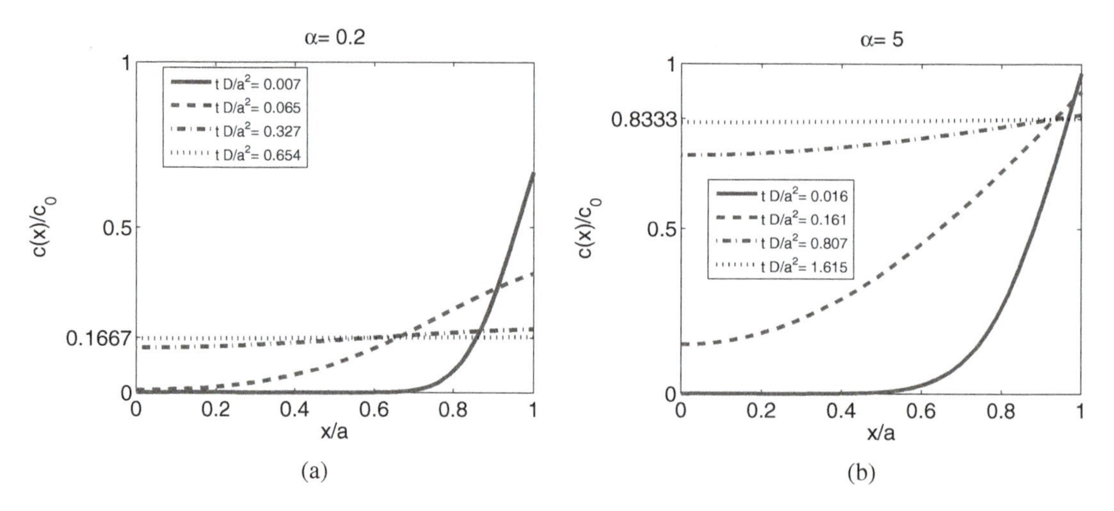

Figure 5.11 Sketch of the solution to Wilson's problem for two values of $\alpha = L/a$. Here $c(x) = c_A/c_0$ is the dimensionless concentration. (a) $\alpha = 0.2$. (b) $\alpha = 5$.

The amount of mass of solute absorbed into the membrane is given by $M(a, t)$ and

$$\frac{M(a,t)}{M_\infty} = 1 - \sum_{n=1}^{\infty} \frac{2\alpha(\alpha+1)}{1+\alpha+\alpha^2 q_n^2} e^{-\beta q_n t} \tag{5.118}$$

where $\alpha = L/a$ and $\beta = D_A/a^2$, and

$$M_\infty = \frac{2aLc_0}{a+L} \tag{5.119}$$

obtained by evaluating $2M(a, \infty)$. Note that the amount of solute absorbed by the membrane is dependent on the dimensions of the reservoir and sheet and on the initial concentration. In the limit as $a/L \to 0$, we have

$$M_\infty = 2ac_0 \tag{5.120}$$

and $q_n = (n + 1/2)\pi$.

Figure 5.11 depicts results for two different values of $\alpha = L/a$. The value of the dimensionless steady state concentration is $c(x) = L/L + a$; at $tD/a^2 \sim 1.6$, the solution for $\alpha = 5$ has reached steady state, while the result for $\alpha = 0.2$, steady state is reached much more quickly.

As an example, consider the diffusion of glucose into a nanopore membrane. A typical reservoir size may be $L = 40$ mm and membrane length $a = 1$ μm. For a solvent concentration of $c_0 = 4$ mole/L, since $L/a \gg 1$,

$$M_\infty = 2ac_0 = 8 \times 10^{-3}\,\frac{\text{mole}}{\text{m}^2}$$

Note that the amount of solute absorbed into the membrane is independent of the diffusion coefficient; however, the rate at which the process occurs is dependent on the diffusion coefficient through the characteristic time scale $\tau = a^2/D_A$. For glucose, $D_A \sim 10^{-9}$, and so $\tau \sim 1$ msec.

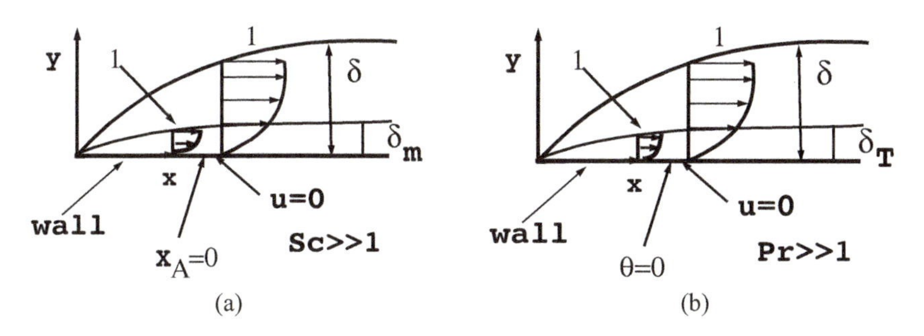

Figure 5.12 Sketch of the boundary layer thicknesses for the flow and heat and mass transfer from a flat plate; the coordinate system (x, y) is shown. For a liquid $\delta \gg \delta_T \gg \delta_m$, and the values of the scaled variables θ and X_A at the wall and the boundary layer edge are shown.

5.10 Temperature and concentration boundary layers

So far, we have considered internal flows exclusively. Yet the flow, heat, and mass transfer in the entrance to a pipe or channel is similar to, though not quite the same as, that in an external boundary layer flow. However, the external boundary layer solution, as with the fluid dynamics problem, is a good first approximation to the heat and mass transfer problems in the developing region.

Consider the steady flow past a flat plate unbounded on one side so that the boundary layer on the plate is thin. This means that the Reynolds number is nominally $Re \gg 1$. Assuming a single length scale, say, L, the governing equations for the mole fraction of a species A and the temperature are

$$u\frac{\partial X_A}{\partial x} + v\frac{\partial X_A}{\partial y} = \frac{1}{ReSc}\frac{\partial^2 X_A}{\partial y^2} \tag{5.121}$$

$$u\frac{\partial \theta}{\partial x} + v\frac{\partial \theta}{\partial y} = \frac{1}{RePr}\frac{\partial^2 \theta}{\partial y^2} \tag{5.122}$$

Assume constant temperature and mass fraction boundary conditions so that, in dimensionless form,

$$X_A = \theta = 0 \ \ y = 0 \tag{5.123}$$

$$X_A = \theta = 1 \quad \text{at } y = \delta_m, \delta_T \tag{5.124}$$

$$X_A = \theta = 1 \ \ x = 0 \tag{5.125}$$

where δ_m and δ_T correspond to the boundary layer edge, as in Figure 5.12. Here, for example, the dimensionless temperature distribution is defined by

$$\theta = \frac{T - T_w}{T_0 - T_w} \tag{5.126}$$

with a corresponding expression for X_A, where T_0 and T_w are the temperature at $x = 0$ and at the wall, respectively; here it is assumed that $T_0 = T_\infty$, the temperature far from the plate. The velocity is a uniform free stream, and there is no pressure gradient.

In the problems of interest here for liquids, $Re \ll Pe_T \ll Pe_m$, where $Pe_T = RePr$ is the thermal Peclet number and $Pe_m = ReSc$ is the mass transfer Peclet number. Note that for air and other perfect gases, $Pr \sim 1$ and $Sc \sim 1$. The Prandtl number for water is $Pe \sim 5$–15, depending on temperature, and for aqueous solutions, the Schmidt number is $Sc \sim 1000$. Thus the mass transfer boundary layer is much thinner than the thermal boundary layer, which is just a bit thinner than the velocity boundary layer. Because the Reynolds and Peclet numbers are large, the temperature and mass fraction vary rapidly near the wall. On the basis of this realization, and recalling the Blasius boundary layer problem of the previous chapter, a boundary layer variable can be defined as $\hat{y} = y/Re^{-1/2}$. Clearly the same similarity analysis that was done for the Blasius boundary layer problem applies to the heat and mass transfer problems. Thus, using the same similarity analysis, it is found that θ and X_A satisfy

$$\theta'' + Prf\theta' = 0 \tag{5.127}$$

$$X_A'' + ScfX_A' = 0 \tag{5.128}$$

where the prime refers to differentiation with respect to $\eta = \hat{y}/\sqrt{2x}$, the Blasius boundary layer variable, and f is the Blasius boundary layer function, defined in the previous chapter.

These two equations may be integrated analytically for arbitrary Prandtl and Schmidt numbers, and with the outer boundary condition applied as $\eta \to \infty$, the solution for the temperature is

$$\theta = \frac{\int_0^\eta e^{-Pr\int_0^\eta f ds} d\eta}{\int_0^\infty e^{-Pr\int_0^\infty f ds} d\eta} \tag{5.129}$$

and similarly for the mole fraction.

For large values of the Prandtl and Schmidt numbers, the thermal and mass transfer boundary layers are much thinner than the fluid boundary layer, and so the value of the function f near the wall may be used in place of the full numerical function obtained from the Blasius solution. It is shown from the numerical solution of the Blasius problem that $f \sim 0.664\eta$ near the wall with the present variables, and so, for example,

$$\theta = \frac{\int_0^\eta e^{-CPr\eta^3} d\eta}{\int_0^\infty e^{-CPr\eta^3} d\eta} \tag{5.130}$$

where $C = 0.332/3$. Equation (5.130) indicates that for large Prandtl number, a new boundary layer variable may be defined as $\hat{\eta} = \eta/Pr^{-1/3}$, which will be seen to be the origin of the $Pr^{1/3}$ dependence on the Nusselt number.

The large Prandtl number dependence of the boundary layer variable would suggest that the ratio of the boundary layer thicknesses should be

$$\frac{\delta_T}{\delta} \sim Pr^{-1/3} \tag{5.131}$$

with the same dependence of δ_m on the Schmidt number. However, the structure of each of the equations is a bit different from the Blasius equation, and numerical evaluation of the solution to the preceding temperature problem indicates that

$$\frac{\delta_T}{\delta} \sim Pr^{-0.4} \tag{5.132}$$

The heat flux at the wall is defined by

$$q_w'' = -k\frac{\partial T}{\partial y}|_{y=0} = -k(T_w - T_0)\theta'(0)\frac{\partial \eta}{\partial y} \tag{5.133}$$

where $\eta = \hat{y}/\sqrt{2x}$. The Nusselt number is thus given by

$$Nu_x = \frac{q_w x}{k(T_w - T_0)} = \frac{h_T x}{k} = -\frac{\theta'(0)}{\sqrt{2}}Re_x^{1/2} \tag{5.134}$$

where

$$\theta'(0) = \frac{-1}{\int_0^\eta e^{-Pr\int_0^\infty f ds} d\eta} \tag{5.135}$$

The value of $\theta'(0)$ can be evaluated from the integral numerically in the general case, and according to White (2006), a good curve fit valid for $0.1 < Pr < 10{,}000$ is

$$\frac{\theta'(0)}{\sqrt{2}} = 0.332Pr^{1/3} \tag{5.136}$$

so that

$$Nu_x = 0.332Re_x^{1/2}Pr^{1/3} \tag{5.137}$$

The Sherwood number is the analogue of the Nusselt number for mass transfer, and the result follows easily from the result for the Nusselt number and

$$Sh_x = \frac{h_m L}{D_{AB}} = \frac{X_A'(0)}{\sqrt{2}}Re_x^{1/2} \tag{5.138}$$

with the similar result that

$$Sh_x = 0.332Re_x^{1/2}Sc^{1/3} \tag{5.139}$$

5.11 Chapter summary

Heat and mass transfer share a number of common themes, and in this chapter, this relationship is apparent. First, simple fully developed temperature profiles are presented and analogies of these simple temperature distributions with Poiseuille and Couette flow are identified. Next, it is shown that the mass transfer and thermal entry lengths have the same form and are straightforward extensions of the hydrodynamic entrance length. The definition of a fully developed temperature distribution in a channel is presented and discussed for the two limiting cases of constant heat flux and constant temperature at the bounding surface. However, because of the magnitude of the Schmidt number, internal flow, mass transfer in liquid flows is almost never fully developed.

What follows next is a survey of several problems that have analytical, exact or nearly-exact solutions. These problem include:

- The Graetz problem: the axially developing temperature distribution in a fully developed fluid flow in a rectangular channel, valid at large thermal Peclet number
- Mass transfer in a thin liquid film for small diffusion coefficient
- Taylor-Aris dispersion, the spreading of a solute in a fully developed pressure driven flow
- Unsteady mass transport in an uncharged membrane
- Heat and mass transport in high Reynolds number boundary layers.

It is also shown in this chapter that the diffusion coefficient can be interpreted stochastically as the variance of the probability that a particle could originate from a position $x - b$ at time t given the position at a later time $t + \tau$. This interpretation of diffusion reinforces the concept that indeed, mass diffusion is a random process.

Chapters 4 and 5 provide an overview of the disciplines within the thermal sciences, called fluid mechanics and heat and mass transfer. In the next three chapters we leave this discipline to study electrostatics, electrochemistry and molecular biology all of which are necessary for the study of electrokinetic phenomena to follow.

EXERCISES

5.1 Determine the solution for the velocity of the superposition of Couette flow and Poiseuille flow, and identify the parameters for which the resulting expression is equivalent to the superposition of one dimensional thermal conduction and a duct with constant heat generation. Is there a reasonable set of parameters such that the two dimensionless distributions have the same value?

5.2 The dimensions of a typical channel used in microfluidic applications are height $h = 20$ nm, length in the flow direction $L = 3$ μm, and width $W = 40$ μm. Show that the concentration distribution will not be fully developed for a flow rate of $Q = 10^{-12}\ \frac{m^3}{\text{sec}}$.

5.3 Write equation (5.14) in dimensionless form, and identify the appropriate temperature scale.

5.4 Integrate the expressions for the Nusselt number and the Sherwood number for the boundary layer over the length of the plate and thus define averaged parameters. Calculate the ratio of these parameters for air and water.

5.5 Beginning with the dimensional form of the energy equation, verify by the proper nondimensionalization the dimensionless equation and boundary conditions for the Graetz problem. Identify the temperature scale.

5.6 Calculate the solution to the Graetz problem discussed in the text for $\mathrm{Pr} \gg 1$ in a cylindrical tube.

5.7 In electro-osmotic flow, the velocity in a channel can be a plug flow for thin electrical double layers. Using the boundary conditions appropriate for the Graetz problem in section 5.5, calculate the solution for the temperature in dimensionless form for a wide channel, assuming constant properties and $u = U_0 =$ constant. How would the solution for a circular tube be different?

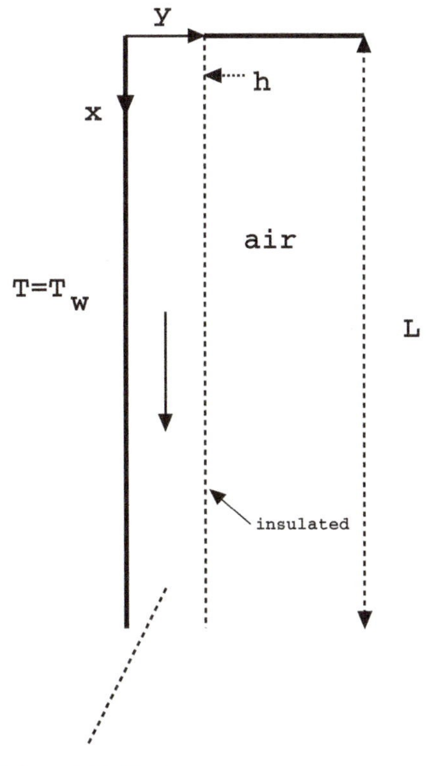

Exercise 5.8

5.8 Water is flowing vertically in the direction of gravity down a plane wall, as shown. Outside the liquid film is air. The energy balance is between conduction and convection. Calculate the temperature of the water inside the liquid film if the free surface is insulated (the thermal conductivity of air is much smaller than that of water) and the wall temperature is constant $T = T_w$ at $y = 0$ and $T = T_e$ at the entrance $x = 0$. Write the expression in dimensionless form, and identify the appropriate temperature scale. Compare with the associated mass transfer problem.

5.9 The mass transfer Graetz problem for flow in a channel having solid walls is

$$\epsilon Pe y(1-y)\frac{\partial \omega}{\partial x} = \frac{\partial^2 \omega}{\partial y^2} + \epsilon^2 \frac{\partial^2 \omega}{\partial x^2}$$

subject to $\omega = \omega_0$ at $x = 0$ and

$$\frac{\partial \omega}{\partial y} = 0 \quad \text{at } y = 0, 1$$

Here $\epsilon = h/L$ and $Pe = Uh/D_A$. Show explicitly that the only solution to this problem for arbitrary Peclet number $Pe = Pe_m = \text{ReSc}$ is $\omega = \omega_0$ and thus there is no dispersion. Does this result contradict the Taylor-Aris solution?

5.10 The solution for an initially concentrated pulse of solute having mass $\langle m_{A0} \rangle$ in an infinite medium is given by

$$\langle \omega_A \rangle = \frac{\langle m_{A0} \rangle}{2\pi a^2 \rho \sqrt{\pi K t}} e^{\left(-\frac{(z-\langle w \rangle t)^2}{4Kt}\right)}$$

Show that this distribution satisfies equation (5.88). Calculate

$$\frac{\partial \langle \omega_A \rangle}{\partial \bar{z}}$$

Plot the solution at several fixed points $\eta = z - \langle w \rangle t$ as a function of time. Write the solution in dimensionless form. Is it easier to plot?

5.11 Repeating the Aris procedure for mass dispersion in a tube, show that the dispersion coefficient for a wide channel of height $2h$ is given by

$$K = D_A \left(1 + \frac{2h^2 \langle u \rangle^2}{105 D_A^2}\right)$$

You might want to solve the problem in dimensionless form. See the discussion of dispersion in wide channels $W/h \to \infty$ in the paper by **?** the limit of which does not correspond to this result.

5.12 Extend the Wilson problem to the case where the reservoirs are not the same size and the symmetry condition at $x = 0$ is no longer appropriate. Is an analytical solution possible?

5.13 Certain types of cells are spherical and assume that the concentration of a solute outside of the cells satisfies

$$\frac{\partial c_A}{\partial t} = D_A \frac{1}{r^2} \frac{\partial}{\partial} \left[r^2 \frac{\partial c_A}{\partial r} \right]$$

with boundary conditions

$$c_A(r,t) = c_0 \quad \text{for } r \geq a \quad \text{at } t = 0$$

$$c_A(r,t) = \quad \text{at} \quad r = a \quad \text{for } t > 0$$

$$c_A(r,t) = c_0 \quad \text{for } r \to \infty \quad \text{for } t > 0$$

Find the solution for the concentration c_A, and show that the steady state solution is

$$\frac{c_A}{c_0} = 1 - \frac{a}{r}$$

Hint: Make a change of variable $C = rc_A$, and show that C satisfies the planar equation.

5.14 A solute used as a drug is confined to a thin band of width $2a$ at time $t = 0$ so that

$$\frac{\partial c_A}{\partial t} = D_A \frac{\partial^2 c_A}{\partial x^2}$$

This is the opposite of Wilson's problem. Using Wilson's procedure, find the solution for the concentration in terms of the total amount of mass of solute, M. Here the initial condition is

$$M = 0 \quad | x | > a$$

$$M = c_0 a \quad | x | \leq a$$

How would you find the solution for $x \geq a$ or $a \to 0$?

5.15 A solute used for drug delivery diffuses through tissue according to the equation

$$\frac{\partial c_A}{\partial t} = D_A \frac{\partial^2 c_A}{\partial x^2} - k_e c_A$$

subject to a maintained concentration at $x = 0$

$$c_A(x,t) = c_0 \quad \text{for } x = 0 \quad \text{for } t > 0$$

$$c_A(x,t) = 0 \quad \text{for } x \to \infty \quad \text{for } t > 0$$

$$c_A(x,t) = 0 \quad \text{for } t = 0$$

Here k_e is an elimination constant and has units of sec^{-1}. Show that the steady state solution is

$$\frac{c_A}{c_0} = e^{-x\sqrt{\frac{k_e}{D_A}}}$$

Write the solution in dimensionless form. Plot the solution for $D_A = 1 \times 10^{-7}$ cm^2/sec and $k_e = 10^{-8} - 10^{-4}$ sec^{-1}. Calculate the effectiveness of drug delivery defined as

$$\eta = \frac{\bar{c}_A}{c_0}$$

where

$$\bar{c}_A = \frac{\int_0^L c_A(x)dx}{L}$$

is the average concentration in the tissue over a length L.

5.16 A solute satisfies

$$D_A \frac{\partial^2 c_A}{\partial x^2} - k_e c_A = 0$$

subject to a maintained concentration at $x = 0$

$$c_A(x) = c_0 \quad \text{for } x = 0 \quad \text{for } t > 0$$

$$N_A = -D_A \frac{dc_A}{dx} = 0 \quad \text{for } x = L$$

Solve this problem, and calculate the average concentration

$$\bar{c}_A = \frac{\int_0^L c_A(x)dx}{L}$$

Find an expression for the flux at $x = 0$. Take $D_A = 1 \times 10^{-7}$ cm^2/sec and $k_e = 10^{-8} - 10^{-4}$ sec^{-1}, $L = 1$ cm.

6 Introduction to Electrostatics

6.1 Introduction

Electrostatics and electrodynamics are essential features of micro- and nanofluidics. At the macroscale pumps, compressors and other fluid-handling devices move fluid using a pressure difference, as was noted in Chapter 4. However, as was demonstrated in Chapter 1, this is often not feasible at the micro- and nanoscale because very large pressure drops on the order of atmospheres are required for flows through channels under about 0.1 μm, channels that are often used in applications.

The objective of this chapter is to introduce, in a general way, the concept of an electric field at the advanced undergraduate level, typically associated with the material presented in an introductory physics class for physics majors; for those students already familiar with the principles of electrostatics, this chapter may be scanned or skipped.

In this chapter, the basics of electrostatics are presented, paying particular attention to the electric field associated with one or more charged spheres, planes, or cylindrical surfaces. In the course of discussion, the concept of a *surface charge density* is defined, which is a concept central to electrochemistry and micro- and nanofluidics. This is followed by a discussion of the fundamental law of Gauss relating the electric field to the charge density of a given medium. The concept of electrical potential is then introduced, and the similarities between the velocity potential of fluid mechanics are noted.

Next, the concept of an electric dipole is introduced, followed by the derivation of the Poisson equation of electrostatics, which is solved for several different situations. Current and current density are then defined, and all these topics will lead into a detailed discussion of the electric double layer in an electrolyte mixture, presented in the next chapter.

6.2 Coulomb's law: The electric field

Most of the physical laws bear the name of their originator and Coulomb's Law is no different. **Charles Augustin de Coulomb** (1736–1806) was a French physicist from a rather aristocratic family who performed experiments that resulted in the law that bears his name. Actually while Coloumb's Law bears his name it was a contemporary, Joseph Priestley (1733–1804), a chemist and natural philosopher who anticipated his results in a comprehensive treatise on electricity (Priestley, 1767) twenty years before Coulomb's demonstration. Priestley is best known as the discoverer of oxygen. Other of his books include *Essay on the First Principles of Government*. A Renaissance man indeed.

Electrically conducting fluids are the norm in micro- and nanofluidics. To begin to discuss how such electrically conducting fluids may be made to move under the action of an electric field, it is useful to discuss the underlying theory of electrostatics in some detail. Recall from the introduction of the concept of an electric field in Chapter 1 that the force acting between two particles of charge q and q' is

$$F = \frac{qq'}{4\pi \epsilon_e r^2} \tag{6.1}$$

where ϵ_e is the electrical permittivity of the medium and r is the distance between the two charges; if the charged particles are atoms or ions, $|q| = 1.602 \times 10^{-19}$ C, the charge of one proton.

The electric field at any point is defined as the force per unit charge acting on a single charge at that point, or

$$E = \frac{F}{q'} = \frac{q}{4\pi \epsilon_e r^2} \frac{N}{C} \tag{6.2}$$

where r is the distance from the point charge. This equation is known as *Coulomb's law*. Note that q is positive and that the electric field is directed outward for a positive charge in the radial direction (Figure 6.1) and is directed radially inward if the charge is negative.

If a number of charges are present, the force at charge q_i, due to all the other charges, is given by

$$\vec{F}_i = \sum_{j \neq i} \frac{q_i q_j (\vec{r}_i - \vec{r}_j)}{4\pi \epsilon_e \mid \vec{r}_i - \vec{r}_j \mid^3} \tag{6.3}$$

where $\vec{r}_i$, for example, is the vector distance of the charge q_i from the origin of the coordinate system. The electric field at the point q_i, due to all the other charges, is the vector sum of the field produced by all the charges:

$$\vec{E}(\vec{r}_i) = \frac{1}{4\pi \epsilon_e} \sum_{i \neq j} \frac{q_j (\vec{r}_i - \vec{r}_j)}{\mid \vec{r}_i - \vec{r}_j \mid^3} \tag{6.4}$$

The electric field lines for two charges are shown in Figure 6.2.

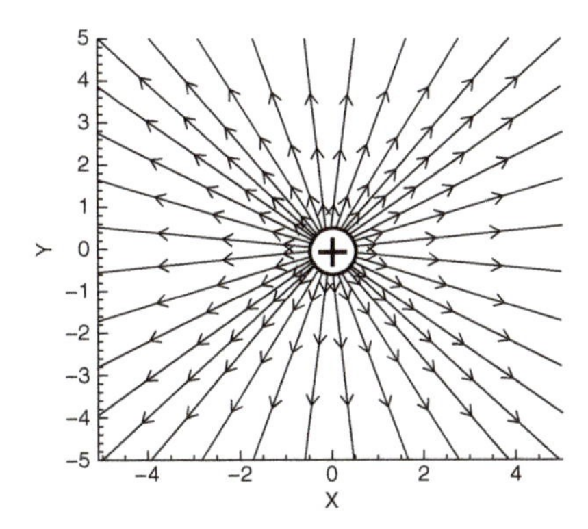

Figure 6.1 Lines of constant electric field or electric field lines from a point charge in free space.

As an indication of the magnitude of an electric field, suppose a 100 volt (recall that 1 volt = 1 Nm/Coul) battery consisting of two parallel plates 1 cm apart generates an electric field of $E = 10^4$ N/C, where C stands for $Coul$. Then the force on an electron in this field is given by

$$F_e = qE = 1.6 \times 10^{-19} C \times 10^4 N/C = 1.6 \times 10^{-15} N \tag{6.5}$$

Compare this value with the gravitational force,

$$F_g = \text{mg} = 9.1 \times 10^{-31}\ \text{kg} \times 9.8\ \text{N/kg} = 8.9 \times 10^{-30}\ \text{N} \tag{6.6}$$

and note that the gravitational force is much smaller than the force generated by the electric field.

In reality, electric fields are set up by charges distributed over linear length; surfaces or volumes of conductors of finite size, that is, along a wire; the walls of a channel; the surface or volume of a sphere; and the region between the walls of a channel. In the general case, when charges are distributed over a volume, the sum in the preceding equation becomes an integral, and

$$\vec{E}(\vec{r}) = \frac{1}{4\pi\epsilon_e} \int_{\mathcal{V}} \frac{\rho_e(\vec{r}')(\vec{r} - \vec{r}')}{\mid \vec{r} - \vec{r}' \mid^3} d\mathcal{V}' \tag{6.7}$$

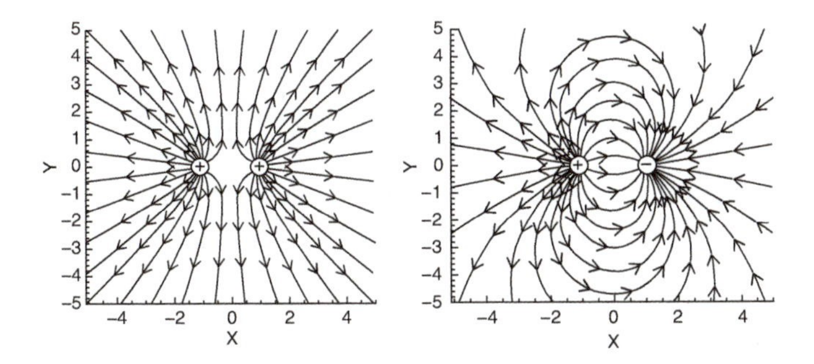

Figure 6.2 Electric field lines for two charges in the given configuration.

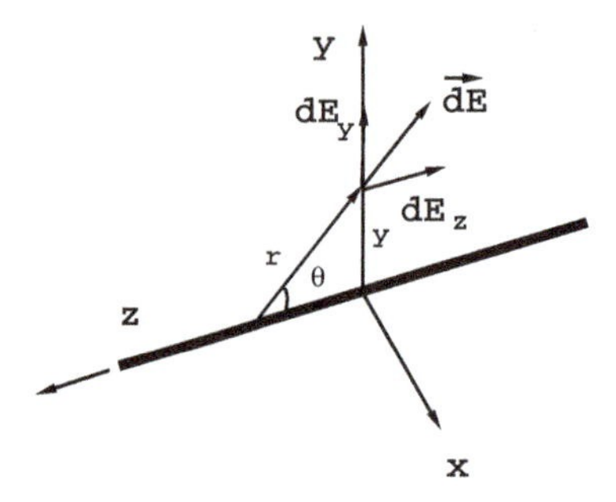

Figure 6.3 Geometry for calculating the electric field due to a long charged wire.

where ρ_e is the volume charge density. Equation (6.7) is one of the many forms of the *Bio–Savart law*, a form similar to that which appears in fluid mechanics.

Consider the case in which electrical charges are distributed over a surface. Then, in the limit as $\Delta q_j \to 0$, the result is just the integral, and writing $dq = \sigma dA$;

$$\vec{E} = \frac{1}{4\pi\epsilon_e}\int \frac{\hat{r}\,dq}{r^2} = \frac{1}{4\pi\epsilon_e}\int_{A'} \frac{\hat{r}\sigma\,dA'}{r^2} \tag{6.8}$$

where σ is the surface charge density having units C/m^2. Here we have written the result in terms of the unit vector $\hat{r}$, where

$$\hat{r} = \frac{(\vec{r} - \vec{r}')}{|\vec{r} - \vec{r}'|}$$

and $r^2 = |\vec{r} - \vec{r}'|^2$. The surface charge density plays an important role in electrokinetic phenomena, and the electrochemistry of surface charge density will be discussed in the next chapter.

6.3 The electric field due to an isolated large flat plate

To begin to derive the expression for the electric field due to a large, flat surface, consider the field induced by a long charged wire, as depicted in Figure 6.3. This is a classical problem that appears in many textbooks on electrostatics, and the present treatment closely follows the approach of Sears and Zemansky (1964). Assume that the wire is very long so that the integral for the electric field is

$$\vec{E} = \frac{1}{4\pi\epsilon_e}\int_{\text{wire}} \frac{\hat{r}\,dq}{r^2} = \frac{1}{4\pi\epsilon_e}\int_{-\infty}^{\infty} \frac{\hat{r}\lambda dz}{r^2} \tag{6.9}$$

where, in this section, we use λ to denote the charge density per unit length. By symmetry, there are two components to the electric field,

$$E_z = -\frac{1}{4\pi\epsilon_e}\int_{-\infty}^{\infty} \frac{\lambda\cos\theta\,dz}{r^2} \tag{6.10}$$

$$E_y = \frac{1}{4\pi\epsilon_e}\int_{-\infty}^{\infty} \frac{\lambda\sin\theta\,dz}{r^2} \tag{6.11}$$

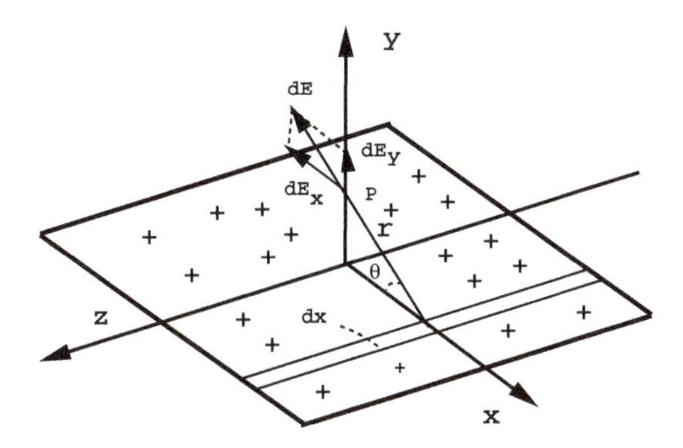

Figure 6.4 Geometry for calculating the electric field due to a single charged plate.

where we have used the relations $r^2 = y^2 + z^2$, $z = -r\cos\theta$, and $y = -r\sin\theta$ so that $r = -y\csc\theta$. Thus $\frac{y}{z} = -\tan\theta$ so that $z = -y\cot\theta$ and thus $dz = y\csc^2\theta d\theta$. Then, assuming that the line charge density λ is constant,

$$\int_{-\infty}^{\infty} \cos\theta \frac{dz}{r^2} = \int_0^{\pi} \frac{y\cos\theta\csc^2\theta}{y^2\csc^2\theta} d\theta = \frac{1}{y}\int_0^{\pi} \cos\theta d\theta = 0 \tag{6.12}$$

Thus $E_z = 0$. Similarly, for E_y, we have

$$E_y = \frac{\lambda}{4\pi\epsilon_e r}\int_{-\infty}^{\infty} \sin\theta \frac{dz}{r^2} = \frac{\lambda}{4\pi\epsilon_e y}\int_0^{\pi} \sin\theta d\theta = 2\frac{\lambda}{4\pi\epsilon_e y} \tag{6.13}$$

Note that the electric field is oriented in a direction normal to the wire and decays as $1/y$ as $y \to \infty$.

The expression for the electric field induced by a charged wire can be used to calculate the electric field due to a large flat plate, as also discussed in Sears and Zemansky (1964). Assume that the plate is very large in the x and z directions of Figure 6.4 and that the distance of the point P to the surface is much smaller than the plate dimensions so that edge effects can be neglected, and in addition, assume that the electric field does not vary in the z direction. Then the electric field will be in a direction perpendicular to the plate, and $dA = Ldx$ and $dq = \sigma Ldx$, where L is the length of the plate in the z direction. Then, in terms of the charge density per unit area,

$$dE_y = dE\sin\theta = 2\sigma\frac{dx}{4\pi\epsilon_e r}\sin\theta \tag{6.14}$$

where $r^2 = x^2 + y^2$ and with $\sin\theta = \frac{y}{r}$. Thus

$$E_y = 2k\sigma\int_{-\infty}^{\infty} \frac{y}{r}\frac{dx}{r} \tag{6.15}$$

Assuming that the surface charge density is independent of position on the plate,

$$E_y = \int_{\text{plate}} 2\sigma\sin\theta\frac{dx}{4\pi\epsilon_e r} = 2\frac{\sigma y}{4\pi\epsilon_e}\int_{-\infty}^{\infty}\frac{dx}{y^2+x^2}$$

$$= 2\frac{\sigma y}{4\pi\epsilon_e}\frac{1}{y}\tan^{-1}\frac{x}{y}\Big|_{-\infty}^{\infty} = \frac{\sigma}{2\epsilon_e} \tag{6.16}$$

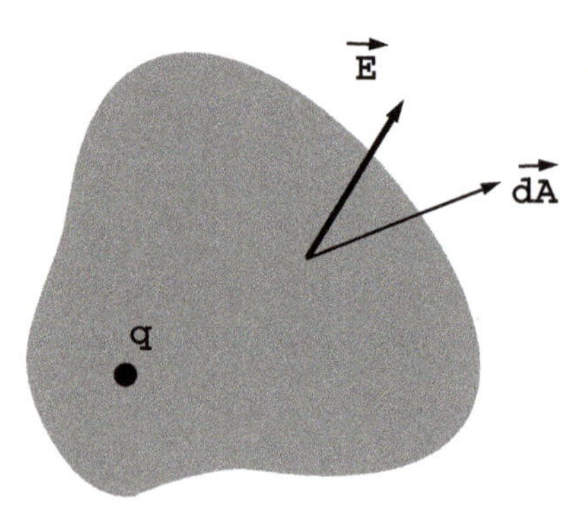

Figure 6.5 A charged surface showing outward normal $d\vec{A}$ and the direction of the electric field.

This result is constant and independent of position. In addition, as long as the dimensions of the plate are very large compared to the distance of the field point from the surface, this result is independent of the shape of the surface. Moreover, near any surface, the electric field is always normal to the surface.

The concept of a charged surface is critical to the transport of fluids at nanoscale. Charged surfaces provide the driving mechanism for the generation of electro-osmotic flow discussed in Chapter 9.

6.4 Gauss's law

Carl Friedrich Gauss was a German mathematician who made significant contributions to number theory. In the spirit of Fourier he could be described as a physicist and engineer as well as a mathematician. Gauss's law was formulated around 1835 but not published until 1867.

Consider a single point charge in a dielectric material; then, taking the integral of the charge over a surface surrounding it (Figure 6.5),

$$4\pi \iint_S \epsilon_e \vec{E} \bullet d\vec{A} = \iint_S \frac{q}{r^2} \cos\theta dA = q \iint_S d\Omega = 4\pi q \tag{6.17}$$

where $d\Omega$ is the solid angle. Thus

$$\iint_S \epsilon_e \vec{E} \bullet d\vec{A} = q \tag{6.18}$$

for a single charge. For a finite number of discrete charges,

$$\iint_S \epsilon_e \vec{E} \bullet d\vec{A} = \sum_j q_j \tag{6.19}$$

and for a continuous distribution of charges over a volume,

$$\iint_S \epsilon_e \vec{E} \bullet d\vec{A} = \iint \int_{\mathcal{V}} \rho_e d\mathcal{V} \tag{6.20}$$

where ρ_e is the volume charge density. Equation (6.20) is the integral form of Gauss's law for a continuous distribution of charges in a volume.

Using Gauss's law, the electric field of a uniformly charged sphere of total charge q over an area $A = 4\pi r^2$ at a given radius r is given by

$$\epsilon_e \iint_S \vec{E} \bullet d\vec{A} = \epsilon_e E A = q \tag{6.21}$$

and so

$$E = \frac{q}{4\pi\epsilon_e r^2} \tag{6.22}$$

as obtained from the definition of the electric field.

Similarly, for a single charged flat plate, the result is

$$\epsilon_e \textstyle\iint_S \vec{E} \bullet d\vec{A} = \sigma A \tag{6.23}$$

$$2\epsilon_e E A = \sigma A \tag{6.24}$$

or

$$E = \frac{\sigma}{2\epsilon_e} \tag{6.25}$$

also as derived previously.

6.5 The electric potential

Consider the electric field given by a single point charge in a medium of permittivity ϵ_e at a position r, which we have given as

$$E = \frac{q}{4\pi\epsilon_e r^2} \tag{6.26}$$

Integrating from a point a to point b,

$$\begin{aligned} \int_a^b E dr &= \int_a^b \frac{q dr}{4\pi\epsilon_e r^2} \\ &= \frac{q}{4\pi\epsilon_e}\left(\frac{1}{r_a} - \frac{1}{r_b}\right) \end{aligned} \tag{6.27}$$

If the integral is taken from b to a, then the result is the negative of the preceding result. Thus, around any closed path, the integral of the electric field satisfies

$$\oint \vec{E} \bullet d\vec{s} = 0 \tag{6.28}$$

This property of the electric field is the same condition that defines a thermodynamic property. This turns out to be a general condition, and using Stokes's theorem,

$$\oint \vec{E} \bullet d\vec{s} = \int_A \nabla \times \vec{E} \bullet \hat{n} dA \tag{6.29}$$

it is seen that the electric field is irrotational: $\nabla \times \vec{E} = 0$.

We then define the *potential difference* as

$$\phi_a - \phi_b = \int_1^2 \vec{E} \bullet d\vec{s} \tag{6.30}$$

The *electric potential* is defined as the work done in moving a unit charge:

$$\phi = -\int_a^b \vec{E} \bullet d\vec{s} \tag{6.31}$$

This formula is the analogue of the formula for mechanical work given by

$$W = -\int_a^b \vec{F} \bullet \vec{ds} \tag{6.32}$$

The units of the electric potential are Nm/C = 1 volt = 1 V.

The potential for a sphere of a single charge in an unbounded medium is given by

$$\phi = \int_r^\infty E dr = \frac{q}{4\pi\epsilon_e r} \tag{6.33}$$

It is perhaps surprising that this distribution is virtually identical to the radial velocity in potential flow for a point source or sink, which is given by

$$u_r = \frac{m}{r} \tag{6.34}$$

where m is the strength of the source or sink.

Note that equation (6.33) implies that for this simple geometry,

$$E = -\frac{\partial \phi}{\partial r} \tag{6.35}$$

so that the electric field is the radial derivative of the potential. In general, in differential form,

$$d\phi = -\vec{E} \bullet d\vec{l} \tag{6.36}$$

Conversely, the directional derivative of the potential along $\vec{l}$ is

$$d\phi = \nabla\phi \bullet d\vec{l} \tag{6.37}$$

and thus

$$\vec{E} = -\nabla\phi \tag{6.38}$$

and it is noted that this relationship occurs often in mechanics; for inviscid and irrotational flow and potential flow, the fluid velocity may be defined by a velocity potential:

$$\vec{u} = -\nabla\phi \tag{6.39}$$

It will turn out that this relationship implies that in the absence of charges in a medium, $\vec{u} \propto \vec{E}$; other situations in which the velocity is identical to the electric potential arise in electro-osmotic flow where there is net charge near a surface, and this point is discussed in Chapter 9.

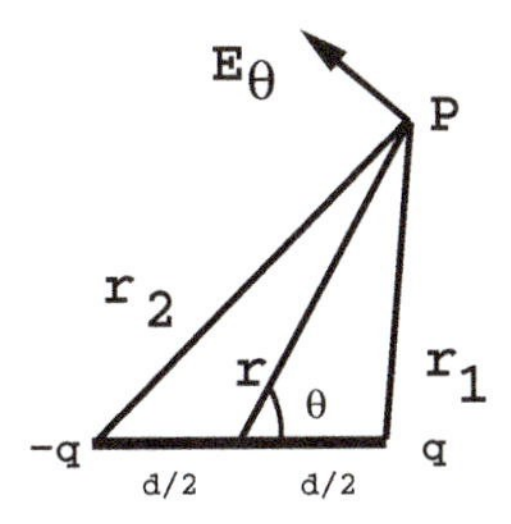

Figure 6.6 Geometry for calculating the potential field due to a dipole.

6.6 The electric dipole and polar molecules

An electric dipole is set up as a combination of a positive and negative charge, and the expression for the potential is a linear superposition of each, according to

$$\phi = \frac{q}{4\pi\epsilon_e}\left(\frac{1}{r_1} - \frac{1}{r_2}\right) \tag{6.40}$$

From Figure 6.7, using the law of cosines, $r_1 = r - (d/2)\cos\theta$ and $r_2 = r + (d/2)\cos\theta$ so that

$$\phi = \frac{q}{4\pi\epsilon_e}\frac{d\cos\theta}{r^2 - \frac{d^2}{4}\cos^2\theta} \tag{6.41}$$

Now, for $d \ll r$,

$$\phi = \frac{qd\cos\theta}{4\pi\epsilon_e r^2} \tag{6.42}$$

The quantity $p = qd$ is called the *dipole moment* and the unit of dipole moment is 1 debye = 1 D = 3.336×10^{-30} Cm. As an example, the dipole moment associated with two charges separated by 0.5 nm is $p = 1.602 \times 10^{-19}\,\mathrm{C} \times 0.5 \times 10^{-9}\,\mathrm{m} = 0.8 \times 10^{-28}\,\mathrm{Cm} = 24$ D. Small molecules have dipole moments on the order of 1 D; for water vapor, $d = 1.85$ D (Israelachvili, 1992). Again, note the similarity to the radial velocity induced by a potential doublet, the analogue of the electric dipole in fluid mechanics,

$$u_r = -m\frac{\cos\theta}{r^2} \tag{6.43}$$

The electric field lines, or lines of constant electric field due to a dipole, are shown in Figure 6.6. Note that the lines flow outward from the positive charge and inward to the negative charge.

Molecules in a dielectric are classified as being *polar* or *nonpolar*. A nonpolar molecule is one in which the center of mass of the positive nuclei and the electrons coincide, while a polar molecule is one in which they do not. Symmetrical molecules such as H_2, O_2, and N_2 are nonpolar whereas water is polar; the polar nature of water is illustrated in Figure 6.7. Under the influence of an electric field, nonpolar molecules can be polarized; in this case, we call them *induced dipoles*.

The polarization field strength is defined by

$$\vec{P} = N e \vec{d} \tag{6.44}$$

where N is the number of dipoles per unit volume and $\vec{d}$ is the vector connecting the two charges oriented in a direction from the negative charge to the positive charge. For most materials $\vec{P}$, the polarization vector is in the direction of the electric field and is written in terms of a *susceptibility* χ, so $\vec{P} = \chi \epsilon_e \vec{E}$.

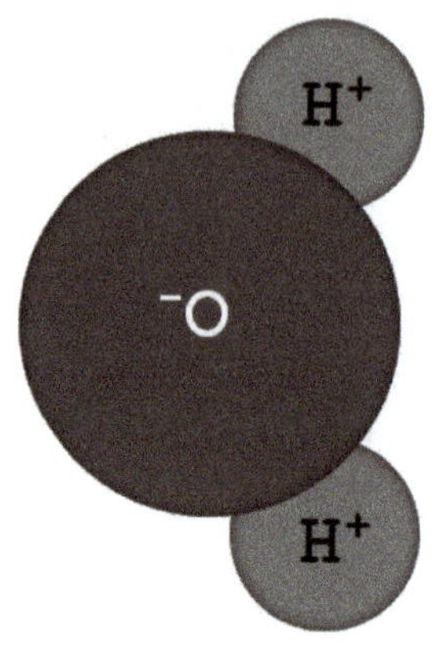

Figure 6.7 A water molecule, illustrating its polar nature.

The behavior of polar molecules in the absence of an electric field and in the presence of an electric field is depicted in Figures 6.8a and 6.8b, respectively. In the presence of an electric field, the polar molecules line up with the electric field lines. The behavior of a nonpolar molecule in the absence of an electric field is depicted in Figure 6.9a; in the presence of an electric field, these molecules will orient in the direction of the polarization field, as in Figure 6.9b, which is very similar to the polar molecule.

6.7 Poisson's equation

Using Gauss's law, along with the divergence theorem from calculus, it follows that

$$\epsilon_e \int_{\mathcal{V}} \nabla \bullet \vec{E} d\mathcal{V} = \int_{\mathcal{V}} \rho_e d\mathcal{V} \tag{6.45}$$

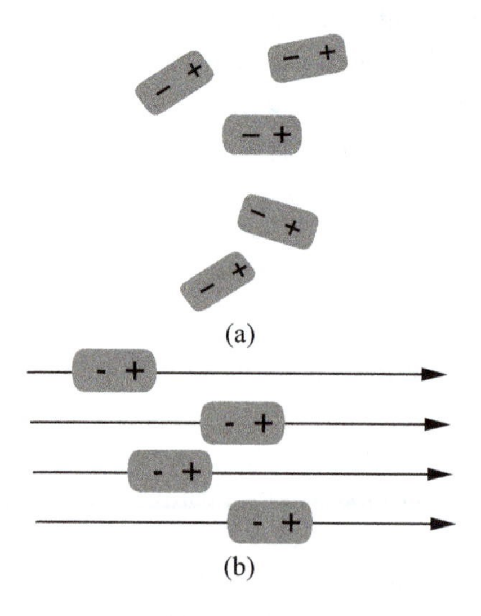

Figure 6.8 Polar molecules (a) in the absence of an electric field and (b) in the presence of an electric field.

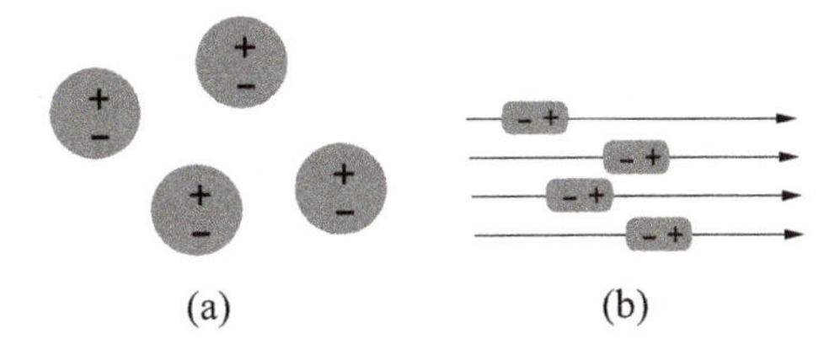

Figure 6.9 Nonpolar molecules (a) in the absence of an electric field and (b) in the presence of an electric field: Induced dipoles.

Because the limits of integration are the same for each side of the equation, we may equate integrands, and so

$$\nabla \bullet \vec{E} = \frac{\rho_e}{\epsilon_e} \tag{6.46}$$

This is the differential form of Gauss's law. Because the electric field is the gradient of the potential, we have

$$\nabla^2 \phi = -\frac{\rho_e}{\epsilon_e} \tag{6.47}$$

where ρ_e is the volume charge density and, in general, is a function of (x, y, z). This equation is Poisson's equation for the electrical potential, as was derived in Chapter 3.

As an example, let us calculate the potential field between two large plates between which is fluid having volume charge density ρ_e (Figure 6.10). Consider first the case of $\rho_e = 0$, which may be the case for deionized water. If the plates are very long in both the x and z directions, the potential will satisfy

$$\frac{\partial^2 \phi}{\partial y^2} = \frac{d^2 \phi}{dy^2} = -\frac{\rho_e}{\epsilon_e} \tag{6.48}$$

Thus, for $\rho_e = 0$,

$$\phi = Ay + B \tag{6.49}$$

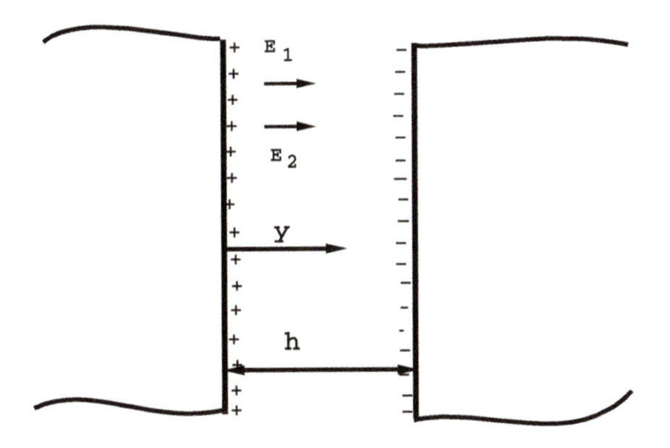

Figure 6.10 Geometry for calculating the potential field between two large plates.

where A and B are constants. To find these constants, the potential or its normal derivative at the two walls is required, as discussed in Chapter 3. At the two surfaces, set $\phi = \zeta_0$ at $y = 0$ and $\phi = \zeta_1$ at $y = h$. Then

$$\phi = (\zeta_1 - \zeta_0)\frac{y}{h} + \zeta_0 \tag{6.50}$$

so that the electric field is constant and

$$E = -\frac{d\phi}{dy} = -\frac{(\zeta_1 - \zeta_0)}{h} = \frac{\sigma}{\epsilon_e} \tag{6.51}$$

Note that if $\zeta_1 = \zeta_0$, the electric field vanishes, which is the case for a conductor.

The surface charge densities are thus defined by

$$\sigma_0 = -\epsilon_e \frac{d\phi}{dy}|_{y=0} = -\epsilon_e \frac{(\zeta_1 - \zeta_0)}{h} \tag{6.52}$$

$$\sigma_1 = \epsilon_e \frac{d\phi}{dy}|_{y=h} = \epsilon_e \frac{(\zeta_1 - \zeta_0)}{h} \tag{6.53}$$

There is a plus sign in the second equation because the outward unit normal to the surface is in the negative y direction. Note that $\sigma_0 + \sigma_1 = 0$ to keep the system electrically neutral. For water with $\zeta_0 = 1$ volt, $\zeta_1 = 0$, and $h = 100$ nm,

$$\sigma_0 = \frac{78.54}{36\pi} \times 10^{-9} \frac{\mathrm{C}^2}{\mathrm{Nm}^2} \times 1 \frac{\mathrm{Nm}}{\mathrm{C}} \times \frac{10^7}{\mathrm{m}} = .00125 \frac{\mathrm{C}}{\mathrm{m}^2} \tag{6.54}$$

Also, if both walls are of the same charge, the potential is constant and the electric field $E = 0$.

For nonzero volume but constant charge density, the electric potential is given by

$$\phi = \frac{-\rho_{e0}}{2\epsilon_0} y^2 + Ay + B \tag{6.55}$$

and applying the boundary conditions,

$$\phi = \frac{-\rho_{e0}}{2\epsilon_e}(y^2 - yh) + \frac{\zeta_0 - \zeta_1}{h} y + \zeta_0 \tag{6.56}$$

Note that for equal potentials on the two walls, which occurs often, the electric potential is of the same parabolic form as the velocity in Poiseuille flow.

The electric field is

$$E = -\frac{d\phi}{dy} = \frac{\rho_{e0}}{2\epsilon_e}(2y - h) + \frac{\zeta_0 - \zeta_1}{h} \tag{6.57}$$

Integrating the potential equation once,

$$\frac{d\phi}{dy}|_{y=h} - \frac{d\phi}{dy}|_{y=0} = \frac{-\rho_{e0} h}{\epsilon_e} \tag{6.58}$$

or

$$\sigma_1 + \sigma_0 + \rho_{e0} h = 0 \tag{6.59}$$

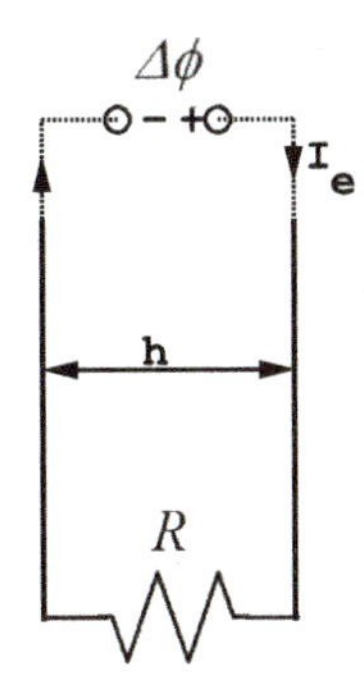

Figure 6.11 Current flows through an electric circuit, in this case, a resistor.

This means that to be electrically neutral,

$$\sigma_0 = -\epsilon_e\left(\frac{\rho_{e0}h}{2\epsilon_e} + \frac{\zeta_1 - \zeta_0}{h}\right) \tag{6.60}$$

$$\sigma_1 = -\epsilon_e\left(\frac{\rho_{e0}h}{2\epsilon_e} - \frac{\zeta_1 - \zeta_0}{h}\right) \tag{6.61}$$

The present example may be of theoretical interest only because the volume charge density is rarely, if ever, a nonzero constant in a given region.

6.8 Current and current density

In an electrical conductor, the charged species will "flow" in the direction of a decreasing potential. The current density in the y direction is defined by

$$J = -\sigma_e \frac{d\phi}{dy} \tag{6.62}$$

where σ_e is termed the electrical conductivity. In terms of the electric field, $J = \sigma_e E$. Equation (6.62) is *Ohm's law*; J is termed the *omhic current*; and if the potential is linear in a given direction, as is the case in the previous section with $\rho_e = 0$, then

$$J = -\frac{\sigma_e \Delta\phi}{h} \tag{6.63}$$

where h is the length over which the potential changes (Figure 6.11).

The *current* is defined as $I_e = JA$, where A is the cross-sectional area. Defining the *electrical conductance* as

$$L_e = \frac{\sigma_e A}{h} \tag{6.64}$$

the current is given by $I_e = L_e \Delta\phi$. The units of current density and current are amp/m^2 and amp, respectively. The *resistance* of a conductor is defined as

$R = 1/L_e$, and so Ohm's law is written as

$$\Delta\phi = I_e R \tag{6.65}$$

a familiar form in many physics and chemistry texts. Many of these texts use the form $V = I_e R = IR$, where V is actually the potential drop across the circuit. The unit of resistivity is the ohm, and the unit of conductivity is ohm^{-1}m^{-1} = siemens/m.

In the example in the previous section that investigates the electric field between two large plates,

$$E = -\frac{d\phi}{dy} = -\frac{(\zeta_1 - \zeta_0)}{h} = \frac{\sigma}{\epsilon_e} \tag{6.66}$$

so that the current density is

$$J = \frac{\sigma_e \sigma}{\epsilon_e} \tag{6.67}$$

6.9 Maxwell's equations

When a current is set up in an electrically conducting medium by an electric field, there will also be a magnetic field, a result of the forces induced by the moving charges that make up the current. When the electric and magnetic fields are coupled, and the resulting equations are *Maxwell's equations*.

Maxwell's equations, which describe the flow of charge and the electric and magnetic fields in moving media, are

$$\nabla \bullet \vec{B} = 0 \quad \text{Gauss's law for magnetism} \tag{6.68}$$

$$\nabla \times \vec{E} = -\frac{\partial \vec{B}}{\partial t} \quad \text{Faraday's law} \tag{6.69}$$

$$\nabla \times \vec{B} = \mu_M \left(\vec{J} + \frac{\partial \vec{D}}{\partial t} \right) \quad \text{Ampere's law} \tag{6.70}$$

$$\nabla \bullet \vec{D} = \rho_e \quad \text{Gauss's law} \tag{6.71}$$

Here $\vec{H}$ is the intensity of the magnetic field, $\vec{E}$ is the electric field, $\vec{D}$ is the displacement field, $\vec{B}$ is the magnetic induction field, $\vec{J}$ is the current density, and ρ_e is the charge density.

If the material is an isotropic, permeable dielectric, then

$$\vec{D} = \epsilon_e \vec{E}, \quad \vec{B} = \mu_M \vec{H} \tag{6.72}$$

and Ampere's law becomes

$$\nabla \times \vec{B} = \mu_M \left(\vec{J} + \epsilon_e \frac{\partial \vec{E}}{\partial t} \right) \tag{6.73}$$

where the *magnetic permeability* μ_M is related to the electrical permittivity by $c = (\mu_M \epsilon_e)^{-1/2}$ and c is the speed of light. The unit of the magnetic induction field is the tesla, 1 tesla = 1 volt sec/m^2. In a vacuum, $\mu_M = 4\pi \times 10^{-7}$ H/m, where H stands for henry. The magnitude of magnetic induction fields that are measured can be on the order of 1 tesla. Even when the electric field is time varying, the magnetic field term in Faraday's law is negligible; for example, an electric field of 1 V over a 10 μm channel will require a very short time scale of about 10^{-10} sec for a magnetic field of $B = 1$ T to balance Faraday's law. Similarly, the left side of Ampere's law dominates, unless the electric field is very large. Thus, in most cases of practical interest for this book, the electric and magnetic fields can be decoupled.

6.10 Chapter summary

In this chapter, the basic principles of electrostatics have been presented. The two main dependent variables in electrostatics are the electric field, the force per charge that exists between two charged bodies, and the electrical potential, the gradient of the electric field. In Chapter 3, it was shown that the fluid velocity is equivalent to the electric potential under rather general assumptions of the flow in pipes and channels.

Several canonical solutions for the electric field and potential relevant to micro- and nanofluidics have been derived, including the electric field due to a long wire and a large flat plate. The concepts of surface and volume charge density have been introduced and will be used extensively in the following chapters. The differential form of Gauss's law relates the electric field, and hence the potential, to volume charge density and leads to a Poisson equation for the electrical potential.

The potential between two charged plates has been discussed, and the current density has been defined. Both concepts will be used extensively in the following chapters.

Finally, Maxwell's electromagnetic equations are presented in SI units, and it is shown that in micro- and nanofluidics, the magnetic and electric fields can be decoupled – a significant simplification.

EXERCISES

6.1 Calculate the electric field and the electric potential due to a point charge of magnitude $e = 10^{-10}$ C (C = coulomb) at a distance $r = 10$ nm from the charge in air. Repeat for water.

6.2 A finite-sized sphere has a surface charge density of magnitude $\sigma = -0.01$ C/m^2. Show, using Gauss's law, that the electric field is given by

$$E = \frac{Q}{4\pi \epsilon_e r^2}$$

where $Q = 4\pi\sigma a^2$ is the total charge on the sphere in C. Calculate the electric field at $r = 100$ nm from the surface of the sphere of radius $a = 1$ μm for a charge density $\sigma = -0.01$ C/m^2 in water and methanol.

6.3 Find the electric field using Gauss's law for a conductor that has a surface charge density on a wall of $\sigma = \sigma_0$ for $y = 0$ and a volume charge density $\rho = \rho_0 e^{-\beta y}$ for $y > 0$. **Hint:** Note that there is only one component to the electric field, and this is in a direction normal to the surface. Calculate the integral

$$\oint_A \vec{E} \bullet \vec{n} dA$$

over the two appropriate surfaces.

6.4 Calculate the work done in charging a sphere to a surface charge density $\sigma = -0.05$ C/m^2 if the sphere diameter is $D = 1$ μm. Repeat for $D = 10$ nm.

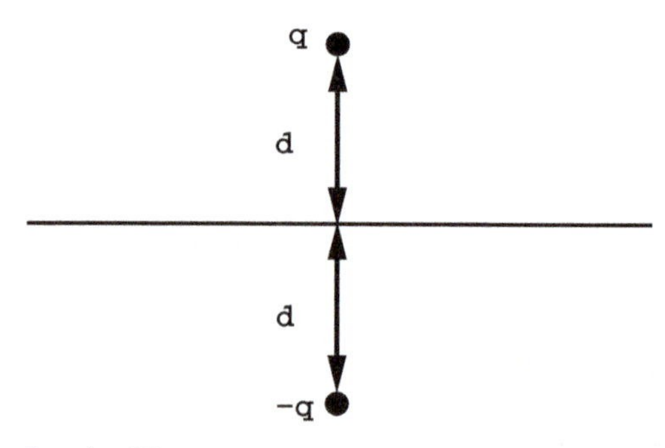

Exercise 6.5

6.5 A charge of magnitude q sits a distance d above a grounded plane wall (i.e., a wall having potential $\phi = 0$) as shown. Show that the potential can be represented as a superposition of the indicated charge and an image charge placed a distance d below the plane wall. Neglect edge effects. Calculate the surface charge density on the wall, assuming a two-dimensional distribution. Show that

$$\int_{-\infty}^{\infty} \sigma(x) dx = -q$$

Note that the same method of images is used in fluid mechanics to satisfy the condition that the velocity normal to the surface is zero.

6.6 In the Helmholtz view of the electrical double layer, the volume charge density near a charged wall is given by

$$\rho_e = a + by \quad 0 \le y \le d$$

$$\rho_e = 0 \quad y \ge d$$

Using the boundary conditions

$$\frac{d\phi}{dy} = \frac{\sigma}{\epsilon_e} \quad \text{at } y = 0$$

$$\phi = 0 \quad y = d$$

integrate the Poisson equation to find the potential for $0 \le y \le d$.

ε_{e1}

y=0

ε_{e2}

Exercise 6.7

6.7 A boundary separates two media having different dielectric properties. The volume charge density is given by

$$\rho_{ei} = k_i \phi$$

Find the potential in each region, and by integrating the governing equation for the potential over a suitable region, show that the surface charge density is given by

$$\sigma = \epsilon_{e1} \frac{d\phi}{dy}|_1 - \epsilon_{e2} \frac{d\phi}{dy}|_2$$

Suppose that region 1 is water and region 2 is silica. Calculate the surface charge density for $k_1 = k_2 = .01\ 1/\text{m}^2$. How far from the wall in k units is the potential $\phi = 0.01\zeta$, where ζ is the potential at $y = 0$?

7 Elements of Electrochemistry and the Electrical Double Layer

7.1 Introduction

A working definition of the field of *electrochemistry* is the study of charged chemical and biological material. The field of electrochemistry appears to have begun over 200 years ago, with the discovery by Galvani that a frog's limbs responded to the touching of its nerves by a scalpel with a severe twitch when the scalpel was electrically charged by a nearby source (Bockris & Reddy, 1998).

Much of electrochemistry involves the study of the properties of ionic solutions first studied by P. Debye and E. Hückel in the 1920s. The modern field of electrochemistry is separated into two related but distinct aspects: ionic solutions and electrode kinetics. Of course, in the problems of interest in this book, the two are related; however, Bockris and Reddy (1998) point out that the two fields today are significantly different, with the field of what they call "electrodics" branching out to applications in the automobile industry, the study of batteries, and fuel cells.

In this chapter, the essential principles of electrochemistry are presented, enabling a thorough understanding of the chemical principles involved in the study of electrokinetic phenomena to follow. Many biofluids are aqueous mixtures of *electrolytes*, which are species that are either positively or negatively charged. The mixtures of interest are aqueous and are usually dilute in the electrolyte species. An example of a common electrolyte mixture is phosphate buffered saline (PBS), which contains five different ionic species in an aqueous solution.

The chapter is divided into four parts. It begins with a presentation of the structure of water and then proceeds to a qualitative discussion of the phenomenon of hydration, in which water molecules can surround an ion due to the polar nature of the water molecule.

The second portion of the chapter introduces the concept of the chemical and electrochemical potential and the Gibbs function and their roles in chemical equilibrium. In this portion of the chapter, the definition of acids and bases, central to the field of modern electrochemistry, is discussed, leading to the characterization of surfaces exposed to an aqueous solution. Many solids acquire a negative charge when they come in contact with an aqueous medium (Hunter, 1981; Probstein,

1989). The development of electrical charge on the walls is a combination of physical and chemical processes. The surface charge is acquired through a variety of mechanisms such as ion adsorption, exposure to charged crystal surfaces, and ionization of surface groups (Masliyah and Bhattacharjee, 2006). Among these, the ionization of surface groups plays a dominant role. For devices with surfaces made of glass or silicon, surface silanol groups (SiOH) undergo deprotonation, which results in development of negative charge on the walls. Thus the mechanics of how this process occurs is addressed, leading to the definition of the surface charge density discussed in Chapters 3 and 6, and the relationship between the surface charge and the ζ potential is derived.

If the surface is negatively charged, the positively charged ions (called the counterions) in the vicinity of the surface tend to collect near the charged surface, and an inner layer of counterions is attached to the surface. Outside this pinned layer, ions float within the electrolyte solution near the surface. The total system, including the pinned and floating ions or diffuse layer, is called the *electrical double layer* (EDL) (Hunter, 1981).

The third portion of the chapter deals with the analysis of the electrical double layer on the liquid side and determining the potential and species distribution within the EDL, which is dependent on the value of the surface charge density and hence the ζ potential on the wall. A relationship between the surface charge density and ζ potential is derived based on linking the solid-side definition of these quantities with that derived from the liquid definition. This leads to the result that the surface charge density $\sigma = \sigma(\zeta, pH, K_i)$, where K_i are the equilibrium constants associated with the solid-side surface chemical reactions. In this context, solutions for the electrical potential on spherical and cylindrical surfaces are also calculated, and the interaction between a solid particle and a plane wall is also discussed.

The chapter concludes with a discussion of electrical conductivity, semipermeable membranes that are characteristic of batteries and a discussion of the Derjaguin approximation and its extension for the determination of the interaction energy between two surfaces.

7.2 The structure of water and ionic species

Water is the most abundant liquid on earth and, from a biological perspective, makes up about 70 percent of a human cell's weight (Alberts *et al.*, 1998; Alberts *et al.*, 1994). A sketch of the structure of water is depicted in Figure 6.7; the angle between the two hydrogen atoms and the oxygen atom is about 110°, and the two hydrogen atoms are linked to the oxygen atom by a covalent bond in which the different atoms share pairs of electrons. In the case of water, the oxygen atom is surrounded by eight electrons. A single water molecule has a nominal diameter of about 3 Å.

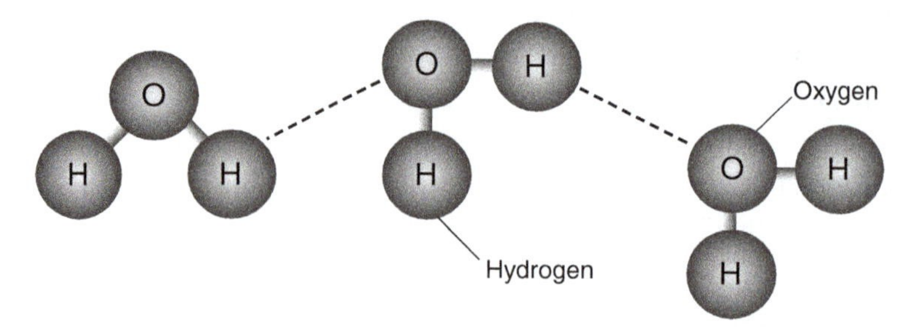

Figure 7.1 Sketch of a hydrogen bond. Image courtesy of the National Institute of General Medical Sciences of the National Institutes of Health.

Because water is polar, when two water molecules approach each other, the negative end of the one water molecule may closely approach the positive end of the other water molecule. This rather weak electrostatic attractive force (Coulombic interaction, defined by Coulomb's law) can result in the formation of a hydrogen bond (Figure 7.1). This type of bond is much weaker than the covalent bond and can easily be broken by the random motion of the surrounding molecules. Indeed, hydrogen bonds are extremely short term, forming and breaking very fast on the order of picoseconds. However, in the bulk, the many hydrogen bonds are the reason that water is a liquid at room temperature and not a gas (Daune, 1993). Hydrogen bonds can occur even between differently charged portions of the same molecule.

In the simple point charge (SPC) model of water, the distance between the oxygen and hydrogen molecules in the H_2O molecule is 0.1 nm and the HOH angle is 109.47°, and the dipole moment of this SPC model for water is $d = 2.27$. There are many other models (Karniadakis *et al.*, 2005; Sadus, 1997; Becker & Karplus, 2006; Franks, 1972) of the structure of water, and in the extended SPC model, the charges on the O and H molecules are assumed to be $q_O = -0.8476$ and $q_H = 0.4238$. The dipole moment for the extended SPC model is $d = 2.35$.

The localization of charge, as shown in Figure 6.7 allows the two regions to bond to ions and other charged molecules, thus making the water molecule highly polar, with a positive charge located on one end and a negative charge on the other. In general, polar molecules have a high dielectric constant and are good electrical conductors. Conversely, hydrocarbon solvents, such as methanol, that are nonpolar have lower dielectric constants and are poor conductors.

In general, polar molecules that can form hydrogen (anions) and ionic (cations) bonds dissolve readily in water. These types of molecules are called *hydrophilic*, or water loving. Ions are an example of a hydrophilic molecule because they can easily be surrounded by one or more water molecules; in this case, the ions are said to be *hydrated*. More will be said about hydration later. Conversely, uncharged molecules, such as the hydrocarbon family, that contain many C–H bonds are water hating or *hydrophobic*. Cell membranes contain many of these types of bonds, and thus portions of cell membranes are hydrophobic. The structure of cell membranes is discussed in the next chapter.

7.3 Chemical bonds in biology and chemistry

Molecules, and indeed all matter are made of atoms, which are arranged in a specific way to form molecules. The individual atoms are held together by chemical bonds. Each atom is composed of a positively charged nucleus consisting of protons and neutrons surrounded by a cloud of negatively charged electrons. An atom as a whole must be electrically neutral.

The electrons are arranged in a rigid series of at most five shells, with the electrons in the innermost shell bound most tightly to the nucleus. An atom is most stable when each of the possible shells (depending on the atom) is filled. The innermost shell can hold at most 2 electrons, the second and third 8 electrons, and the fourth and fifth 18 electrons. In the biological world, atoms with more than four shells are rare (Alberts *et al.*, 1998).

How atoms combine to form molecules is determined by the number of electrons in the outermost shell. For example, hydrogen has one electron in its only shell, and thus it can combine with other atoms that can provide the electron to fill its outer shell; that is, hydrogen is highly reactive. The *valence* is defined as the number of electrons an atom must gain or lose to have a fully populated outer shell. Virtually all the elements in the living world have unfilled outer shells. Those elements having filled outer shells are chemically unreactive; two examples are helium and argon.

Atoms with unfilled outer shells exchange electrons. How they do this leads to the existence of a chemical bond between two or more atoms. When two atoms share a pair of electrons, the bond between the two atoms is called a *covalent bond*. Similarly, an *ionic bond* is formed when an electron is transferred from one atom to the other.

There are several different types of covalent bonds. Most covalent bonds share two electrons and are called *single bonds*. When four electrons are shared, two between each atom, the bond is called a *double bond*. The *bond length* of a double bond is shorter, and the double bond is stronger than the single bond.

The *van der Waals force* is an attractive–repulsive force that is not due to elecrostatic or covalent bonding. This force in a gas is characterized by the van der Waals equation of state (Moran & Shapiro, 2007), which is a modification of the perfect gas law to account for molecular size and intermolecular forces. The effect of van der Waals forces in a liquid is about 10^6 times that in a gas because of the proximity of individual molecules, making the influence of intermolecular forces that much greater.

Chemical reactions play an important role in determining properties of bounding surfaces in nanopore membranes; for example, the surface charge on a silica surface is determined by a reaction between the uncharged silica solid and the hydrogen atom in water. In particular, the nature of the silica surface is discussed in detail later in this chapter.

7.4 Hydration of ions

As has already been mentioned, ions may become hydrated, that is, surrounded by water molecules (Figure 7.2). A molecule is a discrete set of atoms held together by chemical bonds. Inherent in the discussion of hydration phenomena is the concept of solubility. In general, like dissolves like; that is, polar molecules, such as ions, are more soluble in a polar solvent than in a nonpolar solvent. Thus the ion–water interaction is stabilizing; that is, it is more energetically favorable.

Albert Einstein and Hydration

Albert Einstein was writing his dissertation at a time when even the existence of atoms and molecules was being questioned. He showed his fluid dynamics prowess with the derivation of the Stokes-Einstein formula for the diffusion coefficient as discussed in Chapter 2. But also in his dissertation, in talking about the viscosity of an aqueous sugar solution, he derived the formula for the viscosity of the solution as

$$\frac{k^*}{k} = 1 + \Phi$$

where Φ is the volume of the sugar molecules per unit volume of the water. Stachel (1998) points out an error in this equation that Einstein corrects in a publication in 1922. The correct equation is

$$\frac{k^*}{k} = 1 + 2.5\Phi$$

Einstein quotes a value of $\Phi = 0.0245$ "from Burkard's observations (Landolt and Börnstein's Tables)". In a simple calculation, he found that the density of a 1% aqueous sugar solution increased roughly linearly with sugar concentration by a factor of 1.0061. Thus the increase in viscosity is four times that of the increase in density! He concludes:

> Thus, while the sugar solution behaves like a mixture of water and solid sugar with respect to its density, the effect on viscosity is four times larger than what would result from the from the suspension of the same amount of sugar. It seems to me that, from the point of view of molecular theory, this result can only be interpreted by assuming that a sugar molecule in solution impedes the mobility of the water in its immediate vicinity, so that an amount of water that is three times larger than the volume of the sugar molecule is attached to the sugar molecule.

Einstein deduced that a sugar molecule must be hydrated! Such a phenomenon was the subject of heated debate at the time (Stachel, 1998).

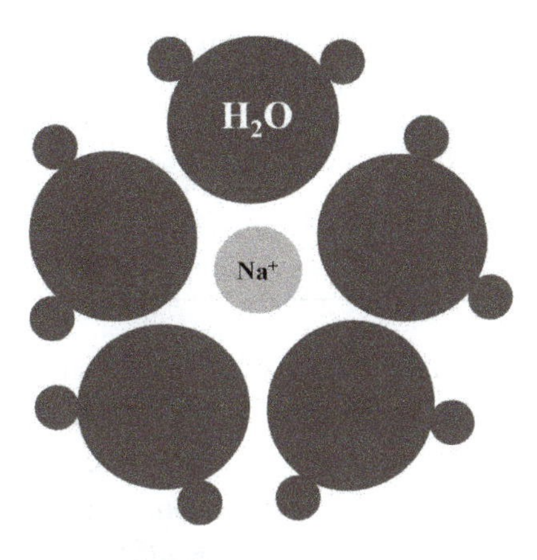

Figure 7.2 Sketch of the hydration of a sodium ion by water molecules.

On the basis of the analysis of the the electrical double layer (EDL) surrounding an ion, as discussed in Chapter 1, the local electric field can be on the order of 10^{10} V/m. Under the influence of this large electric field, the polar ends of water molecules can electrically bond to the ion, creating a "molecule" whose effective radius is larger than the unhydrated ion. A sketch of a hydrated ion is depicted in Figure 7.2. Values of the hydrated ionic radius can range from 3 to 5 times the unhydrated ion radius. For example, the ratio $H = a_H/a$, where a is the unhydrated radius and a_H is the hydrated radius, is typically $H = 1.83$ for Cl^- and $H = 3.91$ for Na^+ (Conway, 1981). Multivalent ions typically have hydration radii larger than monovalent ions because of the stronger electric fields around the multivalent ion.

The hydrated radius of ions is often calculated from the Gibbs free energy (Daune, 1993). One way to analyze the hydration process is to calculate the work required to bring an ion from a vacuum with dielectric constant $\epsilon_{e1} = 1$ to the water medium having a dielectric constant of $\epsilon_{e2} \sim 78$–80. This work is given by integrating the electrical force over all charges, or

$$W_H = \frac{q^2}{8\pi\epsilon_0 r}\left(\frac{1}{\epsilon_{e1}} - \frac{1}{\epsilon_{e2}}\right) \tag{7.1}$$

where r is the radius of hydration; this formula was first given by Born (Daune, 1993). This work value is equivalent to the Gibbs free energy of hydration $W_H = G_H$, and the radius of hydration can then be calculated from measured values of this work.

Several radii of hydrated ions are depicted in Table 2.6. Care should be taken in interpreting the radii depicted in Table 2.6. The ionic radius is that radius of the ion as a crystalline solid, assuming spherical symmetry. The higher the cation charge, the larger the hydrated radius, and the radius of a cation is nearly its van der Waals radius (Daune, 1993), the radius of a molecule modeled as a single hard sphere. Moreover, owing to Brownian motion, the molecules will be vibrating at a very high frequency, and as such, the radii depicted in Table 2.6 must be considered time averages over a large number of ions.

Hydration is also measured by the coordination number, the number of solvent molecules around an ion, and the solvation number, a measure of the dynamic state of the ion–solvent complex as they move in solution. This subject is far beyond the scope of this book, and the details of the definitions of these two numbers are given in Bockris and Reddy (1998).

For the purposes of this book, the most important property of an ion is its hydrated size, and in addition to the experimental data depicted in Table 2.6, the solvated radius can be obtained from *ab initio* molecular simulations.

7.5 Chemical potential

The chemical potential is an important quantity for establishing under what conditions equilibrium will exist. The term *equilibrium* is often used, and it is useful to discuss what the term means in the context of a chemical reaction. Suppose a chemical reaction takes place until completion, and no other changes can be observed. Then the system is said to be in *equilibrium* if the reverse reaction takes place and the original starting state is recovered (Butler, 1998).

The extensive Gibbs function in thermodynamics, defined in units of joules for a pure fluid, is defined as (Moran & Shapiro, 2007)

$$G = H - TS, \tag{7.2}$$

where H is the extensive enthalpy in joules = J or kJ, S is the extensive entropy, and T is temperature. The Gibbs function, or Gibbs free energy, is a property of the state of the system and taking the differential

$$dG = dH - TdS - SdT \tag{7.3}$$

From the first and second laws of thermodynamics, $TdS = dH - Vdp$, and so

$$dG = Vdp - SdT \tag{7.4}$$

Thus, at constant temperature and pressure, equilibrium requires that $dG = 0$.

For a fluid mixture, the Gibbs function is also a function of the number of moles of each species present in the mixture. Thus $G = G(T, p, n_1, n_2, \ldots)$, and so, in the general case,

$$dG = Vdp - SdT + \sum_{i=1}^{m} \mu_i dn_i \tag{7.5}$$

where μ_i is the *chemical potential* and n_i is the number of moles of species i. From the definition of the differential of a function, the *chemical potential* μ_i is thus defined as

$$\mu_i = \frac{\partial G}{\partial n_i}_{T,p,n_j} \tag{7.6}$$

and so the units of chemical potential are J/mole.

The chemical potential is what is termed a *partial molar quantity*. For any extensive property $Y = Y(T, P, n_i)$, a partial molar quantity is defined by taking the partial derivative of Y with respect to n_i, keeping the temperature, pressure, and n_j fixed. For example, the partial molar volume is defined by

$$V_i = \frac{\partial \mu_i}{\partial n_i}|_{T,p,n_j} \tag{7.7}$$

For any extensive property $Y = Y(T, p, n_i)$ and for any constant α,

$$\alpha Y(T, p, n_i) = Y(T, p, \alpha n_i) \tag{7.8}$$

Differentiating with respect to α, we find that

$$Y(T, p, n_i) = \sum_{i=1}^{m} \frac{\partial Y}{\partial(\alpha n_i)} n_i, \tag{7.9}$$

which holds, in particular, for $\alpha = 1$. This means that the extensive Gibbs function is also defined by

$$G = \mu N = \sum_{i=1}^{m} \mu_i n_i \tag{7.10}$$

for a mixture of m components, where N is the total number of moles in the system.

For a single component, in molar units, $G = \mu N$ or $\mu = G/N = \bar{g}$, where $\bar{g}$ is the specific molar Gibbs function in units of J/mole. For example, assuming a perfect gas, the chemical potential of air, taken to be a pure fluid, is

$$\mu = \frac{G}{N} = \bar{g} = \bar{h} - T\bar{s} = T(\bar{c}_p - \bar{s}) \tag{7.11}$$

in molar units. In mass units,

$$\mu = \frac{G}{M} = g = h - Ts = T(c_p - s) \tag{7.12}$$

Most often, as will be seen, changes in the chemical potential are important rather than the specific value of the chemical potential. For air, using property values from thermodynamic tables, the change in chemical potential from $T = 290$ K to $T = 310$ K at constant pressure in mass units is to three digit accuracy

$$\begin{aligned} \Delta\mu &= 310(1.005 - 1.735) - 290(1.005 - 1.668) \\ &= -226.300 + 192.270 = -34.030 \frac{\text{kJ}}{\text{kg}} \end{aligned} \tag{7.13}$$

To obtain the same value on a molar basis, the mass result is multiplied by the molecular weight of air $M_w = 29$ kg/kmole.

As another example, consider the change in chemical potential as liquid water becomes vapor at $T = 100°$C. Using thermodynamic tables for the entropy values,

$$\Delta\mu = 2257 - 373 \times (7.3549 - 1.3069) = 2257 - 2256 \sim 0 \frac{\text{kJ}}{\text{kg}} \tag{7.14}$$

To the accuracy that the tabulated properties of water are measured, the change in chemical potential is zero. In fact, this is a statement of phase equilibrium: that a liquid in equilibrium with its vapor has the same chemical potential, or

$$\mu_{\text{vap}} = \mu_{\text{liq}} \tag{7.15}$$

This relationship also holds for any component in a mixture:

$$\mu_{i,\text{vap}} = \mu_{i,\text{liq}} \tag{7.16}$$

For a mixture, the differential change in Gibbs function from equation (7.10) is given by

$$dG = \sum_{i=1}^{m} \mu_i dn_i + \sum_{i=1}^{m} n_i d\mu_i \tag{7.17}$$

Consequently, equating the two expressions for dG readers

$$\sum_{i=1}^{m} n_i d\mu_i = Vdp - SdT \tag{7.18}$$

Equation (7.18) is called the Gibbs–Duhem equation and provides the necessary relationship between temperature, pressure, and chemical potential changes. Thus, at constant temperature and pressure, the chemical potential of each species is constant.

Consider now the case of a perfect gas mixture of m components. Then the chemical potential μ_i is defined as

$$\mu_i = \mu_i^o(T) + RT \ln X_i \quad \text{for } i = 1, \ldots, m, \tag{7.19}$$

where R is the *universal* gas constant and X_i is the mole fraction. The quantity μ_i^o is the chemical potential of the gas (or vapor) at a suitably defined standard state. For a gas, the quantity μ_i^o is usually independent of pressure, at least if the gas is perfect.

For example, air, though, for practical purposes, it is considered a pure substance, is about 79 percent nitrogen on a molar basis, and the quantity

$$\mu_{N_2} - \mu_{N_2}^o = 8.314 \frac{\text{kJ}}{\text{kmole K}} \times 298 \text{ K} \ln 0.79 = -584.02 \frac{\text{kJ}}{\text{kmole}} \tag{7.20}$$

at $T = 298$ K. Because the molar mass of nitrogen is $M_{N_2} = 28.01$, $\mu_{N_2} - \mu_{N_2}^o = -584.02/28.01 = -20.85$ kJ/kg.

For a process in which a given species in a gas changes composition from $X_{i,1}$ to $X_{i,2}$, the change in chemical potential at constant temperature for a perfect gas is given by

$$\Delta\mu_i = RT \ln \frac{X_{i,2}}{X_{i,1}} \tag{7.21}$$

where 1 and 2 refer to the initial and final states, respectively. Thus

$$X_{i,2} = X_{i,1} e^{-\frac{\Delta\mu}{RT}} \tag{7.22}$$

which is the classical Boltzman distribution.

For a liquid mixture, the chemical potential is defined by (Denbigh, 1971)

$$\mu_i = \mu_i^*(T, p) + RT \ln a_i \quad \text{for} \quad i = 1, \ldots, m, \tag{7.23}$$

where now the reference chemical potential $\mu_i^*(T, p)$ is a function of temperature and pressure, but is in general unknown, and a_i is the activity. The activity coefficient is defined as $\gamma_i = a_i/X_i$, where x_i is the mole fraction of species i in solution. The activity coefficient is, in general, a function of temperature, pressure, and mole fraction. A liquid mixture is said to be an *ideal solution* if $\gamma_i = 1$ so that the activity may be identified with the mole fraction of species i.

Consider now a nonideal or concentrated binary mixture. Then, using the definition of the intensive Gibbs function in equation (7.10), the intensive Gibbs function on a molar basis is given by

$$g(T, p, x_A) = X_A\mu_A^* + (1 - X_A)\mu_B^* + RT\left(X_A \ln X_A + (1 - X_A)\ln(1 - X_A)\right) + g_E, \quad (7.24)$$

where $g_E(T, p, X_A)$ is called the *excess Gibbs function* and is nonzero for nonideal mixtures. Using the definition of the activity coefficient, the excess Gibbs function is

$$g_E(T, p, X_A) = RT(X_A \ln \gamma_A + X_B \ln \gamma_B). \quad (7.25)$$

In concentrated, nonideal solutions, models for the excess Gibbs function are required. A detailed analysis of such models is beyond the scope of this book, and the solutions of interest in micro- and nanofluidics are most often ideal. It is worthwhile to mention several of the most popular methods; these are the Margules, Val Laar, Wilson, and non-random two-liquid (NRTL) models. These models are discussed in great detail in Taylor and Krishna (1993) and, in general, are derived from experimental data.

In the case of ionic species, the chemical potential takes the form

$$\mu_B = \mu_B^* + RT \ln a_\pm \quad (7.26)$$

where $a_\pm = \gamma_\pm X_\pm$, where $X_\pm$ is the mole fraction and $\gamma_\pm = \sqrt{\gamma_+\gamma_-}$ is called the mean activity coefficient. In the Debye–Hückel limit, the activity coefficient is given by

$$-\ln \gamma_\pm = \frac{A_{DH} z_+ z_- \sqrt{I}}{1 + B_{DH} a \sqrt{I}} \quad (7.27)$$

where I is the *classical* ionic strength for a binary mixture, $I = (1/2)\sum c_i z_i^2$ (Fawcett, 2004), and a is the radius of the ion. The constants A_{DH} and B_{DH} are defined by

$$A_{DH} = \frac{N_A e^2 F}{8\pi}\left(\frac{2000}{(\epsilon_e RT)^3}\right)^{1/2} \quad (7.28)$$

$$B_{DH} = F\left(\frac{2000}{\epsilon_e RT}\right)^{1/2} \quad (7.29)$$

The factor of 1000 is inserted to convert mole/liter to mole/m^3 because Faraday's constant and the electrical permittivity include a meter in their units. Fawcett (2004) provides values for these constants and for a 1 mM solution at 25°C, for univalent electrolytes $\gamma_\pm = 0.96$, neglecting the effect of the constant B_{DH} since a is small.

As pointed out by Denbigh (1971), the definition of the standard state quantities μ_i^* and μ_i^o is not always standard, and care must be taken in examination of the experimental data for these quantities. The same is true of the activity coefficient.

7.6 The Gibbs function and chemical equilibrium

As noted in the previous section, the Gibbs function plays a crucial role in the establishment of chemical equilibrium. Consider the extensive form of the first law of thermodynamics for an open system:

$$\delta Q - \delta W = dH = dG + TdS \tag{7.30}$$

where the process is assumed to occur at constant temperature. From the second law of thermodynamics,

$$TdS = \delta Q + Td\sigma \tag{7.31}$$

where $\sigma \geq 0$ is the entropy production in units of, say, kJ/K. Substituting into the first law

$$-\delta W = dG + Td\sigma \tag{7.32}$$

or

$$dG = -Td\sigma - \delta W \tag{7.33}$$

Thus, whatever the sign of the work done, $dG \leq \delta W_{\text{max}}$.

Now suppose $\delta W = 0$. Then it is clear that $dG \leq 0$ since $d\sigma \geq 0$. Note that $dG > 0$ is impossible, and so this restriction on the Gibbs function follows from the limitations on the entropy change for a given process. Thus the value of the Gibbs function must decrease during a process until equilibrium is reached when $dG = 0$, as we saw from the Gibbs–Duhem equation. Note that constant temperature and pressure have been assumed.

Suppose a general reaction is represented by the equation

$$\nu_A A + \nu_B B \rightleftharpoons \nu_E E + \nu_F F \tag{7.34}$$

Then, if the reaction proceeds at constant temperature and pressure, the condition of equilibrium is given by

$$dG_{T,P} = \sum_{i=1}^{m} \mu_i dn_i = \mu_A dn_A + \mu_B dn_B + \mu_E dn_E + \mu_F dn_F = 0 \tag{7.35}$$

For example, for the forward reaction, $dn_A = -k\nu_A$ and $dn_E = +k\nu_E$, with similar expressions for the other two species. Here k is a proportionality constant, depending on the amount of the reactants, and is normally small. Because $dG_{T,P} = 0$, it must be that

$$\nu_E \mu_E + \nu_F \mu_F - \nu_A \mu_A - \nu_B \mu_B = 0 \tag{7.36}$$

Equation (7.36) is known as the *equation of reaction equilibrium*.

If a liquid mixture is ideal, it has been seen that the chemical potential is defined as

$$\mu_i = \mu_i^*(T, p) + RT \ln X_i \quad \text{for } i = 1, \dots, m \tag{7.37}$$

and substituting into the equation of reaction equilibrium

$$\nu_E(\mu_E^*(T, p) + RT \ln X_E) + \nu_F(\mu_F^*(T, p) + RT \ln X_F) \\ - \nu_A(\mu_A^*(T, p) + RT \ln X_A) - \nu_B(\mu_B^*(T, p) + RT \ln X_B) = 0 \tag{7.38}$$

where it is recognized from the previous section that the chemical potential $\mu_i^*(T, p) = g_i^*$, the intensive Gibbs function. Rearranging the terms of this equation, it is readily seen that

$$-\Delta G^*(T, p) = RT \ln \left(\frac{(X_E)^{\nu_E}(X_F)^{\nu_F}}{(X_A)^{\nu_A}(X_B)^{\nu_B}} \right) \tag{7.39}$$

where $\Delta G^*(T, p)$ is defined by

$$\Delta G^*(T, p) = \nu_E \mu_E^*(T, p) + \nu_F \mu_F^*(T, p) - \nu_A \mu_A^*(T, p) - \nu_B \mu_B^*(T, p) \tag{7.40}$$

Note that the stoichiometric coefficients ν_i are given from the reaction. The *equilibrium constant* is then defined as

$$K = \ln \left(\frac{(X_E)^{\nu_E}(X_F)^{\nu_F}}{(X_A)^{\nu_A}(X_B)^{\nu_B}} \right) \tag{7.41}$$

and so in terms of the change in Gibbs function for the reaction,

$$K = e^{-\frac{\Delta G^*(T,p)}{RT}} \tag{7.42}$$

where, in all of this analysis, R is the universal gas constant. Of course, this analysis can be extended for any number of reacting species and for simultaneous reactions and ionization processes.

The equilibrium constants are, in most cases, determined from experiments. The equilibrium mole fractions in the case of an ideal liquid mixture are then determined from the solution of a set of nonlinear equations (Moran & Shapiro, 2007). This procedure also applies to ionic mixtures, as will be described for reactions that occur on silica surfaces in Section 7.9.

To illustrate the process, consider the reaction of gaseous carbon monoxide with water vapor. The theoretical chemical reaction, the completed reaction is given by

$$CO_2 + H_2 \rightleftharpoons CO + H_2O \tag{7.43}$$

In actuality, there may be some reactants left in the products, and so the actual reaction may be

$$CO_2 + H_2 \rightleftharpoons wCO + xCO_2 + yH_2O + zH_2 \tag{7.44}$$

It is common to use the number of moles of each species in the definition of the equilibrium constant, which is

$$K = \frac{wy}{xz}\left(\frac{p}{p_{ref}N_T}\right)^{1+1-1-1} \tag{7.45}$$

where $N_T = w + x + y + z$ is the total number of moles in the product gases. Note that the reverse reaction has an equilibrium constant of $1/K$. Equation (7.44) needs to be balanced by conserving the moles of C, O, and H. These equations are

$$2 = w + 2x + y \quad \text{O balance} \tag{7.46}$$

$$1 = w + x \quad \text{C balance} \tag{7.47}$$

$$2 = 2y + 2z \quad \text{H balance} \tag{7.48}$$

Equation (7.45) becomes

$$K = \frac{wy}{xz} \tag{7.49}$$

These are four equations in four unknowns for x, z, w, and y that may be solved in Matlab or some other software package. However, the equilibrium constant K must be known; these values are given in many different places, including in undergraduate textbooks on thermodynamics and in the National Institute of Standards (NIST) database. Its value depends on the temperature of reaction.

At room temperature, $T = 298°$, $\log_{10} K \sim -0.5$ or $K = 10^{-5}$, and the reaction does not proceed. Conversely, at $T = 1600°$, $\log_{10} K = 0.474$, $K = 2.979$, and for this value of the equilibrium constant, $x = 0.367$, $y = 0.633$, $z = 0.367$, and $w = 0.633$. Note that because of the nonlinearity of the equilibrium constant equation, there may be multiple solutions, with only one being relevant. Note also that the total number of moles has been conserved. One can plot the yield of this reaction as a function of temperature by plotting the results for a number of values of K (Moran & Shapiro, 2007). Can you combine these four equations into a single nonlinear equation for one of the variables?

Various salts, acids, and bases dissociate when in contact with water. For example, in aqueous solution, sodium chloride dissociates according to

$$\text{NaCl(solid)} \rightleftharpoons \text{Na}^+\text{(aq)} + \text{Cl}^-\text{(aq)} \tag{7.50}$$

The same methodology leads to the definition of the equilibrium constant as

$$K = \frac{x^2}{1-x}\frac{1}{1+x} = \frac{x^2}{1-x^2} \tag{7.51}$$

where, for example, $N_+ = N_- = x$ is the number of moles of Na^+. Thus

$$x = \sqrt{\frac{K}{1+K}} \tag{7.52}$$

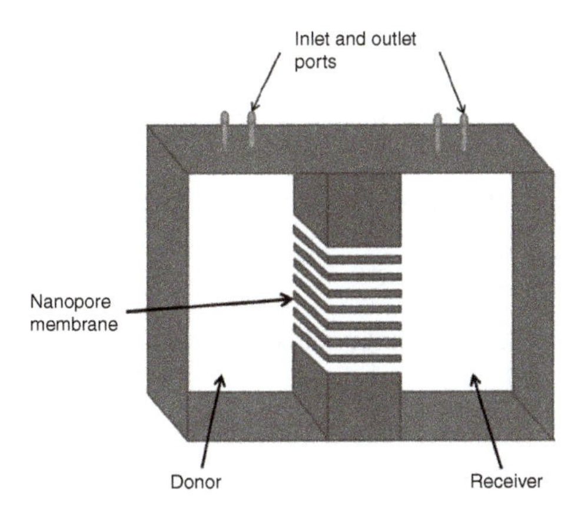

Figure 7.3 Depiction of donor and receiver regions separated by a nanopore membrane.

In $NaCl$ dissociation, the equilibrium constant for this ionic reaction is large, and so the dissociation is complete, and $x = 1$. Multiple reactions may occur in many situations, and this case will be considered later in this chapter, in the section on silica surfaces.

7.7 Electrochemical potential

Analgous to the chemical potential, the electrochemical potential in J/mole is defined by

$$\mu_i = \mu_i^* + RT \ln a_i + z_i F\phi \tag{7.53}$$

where $R = N_A k_b$ is the universal gas constant; F is Faraday's constant, $F = N_A e = 96{,}500$ C/mole; and ϕ is the electrical potential. For the dilute solutions of interest, here the activity is assumed to be equal to the mole fraction, $a_i = x_i$. Thus

$$\mu_i = \mu_i^* + RT \ln X_i + z_i F\phi \tag{7.54}$$

In the same way that the chemical potential is constant as a condition of equilibrium for uncharged species, the electrochemical potential is constant for charged species (ions). Note that at $T = 300$ K, $RT = 2494$ J/mole for $R = 8.314$ J/mole K. For a monovalent ion and a potential of $\phi = 1$ V $= 1$ J/C, $F\phi = 96{,}500$ J/mole and so often at high potentials, the electrostatic term can dominate.

To illustrate the use of the electrochemical potential, consider a charged species in an electrolyte solution in an upstream and downstream reservoir separated by a porous membrane of some type, as depicted in Figure 7.3. At equilibrium for constant temperature and pressure, the chemical potentials in the donor and receiver must be equal so that $\mu_D = \mu_R$, where D stands for donor reservoir and R stands for receiver reservoir. The chemical potential scale μ_i^* can be chosen as equal in the two reservoirs, and so

$$\mu_R - \mu_D = RT \ln \frac{X_R}{X_D} + zF\Delta\phi = 0 \tag{7.55}$$

where $\Delta\phi = \phi_R - \phi_D$. Thus the mole fractions in the two regions are related by

$$X_R = X_D e^{-zF\Delta\phi/RT} \tag{7.56}$$

which is recognized again as the Boltzmann distribution. Note that $X_R < Xx_D$ if $\Delta\phi > 0$ and $X_R > X_D$ if $\Delta\phi < 0$.

Putting equation (7.56) in terms of the potential leads to the *Nernst equation* (Hunter, 1981), which is given by

$$\Delta\phi = \frac{RT}{zF} \ln \frac{X_R}{X_D} \tag{7.57}$$

The fact that at equilibrium, the electrochemical potentials are constant in the donor and receiver regions is called *Donnan equilibrium*.

7.8 Acids, bases, and electrolytes

A *solution* is defined as a mixture of different components of different chemical structure. Under normal conditions, sodium chloride dissociates into Na^+ and Cl^- in an aqueous solution, and because of their charge, these ions are surrounded by water molecules. This complex thus has a length scale that is larger than its so-called ionic radius. Blood serum is similar in composition to PBS. In electrochemistry, a *buffer* consists of a weak acid and a conjugate base so that when added to the solution, the pH of the solution does not change; sodium phosphate and potassium phosphate are the buffers in PBS, both containing a hydrogen atom. Generally, the biofluids of interest here are dilute in the charged species, and so the properties of such fluids, density and viscosity, for example, can be taken to be those of water.

Milk of magnesia is an antacid, a base that acts to neutralize acids in the stomach. The active base is magnesium hydroxide: $Mg(OH)_2$.

An *acid* is a proton (H^+) donor, and a *base* is a proton acceptor, according to the Bronsted–Lowry terminology (Raymond, 2007). Acids and bases are linked by the general proton transfer reaction

$$HA + B \rightleftharpoons A^- + BH^+ \tag{7.58}$$

and the extent of the dissociation is measured by an *equilibrium constant* defined by

$$K = \frac{[A^-][BH^+]}{[HA]} \tag{7.59}$$

The limit $K \to 0$ means little dissociation has occurred and defines a weak acid or base. The double arrow in equation (7.58) means that some of the reactant HA remains, and the equilibrium constant is the ratio of the constants associated with the forward and backward reactions: $K = k_f/k_b$. The concept of an equilibrium constant described here is the same as is customarily used in the combustion literature, another example of a type of chemical reaction (Moran & Shapiro, 2007).

Notation alert: The square brackets denote the activity (which is dimensionless) of the given species, as is common in the chemistry literature. This notation eliminates the need for cumbersome subscripts on the concentrations such as $[A^-] = a_{A^-}$.

Acids and bases can be strong or weak, depending on the degree of ionization. For example, a strong acid or base completely dissociates; sodium hydroxide (NaOH), also known as lye, is a strong base such that

$$NaOH \to Na^+ + OH^- \tag{7.60}$$

In chemistry, the notation pC is defined as $pC = -\log_{10} C$. In this context, a strong base is defined as that having $pOH = -\log_{10}[OH^-] \sim 0$ and increasing pH, while a strong acid is defined as $pH = -\log_{10}[H^+] \sim 0$ and increasing pOH.

For example, ammonia (NH_3) is a weak base in water because very little dissociates, having pH ~ 9. Its reaction with water is governed by the equation

$$NH_3 + H_2O \rightleftharpoons OH^- + NH_4^+ \tag{7.61}$$

with the equilibrium constant defined by

$$K = \frac{[OH^-][NH_4^+]}{[NH_3]} \sim 2 \times 10^{-5} \tag{7.62}$$

Water itself is also an acid and base; for the ammonia reaction, the water serves as an acid because it donates a proton to the ammonia. The equilibrium constants for many weak acids range from $K_a \sim 10^{-1}$ to 10^{-5} and for weak bases from $K_b \sim 10^{-5}$ to 10^{-9}. These values can be found in most undergraduate analytical chemistry textbooks and in Butler (1998).

Water itself can ionize; it does so according to the reaction

$$2H_2O \rightleftharpoons H_3O^+ + OH^- \tag{7.63}$$

with

$$K = \frac{[OH^-][H_3O^+]}{[H_2O]^2} \tag{7.64}$$

In the chemistry literature, the liquid water reactant is often dropped, and the equilibrium constant for this reaction is sometimes written

$$K_w = [OH^-][H^+] \tag{7.65}$$

for a single H_2O molecule transfering its proton. The equilibrium constant for the ionization of water at $T = 25°C$ is $K_w = 1 \times 10^{-14}$. Because pure water must be electrically neutral, $[H^+] = [OH^-] = 1 \times 10^{-7}$. Thus the pH of neutral water is pH $= 7$ and is termed the *neutral pH*. The value of K in the ionization of water decreases with decreasing temperature.

In the same way that we speak of strong and weak acids and bases, we speak of strong and weak electrolytes. For example, as noted earlier, sodium chloride dissociates in water according to

$$\text{NaCl (solid)} \rightarrow \text{Na}^+\text{(aq)} + \text{Cl}^-\text{(aq)} \tag{7.66}$$

with the equilibrium constant defined by

$$K = \frac{[Cl^-][Na^+]}{[NaCl]} \tag{7.67}$$

and $K \rightarrow \infty$, indicating complete dissociation (no NaCl in the products) at normal temperatures as described in Section 7.6; thus sodium chloride is a strong electrolyte and you sense the different approaches taken in the engineering community, from this, the chemistry presentation. The limit $K \rightarrow 0$ defines a weak electrolyte. Generally speaking, the electrolytes of interest in micro- and nanofluidics are strong. Water itself is a weak electrolyte.

7.9 Site-binding models of the silica surface

In Chapter 3, the surface charge density was defined in terms of the normal derivative of the electrical potential with the proportionality constant being the electrical permittivity. It was revisited in Chapter 6 in the definition of the electric field due to a charged surface.Thus the electrical potential was defined in terms of the potential in the liquid adjacent to the solid wall. In this section, a complementary view of the surface charge density based on the properties of the solid silica (i.e., silicon family) surface is presented; this will lead to a relationship between the surface charge density and the ζ potential that incorporates the effect of pH. Recall that the electrical double layer was introduced in Section 1.10.

Silicon (chemical symbol Si) is one of the most abundant elements on earth. Silicon may bond with oxygen to form silicon dioxide or silica, which has chemical symbol SiO_2, found commonly in the form of a crystalline mineral. Recently silicon or silica nanoporous membranes have received attention as candidates for implantable devices, specifically in the design of artificial kidneys (Fissell, 2006; Fissell & Humes, 2006) and for glucose sensors. Silicon may be made biocompatible by treating the surface with polyethylene glycol (PEG), which also can reduce the chances of biofouling at the surfaces of the membrane.

The charge on the silica surface is regulated by the availability of potential determining ions (PDI) (Hunter, 1981), that is, ions that allow the surface to acquire charge through electrochemical reactions. For a silica surface in the pH range of interest (pH = 5–9), the hydrogen ion H^+ is considered the primary PDI (Iler, 1979). The pH-dependent surface charge acquired by the surface in turn affects the distribution of ions in the electrolytes that form the electrical double layer near the surface (Elimelech, 1998). To understand the effect of pH on the ζ potential, the predictions of the surface charge from a solid-side model of adsorption on the silica surface will be combined with a liquid-side model of the EDL, to be discussed later in this chapter.

At pH $\sim$ 4, protons desorb from the SiOH to form the SiO^- on the surface so that at pH $>$ 4, the silica surface is negatively charged (Icenhower & Dove, 2000). For pH $<$ 6, the SiOH complexes make up virtually all of the surface sites on a quartz surface, while even at a pH = 7, they account for $\sim$98 percent of the sites. Only at large pH $\sim$ 12 is there a significant fraction of the complex $SiO–Na^+$, on the surface, the fraction being just over 30 percent.

The *isoelectric point* of the surface, or point of zero charge, occurs at pH = 2.4, meaning the reaction is in equilibrium and the surface has no charge. If pH $>$ 2.4, the process of deprotonating SiOH will overwhelm the protonation of SiO, driving the surface charge density strongly negative. If pH $>$ 4, almost all SiOH molecules are deprotonated so that the silicon surface is highly negatively charged and the charge density can be treated as a constant. For pure water, pH = 7, and the pH of most physiological fluids, such as blood serum and saline, is pH $\sim$ 7.4.

The family of surfaces made of silicon, Si, and its derivatives, among them silica, SiOH, and silicon dioxide, SiO_2, can be crystalline (rigidly ordered) or amorphous (disordered). SiO_2 and SiOH are also known as two of many forms of what is termed the *quartz group*. As noted in Section 7.1, when exposed to water, for example, SiOH dissociates according to (Israelachvili, 1992)

$$\mathrm{SiOH\ (solid)} \rightleftharpoons \mathrm{SiO^-(solid)} + \mathrm{H^+(aqu)} \tag{7.68}$$

Thus the silica walls having SiO^- ions (Figure 7.4) will attract the positive ions that are in solution, a process termed *adsorption*, forming the EDL.

The equilibrium constant for the reaction is defined by

$$K_1 = \frac{[\mathrm{SiO^-}][\mathrm{H^+}]_s}{[\mathrm{SiOH}]} \tag{7.69}$$

which has the value $K_1 = 10^{-6.8}$. The subscript s denotes that the H^+ adsorbs to the surface from the liquid side. In dilute electrolyte solutions, the activity and concentration are approximately equal. In both these equations, the square brackets denote mole fraction, and the equilibrium constant K_1 is scaled by 1 mole/m^3

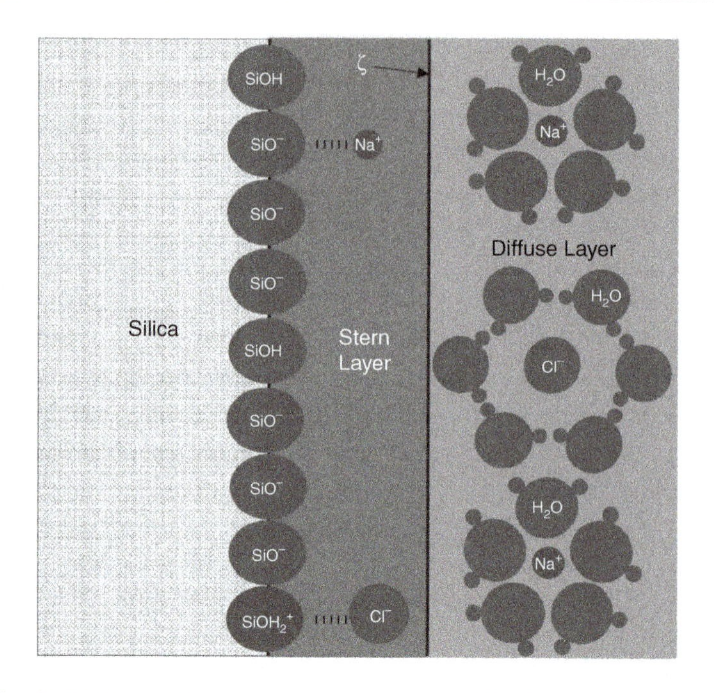

Figure 7.4 Sketch showing the ions that make up the solid side of the silica surface, predominantly SiO^- in the pH range of interest. Adsorption of Cl^- in the pH range of interest is rare and the surface consists of predominantly SiO^- in the pH range of interest.

so that, for example, $K_1 = K_1^*/1\ \text{mole/m}^3$, so that K_1 is dimensionless. As noted previously, this point is not often very clear in textbooks and research papers.

The number density of surface silanols (N_s) in molecules/m^2 is constant for a particular silica surface in its pristine state, meaning that the surface is immersed in a vacuum. The number of molecules on the surface is given by

$$N_s = ([\text{SiO}^-] + [\text{SiOH}])N_A \tag{7.70}$$

where N_A is Avogadro's number. The quantity N_s can be measured using a variety of experimental methods, for example, chromatography with tracer molecules followed by detection with accurate weight gain measurements; infrared spectroscopy and nuclear magnetic resonance (Iler, 1979; Kirby & Hasselbrink, 2004a); and from experimental data, $N_s = 8.3 \times 10^{-6}\ \text{mole/m}^2 N_A$, according to Papirer (2000).

The surface charge density σ that must neutralize the excess charge in the diffuse layer of the EDL is then defined by

$$\sigma = -[\text{SiO}^-]F \tag{7.71}$$

where F is Faraday's constant and the units of surface charge density are C/m^2. As mentioned previously, the first term in equation (7.71) is usually much smaller than the second term.

It is desireable to express equation (7.71) in terms of the equilibrium constants because they are usually known from experiments. Using equations (7.70) and

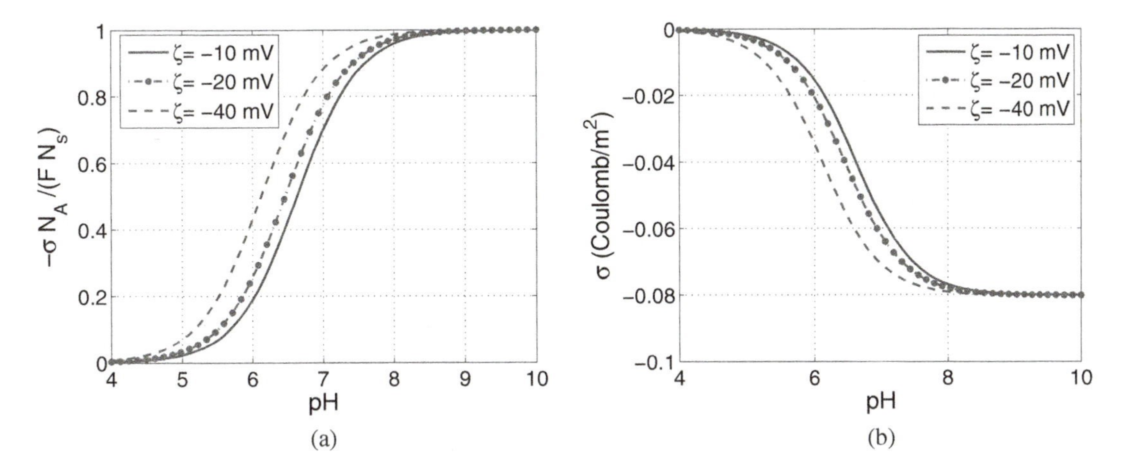

Figure 7.5 Surface charge density for various values of the ζ potential using equation (7.72). (a) Dimensionless form. (b) Dimensional form. Here $N_s = 5 \times 10^{17}/\text{m}^2$ and $K_1 = 10^{-6.8}$.

(7.71) and the definitions of the equilibrium constants,

$$\sigma = -\frac{N_s F}{N_A} \frac{K_1}{K_1 + [\text{H}^+]_s} \tag{7.72}$$

The activity or mole fraction of $[\text{H}^+]_s$ at the surface is essentially unknown; however, its value in the bulk can be calculated and related to the surface value by the Boltzmann distribution, and if the potential at the Stern plane is taken to be the ζ potential, then (see Section 7.12) $[\text{H}^+]_s = b[\text{H}^+]_{bf}$, where $b = e^{-\zeta F/RT}$ and bf stands for "bulk fluid." Equation (7.72) is the solid-side definition of the surface charge density.

The general picture of the relationship between surface charge density and ζ potential is depicted in Figure 7.5. Note that the dimensionless surface charge density $\sigma N_A / F N_s \to 1$ as pH $\to$ 9–10, and for these values of K_1 and N_s, $\sigma \to -0.08$ C/m^2 as $pH \to$ 9–10.

From this analysis, it is apparent that the surface charge density $\sigma = \sigma(\zeta, pH, K_i)$ and that the chosen value of N_s is a parameter. As is seen here, the precise form of the expression for the surface charge density (here equation (7.72)) depends on the specific model of the EDL. These EDL models are summarized by Venema *et al.* (1996), including the Nernstian–Stern layer model, which requires no surface binding sites. They specifically considered cadmium ion adsorption and found that none of the five models they tested, including the triple layer model, could explain all the data. In that sense, this area is still fertile ground for research.

7.10 Polymer surfaces

The term *polymer* is used to describe large molecules that are formed by the chemical union of five or more molecules of a single compound, called a *monomer*. A

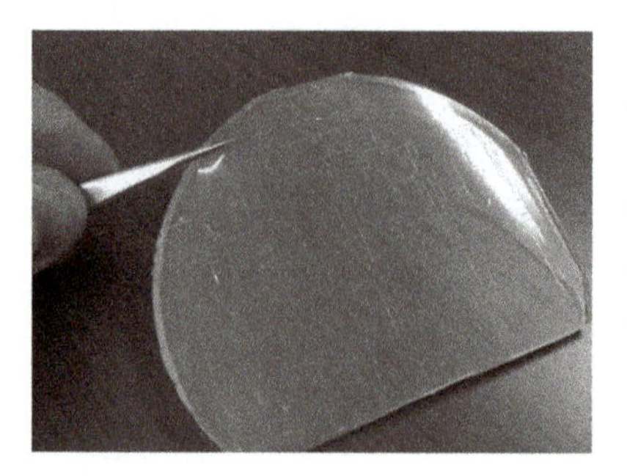

Figure 7.6 Image of PDMS contributed by Professor Derek Hansford.

combination of two monomers is called a *dimer*, whereas the combination of three and four monomers are called *trimers* and *tetramers*, respectively. Proteins and nucleic acids containing carbon are organic *biopolymers* and will be discussed in detail in the next chapter. The reaction mechanism for the formation of the polymer as a chain of monomers is called *polymerization*.

Two of the more common polymers that are often used as materials for nanopore membranes are polydimethylsiloxane (PDMS) (Figure 7.6) and polymethyl methacrylate (PMMA). Both materials are of the plastics family and are used as substitutes for glass and the silica family of surfaces. PDMS is used in contact lenses and for caulking, to name just two applications. Both PDMS and PMMA are softer than the silicas, and both are easily scratched. PMMA is sold under the trade name Plexiglas (Kirby & Hasselbrink, 2004b).

The organic polymers PDMS and PMMA have lower surface charge density and are more hydrophobic than other surfaces. The result of lower surface charge density is a smaller electro-osmotic flow rate; conversely, the hydrophobic nature of these two polymer surfaces may lead to slip, resulting in a higher electro-osmotic velocity. The ζ potential of organic polymers is more sensitive to buffer additives with hydrophobic groups in comparison to silica because of their higher surface hydrophobicity (Kirby & Hasselbrink, 2004C). The properties of the substrates summarized in Table 7.1 are obtained from the review by Kirby and Hasselbrink (2004c), unless otherwise mentioned.

Data for PDMS from a number of different sources suggest that there is a linear dependence of the ζ potential on pH over a wide range of mole fractions of the bulk mixture; Kirby and Hasselbrink (2004c) suggest from the data that

$$\zeta \cong -(2 + 7(\text{pH} - 3)) \log_{10} C \tag{7.73}$$

where C is a reference counterion concentration.

Table 7.1. Properties of silica, PDMS, and PMMA surfaces relevant to microfluidic applications, based on Kirby and Hasselbrink (2004c)

Substrate	Repeating unit	Dissociable surface group	Surface charge	$-\frac{d\zeta}{d(pC)}$ (mV)	Hydrophilic or hydrophobic
Silica	Silicon dioxide	silanol	high	20	hydrophilic
PDMS	Dimethylsiloxane	silanol	low	30	hydrophobic
PMMA	Methylmethacrylate	carboxyl	low	10	hydrophobic

The fifth column tabulates the approximate reduction in the magnitude of zeta potential per 10-fold increase in the bulk solution concentration C of a 1:1 symmetric salt at near-neutral pH. Here $pC = -\log_{10}(C)$. The PDMS surface has fewer surface silanol and therefore typically lower surface charge density than silica surfaces under the same conditions (Kirby & Hasselbrink, 2004C).

Experimental results for PMMA are less clear, and there is a wide range of scatter in the ζ potential as a function of pH. In this case, only the dependence on the concentration is clear, being $\zeta = a_0 - a_1 \log_{10} C$ for univalent electrolytes. The values for the constants $a_0 = -4.06$ mV and $a_1 = -12.57$ mV are valid over the range $7 < \text{pH} < 8$.

7.11 Qualitative description of the electrical double layer

A charged surface in contact with an aqueous electrolyte is shown in Figure 7.9. If a surface is negatively charged, the positively charged ions in the vicinity of the surface tend to collect near the charged surface, a process often called *charge polarization*, a phenomenon that occurs in several contexts (Chang and Yeo, 2010). This process forms a thin layer of positively charged ions near the surface, so-called *counterions*. The thickness of this layer of "pinned" ions near the surface is of the order of the ionic diameter of hydrated counterions, and the edge of this layer is called the *Stern layer*, and the edge of this layer is the *Stern plane*. The electrical potential at the Stern plane is called the ζ *potential*.

The *Stern plane* is the plane of closest approach of an ion, and the precise position is dictated by the size of the ions at the surface. In the classical theory of the EDL, the ions outside the Stern layer are distributed according to the equilibrium *Boltzman distribution*. The layer outside the Stern layer is called the diffuse layer. The total system, including the Stern layer and the diffuse layer, is called the EDL. In the continuum model of the EDL, the ions in the diffuse layer are most often modeled as point charges.

Because the electrical double layer is so thin, being on the order of nanometers to hundreds of nanometers at the most, its structure has not been probed experimentally. However, the analogy between the velocity and the potential in electro-osmotic flow discussed in Chapter 3 means that velocity measurements within the EDL can be used to extract electrical potential values (Sadr *et al.*, 2006). This point has already been discussed in Sections 3.13 and 3.17. The concentration of the cations and anions near a charged surface has traditionally been obtained by analysis (e.g., molecular dynamics simulations) and/or qualitative and descriptive models of the EDL near a charged surface.

The ion distribution within the EDL can be described by using the number density, concentration, or mole fraction. Engineers usually prefer the dimensionless mole fraction, whereas chemists usually use concentration or number density. In this section, mole fraction, will be used, and for a binary mixture, these terms will be denoted by g for the cation and f for the anion.

The simplest model for the EDL was originally given by Helmholtz (1897) long ago, and he assumed that electrical neutrality was achieved in a layer of fixed length. For a negatively charged surface, he assumed that the distributions of the anions and the cations are linear with distance from the surface, as depicted in Figure 7.7.

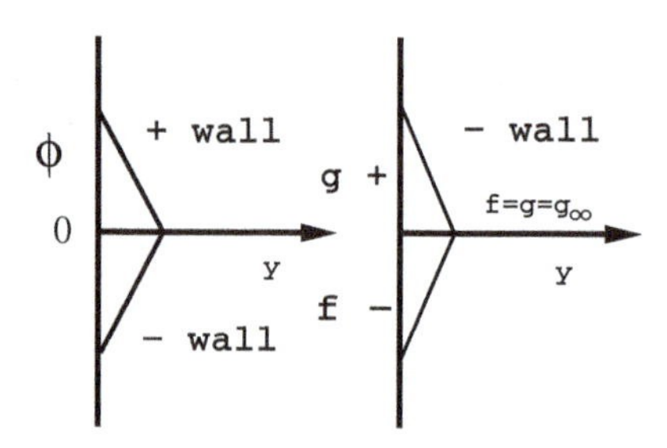

Figure 7.7 Potential (left) and mole fractions near a negatively charged surface according to the ideas of Helmholtz (1897). Here g denotes the cation mole fraction and f denotes the anion mole fraction.

Almost 30 years later, the Debye–Hückel picture of the EDL (Debye & Hückel, 1923) was postulated, in which the influences of the ionic species are equal and opposite but nonlinear functions of distance from the surface so that the surface mole fractions of the coion (anion for a negatively charged surface), the magnitude of the surface mole fractions f^0, and counterion g^0 are symmetric about their asymptotic values far from the surface; see Figure 7.8a.

The Gouy–Chapman (Gouy, 1910; Chapman, 1913) model of the EDL allows for more counterions to bind to the surface charges than coions and is one example of charge polarization. In ion-exchange membranes, the concentration of the counterion may be driven to zero (see Section 7.18). This means that for a negatively charged surface, g^0 can be much larger than its asymptotic value, whereas f^0 is not much lower than the asymptotic value in the core. This situation is depicted in Figure 7.8b. Whether the Debye–Hückel picture or the Gouy–Chapman model of the EDL is relevant depends on the surface charge density, with the Debye–Hückel picture occuring at low surface charge densities and the Gouy–Chapman model occuring for higher surface charge densities.

The double layer model described can be further refined by delineating the intrinsic surface charges SiOH from the adsorbed ions, say, Na^+ and Cl^-, leading to what is now termed the *triple layer model* (James, 1981; Persello, 2000). The structure of the triple layer model is depicted in Figure 7.9. On a silica surface, the protons and the hydroxide ions from the surface charge layer compose the solid-side, innermost layer, with the adsorbed electrolytes pinned in a second

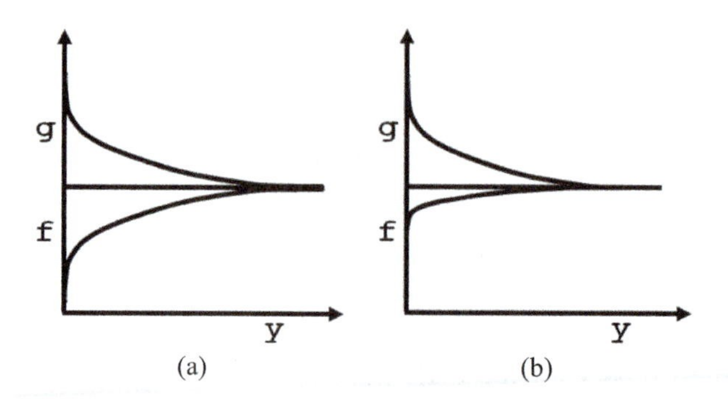

Figure 7.8 (a) Debye–Hückel (Debye & Hückel, 1923) picture of the EDL. Here g denotes the cation mole fraction and f denotes the anion mole fraction. This model occurs for low surface charge densities. (b) Gouy–Chapman model (Gouy, 1910; Chapman, 1913) of the EDL valid at higher surface charge densities.

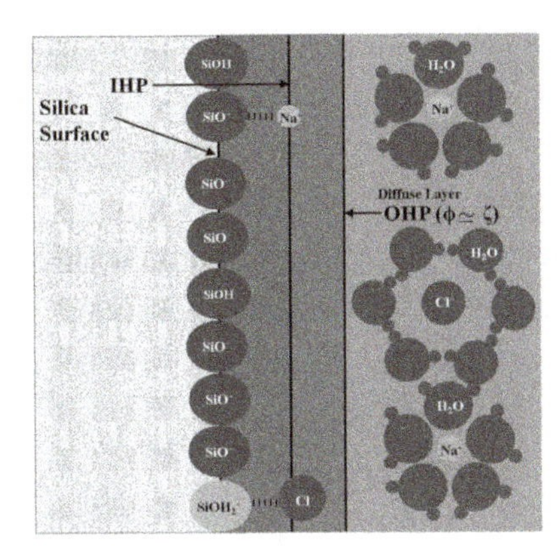

Figure 7.9 The triple layer model of the EDL differentiates layers at the surface according to hydroxyl group surface charges and charge generated by adsorbed ions. The ζ potential takes its value at the outer Helmholtz plane (OHP).

layer adjacent to the surface layer, called the *inner Helmholtz plane* (IHP). In this model, the H^+ and OH^- ions react at the surface of the solid, while the center of the electrolyte ions resides approximately one hydrated ion radius from the surface plane. The remaining electrolyte ions constitute the diffuse layer.

In the triple layer model, there are surface charge densities associated with the reaction with a proton or hydroxyl group (the OH group) and associated with the reactions involving adsorbed ions. This forms what is normally called the surface charge density in the classical EDL picture. Despite the dichotomy between the surface and adsorbed layers, the triple layer model leads to a surface charge density equivalent to that of the double layer model.

Stern (1924) recognized that there are a number of other assumptions embedded in these qualitative and simple models. In the models discussed so far, the ions have been assumed to be point charges, and the solvent is not modified by the presence of the charges. He proposed that the finite size of the ions affects the value of the potential at the surface, and this is called the *excluded volume effect*.

7.12 Electrolyte and potential distribution in the electrical double layer

In the previous two sections, the electrochemistry of the solid side of a charged surface, such as silica, was described. In this section, we discuss the nature of the liquid side of such a surface.

Consider a stagnant electrolyte solution bounded by a single charged flat plate, as in Figure 1.18. As we know, at equilibrium, the electrochemical potential is constant. Thus, if y is the distance measured from the wall, assuming that the wall is long in both the x and z directions, the electrochemical potential of each species is constant in the y direction, and

$$\frac{d\mu_i}{dy} = 0 \tag{7.74}$$

In the general case, where the electrochemical potential may depend on all three spatial variables, $\nabla \mu_i = 0$. Thus, in one dimension,

$$\frac{1}{c_i}\frac{dc_i}{dy} = -\frac{z_i F}{RT}\frac{d\phi}{dy} \tag{7.75}$$

and the concentration of each species is obtained by integrating;

$$c_i = c_i^0 e^{\frac{-z_i F\phi}{RT}} \tag{7.76}$$

the Boltzmann distribution where here c_i^0 is the mole fraction far from the wall, where the potential is defined to be zero. The mole fraction is defined as the concentration of a given species divided by the total concentration, including the solvent.

Conversely, we could define

$$c_i = c_i^0 e^{\frac{-z_i F(\phi-\zeta)}{RT}} \tag{7.77}$$

in which case, c_i^0 is the concentration of species i at the wall, where $\phi = \zeta$. Because the concentration is given by $c_i = n_i/N_A$, the Boltzman distribution could also be expressed as a number density:

$$n_i = n_i^0 e^{\frac{-z_i F\phi}{RT}} \tag{7.78}$$

Consider the case of N species of valence z_i. Note that counterions will collect on the wall in greater numbers than coions, which are repelled by the wall. Then the equation for the potential becomes

$$\nabla^2\phi = \frac{d^2\phi}{dy^2} = -\frac{\rho_e}{\epsilon_e} = -\frac{e}{\epsilon_e}\sum_{i=1}^{N} c_i^0 z_i e^{\frac{-z_i F\phi}{RT}} \tag{7.79}$$

where ρ_e is the volume charge density. It is clear from the argument of the exponential in the volume charge density that the potential scale for nondimensionalization should be taken as $\phi_0 = RT/F$ so that the potential scale is $\phi_0 = 26$ mV.

For $N = 2$, and for a pair of monovalent ions of equal but opposite valence, with ϕ scaled on ϕ_0,

$$\nabla^2\phi = \frac{d^2\phi}{dy^2} = \frac{1}{\kappa^2}\sinh\phi \tag{7.80}$$

where

$$\kappa^2 = \frac{2F^2c^0}{\epsilon_e RT} \tag{7.81}$$

has units of m^{-2} and so $1/\kappa^2 = \lambda^2$ has units of m^2 and λ is the Debye length. Here c^0 is the concentration of the two electrolytes in the bulk, where their concentrations are equal. Defining $Y = \kappa y = y/\lambda$, in dimensionless form,

$$\frac{d^2\phi}{dY^2} = \sinh\phi \tag{7.82}$$

This equation is recognized as dimensionless because of the absence of all of the dimensional parameters present in equation (7.80). For a pair of multivalent ions of valence z and $-z$, the coefficient is just multiplied by the absolute value of the valence, and in dimensionless form,

$$\frac{d^2\phi}{dY^2} = z \sinh z\phi \tag{7.83}$$

If the boundary condition at the surface is assumed to be given by

$$\phi = \phi_s \quad \text{at } Y = 0 \tag{7.84}$$

and at large distances, $\phi \to 0$ as $Y \to \infty$, an analytical solution may be found. The quantity ϕ_s can be identified as the dimensionless ζ potential.

Consider first a pair of monovalent ions. To find the solution, we can integrate once after multiplying by 2 $(d\phi/dY)$ to obtain

$$\frac{d\phi}{dY} = -2\sqrt{\frac{(\cosh\phi - 1)}{2}} \tag{7.85}$$

Now, using the identity

$$\sqrt{\frac{\cosh\phi - 1}{2}} = \sinh\frac{\phi}{2} \tag{7.86}$$

it is possible to integrate again, noting that

$$\int \frac{d\phi}{\sinh\frac{\phi}{2}} = \ln\left(\tanh\frac{|\phi|}{4}\right) \tag{7.87}$$

and so

$$\frac{\tanh\frac{\phi}{4}}{\tanh\frac{\phi_s}{4}} = e^{-Y} \tag{7.88}$$

The solution is thus

$$\phi = 4\tanh^{-1}\left(\tanh\frac{\phi_s}{4}e^{-Y}\right) \tag{7.89}$$

For an arbitrary value of the valence, we have

$$\phi = \frac{4}{z}\tanh^{-1}\left(\tanh\frac{z\phi_s}{4}e^{-zY}\right) \tag{7.90}$$

Since

$$\tanh^{-1}x = \frac{1}{2}\frac{1+\tanh x}{1-\tanh x} \tag{7.91}$$

this solution can also be expressed as

$$\phi = \frac{2}{z}\ln\left(\frac{1 + e^{-zY}\tanh(\frac{z\phi_s}{4})}{1 - e^{-zY}\tanh(\frac{z\phi_s}{4})}\right) \tag{7.92}$$

The combination of equations (7.76) or (7.78) and (7.92) is called the Poisson–Boltzmann solution for the classical EDL. Here ϕ_s is identified as the dimensionless ζ–potential.

The solution for the potential is depicted in Figure 7.10; the Debye–Hückel approximation valid for small potentials is also shown. A negative value of the

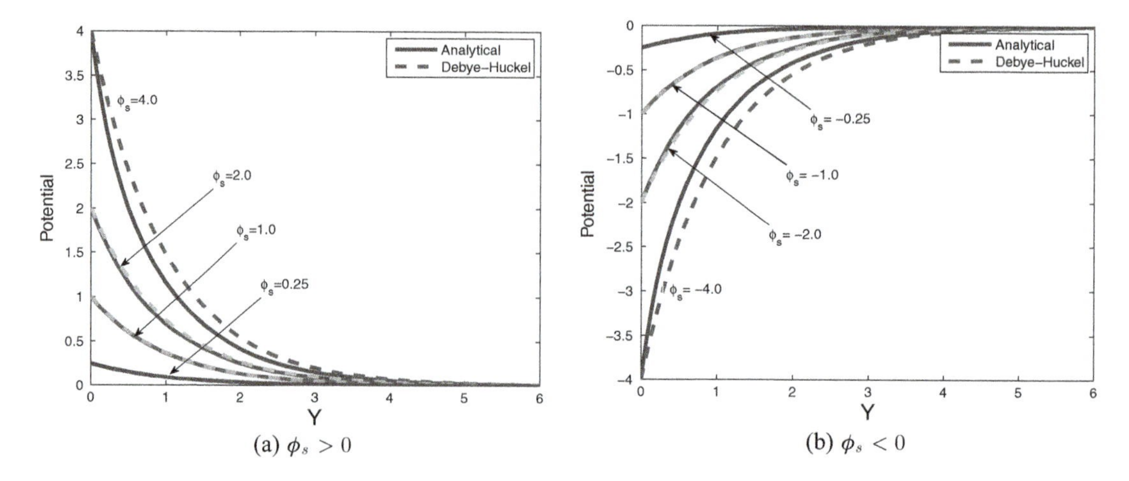

(a) $\phi_s > 0$ (b) $\phi_s < 0$

Figure 7.10 Dimensionless potential for several different values of the surface potential for a 1:1 electrolyte ($z_1 = 1$, $z_2 = -1$) at a bulk ionic strength of $I = 0.3$ M $= 0.15$ Na $+ 0.15$ Cl.

surface potential implies that the wall is negatively charged, whereas a positive value implies a positively charged wall. Note that the solutions are mirror images of each other. The EDL thickness is indicative of the region near the wall over which there is a surplus of cations over anions for a negatively charged wall or vice versa for a positively charged wall.

For small dimensionless potentials, equation (7.83) may be linearized because $\sinh\phi \sim \phi + \cdots$ as $\phi \to 0$, and thus for $z = 1$

$$\frac{d^2\phi}{dY^2} = \sinh\phi = \phi + \cdots \tag{7.93}$$

The solution to this equation is simply

$$\phi = \phi_s e^{-Y} \tag{7.94}$$

and the linearization process is called the Debye–Hückel approximation; this solution is also plotted in Figure 7.10. The thickness of the EDL for the DH approximation can be determined analytically, and

$$0.01 = e^{-\frac{\delta_e}{\lambda}} \tag{7.95}$$

or

$$\delta_e = \lambda \ln \frac{1}{0.01} \tag{7.96}$$

or $\delta_e = 4.6\lambda$, a formula that is similar to the definition of the boundary layer thickness in fluid mechanics.

In the general case where a mixture contains N ions, the Debye length is given by $\lambda^2 = 1/\kappa^2 = \epsilon_e RT/F^2 I$, where $I = \sum_{i=1}^{N} c_i^0 z_i^2$. For a pair of monovalent dissociated ions such as NaCl, $I = c_1 + c_2 = 2c^0$, where c^0 is the concentration of each species in the region where the mixture is electrically neutral.

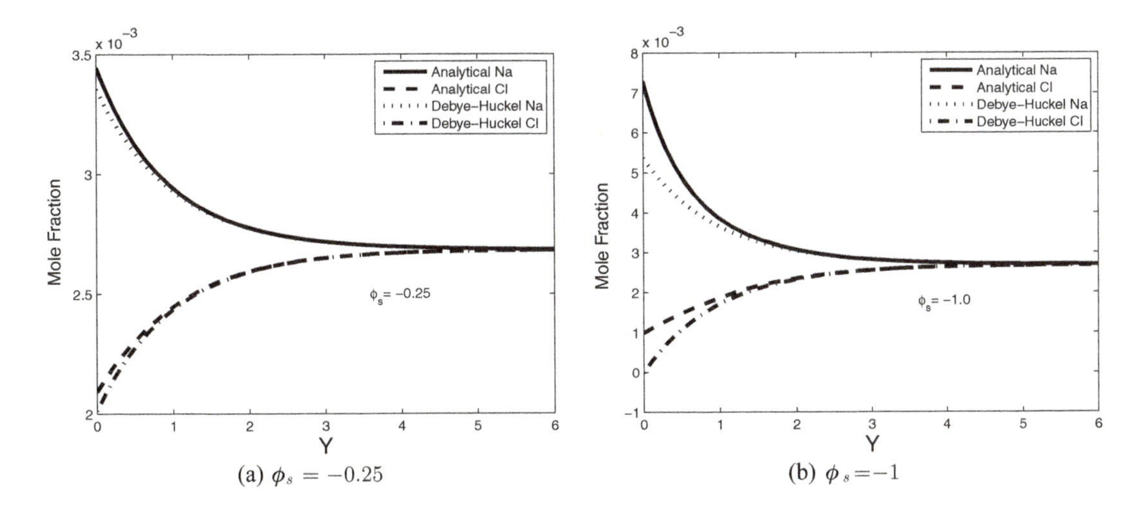

(a) $\phi_s = -0.25$ (b) $\phi_s = -1$

Figure 7.11 Mole fractions for two different values of the surface potential for a 1:1 electrolyte ($z_1 = 1$, $z_2 = -1$) for a negatively charged wall at a bulk ionic strength of $I = 0.3$ M.

Notation alert: This is just a reminder that chemists define the ionic strength as $I = 1/2 \sum_{i=1}^{N} c_i^0 z_i^2$ so that there will be a 2 in the definition of the Debye length. This definition is not appropriate for mixtures containing multivalent ions.

As a numerical example, consider an aqueous electrolyte solution with 0.015 M Na^+ and 0.015 M Cl^-. Then, at $T = 300$ K,

$$\lambda = \sqrt{\frac{8.315 \frac{\text{J}}{\text{mole K}} \times 300 \text{ K} \times 78.54 \times \frac{1}{36\pi} \times 10^{-9} \frac{C^2}{\text{Nm}^2}}{30 \frac{\text{moles}}{\text{m}^3} \times (96500 \frac{\text{C}}{\text{moles}})^2}}$$
$$= 2.5 \times 10^{-9} \text{ m} = 2.5 \text{ nm} \tag{7.97}$$

For an ionic strength of 0.3 M, $\lambda = 0.8$ nm.

Note in Figure 7.10, the surprising result that the DH approximation gives a good representation of the full nonlinear solution for values of the potential much higher than expected. It is expected that the approximation should break down for potentials $\phi_s \sim 0.2$, but this does not occur. However, it is seen that the DH approximation is much worse for the mole fractions, for which

$$X_i = X_i^0 e^{-z_i \phi} \sim X_i^0 (1 - z_i \phi + \cdots) \tag{7.98}$$

as depicted in Figure 7.11. Adding more terms to the Taylor series approximation for the exponential only marginally improves the situation.

Let us now consider the EDL on channel walls, as depicted in Figure 7.12. Now there are two length scales λ and the height of the channel h. Moreover, if the Debye length is large enough, or the channel height is small enough, the potential may not be zero over most of the core of the channel. In anticipation of this possibility, if we nondimensionalize the y coordinate on h, then

$$\epsilon^2 \frac{d^2 \phi}{dY^2} = z \sinh z\phi \tag{7.99}$$

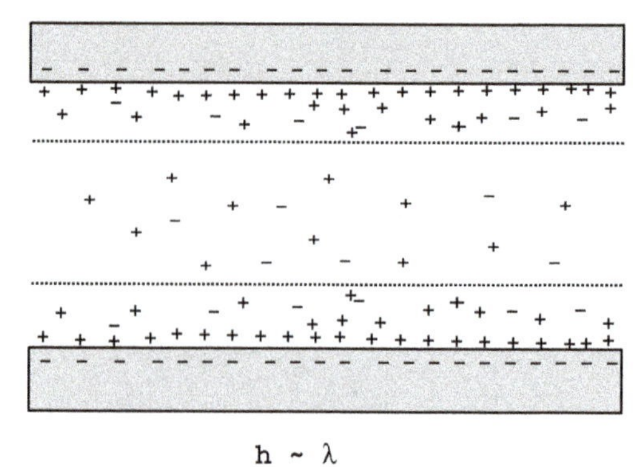

Figure 7.12 The classical EDL on two plates.

where ϕ is the dimensionless potential and $Y = y/h$ and $\epsilon = \lambda/h$. The boundary conditions now become

$$\phi = \phi_s \quad \text{at } Y = 0, 1 \tag{7.100}$$

If both of the EDLs are thin, then the solution for the potential is as given earlier, with $Y = y/\lambda = y/\epsilon h$ near $y = 0$ and $Y = h - y/\lambda = h - y/\epsilon h$ near $y = h$. This situation is depicted in Figure 7.13 and is the straightforward extension to the case of a single plate.

If $\epsilon = O(1)$, the double layers are said to be overlapping, and the integration procedure for the overlapped case cannot be performed as before because the boundary condition must be applied at the walls $Y = 0, 1$. The integration

$$\int_0^Y \frac{d\phi}{\sqrt{\cosh z\phi - \cosh z\phi_s}} = \pm\sqrt{2}Y \tag{7.101}$$

results in a generalized hypergeometric series, which must be evaluated numerically. It is thus easier to numerically calculate the result directly, and these results are shown in Figure 7.13 for several values of the surface potential, the results

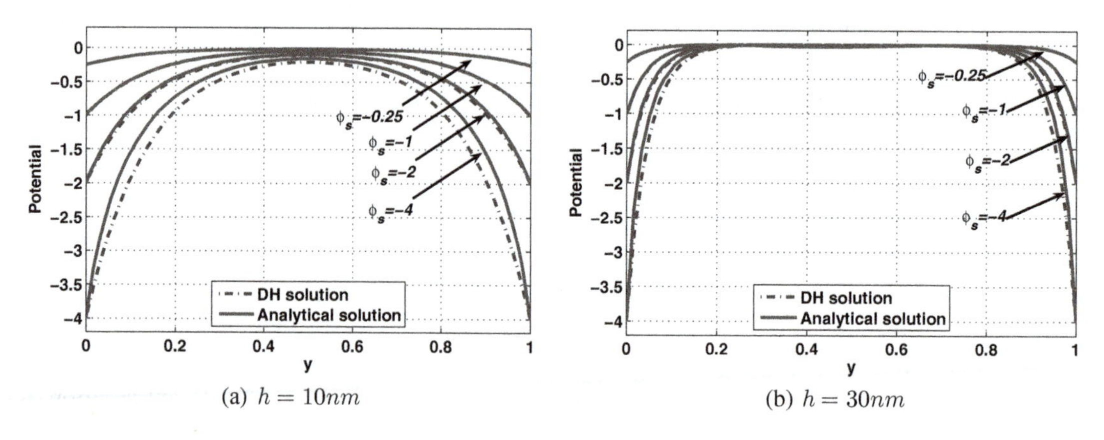

Figure 7.13 Dimensionless potential for several different values of the surface potential for a 1:1 electrolyte ($z_1 = 1$, $z_2 = -1$) for a positively charged wall in a channel of height $h = 10, 30$ nm at a bulk ionic strength of 0.1 M (0.05 + 0.05 M).

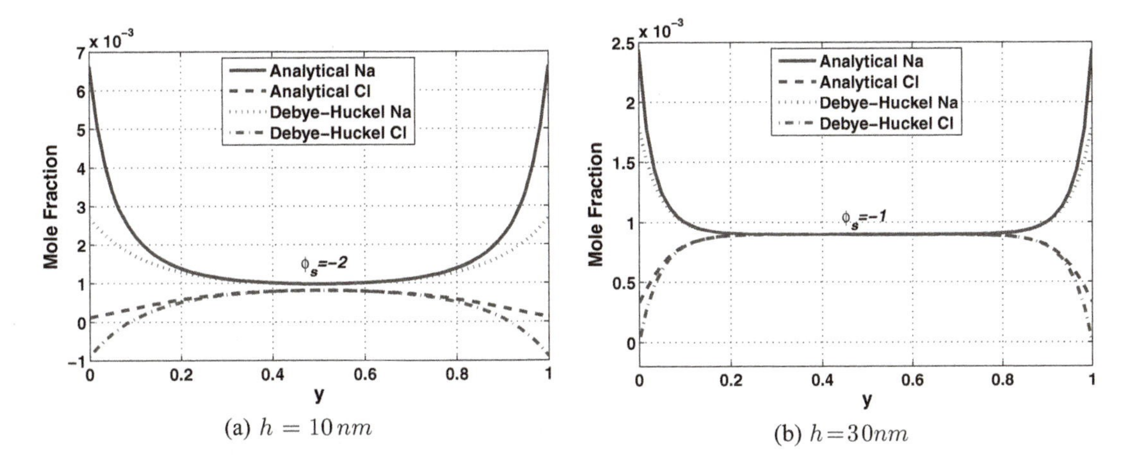

Figure 7.14 Mole fractions for two different values of the surface potential for a 1:1 electrolyte ($z_1 = 1$, $z_2 = -1$) for a positively charged wall in a channel of height $h = 10$, 30 nm at a bulk ionic strength of 0.1 M.

for the mole fractions are depicted in Figure 7.14. In the Debye–Hückel limit, the solution in the overlapping case is given by

$$\phi = \phi_s \frac{\cosh \frac{y-1/2}{\epsilon}}{\cosh \frac{1}{2\epsilon}} \tag{7.102}$$

It is left as an exercise to calculate the solution for the potential when the surface charge density rather than the ζ potential is known.

7.13 Multivalent asymmetric mixtures

In the previous section, a binary mixture of monovalent ions was considered exclusively. Grahame (1953) was apparently the first to investigate the case of asymmetric multivalent mixtures. Using the methods described in the previous section results in the generalized dimensionless expression after one integration of the Poisson equation ($Y = y/\lambda$) is

$$\left(\frac{d\phi}{dY}\right)^2 = \left(\sum_{i=1}^{N} X_i^0 z_i e^{-z_i\phi} - \sum_{i=1}^{N} X_i^0\right) \tag{7.103}$$

where X_i^0 are the dimensionless concentrations scaled on the ionic strength in the bulk; the potential scale is RT/F, as before.

Unfortunately, the integration can be carried out only for specific values of the valences. Grahame (1953) considered explicitly the case of a 2:1 ($z_1 = 2; z_2 = -1$) electrolyte and a 1:2 electrolyte. In his notation, the Debye length is based on the bulk concentration of the less abundant species, and for a 2:1 electrolyte, this results in the dimensionless equation

$$\left(\frac{d\phi}{dY}\right)^2 = e^{-2\phi} + e^{\phi} - 3 \tag{7.104}$$

Using the identity

$$(e^{-\phi} - 1)(1 + 2e^{\phi})^{1/2} = (e^{-2\phi} + e^{\phi} - 3)^{1/2} \tag{7.105}$$

equation (7.104) can be integrated again to obtain ($\phi_s = \zeta$)

$$\int_0^Y dY = \int_\zeta^\phi \frac{d\phi}{(e^{-\phi} - 1)\sqrt{1 + 2e^{\phi}}} \tag{7.106}$$

This integral can be evaluated by making the transformation $\xi = (1 + 2e^{\phi})^{1/2}$ or from integral tables for $\phi < 0$ to obtain

$$Y = \frac{2}{\sqrt{3}}\left[\tanh^{-1}\left[\frac{1 + 2e^{\phi}}{3}\right] - \tanh^{-1}\left[\frac{1 + 2e^{\zeta}}{3}\right]\right] \tag{7.107}$$

which is an equation that gives the potential ϕ as an implicit function of Y. Here $\tanh^{-1}$ is the inverse hyperbolic tangent function.

For $\phi > 0$, the result is given by

$$Y = \frac{1}{\sqrt{3}}\left(\ln\left[\frac{(1 + 2e^{\phi})^{1/2} + \sqrt{3}}{(1 + 2e^{\phi})^{1/2} - \sqrt{3}}\right] - \ln\left[\frac{(1 + 2e^{\zeta})^{1/2} + \sqrt{3}}{(1 + 2e^{\zeta})^{1/2} - \sqrt{3}}\right]\right) \tag{7.108}$$

It is left to the exercises to evaluate these two equations and plot the potential as a function of Y.

7.14 The ζ potential and surface charge density: Putting it all together

Having determined the electrical potential and the concentrations on the liquid side of the surface, we can now develop a relationship that relates the surface charge density, the ζ potential, and pH.

7.14.1 The classical liquid-side view for a symmetric electrolyte

Recall from Figure 1.18 that the ζ potential is the electrical potential at the Stern plane and is directly related to the surface charge density. An explicit expression for surface charge density as a function of ζ potential can be obtained for a symmetric binary $z{:}z$ electrolyte using the liquid-side analysis. Then, using the previous site-binding analysis on the solid side, we can tie it all together.

The derivative of the potential normal to the wall is given by

$$\frac{d\phi}{dy} = \mp\sqrt{\frac{2k_bT}{\epsilon_e}\left[\sum_{i=1}^{N} n_i^0\left(e^{\frac{-z_ie\phi}{k_bT}} - 1\right)\right]} \tag{7.109}$$

where we choose the negative sign if the potential is positive on the wall and the positive sign if the potential is negative at the wall. For a symmetrical pair of ions, the number density of the ions $n_1^0 = n_2^0 = n^0$ in the bulk, and

$$\frac{d\phi}{dy} = \mp\sqrt{\frac{4k_bTn^0}{\epsilon_e}}\left(\cosh\left(\frac{ze\phi}{k_bT}\right) - 1\right) = \mp\sqrt{\frac{4RTc^0}{F\epsilon_e}}\left(\cosh\left(\frac{zF\phi}{RT}\right) - 1\right) \tag{7.110}$$

Using the identity given by equation (7.86), for a pair of symmetric electrolytes,

$$\frac{d\phi}{dy} = \mp\sqrt{\frac{8k_bTn^0}{\epsilon_e}}\sinh\left(\frac{ze\phi}{2k_bT}\right) \tag{7.111}$$

Specifically, at the wall,

$$\left.\frac{d\phi}{dy}\right|_0 = \mp\sqrt{\frac{8k_bTn^0}{\epsilon_e}}\sinh\left(\frac{ze\zeta}{2k_bT}\right) \tag{7.112}$$

where it is noted that here ζ is dimensional.

The total charge of the cations and anions in the gap must be offset by the total surface charge so that the system is electrically neutral. Consider the case of a thin double layer $\epsilon = \lambda/h \ll 1$ outside of which is a neutral aqueous solution. Then, if σ_0 denotes the surface charge density on the liquid side,

$$\sigma_0 + \int_0^\infty \rho_e dy = \sigma_0 - \epsilon_e \int_0^\infty \frac{d^2\phi}{dy^2} dy = 0 \tag{7.113}$$

Substituting from earlier,

$$\sigma_0 = -\epsilon_e \left.\frac{d\phi}{dy}\right|_0 = \sqrt{8\epsilon_e k_b T n^0}\sinh\frac{ze\zeta}{2k_bT} \tag{7.114}$$

and solving for the ζ potential,

$$\zeta = \frac{RT}{zF}\sinh^{-1}\left(\frac{\sigma_0 F\lambda}{\epsilon_e RT}\right) \tag{7.115}$$

This is a fundamental relationship between the surface charge density, the Debye length, λ, and the ζ potential. In the Debye–Hückel limit of small potential,

$$\sigma_0 = \epsilon_e \left.\frac{d\phi}{dy}\right|_0 = \sqrt{8\epsilon_e k_b T n^0}\frac{ze\zeta}{2k_bT} \tag{7.116}$$

so that

$$\sigma_0 = z\epsilon_e\kappa\zeta\frac{F}{RT} \tag{7.117}$$

and this equation is called *Grahame's equation*. Note that the quantity

$$\frac{\sigma_0}{\epsilon_e\kappa} = z\zeta\frac{F}{RT} \tag{7.118}$$

is dimensionless.

For a finite channel with overlapped double layers having the same surface charge,

$$2\sigma_0 - \epsilon_e \int_0^1 \frac{d^2\phi}{dy^2} dy = 0 \tag{7.119}$$

so that

$$\sigma_0 = 2z\epsilon_e \kappa \zeta \frac{F}{RT} \tanh \kappa h \tag{7.120}$$

in the Debye–Hückel limit.

7.14.2 The solid-side view and connection to the liquid side

The preceding simple formula for the relationship between the surface charge density and ζ potential valid in general for a z:z electrolyte cannot distinguish the effect of pH since pH defines the adsorption of H^+ ions on the surface, as discussed in Section 7.9. For the case of a simple mixture such as Na^+–Cl^-–water, a straightforward extension of equation (7.72) for the negatively charged surface leads to

$$\sigma_0 = -\frac{N_s F}{N_A} \left(\frac{K_1}{K_1 + b \left(K_{Na}[\mathrm{Na}^+] + [\mathrm{H}^+] \right)} \right) \tag{7.121}$$

where $K_1 \sim 10^{-7}$ is the equilibrium constant of the silanol reaction given by equation (7.68), K_{Na} is the equilibrium constant of surface adsorption of sodium ions, and N_s is the total number density of surface silanols. The adsorption of chloride ions, the coions, has been neglected. Here b is the Boltzmann factor $b = e^{-\zeta F/RT}$, defined previously. In equation (7.121), it is assumed that the silica surface has sufficient negative charge (pH $\gg 3$) so that protonation of neutral silanol groups and adsorption of chloride ions can be neglected. Note that for $\zeta < 0, b > 0$ and in equation (7.121), the square brackets denote bulk or reservoir mole fractions.

To connect the solid side with the liquid side, squaring equation (7.109) leads to

$$\sum_{i=1}^{N} n_i^0 (b^{z_i} - 1) - \frac{\sigma_0^2}{2\epsilon_e RT} = 0 \tag{7.122}$$

To obtain the ζ potential as a function of the concentration of the various species in the system, including hydrogen ions, which reveal the pH dependence of the ζ potential, we equate the liquid- and solid-side expressions for the surface charge density, $\sigma = \sigma_0$, which results in a nonlinear algebraic equation in the Boltzman factor, b requiring numerical solution.

Consider the situation in which all electrolytes (neutral molecules, such as NaCl, KNO_3, HCl, and H_2O, capable of dissociation) used to formulate the solution in the reservoir are symmetric and dissociate into an equal number of z valent cations and z valent anions, where z is taken as a positive integer. In this situation, the total number of species N can only be an even number, and there are only $N/2$ distinct n_i^0 if the ions are modeled as point charges. In this case, the summation in equation (7.122) for a binary electrolyte is given by

$$n^0(b^z + b^{-z} - 2) = n^0(b^{z/2} - b^{-z/2})^2 \quad (7.123)$$

Substituting into equation (7.109) and using the definition of Debye length,

$$\sigma_0 = -\frac{\epsilon_e RT}{zF\lambda}(b^{z/2} - b^{-z/2}) \quad (7.124)$$

where the minus sign in front of the expression on the left emphasizes that the ζ potential and surface charge have the same sign. This equation is an alternate form of equation (7.114), only it is more general in the sense that it is valid for multicomponent mixtures containing only $+z$ and $-z$ valent species.

In the limit of large ζ potential such that $b \gg 1$, equation (7.124) becomes

$$\sigma_0 = -\frac{z\epsilon_e}{\lambda}\frac{RT}{F}b^{z/2} \quad (7.125)$$

It is uncertain whether the adsorption of sodium plays a significant role in the current context when compared with H^+ adsorption (Kirby & Hasselbrink, 2004a). Thus the solid-side equation (7.121) is rewritten, neglecting sodium ion adsorption to the silica surface (setting $K_{Na} = 0$) as

$$\sigma_0 = -\frac{N_s F K_1}{N_A(K_1 + b[H^+])} \quad (7.126)$$

Recall that K_1 is the equilibrium constant for the SiOH dissociation and is estimated to be $K_1 = 10^{-6.8}$. In most cases, the second term in the denominator of equation (7.126) is dominant under conditions when $-\zeta F/RT \gg 1$ and $[H^+]$ is not too small.

When $-\zeta F/RT \gg 1$, we can combine equation (7.125), calculating σ_0 from the liquid side, with equation (7.126), calculating the same quantity from the solid side, to obtain

$$\exp\left(-\frac{3z\zeta F}{2RT}\right) = \frac{A}{[H^+]} \quad (7.127)$$

where the dimensionless number

$$A = \frac{zF^2 N_s F K_1 \lambda}{N_A \epsilon_e RT} > 0 \quad (7.128)$$

Taking natural logarithms of both sides of equation (7.127) and rearranging, it is found that

$$-\frac{\zeta F}{RT} = \frac{2}{3}(\ln A - \ln[H^+]) \quad (7.129)$$

In the chemistry literature, it is common to convert to logarithms to the base 10 using log 10 = 2.303, and utilizing the definition of pH, the final result for the ζ potential is (Revil *et al.*, 1999)

$$\frac{\zeta F}{RT} = 1.535(\log_{10} A + \text{pH}) \quad (7.130)$$

where $A = A(\text{pH})$ through the dependence of A on the Debye length.

Equation (7.130) suggests that the zeta potential has an approximately linear dependence with pH at large ζ potential, as discussed elsewhere in the literature (Kirby & Hasselbrink, 2004a). This equation predicts a change in zeta potential of about $1.535RT/F = 40$ mV per pH unit. In practice, the change in ζ potential per pH unit depends on the type of surface. Experimental data on porous silicon substrates suggest an approximate slope of about 30 mV (Thust *et al.*, 1996). For small ζ potential, $b \to 1$, and combining the liquid and solid sides, the result is

$$\zeta = \frac{N_s \lambda}{\epsilon_e N_A}\left(1 + \frac{10^{-\mathrm{pH}}}{K_1}\right) \tag{7.131}$$

In the general case of symmetric electrolytes discussed earlier, the equation for the ζ potential becomes

$$\frac{\epsilon_e RT}{zF\lambda}(b^{z/2} - b^{-z/2}) = \frac{N_s F}{N_A}\left(\frac{K_1}{K_1 + b\left(K_{\mathrm{Na}}[\mathrm{Na}^+] + [\mathrm{H}^+]\right)}\right) \tag{7.132}$$

which is a nonlinear equation that may be solved using a numerical zero-finding technique. Obviously, this procedure may be extended to more complex mixtures of asymmetric electrolytes, as discussed previously.

Consider the calculation of the effect of solution $\mathrm{pH} = -\log_{10}[\mathrm{H}^+]$ on the zeta potential in an aqueous solution of the salt NaCl. The pH of the solution can be varied by adding small quantities of either the acid hydrogen chloride, HCl, to obtain $\mathrm{pH} < 7$ or the base sodium hydroxide, NaOH, for $\mathrm{pH} > 7$. The relative amounts of NaCl and the acid or base for each desired pH value are so chosen to keep the solution ionic strength I of the solution constant. The required concentration of NaCl can be calculated for a given pH from the desired ionic strength I using $I = [\mathrm{Na}^+] + [\mathrm{H}^+] = [\mathrm{NaCl}] + 10^{-\mathrm{pH}}$ for $\mathrm{pH} < 7$ and $I = [\mathrm{Cl}^-] + [\mathrm{OH}^-] = [\mathrm{NaCl}] + 10^{-\mathrm{pH}+14}$ for $\mathrm{pH} > 7$. It may be noted, however, that for the practical pH range of $5 < \mathrm{pH} < 10$ and NaCl concentrations of 1 mM and above, $[\mathrm{OH}^-] \ll [\mathrm{Cl}^-]$ and/or $[\mathrm{H}^+] \ll [\mathrm{Na}^+]$; therefore the ionic strength is dominated by the sodium chloride, or $I \simeq [\mathrm{NaCl}]$.

Because the ionic strength is held constant, the Debye length λ is a constant as the pH is varied. Noting that $z = 1$ for the sodium chloride mixture, multiplying both sides of equation (7.132) by $x = b^{1/2}$ and rearranging, the following fourth order polynomial equation in x results:

$$B(K_{\mathrm{Na}}[\mathrm{Na}^+] + [\mathrm{H}^+])x^4 + B(K_1 - K_{\mathrm{Na}}[\mathrm{Na}^+] - [\mathrm{H}^+])x^2 - K_1 x - BK_1 = 0 \tag{7.133}$$

where $B = \epsilon RTN_A/N_s F^2$ and is independent of the changing pH of the solution. Equation (7.133) can be solved for x through analytical and/or numerical root finding techniques, for example, through the use of the function `roots` in Matlab. Then the zeta potential is calculated from $\zeta = -2RT/F(\ln x)$.

Figure 7.15 shows the effect of pH on the zeta potential for a solution with $I = 0.2$ mM (0.1 MNa, 0.1 MCl), $K_1 = 10^{-6.5}$, $K_{\mathrm{Na}} = 10^{-3.5}$, $N_s = 4.8\ \mathrm{nm}^{-2}$,

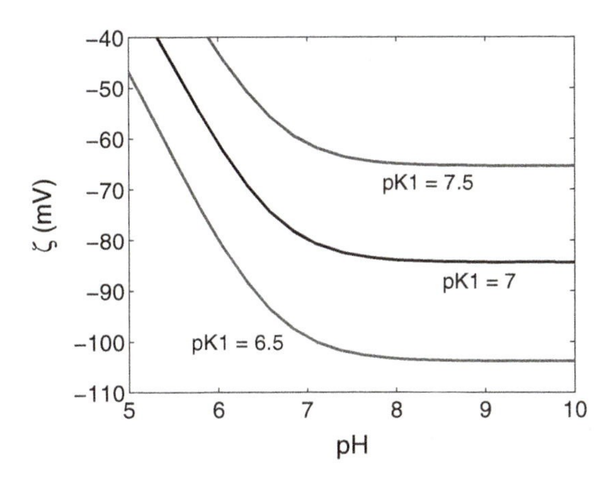

Figure 7.15 The pH dependence of ζ potential calculated using the site-binding model (equation (7.133)) for a sodium chloride solution in contact with a fused silica surface. The parameters used are $I = 0.2$ mM (0.1 MNa, 0.1 MCl), $K_{Na} = 10^{-3.5}$, $N_s = 4.8$ nm^{-2}.

considered to represent a fused silica surface (Revil *et al.*, 1999; Papirer, 2000). For low (acidic) to near-neutral pH values, the magnitude of the ζ potential increases almost linearly with pH because, if the number of available H^+ ions is small, the tendency of the charged SiO^- groups to recombine with H^+ to form neutral SiOH groups is reduced. However, a plateau in the ζ versus pH curve is eventually reached, as can be verified by observing that the right-hand side of equation (7.132) becomes a constant for small values of $[H^+]$. A similar behavior has also been observed in experiments (Kirby & Hasselbrink, 2004a).

7.15 The electrical double layer on a cylinder

The EDL on a cylinder of radius a is also an important problem because tubular membranes are used in filtration processes, such as water purification and desalination, and other applications. Analogous to the plane wall case, the *dimensionless* governing equation for the potential for a z:z electrolyte is given by

$$\epsilon^2 \left(\frac{\partial^2 \phi}{\partial r^2} + \frac{1}{r} \frac{\partial \phi}{\partial r} \right) = z \sinh z\phi \qquad (7.134)$$

where $\epsilon = \lambda / a$ and λ is the Debye length.

In general, in cylindrical coordinates, integration of equation (7.134) cannot be performed analytically in the general nonlinear case. However, under the Debye–Hückel approximation, an analytical solution may be found. Linearizing this equation for $z = 1$ for a pair of symmetric electrolytes yields

$$\epsilon^2 \left(\frac{d^2 \phi}{dr^2} + \frac{1}{r} \frac{d\phi}{dr} \right) = \phi \qquad (7.135)$$

This is a Bessel equation of order zero, and the solution for the potential in the EDL on the *inside* of the cylinder is given by

$$\phi = \zeta \frac{I_0\left(\frac{r}{\epsilon}\right)}{I_0\left(\frac{1}{\epsilon}\right)} \tag{7.136}$$

Note that this solution is valid for any value of $\epsilon = \lambda/a$ where a is the radius of the cylinder. For small ϵ, the solution reduces to the planar case of a decaying exponential.

7.16 The electrical double layer on a sphere

A spherical microparticle flowing in a channel is a common process in micro and nanofluidics. Thus it is often necessary to solve for the electrical potential in the EDL around the sphere. In this case, the governing equation for the potential is, in dimensionless form,

$$\epsilon^2 \frac{1}{r^2} \frac{d}{dr}\left(r^2 \frac{d\phi}{dr}\right) = z \sinh z\phi \tag{7.137}$$

as in the planar case; the boundary conditions are $\phi = \zeta$ at $r = a$ and $\phi \to 0$ as $r \to \infty$.

As with the cylindrical geometry, the full nonlinear problem cannot be integrated analytically. Under the Debeye–Hückel approximation for small potential, we have the reduced equation

$$\epsilon^2 \frac{1}{r^2} \frac{d}{dr}\left(r^2 \frac{d\phi}{dr}\right) = z^2 \phi \tag{7.138}$$

and the solution is given by

$$\phi(r) = A \frac{e^{-z\frac{r}{\epsilon}}}{r} + B \frac{e^{z\frac{r}{\epsilon}}}{r} \tag{7.139}$$

The constants A and B must be found from the boundary conditions because the potential must remain bounded at $r = \infty$, $B = 0$. Applying the boundary condition on the sphere, we have

$$\phi(r) = \frac{1}{r} \zeta e^{z(1-r)/\epsilon} \tag{7.140}$$

The condition of charge neutrality is similar to that for the flat plate, and so the total charge in coulombs on the surface is given by

$$\Sigma = -4\pi\epsilon_e \int_a^\infty r^* 2\rho_e dr^* = 4\pi\epsilon_e \int_a^\infty \frac{d}{dr^*}\left(r^{*2} \frac{d^2\phi^*}{dr^*}\right) dr^* \tag{7.141}$$

Performing the integration,

$$\Sigma = 4\pi\epsilon_e a^2 \zeta \left(\frac{1}{\lambda} + \frac{1}{a}\right) \tag{7.142}$$

The surface charge density is thus given by

$$\sigma_0 = \frac{\Sigma}{4\pi a^2} = \epsilon_e \zeta \left(\frac{1}{\lambda} + \frac{1}{a} \right) \tag{7.143}$$

As an example, consider a sphere having a radius $a = 10^{-7}$ m $= 100$ nm in a mixture where the number density is $n^0 = 10^{25}$ molecules/m^3. The concentration is

$$c = \frac{n^0}{N_A} = \frac{10^{25}}{6.02 \times 10^{23}} = 17 \frac{\text{moles}}{\text{m}^3} = 0.017 \text{ M} \tag{7.144}$$

with a Debye length of

$$\lambda = \sqrt{\frac{RT\epsilon_r\epsilon_0}{2IF^2}} = \sqrt{\frac{8.315 \times 300 \times 78.54 \times \frac{1}{36\pi} \times 10^{-9}}{2 \times 17 \times 96{,}500^2}} = 2.34 \text{ nm} \tag{7.145}$$

For $\sigma_0 = -0.1\ C/m^2$, the ζ potential at the surface is

$$\zeta^* = \frac{\lambda}{\epsilon_e}\sigma_0 = -0.34 \text{ V} \tag{7.146}$$

7.17 Electrical conductivity in an electrolyte solution

We have already defined the electrical conductivity for a general electrically conducting material in Section 6.8; just as in a solid, electrically conducting media, electric current in an electrolyte solution or plasma, originates from the motion of charge carriers such as ions and charged biomolecules. For constant electrolyte concentration, a given ion in the electrolyte reaches a constant velocity in the direction of the electric field, the *migration velocity*, determined by the action of the electric field and the electrostatic and hydrodynamic interactions with the surrounding solvent molecules. These two effects act in opposite directions, with the electric field tending to increase the migration velocity and the interactions with the surrounding molecules tending to decrease the migration velocity.

We have already seen in Chapter 3 that the flux of a charged species A in an electrolyte solution, in the absence of a velocity field, is given by

$$\vec{N}_A = -D_A \nabla c_A + m_A z_A c_A \vec{E} \tag{7.147}$$

and the current density is defined by

$$\vec{J} = F \sum_{i=1}^{n} z_i \vec{N}_i \tag{7.148}$$

Assuming that the concentration is constant, we have

$$\vec{J} = F \sum_{i=1}^{n} z_i m_i \vec{E} \tag{7.149}$$

where $m_i = z_i \ FD_i/RT$ is the *ionic mobility*. Ohm's law for an electrolyte solution is

$$\vec{J} = \sigma_e \vec{E} \tag{7.150}$$

so that the electrical conductivity σ_e is defined by

$$\sigma_e = F \sum_{i=1}^{n} m_i z_i^2 c_i \tag{7.151}$$

For solutions of 1:1 electrolytes, specifying the ionic strength is equivalent, under electrical neutrality situations, to specifying the concentration of cations and anions. If the ionic strength (without the 1/2, as the chemists define it) is defined by $I = \sum z_i^2 c_i$, there are $I/2$ moles of cations and $I/2$ moles of anions per liter of a 1:1 electrolyte solution. If the mobility of the cation is m^+ and the mobility of the anion is m^-, then equation (7.151) takes the form

$$\sigma_e = \frac{(m^+ + m^-)IF}{2} \tag{7.152}$$

According to the classical theory of electrolyte solutions (Fuoss & Onsager, 1955), the mobilities of the ions are dependent on the diffusion coefficients, which are functions of concentration, and so both m^+ and m^- are dependent on the concentration of the ions in the electrolyte and therefore on the ionic strength I. The dependence of ionic mobilities on ionic strength results in the following expression for the sum of mobilities in equation (7.152) (Fawcett, 2004):

$$m^+ + m^- = m_+^0 + m_-^0 - \left\{ B_1 \left(m_+^0 + m_-^0 \right) + \frac{B_2}{F} \right\} \frac{I^{1/2}}{\sqrt{2}(1 + \kappa a)} \tag{7.153}$$

where m_+^0 and m_-^0 are the mobilities at infinite dilution (in the limit $I \to 0$) for the cation and anion, B_1 and B_2 are experimentally determined constants (Fawcett, 2004), and a is an ionic size parameter.

The mobilities at infinite dilution are most often experimentally determined and often tabulated in standard physical chemistry texts (Castellan, 1983) or electrochemistry texts (Fawcett, 2004; Bockris & Reddy, 1998). At 25°C, $B_1 = 0.2292\,\text{L}^{1/2}\text{mol}^{1/2}$ and $B_2 = 60.57 \times 10^{-4}$ S m^2 $\text{L}^{1/2}\,\text{mol}^{-3/2}$. The SI unit of conductivity is siemens/meter = S/m. In the denominator of the second term, κ is the inverse of the Debye length and is proportional to $I^{1/2}$.

As noted previously in this chapter, a given ion in an electrolyte is always surrounded by a cloud of other ions, consisting predominantly of ions of opposite charge and solvated by polar water molecules and thus *hydrated*. When an ion moves, its ionic cloud moves in the opposite direction and generates an additional viscous drag on the ion. The term containing B_1 in equation (7.154) represents this *electrophoretic effect*, which results in a decrease in the drift velocity of the ion. A given ion moving together with its ionic cloud is also electrically polarized, with the effective center of positive charge slightly leading the center of negative

Table 7.2. The experimental and theoretical conductivities calculated using equation (7.154) of $I_0 = 0.001, 0.01$ and 0.1 M potassium chloride (KCl) solutions

I_0	$\sigma_{e,\text{exp}}$ $\left(\frac{\text{S}}{\text{m}}\right)$	$\sigma_{e,\text{theo}}$ $\left(\frac{\text{S}}{\text{m}}\right)$
0.001 M	0.0147	0.0147
0.01 M	0.1413	0.1413
0.1 M	1.2880	1.2780

Equation (7.154) was used with a = 0.35 nm. The experimental conductivity values are as quoted for the corresponding *KCl* conductivity standards in the product catalog of Aquaspex (http://www.aquaspex.com.au/products/index.php?class=13).

charge in the direction of its motion. This charge asymmetry originates from the finite time it takes for the ionic cloud to respond to the motion of the ion and is known as the *relaxation effect*. The resultant electric field is oppositely directed to the applied field and results in a reduction in mobility of the ion. The relaxation effect is incorporated in the term containing B_2 in equation (7.154). Substituting equation (7.153) into equation (7.152),

$$\sigma_e = I\left[\frac{(m_+^0 + m_-^0)}{2}F - \left\{B_1\frac{(m_+^0 + m_-^0)}{2}F + B_2\right\}\frac{I^{1/2}}{2\sqrt{2}(1+\kappa a)}\right] \tag{7.154}$$

As an illustration of use of equation (7.154), consider the conductivity of a potassium chloride or KCl aqueous solution. The mobilities at infinite dilution for K^+ ions is well known – $m_+^0 = 7.91 \times 10^{-8}$ m²/sV – and that for Cl^- ions is $m_-^0 = 7.62 \times 10^{-8}$ m²/sV (Chang, 2000). The conductivity of an electrically neutral $I_0 = 0.001$ M KCl is calculated using these mobility values, $\sigma_e = F(m_+^0 + m_-^0)c = 0.0147$ S/m, and this value agrees with experiment (see Table 7.2). The magnitudes of the electrical conductivity for water under several conditions are shown in Table 7.3.

Table 7.3. Electrical conductivities for water under several conditions

Fluid	σ_e
Ultra pure water	$\sim 0.06\ \frac{\mu\text{S}}{\text{cm}}$
Distilled water	$\sim 1\ \frac{\mu\text{S}}{\text{cm}}$
Raw water	$\sim 60\ \frac{\mu\text{S}}{\text{cm}}$
0.05% NaCl	$\sim 1\ \frac{\text{mS}}{\text{cm}}$
Seawater	$\sim 50\ \frac{\text{mS}}{\text{cm}}$

In industrial practice, μS/cm is the unit of choice for conductivity.

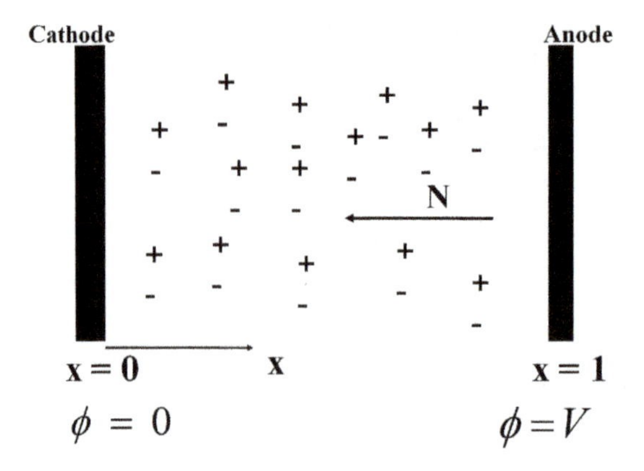

Figure 7.16 Simple model of a cation exchange membrane.

7.18 Semi-permeable membranes

A phenomenon we have mentioned in the context of biological ion channels is the ability of membranes to selectively allow one species to pass through but not another. For example, a battery is a cation exchange membrane in the sense that cations will be allowed to pass from the anode to the cathode freely, while the anions are prevented from passage into the cathode, as depicted in Figure 7.16. This is a subject that could be expanded into an entire chapter, and there is a vast body of literature that I will mention in several open-ended exercises at the end of this chapter. For now, let us consider a simple one-dimensional model for such a membrane. The interested reader should consult the series of papers by Rubinstein referenced in Zaltzman and Rubinstein (2007) and the papers by Bazant *et al.* (2005) and Chu and Bazant (2005).

The governing equations for a one-dimensional analysis are

$$(g_x + g\phi_x)_x = 0 \tag{7.155}$$

$$(f_x - f\phi_x)_x = 0 \tag{7.156}$$

$$\epsilon^2\phi_{xx} = -(g - f) \tag{7.157}$$

where $\epsilon = \lambda/L$ and λ is the Debye length. The subscript x denotes a derivative in the x direction. Here g is the dimensionless cation concentration scaled on the bulk concentration, and f is the corresponding anion concentration. These equations are subject to boundary conditions

$$f_x - f\phi_x = 0 \quad x = 0, 1 \tag{7.158}$$

$$g = g^0 \quad x = 0, 1 \tag{7.159}$$

$$\phi = V \quad x = 1 \tag{7.160}$$

$$\phi = 0 \quad x = 0 \tag{7.161}$$

The potential has been scaled on RT/F, and g, f are scaled on their bulk concentrations. With this scaling, these equations and boundary conditions are

Lithium-Ion Batteries

Batteries are electrochemical cells that produce current by converting stored chemical energy into electrical energy. The primary chemical reaction that drives all commonly used batteries is called an oxidation-reduction reaction. The reactions take place at each of two electrodes. The anode is where ions are oxidized and lose some of their electrons. The cathode is where ions are reduced and gain electrons. Depending on whether the cell is being charged or discharged, either electrode can be the anode or cathode at a given instant. In between each half reaction, the electrons traverse an external circuit and provide useful energy in the form of electricity. A separator that is permeable to ions but not electrons lies between the electrodes to prevent an electrical short circuit. A liquid or gel electrolyte solution enables ion transfer between the electrodes.

As a specific example, consider the lithium-ion battery. This battery chemistry has gained recent popularity in a variety of portable electronics and automotive applications. The typical lithium-ion battery features one graphite electrode and the other made of a metal oxide or phosphate such as $LiCoO_2$, $LiNiO_2$, or $LiFePO_4$. During discharge, lithium ions are transported from the graphite electrode into the electrolyte, described by the forward path of the half reaction

$$Li_yC \leftrightarrow Li^+ + e^- + C$$

where C represents a generic graphite compound and $0 < y < 1$ is the amount of lithium stored in the graphite electrode divided by its saturation value.

The ions diffuse through the separator to the cathode where they intercalate or pass into the solid material. This process is described by the forward path of the cathode half reaction

$$Li^+ + e^- + M \leftrightarrow Li_zM$$

where M represents a generic metal compound and $0 < z < 1$ is the quantity analogous to y in the cathode.

Transport of lithium ions into the cathode solid material completes the oxidation-reduction reaction. During charging, the same process is repeated in reverse and the reactions follow the backward paths of the aforementioned reactions. The schematic below illustrates the path of electrons and lithium ions for charge and discharge.

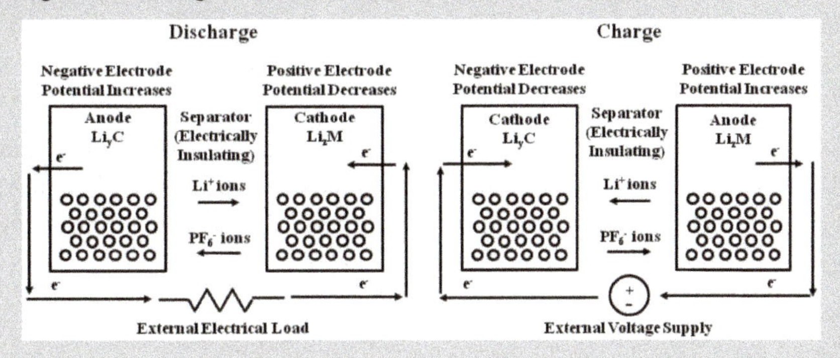

Contributed by Jim Marcicki.

supplenented by the condition that the number of anions remains fixed so that

$$\int_0^1 f(x)dx = 1 \tag{7.162}$$

where the concentration f has been scaled on its bulk value, where the anion concentration is equal to the cation concentration. In a battery, for example, the current is fixed, and only the cations provide current. Thus the current is given by

$$I = N_A = N_g = N \tag{7.163}$$

Thus one integration of the governing equations for g and f yields

$$g_x + g\phi_x = N \tag{7.164}$$

$$f_x - f\phi_x = 0 \tag{7.165}$$

Consider first the outer solution away from the electrodes; adding the equations for g and f results in

$$g_o = f_o = \frac{N}{2}(x + A) \tag{7.166}$$

and subtracting,

$$\phi_x = \frac{1}{x + A} \tag{7.167}$$

so that

$$\phi_o = \ln(x + A) + B \tag{7.168}$$

To obtain A, we use the integral constraint on the anions and get

$$A = \frac{2}{N} - 1/2 \tag{7.169}$$

Thus, in the outer region, where to leading order the problem is electrically neutral,

$$g_o = f_o = \frac{N}{2}\left(x - \frac{1}{2}\right) + 1 \tag{7.170}$$

This is the Rubinstein result. The constant B is still undetermined.

To find B, we need to match the outer solution to the inner solution near the electrodes. Near $x = 0$, we write, in the usual way, $\xi = x/\epsilon$, giving a Boltzmann distribution for both f and, g, and, for example,

$$g(\xi) = g^0 e^{-\phi} \quad x = 0 \tag{7.171}$$

$$g(\xi) = g^0 e^{-(\phi - V)} \quad x = 1 \tag{7.172}$$

where ϕ is the solution for the potential inside the EDL. Matching in the limit as $\xi \to \infty$, we obtain

$$\lim_{\xi \to \infty} \ln(g^0 e^{-\Phi}) = \ln g^0 - \phi = \lim_{x \to 0} \ln\ g = \ln\left(1 - \frac{N}{4}\right) \tag{7.173}$$

so that near $x = 0$, the outer solution for ϕ must satisfy

$$\phi(0) = \ln g^0 - \ln\left(1 - \frac{N}{4}\right) \tag{7.174}$$

and similarly, near $x = 1$,

$$\ln g^0 - (\phi - V) = \ln\left(1 + \frac{N}{4}\right) \tag{7.175}$$

giving

$$\phi(1) = \ln g^0 + V + \ln\left(1 + \frac{N}{4}\right) \tag{7.176}$$

We now find the constant B by equating the limit of the outer solution for ϕ as $x \to 1$ to find

$$B = \ln g^0 + V + \ln\left(1 + \frac{N}{4}\right) - \ln(1 + A) \tag{7.177}$$

Thus the outer solution for the potential is given by

$$\phi = V + \ln\left(\frac{1}{2}N(x - 1/2) + 1)\right) + \ln\frac{g^0}{(1 + \frac{N}{4})^2} \tag{7.178}$$

Note the interesting result that to leading order,

$$\phi_{xx} = -\frac{1}{4}\frac{N^2}{\left(1 + \frac{1}{2}N\left(x - \frac{1}{2}\right)\right)^2} \neq 0 \tag{7.179}$$

so that the volume charge density must be nonzero at $O(\epsilon^2)$; that is, the volume charge density

$$\rho_e = \epsilon^2\left(\frac{1}{4}\frac{N^2}{\left(1 + \frac{1}{2}N\left(x - \frac{1}{2}\right)\right)^2}\right) \tag{7.180}$$

to leading order in ϵ and is induced by the electric field.

The solution for the outer potential needs to be matched with the inner solution at $x = 0$. This matching will yield the current–voltage relationship, and near $x = 0$,

$$\ln g^0 - \ln\left(1 - \frac{N}{4}\right) = \ln\left(1 - \frac{N}{4}\right) + \ln g^0 - 2\ln\left(1 + \frac{N}{4}\right) \tag{7.181}$$

leading to

$$N = 4\left[\frac{1 - e^{-V/2}}{1 + e^{-V/2}}\right] \tag{7.182}$$

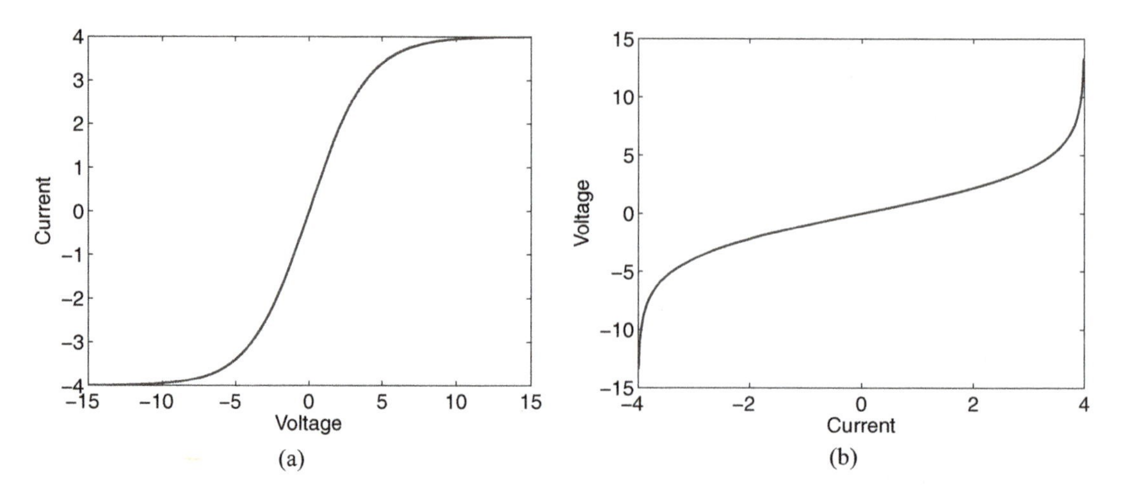

Figure 7.17 Plots of the current-voltage relationship for a cation exchange membrane.

This is the Rubinstein result. Note that as $V \to \infty$, the current $J = N \to 4$; this value is termed the *limiting current*. In terms of the voltage, we have

$$V = 2 \ln \frac{1 + N/4}{1 - N/4} \tag{7.183}$$

The result is depicted in Figure 7.17. The limiting current is reached at a dimensionless voltage of about $V \sim 10$.

If the permselective membrane is selective to cations, its concentration near the electrodes will increase substantially, and the anion concentration will decrease to near zero. A typical result is depicted in Figure 7.18. Note that this result resembles the Gouy–Chapman distribution, for which the concentrations are asymmetric around the mean and are discussed in Section 7.11, although this time, the current is nonzero, resulting in a linear variation of the concentrations.

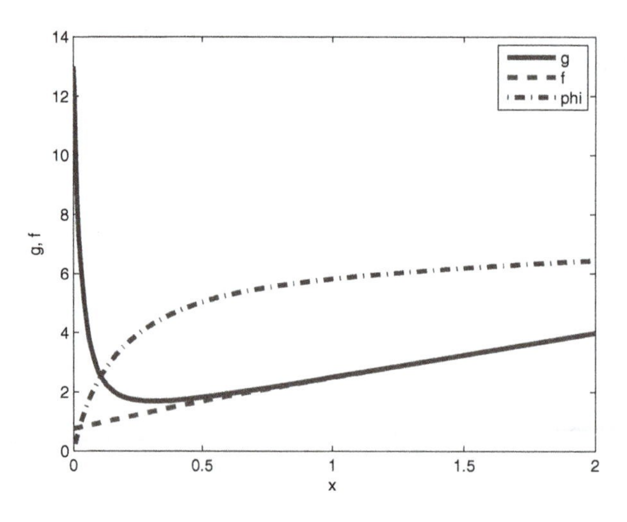

Figure 7.18 Typical result for the scaled concentrations and potential in a cation exchange membrane, near $x = 0$. As $J \to 4$, the limiting current, the anion concentration at the wall, approaches zero.

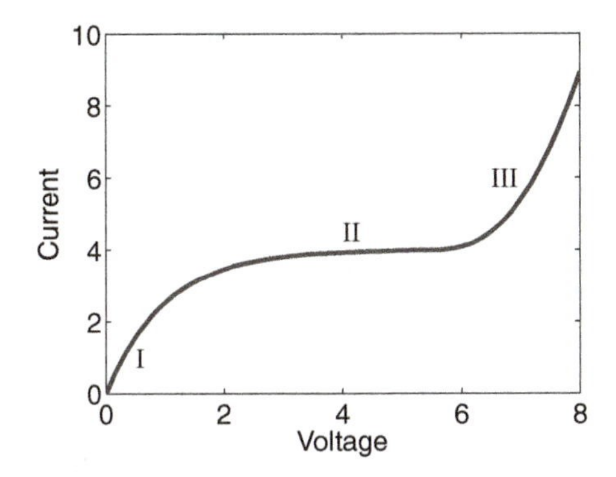

Figure 7.19 Plot of the current-voltage relationship for a cation exchange membrane, illustrating the behavior in the three regions; I, below the limiting current; II, at or near the limiting current; and III, above the limiting current.

The divergence of the cation and anion concentrations near a permselective membrane is called *concentration polarization* (Zaltzman & Rubinstein, 2007).

Bazant *et al.* (2005) and Chu and Bazant (2005) discuss the structure of boundary layers that appear when the limiting current is approached. If $N = J = 4 - O(\epsilon^{2/3})$, a boundary layer of $x \sim O(\epsilon^{2/3})$ appears outside the classical diffuse layer of $O(\epsilon)$. Chu and Bazant (2005) also analyze the regime above the limiting current, as illustrated in Figure 7.19. Above the limiting current, there is a finite region over which the anion concentration vanishes near the cathode, a case of extreme concentration polarization. The concept of a limiting current was posed by Nernst over a century ago (Chu & Bazant, 2005).

The EDL near each wall is fundamentally different from the EDL discussed in Section 7.12 because the structure is generated not by a native surface charge and/or ζ potential but by conducting electrodes with imposed potential that creates a current. No current is passed by silica surfaces and other surfaces that have been discussed previously in this chapter. Thus the EDLs near $x = 0, 1$ are said to be *induced* by the electric field. The possibility of large electric fields can lead to extreme polarization, whereby, in this case, the cation is highly concentrated near $x = 0$, while the anion is severely depleted. This means that the solution may not be dilute and that the electrical permittivity and the diffusion coefficient may depend on the local concentration (Bazant *et al.*, 2005). Moreover, large concentrations may require the consideration of finite ion volume effects and ion–solvent interactions, effects that are beyond the scope of this book. In this problem, no flow field is generated in the x direction because there are no walls to provide the excess charge required for such a flow.

7.19 The Derjaguin approximation

Notation alert: The symbol E in this section denotes interaction energy and not electric field.

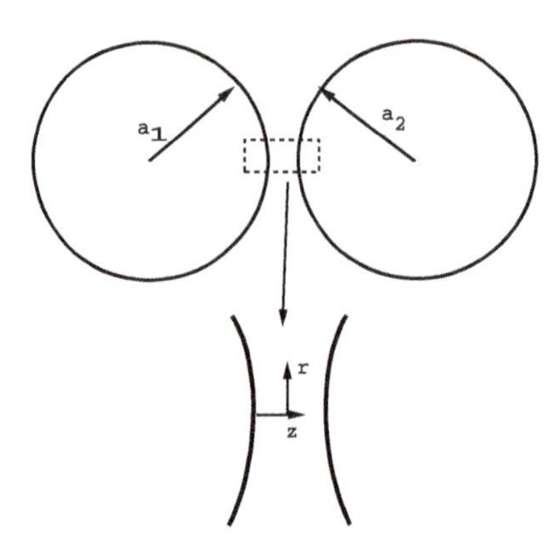

Figure 7.20 The interaction of two speres of radius a_1 and a_2, assuming the interaction is confined to a small area near the point of minimum distance $z = h$, is called the Derjaguin approximation. This assumption is equivalent to the lubrication approximation.

The *Derjaguin approximation* uses the lubrication approximation to determine the interaction between colloidal particles or between colloidal particles and walls (Derjaguin, 1934; Russel *et al.*, 1991; Bhattacharjee & Elimelech, 1997). Derjaguin (1934) recognized that in many cases, the interaction between different bodies, two flat surfaces, two particles, or a particle and a surface is not significant until the particles come very close together. Consider the interaction between two colloidal spheres shown in Figure 7.20. Then the force normal to the surface of the two particles is given by an integral of the pressure over the surface of the particle:

$$F_z = \iint_S p(h)\vec{n} \cdot \vec{e}_z dS \tag{7.184}$$

where h is the minumum distance between the particles. The energy between the particle and the surface is the integral of the total force, or

$$E = -\int_z F_z dz \tag{7.185}$$

where z measures the normal distance between the particles. Let

$$P = -\int_z p(h) dz \tag{7.186}$$

so that P has units of N/m. Thus

$$E = \iint_A P(h) \frac{\vec{n} \cdot \vec{e}_z}{|\vec{n} \cdot \vec{e}_z|} dA \tag{7.187}$$

because $dS = 1/|\vec{n} \cdot \vec{e}_z| dA$, where $dA = dxdy$ from standard vector integration methods; note that the order of the integrations has been switched.

It is difficult to use Cartesian coordinates for the integration, and the Derjaguin approximation for the interaction of two spheres of radii a_1 and a_2 relies on the fact that in a local cylindrical coordinate system as depicted in Figure 7.20 (Russel *et al.*, 1991),

$$z(r) = h + \frac{1}{2} r^2 \left(\frac{1}{a_1} + \frac{1}{a_2} \right) + O\left(\frac{r^4}{a_1^3}, \frac{r^4}{a_2^3} \right) \tag{7.188}$$

where h is the point of minimum distance between the two spheres and thus $dz = rdr\,(1/a_1 + 1/a_2)$. The energy is thus

$$E = 2\pi \int_0^\infty P r dr = \frac{2\pi}{\frac{1}{a_1} + \frac{1}{a_2}} \int_h^\infty P dz \tag{7.189}$$

Note that if $a_1 \gg a_2$, the formula reduces to

$$E = 2\pi a_2 \int_h^\infty P dz = 2\pi a_2 F_z \tag{7.190}$$

which is essentially the Derjaguin formula for the interaction between a plane, wall, and sphere.

The force between the two bodies is obtained by differentiating equation (7.189) with respect to h; the result is

$$F_z = 2\pi \frac{a_1 a_2}{a_1 + a_2} P \tag{7.191}$$

where P is defined earlier.

Bhattacharjee and Elimelech (1997) and Bhattacharjee *et al.* (1998) have gone beyond the Derjaguin approximation and calculated the interaction between two spheres (Bhattacharjee *et al.*, 1998) and a sphere and a plane wall (Bhattacharjee & Elimelech, 1997) using what they call the surface element integration (SEI). They relaxed the assumption that all the interactions are localized, integrating over cylindrical slices on the sphere; for the interaction between a sphere and a plane wall, they obtained

$$E = 2\pi \int_0^a \int_0^a \left[P\left(h + a - a\sqrt{1 - \frac{r^2}{a^2}} \right) - P\left(h + a + a\sqrt{1 - \frac{r^2}{a^2}} \right) \right] r dr \tag{7.192}$$

The case of two spheres is of a similar form.

The use of the Derjaguin approximation depends on the nature of the interactions between the two bodies. For example, the form of energy per area, P, for the plane wall–particle interaction for van der Waals attraction is given by

$$P_{vdW} = -\frac{A}{12\pi z^2} \tag{7.193}$$

where $A \sim 10^{-18} J$ is the Hamaker constant, and so the Derjaguin approximation gives

$$E = -\frac{Aa}{6h} \tag{7.194}$$

The integral using the SEI method can be integrated analytically; the result is

$$E = -\frac{A}{6}\left(\frac{a}{h} + \frac{a}{h + 2a} + \ln\left(\frac{h}{h + 2a} \right) \right) \tag{7.195}$$

It is left as an exercise to plot equations (7.194) and (7.195) to show that the Derjaguin approximation is inaccurate for $h/a > 0.1$.

As the particles get smaller, the Derjaguin approximation results in a larger error because the local curvature is larger. The minimum value of h for which the Derjaguin approximation is valid drops significantly as the particle size decreases. Whether the particle can get that close to the wall becomes an issue, especially if the two surfaces have the same sign charge. Russel *et al.* (1991) suggests that the approximation may fail even for micron-size particles.

7.20 Chapter summary

This chapter provides an introduction to the basic concepts of electrochemistry that are important for the discussion of electrokinetic phenomena to follow. The basic structure of water and the concept of hydration of ions and the definition of the various radii that are used in the electrochemistry are presented. The fundamental definitions of acids and bases are given for the reader to understand ionization processes.

The chemical and electrochemical potential and the Gibbs function are then introduced; the invariance of these potentials determines the equilibrium behavior of the electrical potential and the concentrations of the ionic species within the EDL on the liquid side of the surface.

These subjects lead to an extensive discussion of the EDL, including the presentation of the various phenomenological models of the EDL that have appeared, including the description of the triple layer model of the EDL, a refinement of the basic two layer model. The surface charge density is defined based on the nature of site-binding models on the solid side of a surface.

The potential distribution within the electrical double and the Boltzmann distribution for the ionic species allows the solid-side surface charge density of a site-binding model and fluid site-binding model surface charge density to be linked, yielding a fundamental expression for the ζ potential as a function of pH and the equilibrium constants of the other important adsorption reactions specific surfaces.

The potential distribution in the EDL is derived for both monovalent symmetric mixtures and asymmetric mixtures in the case of a plane wall. The potential distribution on a cylinder or a sphere must be calculated numerically in the nonlinear regime; analytical solutions are derived in the Debye–Hückel limit of small potential.

Next is a discussion of the electrical conductivity of an electrolyte solution, a measure of its ability to conduct electrical current from one electrode to the other. This occurs in a semipermeable membrane that allows one set of ions to pass through the solution and into the electrode(s), while preventing the other ions from doing so. A cation exchange membrane allows cations to pass from the anode to the cathode, while preventing anions from doing so. The concentrations resemble the behavior of a Gouy–Chapman picture of the EDL and is one example of concentration polarization. The chapter ends with an introduction to Derjaguin's approximation and an improvement to that approximation for the interaction between two spheres or a sphere and a plane wall.

Many of these concepts will be revisited in Chapter 9 and Chapter 10 in various forms. In particular, it will be seen that this EDL analysis leads to the analogy between the potential in the EDL and the velocity in electro-osmotic flow, as was already pointed out in Chapter 3.

EXERCISES

7.1 A 0.2 M sodium–chloride–water solution (0.1 MNa, 0.1 MCl) is contained in a region upstream of a nanopore membrane. The potential drop across the membrane is measured to be −5 V. At a temperature of $T = 20°\text{C}$, calculate the activity of sodium in the downstream reservoir.

7.2 Given that the reaction

$$\text{SiOH} + \text{M}^+ \rightleftharpoons \text{SiOHM} + \text{H}^+$$

describes the adsorption of a metal cation, such as sodium, on a silica, with the equilibrium constant defined by

$$K_M = \frac{[\text{H}^+][\text{SiOHM}]}{[\text{SiOH}][\text{M}^+]}$$

show that the surface charge density is given by

$$\sigma = \frac{N_s F}{N_A} \frac{[\text{H}^+]_s^2 - K_1}{K_1 + [\text{H}^+]_s + K_M[M^+]}$$

Calculate the surface charge density for the adsorption of sodium for which $K_{NA} = 10^{-7.1}$, and use the values of $K_1 = 10^{-6.8}$ and $\zeta = -20$ mV at pH = 7.

7.3 Given that the reaction

$$\text{SiOH} + \text{Ca}^{2+} + \text{H}_2\text{O} \rightleftharpoons \text{SiOCaOH} + 2\text{H}^+$$

describes the adsorption of calcium on a silica surface, find an expression for the surface charge density in terms of the appropriate activities or mole fraction in the same form, as in the previous problem for a sodium–chloride–calcium solution.

7.4 A 0.1 M sodium–chloride–water (0.05 Na, 0.05 Cl) mixture fills a channel made of silica surfaces. The surface charge density is known to be $\sigma = -20\frac{C}{m^2}$. The ζ potential is unknown. Using the Debye–Hückel approximation with the Neumann condition at the wall and the proper condition in the bulk, calculate the potential if the EDL is thin. What is the effective width of the EDL? Plot the potential near the surface $y = 0$.

7.5 For the previous problem, find the concentrations of the sodium and chloride at the surface $y = 0$. Evaluate the validity of the Debye–Hückel approximation.

7.6 A 0.01 M sodium–chloride–water mixture fills a channel made of silica surfaces, with a height of $h = 100$ nm. The surface charge density at $y = 0$ is known to be $\sigma = -20\frac{C}{m^2}$, and at $y = h$, the surface charge density is $\sigma =$ 15 mV. Using the Debye–Hückel approximation and the proper condition in the bulk, calculate the potential distribution in the channel. Repeat for $h = 10$ nm.

7.7 For the previous problem, find the concentrations of the sodium and chloride at each of the surfaces. Evaluate the validity of the Debye–Hückel approximation.

7.8 Consider the case of EDLs near each wall in a channel under the Debye–Hückel approximation. Compare the dimensionless solutions for the fixed ζ potential boundary condition and the fixed surface charge distribution at the wall. Conclude that in dimensionless form, the surface charge density and the ζ potential are equivalent. Identify the dimensional form of this equivalence as the Grahame equation.

7.9 Plot the dimensionless potential for a z:z electrolyte of known ζ potential near $y = 0$ for $z = 1, 3$ (equation (7.92)) and dimensionless ζ potential, $\zeta = \phi_s = -2$. What do you notice about the effective EDL thickness (i.e., the number of Debye lengths), defined as when the potential $\phi < 0.01$? Base the Debye length on a bulk ionic strength of $I = 0.3$ M. Perform the calculation at $T = 300$ K. Compare with the Debye–Hückel result.

7.10 In the text, the solution of Grahame for a 2:1 electrolyte was calculated. Plot the solution for the potential for $\zeta = 1$, and compare with the solution for a 1:1 electrolyte. Plot each of the concentrations, and again compare with the 1:1 case. A numerical solution using a zero-finding technique is required.

7.11 Following Grahame, calculate the solution for a 1:2 electrolyte solution, and repeat the steps of the previous problem. In this case, the key identity is given by

$$(e^{\phi} - 1)(1 + 2e^{-\phi})^{1/2} = (e^{2\phi} + e^{-\phi} - 3)^{1/2}$$

7.12 An electrolyte solution resides between two plates having the same ζ potential. Show that the pressure between the two plates in a static fluid satisfies

$$\frac{dp}{dy} + \rho_e \frac{d\phi}{dy} = 0$$

where ρ_e is the volume charge density and $\rho_e = -\epsilon_e \, d^2\phi/dy^2$. For a 1:1 electrolyte, show that the force between the two plates under the Debye–Hückel approximation due to the presence of the electrolyte solution satisfies

$$\frac{F}{A} = \mp\frac{1}{2}\epsilon_e\zeta^2\kappa^2 \quad \text{at } y = 0, 1$$

for $\kappa h \gg 1$, and

$$\frac{F}{A} = p(h) - p(0) = \frac{1}{2}\epsilon_e\zeta^2\kappa^2 \tanh \kappa h \quad \text{for } \kappa h \ll 1$$

What do you conclude about the force between the plates in each limit? Here $\kappa = 1/\lambda$. Note that $dp/dy = 0$ when $\rho_e = 0$. The pressure p in this problem is one form of *osmotic pressure*.

7.13 From the liquid-side point of view, for a $z{:}z$ electrolyte, the ζ potential and the surface charge density are related by equation (7.115). Given that the ζ potential can be measured using the *streaming potential*, plot the surface charge density for $\zeta = 26, 52, 78$ mV at a temperature $T = 35°$C for a 0.15 M solution for $z = 1$. Keeping the molarity fixed, how does the solution change for $z = 2, 3$?

7.14 For a binary symmetric monovalent electrolyte, a Taylor series approximation to the exponential Boltzmann distribution yields the equation for the potential as (Oyanader & Arce, 2005)

$$\epsilon^2 \frac{d^2\phi}{dy^2} = \phi\left(1 + \frac{\phi^2}{3} + \frac{\phi^4}{5} + \frac{\phi^6}{7} + \cdots\right)$$

Solve this equation numerically, and find the potential and mole fraction profiles for electro-osmotic flow in a 20 nm channel. The electrolyte solution is 0.1 M NaCl, and the ζ potential on the walls is -40 mV.

7.15 Plot the ζ potential as a function of pH for a 0.015, 0.15 M binary mixture of monovalent electrolytes.

7.16 In the limit as $\epsilon \to 0$, find the solution for the potential on the outside of the surface of a cylinder for the constant potential boundary condition.

7.17 The governing equations for a cation exchange membrane in the core are

$$g_x + g\phi_x = I = N$$

$$f_x - f\phi_x = 0$$

Defining

$$c = \frac{1}{2}(g + f)$$

and

$$\rho = \frac{1}{2}(g - f)$$

show that the equations can be written in the form

$$c_x + \rho\phi_x = N/2$$

$$\rho_x + c\phi_x = N/2$$

and that conservations of anions reduce to

$$\int_0^1 (c - \rho)dx = 1$$

Away from the limiting current, $\rho \ll$ c; using this approximation, deduce the form of c and ρ. Show, in particular, that using conservation of anions in the outer region with $I = N$,

$$g = f = \frac{I}{2}\left(x - \frac{1}{2}\right) + 1$$

$$\phi = \ln\left(\frac{I}{2}\left(x - \frac{1}{2}\right) + 1\right) + B$$

where B is a constant to be determined by matching.

7.18 This is an open-ended exercise. Write a report on the derivation of the Butler–Volmer law of electrochemistry. Discuss what is it used for, and relate its use to electrochemical cells.

7.19 This is an open-ended exercise. Write a report on the results of the paper by Bazant *et al.* (2005). How is the Butler–Volmer law used? What are the differences between this paper and the paper by Zaltzman and Rubinstein (2007)?

7.20 Compare the results using the Derjaguin and SEI methods for the van der Waals attraction, equations (7.194) and (7.195), as a function of h/a. For a fluid particle of radius $a = 1$ μm, at what value of h is the difference between the two results more than 20 percent?

7.21 Using the methods described by Bhattacharjee *et al.* (1998), show that the interaction energy between two spheres of radii a_1 and a_2 in a van der Waals attraction is given by

$$E_{\mathrm{vdW}} = -\frac{A}{6h}\left(\frac{2a_1a_2}{h^2 + 2a_1h + 2a_2h} + \frac{2a_1a_2}{h^2 + 2a_1h + 2a_2h + 4a_1a_2}\right.$$
$$\left. + \ln\left(\frac{h^2 + 2a_1h + 2a_2h}{h^2 + 2a_1h + 2a_2h4a_1a_2}\right)\right)$$

8 Elements of Molecular and Cell Biology

8.1 Introduction

It may be said that at some level cells are the building blocks of all living things. The most basic of all organisms are made up of single cells, and higher organisms, such as animals and humans, are made of a large number of cells all arranged in a specific way.

Because the nanoscale is the scale of biology, there has been an explosion of new knowledge and new ideas in the biological sciences. This has been the result of significant advances in static measurement techniques such as the surface force apparatus (SFA), the atomic force microscope (AFM), and the scanning tunneling microscope (STM). Moreover, optical techniques can be employed for single molecule detection and analysis. With explosive improvements in computer architecture, molecular dynamics (MD) and its derivatives (nonequilibrium molecular dynamics, NEMD) are now being employed to study the motion of ions and proteins in a flowing solution and the conformation of proteins and other macromolecules in a static medium.

The links between nanotechnology and biology are numerous. Entire devices for the rapid analysis of proteins, drug delivery, biochemical sensing, and other applications are being developed at a rapid pace. Entire books are being published on biomedical applications at the nanoscale (Malsch, 2005; Ciofalo *et al.*, 1999). Artificial and implantable pumping systems employing nanopore membranes are being developed for filtering proteins and ions for use as a renal assist device (Fissell, 2006) (see Figure 4.1).

The main objective of this chapter is to introduce the student to the most common biomolecules used in the applications of interest. These biomolecules, such as proteins and polysaccharides, are soft materials and are deformable when sheared by a fluid flow. That being said, it is extremely difficult to describe the nature of the deformation, though some measurements have been reported in the literature. Indeed, albumin is thought to form the shape of a wedge of cheese as it enters a nanopore membrane (Ferrer *et al.*, 2001). Although a matter of debate in the literature, the most important properties of a biomolecule are its overall size and valence or charge. These two properties determine the primary transport characteristics of the biomolecule.

The Atomic Force Microscope (AFM)

Example of a commercial atomic force microscope.

The atomic force microscope (AFM), also known as the scanning probe microscope (SPM), is a high resolution imaging technique capable of resolving surface features on the order of fractions of nanometers. To obtain a surface image, a cantilever with a sharp tip is scanned on the specimen surface. Interaction forces between the tip and the surface (on the order of nanonewtons), coming from mechanical contact, van der Waals, and capillary forces, lead to deflection of the cantilever. This deflection is measured using optical detection techniques. A piezoelectric tube is responsible for scanning the sample horizontally and vertically, and is attached either to the specimen (as shown in the example above) or to the assembly containing the tip, depending on the microscope configuration. In the static or contact mode of operation, the tip is in contact with the specimen at all times, the tip-surface force is kept constant, and a feedback mechanism allows the tip to trace the surface topography during the scan. In the dynamic or "tapping" mode, the cantilever is vibrated and the oscillation amplitude is maintained constant during scanning. The AFM can be used to probe the structure of biological cells.

In addition to surface imaging, the AFM is also a versatile nanoscale characterization technique. Imaging can be performed either in air or while the tip and specimen are immersed in a liquid medium. Other capabilities of the AFM include the measurement of mechanical forces (such as lateral forces, viscoelasticity, adhesion etc.), electrical properties (such as capacitance, resistance, surface potential), thermal properties, and magnetic forces.

Contributed by Manny Palacio

In this chapter, we introduce the basic concepts of molecular and cell biology, with an emphasis on information that is being exploited by nanotechnology today. This chapter is not meant to be all-inclusive; much more material is found in the excellent book by Alberts *et al.* (1998), which I have found extremely informative and highly recommend to the interested reader.

The chapter begins with a discussion of the chemical structure of nucleic acids, and polysaccharides are described, with an emphasis on the structure of DNA. Proteins and protein binding are described, leading to a discussion of the operation of a cell and the structure of its outer membrane. The chapter concludes with a description of the transport properties of ion channels.

8.2 Nucleic acids and polysaccharides

Nucleic acids are polymers consisting of nucleotides. Those based on a sugar called *ribose* are called ribonucleic acids (RNA), and those based on *deoxyribose* are called deoxyribonucleic acids (DNA). RNA is single stranded, whereas DNA is usually double stranded, although single-stranded DNA (ss-DNA) does exist. These two nucleotides are described in greater detail later.

Early in the 1940s, it was discovered that genetic information is primarily involved with developing instructions for making proteins. Deoxyribonulceic acid (DNA) was identified in the 1940s as the likely carrier of this genetic information (Alberts *et al.*, 1998). Crick and Watson (1953) determined the structure of DNA as a double helix, which immediately made clear how the DNA may be copied, or *replicated*, and began to explain the mechanism by which DNA encodes instructions for making proteins.

Watson, Crick, Franklin, and Wilkins and the Discovery of DNA

James Watson (1928–) and Francis Crick (1916–2004) were colleagues at the Cavendish Laboratory in Cambridge when they began their work leading to the discovery of the primary structure of DNA. Watson's background was in genetics, having a PhD from the University of Indiana. Francis Crick was a physicist having received a bachelor's degree from University College, London. He was already at Cavendish when Watson arrived and together they put together the structure of DNA. Two other colleagues played key roles and the paper Watson and Crick wrote (Crick and Watson, 1953) relied on their colleagues' input. Rosalind Franklin had produced X-ray crystallography images of DNA that Watson and Crick used to formulate their view of the structure of DNA. Unfortunately Franklin died in 1958 before she could be honored with the Nobel prize that was won by Watson and Crick in 1962. Maurice Wilkins was a nuclear physicist who changed fields to work with Franklin in X-ray crystallography. Wilkins had also produced X-ray images of DNA of which Watson and Crick were aware. In reality, many believe that both Franklin and Wilkins should have been co-authoers on the paper describing one of the most important discoveries in the history of the world.

DNA, genes, and chromosomes are interrelated in that based on work in the early twentieth century, it was discovered that genes are carried by chromosomes, which consist of DNA and proteins. In a glossary at the end of his excellent book, Alberts *et al.* (1998) defines a gene as a "region of DNA that controls a discrete hereditary characteristic of an organism, usually corresponding to a single protein or RNA" (p. G-8). In the same way, he defines a chromosome as a "long threadlike structure composed of DNA and associated proteins that carries part or all of the genetic information of an organism (p. G-4). Thus these three structures, DNA, genes, and chromosomes, and their associated proteins work together to keep cells functioning properly and are crucial elements in the development of new cells.

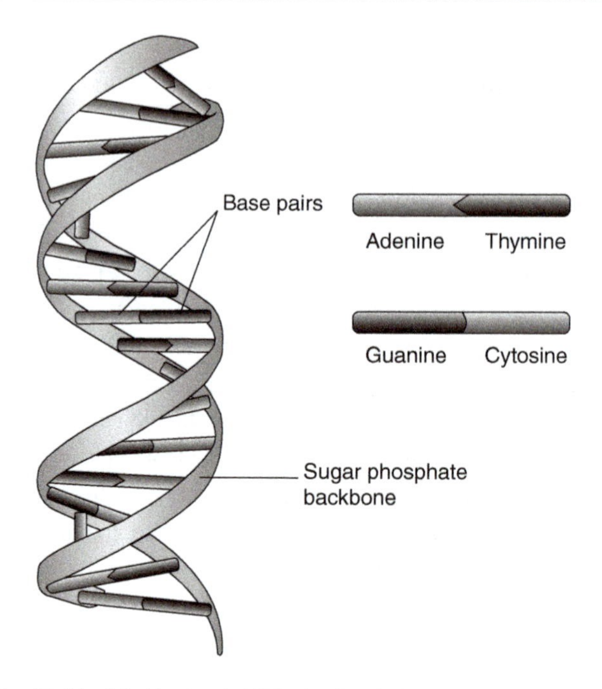

Figure 8.1 Sketch of double stranded DNA showing the four bases adenine (A), cytosine (C), guanine (G), or thymine (T). Image courtesy of the National Library of Medicine.

DNA is composed of two long nucleotide chains held together by hydrogen bonds. Nucleotides are structures containing five-carbon sugars attached to one or more phosphate groups (a phosphorus central atom surrounded by four oxygens) and a base that can be either adenine (A), cytosine (C), guanine (G), or thymine (T). A high-resolution image and a sketch of DNA is depicted in Figure 8.1. Two nucleotides connected by a hydrogen bond is called a base pair (bp). The length of a DNA molecule is measured by its number of base pairs. One base pair is approximately 2 Å long and 3 nm in diameter. A DNA strand can unravel and be as long as several microns.

RNA, or ribonucleic acid, as with DNA, is made up of four nucleotides containing the four bases uracil (U), adenine (A), cytosine (C), and guanine (G). RNA has two very important features: it carries information based on its individual sequence and passes it on by replication, and its unique shape determines its function.

In these DNA–RNA structures, A pairs with T or U and G pairs with C; the A–T(U) pairing contains two hydrogen bonds, and the G–C pairing contains three hydrogen bonds. RNA comes in many forms, and messenger RNA (mRNA) is one of the most important. The production of mRNA from its DNA template is called *transcription*, and mRNA is essential in the synthesizing, or *coding*, of proteins. The coding process is the initial step in the assembly of amino acids into a protein.

Polysaccharides consist of a large number of sugars, many composed of glucose, which can be linear as well as globular in shape; glucose is globular in solution, with a nominal radius of $a = 0.37$ nm. The term *globular* means that the molecule is roughly in the shape of a sphere; polysaccharides can be highly deformable. Glucose alone is a *monosaccharide* with the chemical formula $C_6 H_{12}O_6$. The monosaccharides have a chemical formula of the form $(CH_2O)_n$, where n is usually $n = 3, 4, 5, 6, 7$. The simple sugars are the primary energy sources for cells.

Polysaccharides are held together by covalent bonds and are often very large and almost macroscopic in size. They occur in bacteria cell walls, and plant cells are made exclusively of polyaccharides.

8.3 Proteins

Many biologists consider proteins to be the most important biological element of any living system. All chemical reactions within a living system are catalyzed by enzymes, which are protein molecules that catalyze, or increase the rate of, a biochemical reaction by themselves or in complexes with small molecules. They provide an organism with the essential food elements of carbon, hydrogen,

The Thermodynamics of Molecular Biology

As long ago as the early 1960s, **Terrell L. Hill** was talking about the *Thermodynamics of Small Systems* (Hill, 1963, 1964). The "small systems" he was talking about at the time were colloidal particles and biomolecules in solution. His Part I and Part II monographs were part of the, at that time, W. A. Benjamin Inc. series titled "Frontiers in Chemistry." He was a Professor of Chemistry at the University of Oregon and the monograph appears to be the first in the series of short books on these topics.

Terrell Hill is called "Theoretician Extraordinarius" by Lloyd Ferguson in a short, two-page historical narrative published in the journal *Cell Biochemistry and Biophysics* Vol. 11 in 1987, a volume dedicated to Professor Hill. Hill worked on the Manhattan Project at UC Berkeley and worked at the National Institutes of Health as Chief of Theoretical Molecular Biology in the National Institute of Diabetes, and Digestive and Kidney Diseases. In the preface of Part I, Hill (1963) says

> Small thermodynamic systems should be of interest to three classes of readers (1) experimentalists working with colloidal particles, polymers, or macromolecules; (2) theoreticians concerned with the preceding fields or with the statistical mechanics of any kind of finite system; (3) those with interest in thermodynamics *per se*.

He talks about biolecules in solution and his main objective is to to extend the range of validity of ordinary thermodynamic definitions to include small, nonmacroscopic systems (e.g., a single macromolecule or colloidal particle.)

For example, he writes the Gibbs free energy of a colloidal particle

$$G = Nf(p,T) + a(p,T)N^{2/3} + b(T)\ln N + c(p,T)$$

Here, N is the number of molecules in solution and, as $N \to \infty$, the macroscopic limit is attained. Only the first term describes macroscopic thermodynamics and the last three terms describe the microscopic system. Recall that in the macroscopic sense the chemical potential μ is defined by

$$\mu = \frac{\partial G}{\partial N}\Big|_{p,T}$$

Hill appears to be the first researcher to address these problems in a formal way. As applications he suggests in Volume I

> We may anticipate two main classes of applications of small systems thermodynamics: (1) as an aid in analyzing, classifying, and correlating equilibrium experimental data on "small systems" such as (noninteracting) colloidal particles, liquid droplets, crystallites, macromolecules, polymers, polyelectrolytes, nucleic acids, proteins, etc.; and (2) to verify, stimulate, and provide a framework for statistical mechanical analysis of models of finite (i.e. "small") systems.

Terrell Hill is the forerunner of researchers in the modern era of the thermodynamics of molecular biology.

nitrogen, and sulfur. Also, proteins carry out most of the processes in the cell. These molecules are made up of a combination of 20 different amino acids that are joined together by a covalent peptide bond.

The production of a protein, or *protein synthesis*, begins with a gene on a particular strand of the DNA. A gene governs which amino acids join together; they do this by expelling a molecule of water, a process called *condensation*. A gene is made up of a linear array of bases that contains the specific information or coding for the synthesis of a protein. Thus a gene contains the information required to produce a functional RNA. A gene is said to be *expressed* when its product, the RNA or a protein, is completed.

8.3.1 Protein function

There are seven basic types of proteins classified according to their function, although different authors use different terms to describe each class; see, for example, (Alberts *et al.* 1998, panel 5-1). *Enzymes* are catalysts in biological reactions within the cell. The immune system responds to foreign bacteria and viruses by producing *antibodies* that destroy or bind to the antigen, the foreign agent. The antigen is the catalyst, or *reaction enhancer*, for inducing the immune response: the production of the antibodies.

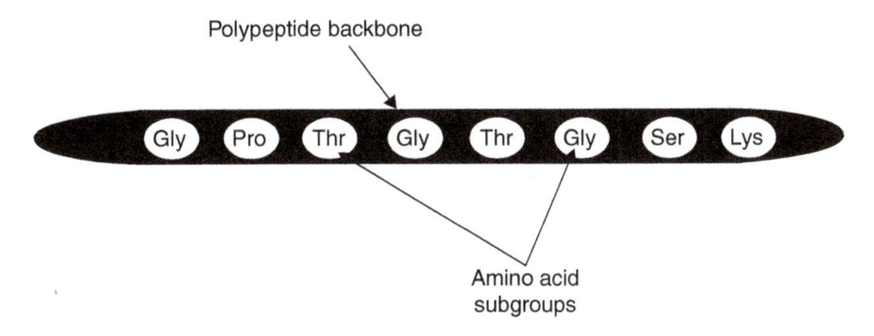

Figure 8.2 Sketch indicating the primary structure of a protein.

Some proteins, called *transport proteins*, carry materials around in the body (e.g., hemoglobin and oxygen), where that material is used by different cells. *Regulatory proteins* control the metabolism and reproduction of the cell by managing aspects such as temperature and pH. Other proteins, *structural proteins*, have mechanical properties that give support to bones, tendons, and skin in man and large animals. The interactions of actin and myosin proteins, which are *movement proteins*, provides the proper mechanism for the contraction and expansion of the heart and the other muscles throughout the body.

Nutrient proteins are important to a growing organism because of their abundance of amino acids. Proteins are present in the solid material of the body as well as in blood serum. For example, actin is one of the three major components of the cytoskeleton. Proteins are dynamic structures that rotate and deform in response to random thermal motions (Alberts *et al.*, 1998).

The two most important properties of proteins are their *structure* and their *function*, the latter of which we just described. We now discuss their structure.

8.3.2 Protein structure

Proteins are complex molecules, and as such, the entire structure cannot be viewed at the same time; moreover, all but the simplest proteins cannot be written in terms of a simple chemical formula. Thus the biology community has defined several levels of structure, described here.

The *primary structure* (or amino acid sequence) is a complete description of the chemical parts of a protein. The sequence is unique to the particular organ or tissue and replicates by cell division. Amino acids are bonded together in a chain by peptide bonds, which are amide bonds formed between the α-carboxylate group of one amino acid and the α-amino group of another. The primary structure of a single polypeptide is shown in Figure 8.2.

The main component of a protein is the *polypeptide backbone*, which is composed of an amino acid sequence to which side chains are attached. The side chains can be polar or nonpolar and charged or uncharged. The polar side chains

Human serum albumin and its cousin bovine serum albumin are proteins of enormous importance in biology. The function of the kidney is to filter out small ions and other small constituents of blood serum while retaining larger biomolecules such as HSA and the red and white blood cells. In addition, altered HSA appears to be an indicator of the progression of heart disease (Sinha *et al.*, 2004). It is well known that albumin binds to metals at its N-terminus, particularly cobalt. Cardiac ischemia, or the reduction of the transport of blood and oxygen to the heart appears to coincide with the reduced ability of albumin to bind to cobalt. Such a situation will change the electrostatic properties of an albumin molecule and hence its transport properties in solution. Ischemia Technologies of Denver, CO commercialized the first device to analyze albumin for indications of cardiac ischemia in 1997.

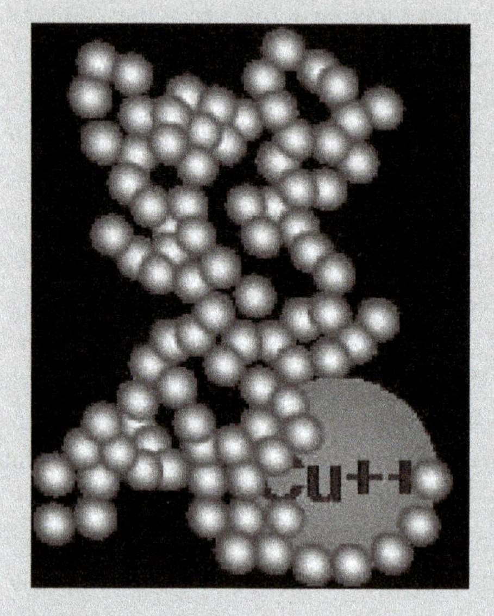

Cobalt bound to albumin at the N-terminus.

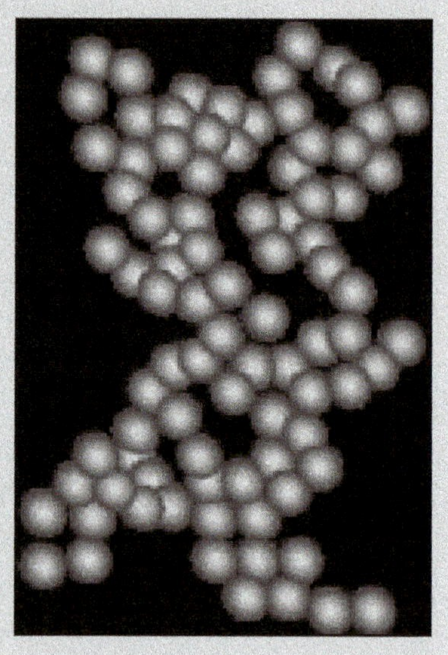

Albumin without bound cobalt.

are often arranged on the outside of the protein, where they can hydrogen-bond with water molecules and other polar solvents. The nonpolar or hydrophobic side chains are often arranged so that they do not disrupt the water molecules; these side chains tend to "hide" from water on the inside of a protein. Amino acids have an amino group on one end (NH_2), called the *N-terminus*, and a carboxyl group (COOH) on the other end, called the *C-terminus*.

The polypeptide backbone of a protein folds back on itself into a shape or *conformation* based on the criterion that the shape be that of minimum Gibbs free energy. A protein can be unfolded under the action of a particular solvent, a state in which the protein is termed *denatured*, meaning the protein loses its biological function capability, and this situation is depicted in Figure 8.3. The difference between the denatured and native state (folded) may be only several

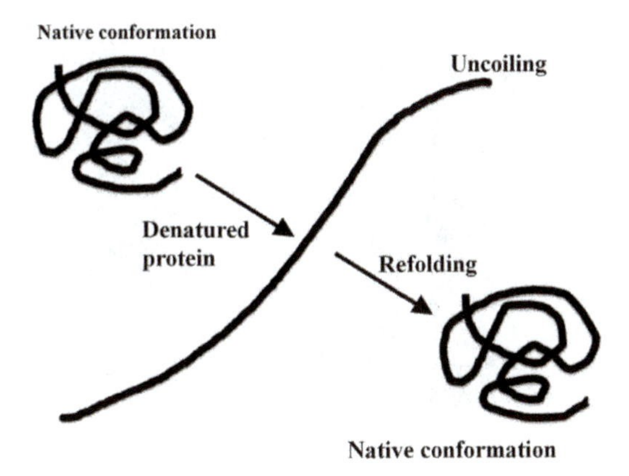

Figure 8.3 Sketch showing protein folding and the unfolding or denatured state in the presence of a particular solvent. Many proteins can be denatured by urea.

hundred hydrogen bonds out of many thousands of bonds, depending on the size of the protein.

The appropriate conformation of a protein in a cell is highly stable and will change only slightly on interaction with other molecules. The way the polypeptide backbone folds is determined entirely by its amino acid sequence. The first protein to have been "sequenced" was *insulin*, whose sequence was decoded in 1955 (Alberts *et al.*, 1998). It should be noted, however, that it is still not possible to predict a protein's shape from its amino acid sequence, and many proteins are much more complex than insulin, some even containing more than one polypeptide backbone.

The *secondary structure* depicts the long amino acid chain by describing its folding pattern. While all proteins have their own unique folding patterns, two of the most common patterns are the α-helix and the β-sheet. Maximizing the hydrogen bonding between amide hydrogen and carbonyl oxygen of the peptide bonds causes these two types of constructions. The α-helix is a helical, right-handed configuration that involves all amide hydrogen (N–H bonds) and carbonyl oxygen (C=O, double bond) of the peptide backbone. The carbonyl oxygen interacts with the amide hydrogen of the amino acid four down from the oxygen.

The nearly completely extended β-sheet also involves the hydrogen bonding of all the amides and carbonyls. The structure can be further classified as either a parallel or antiparallel β-sheet. It is the secondary structure that is visualized in several ways, as depicted in Figure 8.4. Each portion of the protein is color coded in a specific way by the user. The software package ProteinShop is used to produce Figure 8.4; another common rendering program is MidasPlus. Proteins can also be rendered using a wire model and a space-filled model (Alberts *et al.*, 1998).

The *tertiary structure* describes the additional folding of the secondary structure. The three-dimensional twisting and folding of the overall chain is depicted;

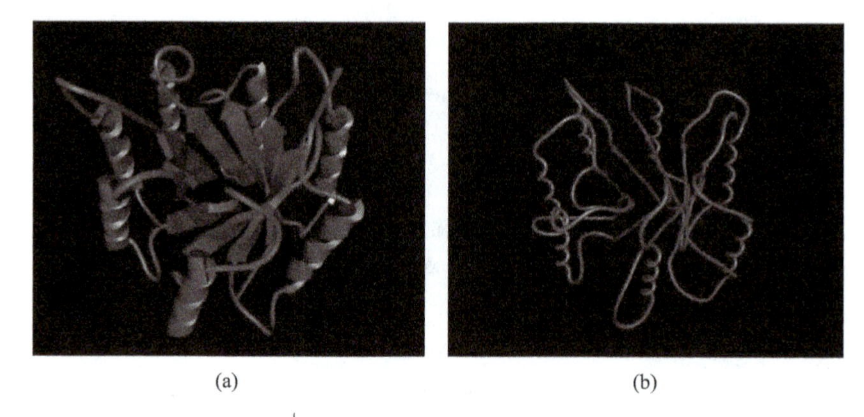

Figure 8.4 Two means of rendering the secondary structure of a protein. (a) Cartoon rendering showing the α-helix helix tubes and the β-sheet shown by the arrows and the coiled regions. This is called a ribbon model. (b) Same as (a), using only tubes, called a backbone model. These diagrams were rendered using the software program ProteinShop. See http://vis.lbl.gov/scrivelli/public/~silvia_page/Pshp.gallery.html.

the forces involved in this further folding are van der Waals forces, hydrogen bonding, ionic bonding, and even covalent bonding. The van der Waals force folding is between the side chain R groups of nonpolar amino acids that are hydrophobic. The hydrogen bonding is between the side chain R groups consisting of one or more organic compounds CH_x. Ionic bond folding is the result of the interaction of a positively charged amino acid and a negatively charged one. Covalent bonding also occurs between thiol-containing amino acids. A *thiol* is a molecule that contains a sulfur hydrogen bond (S-H). The term *quartenary structure* is used if the protein contains more than one backbone. Sketches rendering all these structures are depicted in Figure 8.5.

Bioinformatics is the term given to the creation of databases and computational algorithms for the advancement of our knowledge of the biological world. The structures of proteins are most often determined experimentally, generating a very large amount of information that must be reduced and manipulated to produce the types of pictures seen in this section. More frequently these days, molecular dynamics simulations are used for the same purpose (Figure 8.6).

8.3.3 Some common proteins

Proteins occur in the solid material of the body as well as in blood serum. For example, actin is one of the three major components of the cytoskeleton. The walls of *ion channels*, the primary conduits to and from cells, are made of proteins, which allow certain ions into the cell, depending on the function of the particular channel, and certain waste materials out.

Albumin is a charged protein that exists in many forms. *Human serum albumin* (HSA) is a plasma protein made in the liver that helps maintain chemical equilibrium in the bloodstream. HSA can bind fatty acids and other hydrophobic

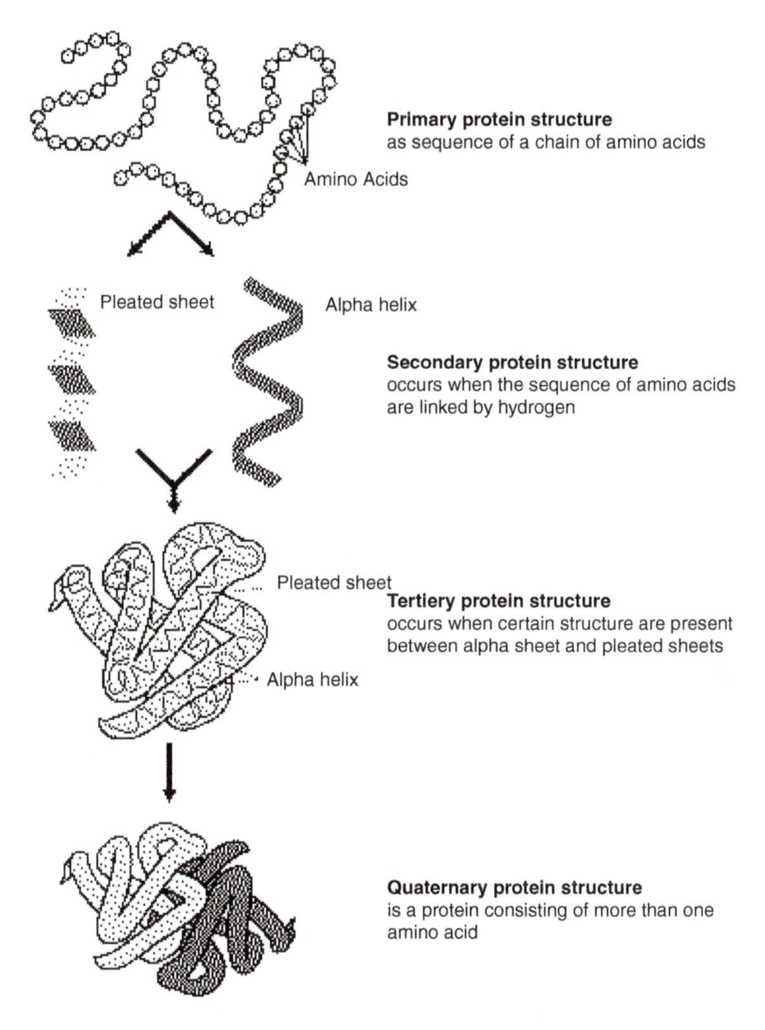

Figure 8.5 Sketches depicting the various structures of a protein. Image courtesy of the National Library of Medicine of the National Institutes of Health.

molecules and can also be used as a drug carrier; there are six distinct sets of binding sites on HSA.

Drugs are often classified according to the extent that they bind to plasma proteins. Calcium is transported by HSA; the maintenance of a stable level of calcium is important in cardiac tissue. Albumin has an average concentration of about 40 mg/ml in serum and has a molecular weight of $M = 66$ kDa. The diffusion coefficient of albumin in blood serum (an electrolyte mixture) is $D = 6.1 \times 10^{-11}$ m^2/sec, and its valence is $z \sim -19$ (Peters, 1996).[1]

[1] Recall that the valence is the number of electrons required to fully populate the outer shell of an atom. To speak of a valence of a protein, then, may not be appropriate because the surface charge of a protein usually varies with position on the outer surface of the protein. A protein may also have a volume charge density. Nevertheless, it is common to speak of the valence of *molecules* such as albumin.

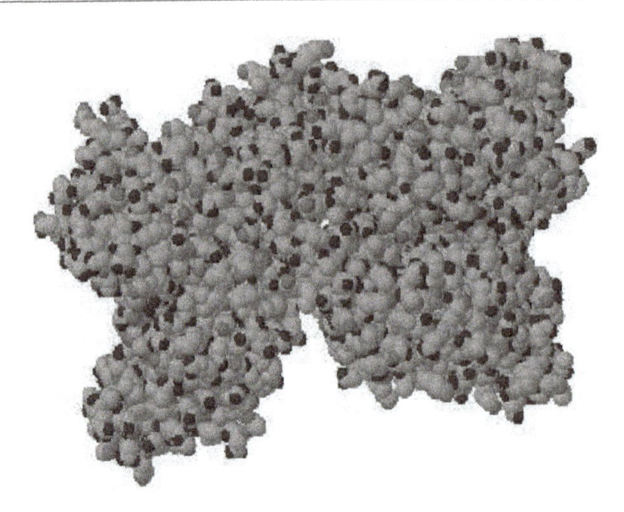

Figure 8.6 Space-filling view of human serum albumin. In a color rendering molecules are coded according to the CPK (Koltun, 1965) color scheme: gray is carbon, oxygen is red, nitrogen is blue, and yellow is sulfur. From the Protein Data Bank, pdb1bm0.

Human IgG is the most abundant of human immunoglobulins, which are more commonly known as antibodies. They recognize antigens on the surface of foreign bodies like bacteria and viruses. Human IgG has a distinctive hinged shape characteristic of antibodies (see Figure 8.8).

Human insulin triggers cells to absorb excess glucose from the blood when the glucose level gets too high. Produced in the pancreas, insulin is soluble in serum at physiological pH $\sim$ 7.4. Its molecular weight is $\sim$6 kDa. If the pancreas cannot produce enough insulin (type 1) or the cells removing glucose casnnot use the insulin that is produced (type 2), *diabetes* is the result.

The key role of *apolipoproteinE* is the metabolism of lipoproteins, which are a macromolecular complex of lipids and proteins that form the walls of the cell. Apolipoprotein activates enzymes that are important in the covalent modification of lipids. They contain receptor ligands that help direct remodeled lipoproteins to specific tissue sites. Apolipoproteins are also involved in development and repair of nervous tissue. There are more than 18 variants of apolipoproteinE, and an imbalance in one of them (called E4) can result in the development of Alzheimer's disease. The molecular weight is $\sim$34.2 kDa.

Lysozyme is a major bacteria-fighting component of the immune system. The small enzyme destroys the carbohydrate cell wall of bacteria. Lysozyme protects areas that are abundant in food from bacteria and is found in tears, mucus, saliva, blood serum, and milk. The molecular weight is around 14.4 kDa.

Human interferon is another major part of the immune defense system. Once activated, interferon slows the growth of a foreign substance, say, bacteria and viruses, by slowing or completely blocking its function. There are three classes of interferon: α, β, and γ. Interferon α–$2a$ may be used to treat a form of leukemia; interferon α–$2b$ may be used to treat hepatitus C, and the molecular weight is $\sim$19 kDa.

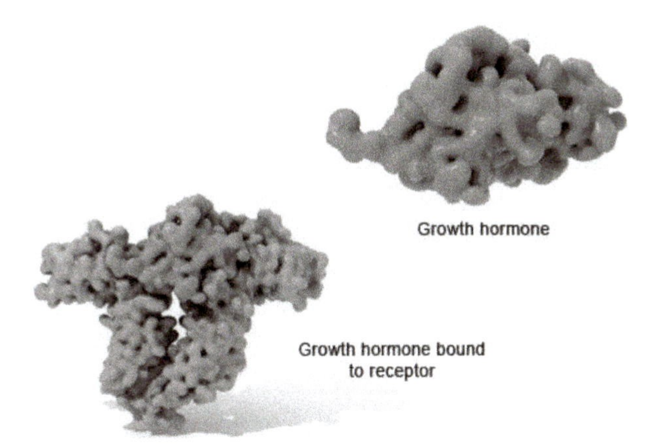

Figure 8.7 Illustration of protein binding, showing a number of weak noncovalent bonds holding the ligand (red) in place.

8.3.4 Few polypeptide chains are useful

Because there are 20 distinct amino acids, there are 20^4 possible combinations for a polypeptide chain four amino acids long. Thus there are 20^n combinations of a polypeptide chain n amino acids long. This is a vast number, but fortunately, many fewer than these very large numbers will be important due to the unique structure and function of each polypeptide chain. Moreover, owing to natural selection, each protein conformation is extremely stable, although even a change of a few atoms in the structure can have disastrous consequences (Alberts *et al.*, 1998).

8.4 Protein binding

One of the consequences of the specificity of a protein's structure and function is the ability of the protein to bind to other molecules. This binding can be either strong and long in duration or weak and short-lived. A particular protein can bind only one or a small number of molecules, and that molecule is called a *ligand*. Where that ligand binds on a given protein is called a *binding site*. Many proteins have several binding sites.

Binding sites on a protein are cavities in the surface of the protein; the ligand fits into the cavity such that a large number of weak noncovalent bonds may form, as shown in Figure 8.7. Note that the ligand is ideally suited to the binding site, and unless this is so, many weak noncovalent bonds do not form, and the binding fails.

As has already been mentioned, immunoglobulins, or antibodies, are the body's response to a foreign substance. They bind to the target called the *antigen* with a remarkable specificity, and there is a one-to-one correspondence between the number of antibodies and the number of antigens. Antibodies are Y-shaped, with two antigen binding sites at the tips of the Y, as shown in Figure 8.8. This

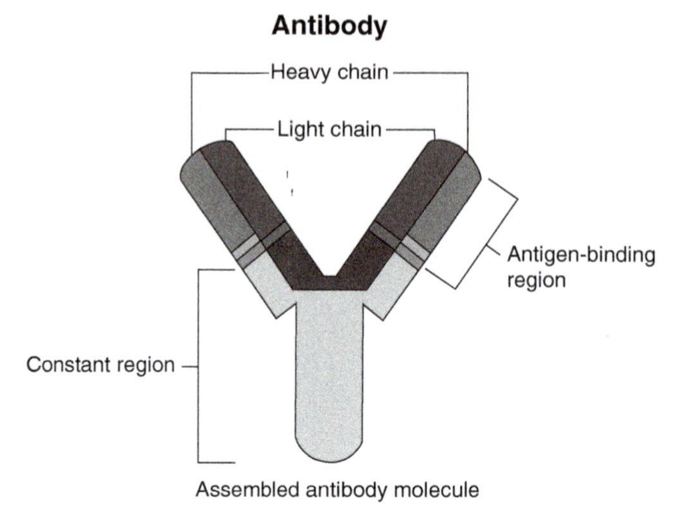

Figure 8.8 Sketch of an antibody.

structure is known to be well suited for grabbing other molecules. Antibody–antigen binding is extremely tight, so the limiting rate constant is the dissociation constant; that is, $k_{\rm off} \sim 0$ (see the chemical reaction boundary condition discussion in Chapter 3).

Two of these chains are said to be "light," and two of them are said to be "heavy," with the terms referring to their relative molecular weights. IgG also contains two binding sites for the antigen, as shown in Figure 8.8, and there are several types of IgG, which are built up from the same basic structure.

IgG occurs naturally in the body but may also be introduced into the body in many different ways (Saltzman, 2001). The controlled delivery of IgG extends its life in the body greatly. However, depending on the methods of delivery, early time results for the fraction of IgG in the plasma (<5 days) can vary widely.

The strength of the binding of a protein to a ligand is characterized by its equilibrium constant, K. The process of binding is considered a chemical reaction, and if the chemical equation is defined by

$$\mathrm{A} + \mathrm{B} \rightleftharpoons \mathrm{AB} \tag{8.1}$$

the equilibrium constant is defined by

$$K = \frac{k_{\rm on}}{k_{\rm off}} = \frac{[\mathrm{AB}]}{[\mathrm{A}][\mathrm{B}]} \tag{8.2}$$

where $k_{\rm on}$ is the *association rate* of the bound complex [AB] and $k_{\rm off}$ is the *dissociation rate* of [AB]. The requirement of conservation of species at equilibrium that dissociation rate = association rate leads to equation (8.2). The greater the equilibrium constant, the greater the strength of the binding; typically, in a cell, in dimensional form, $K \sim 10^8$–10^{10} L/mole.

At equilibrium, the change in Gibbs function for the reaction $dG = 0$, where G is the Gibbs free energy.[2] Then $dG = 0$ requires that $G_f - G_i = 0$, where the subscripts f and i denote the final and initial states, respectively. Initially, the value of the Gibbs free energy is the sum of that for species A and species B:

$$G_i = G_i^0 + RT\ln[\mathrm{A}] + RT\ln[\mathrm{B}] \tag{8.3}$$

After the reaction has taken place,

$$G_f = G_f^0 + RT\ln[\mathrm{AB}] \tag{8.4}$$

requiring

$$-\frac{\Delta G^0}{RT} = \ln\frac{[\mathrm{AB}]}{[\mathrm{A}][\mathrm{B}]} = \ln K \tag{8.5}$$

where ΔG^0 is the Gibbs function change for the reaction $\Delta G^0 = G_f^0 - G_i^0$. Thus, if $K = 10$ at $T = 298$ K, then $\Delta G^0 = -1.36$ kcal/mole $= -5.7$ kJ/mole, and each 10-fold increase in the equilibrium constant corresponds to an increase in free energy of about 1.4 kcal/mole.

As a typical example described by Alberts *et al.* (1998) supposes n molecules of both A and B are present in a cell; let the concentration initially $[A] = [B] = C$. Let $C - x$ be the concentrations of A and B at equilibrium. Because the reaction is given by $[A] + [B] \rightleftharpoons [AB]$, the number of moles at equilibrium of the complex $[AB]$ is x. Then, using the definition of K,

$$\frac{x}{(C - x)^2} = K \tag{8.6}$$

or

$$K(C - x)^2 - x = 0 \tag{8.7}$$

This is a single quadratic equation for x that can easily be solved by the quadratic formula, and the result is

$$x = \frac{2KC + 1 \pm \sqrt{(2KC + 1)^2 - 4K^2C^2}}{2K} \tag{8.8}$$

In this case, only the negative sign is relevant (why?), and for a solution containing 1000 molecules of A and B, which, for a typical cell, corresponds to $[A] = [B] = 10^{-9}M$, and for $K = 10^{10}$ L/mole at equilibrium, there are 270 molecules of A and B and 730 molecules of AB. Of course, as the equilibrium constant decreases, the yield of AB decreases.

[2] The mole fractions should be used here to avoid a dimensional quantity within a function; here square brackets denote concentration, as in Alberts *et al.* (1998), and the end result will not be affected.

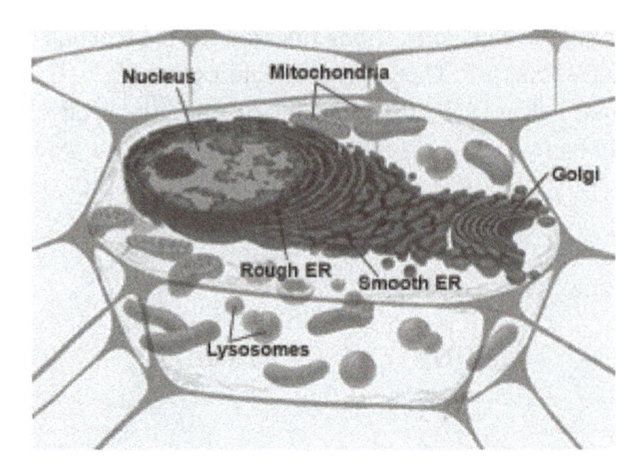

Figure 8.9 Sketch of sliced-open animal eukaryotic cell. Image courtesy of the National Institute of General Medical Sciences of the National Institutes of Health.

8.5 Cells

The cell is the basic unit of all living things. Each cell consists of a membrane surrounding an aqueous solution through which the cell aquires nutrients and removes waste products. Cells are small, of the order of 5–20 μm, and they can be visualized with the aid of a microscope. The various types of microscopes that are used to visualize cells are described by Alberts *et al.* (1998), and it is the invention of the microscope in the seventeenth century that led to the discovery of the cell.

By the nineteenth century, with the aid of research conducted by a number of biologists, a *cell doctrine* (Alberts *et al.*, 1998) maintained that

1. All living things are made up of cells and the products formed from cells.
2. Cells are units of structure and function.
3. All cells arise from preexisting cells.

There are two types of cells: *eukaryotic* cells, which have distinct compartments and are found in the higher mammals, and *prokaryotic* cells, such as bacteria, which have no such compartments (Figure 8.9). Eukaryotic human cells are the subject of this section. It is important to note that all cells are living entities and that viruses are not cells because they cannot replicate.

Cells are composed of an outer flexible cell membrane made of lipids called a *lipid bilayer*. Most of the cell membrane is impermeable; however, it does contain the ion channels that are used to transport material into or out of the cell, depending on its function. This membrane allows the cell to maintain the appropriate level of pH, electrolytes, and other chemicals. Cell death most often occurs when this membrane is ruptured.

The *cytoskeleton* allows the cell to change its shape and move. It consists of a series of rodlike filaments of various diameters (7–25 nm) that anchor the cell to its surroundings. These filaments extend throughout the cell, forming a gridlike

Table 8.1. Composition of a typical eukaryotic cell.

Molecule	% Weight	Types of each molecule
Water	70.0	1
Inorganic ions	1.0	20
Sugars and precursors	1.0	250
Amino acids and precursors	0.4	100
Nucleotides and precursors	0.4	100
Fatty acids and precursors	1.0	50
Other small molecules	0.2	300
Macromolecules	26.0	3000

From Alberts *et al.* (1998). Macromolecules include proteins, nucleic acids, and polysacchrides.

network. The cytoskeleton is made up of microtubules, intermediate filaments, and actin. The intermediate filaments, in particular, are strong and durable and allow the cell to withstand mechanical stress.

Inside the cell, there are specific regions that have specific functions required for the operation of a healthy cell. The *mitochondria* are porous membranes within the cell, in which most of the cell's energy is produced, often from sugars such as glucose. The so-called outer membrane contains a large number of ion channels, some called *transport proteins*, that allow large molecules into the mitochondria.

The *nucleus* is the "command" center (Ethier & Simmons, 2007) and contains the genetic material for molecules that are synthesized by the cell. Proteins and other molecules are synthesized initially by the copying of genetic information coded in DNA into messenger (mRNA), which then leaves the nucleus and enters the *endoplasmic reticulum*, where the proteins and molecules are actually synthesized.

Once the molecules are synthesized in the endoplasmic reticulum, they are moved to the *Golgi apparatus*, where the newly synthesized molecules undergo a series of modifications before they can be biologically active. Any postsynthesis mistakes are automatically discarded in the *lysozomes*.

The *cytosol* (the light region in Figure 8.9) is what surrounds the structures just described. The cytosol consists of a water-based gel in which a number of key biochemical reactions take place. The cytosol is the origin of the manufacturing process of proteins, which are synthesized by molecular machines called *ribosomes*.

Most of a cell is water, and other than water, most of the other constituents of a cell are carbon based. Carbon has four electrons in its outer shell so that carbon can form covalent bonds with four other atoms. This means that carbon is ideal for forming large macromolecules. The carbon compounds that compose the cell, and indeed all molecules that contain carbon, are called *organic molecules*.

The main constituents of a cell are depicted in Table 8.1. From the table it is seen that a cell contains four major classes of organic molecules: sugars, nucleotides, fatty acids, and amino acids. The sugars, especially glucose, act as an energy source for the cell. Nucleotides act as carriers of chemical energy, an important example of which is adenosine triphosphate (ATP). A *fatty acid* is, in general, composed of a long hydrophobic hydrocarbon chain that is not very reactive. At the other end is a hydrophilic carboxyl group that is chemically reactive. Fatty acids act as a food reserve and can produce much more energy than glucose. A fatty acid belongs to a class of materials called *lipids*.

Cells communicate with each other through a variety of molecules such as proteins, nucleotides, fatty acids, and several others. Those molecules used to communicate with other cells are called *hormones*. Insulin, cortisone, and estrogen are examples of hormones. *Insulin* regulates the blood sugar level, for example, and *estrogen* regulates reproductive function.

The destination of the message sent by one cell to another is called the *receptor*. Large hydrophilic signaling molecules cannot cross the cell plasma membrane and so must reside there. Conversely, small hydrophobic molecules, such as *steroid hormones*, can cross the plasma membrane into the interior of the target cell. Perhaps the best known of this class of signaling molecules is *testosterone*. A *steroid* is any one of a class of organic compounds that has a 17 carbon nucleus (Alberts *et al.*, 1998); cholesterol is a steroid.

8.6 The cell membrane

Each cell is surrounded by a thin, fatty film called the *plasma membrane* that is on the order of 10 nm thick. This membrane prevents the contents of the cell from leaking out but also provides the pathway for the cell to receive the necessary nutrients to grow and for the cell to rid itself of waste. Thus the cell membrane contains a network of ion-selective channels and pumps, some of which are made from selected proteins. As the cell grows, the membrane grows as well; thus the cell membrane is a thin, deformable shell and behaves much like a fluid in that it deforms under stress and, if torn, it heals quickly.

The plasma membrane has a two-layer structure comprising lipid molecules that have a hydrophilic head and a hydrophobic tail. The lipid molecules are oriented with their hydrophilic heads in contact with the aqueous environment outside the cell, a configuration that is energetically most favorable (Alberts *et al.*, 1998). The cell membrane is depicted in Figure 8.10. The lipid bilayer is about 5 nm thick, and embedded in the lipid bilayer are the membrane protein channels that facilitate transport to and from the cell. These membrane proteins are usually negatively charged, and some of these channels make up what are called *ion channels*, which regulate the concentration of crucial ions, such as sodium, chloride, calcium, and magnesium, to name a few, inside and outside the cell.

Because of the chemical structure of the lipid bilayer, small, nonpolar molecules, such as gaseous oxygen and carbon dioxide, diffuse through readily. Small, uncharged polar molecules, such as water, can also pass through the lipid

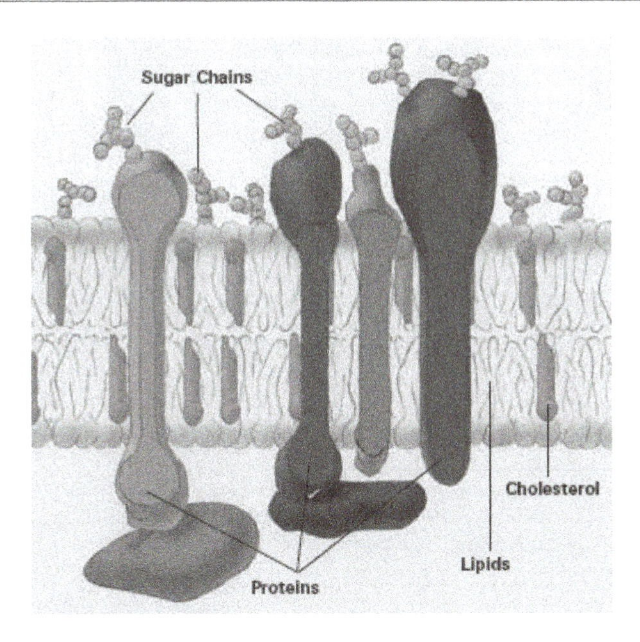

Figure 8.10 Sketch of the structure of a plasma membrane surrounding a cell, including the lipid bilayer. The ion channels are shown as the protein structures they are. Image courtesy of the National Institute of General Medical Sciences of the National Institutes of Health.

bilayer rather easily, but the larger the molecule, the more difficult transport becomes. Lipid bilayers prevent passage of ions and other charged molecules. Because these molecules are attracted to water, they do not want to enter the fatty or hydrocarbon region of the lipid bilayer.

The transport of nonpolar and polar molecules through the lipid bilayer is solely by diffusion. However, for cells to grow, that is, to absorb nutrients such as sugars, ions, and amino acids, among others, and for the cell to release waste material, much faster transport is necessary. The cell provides this faster transport through the use of ion channels; these channels are described in detail in the next section.

8.7 Membrane transport and ion channels

Ion channels play a crucial role in the transport of biofluids to and from cells. To keep the cells functioning properly, there needs to be a continuous flux of ions in and out of the cell and the cell components. In this section, a qualitative picture of ionchannels is described, in keeping with the dominant theme of this chapter. Approximate concentrations of selected ions inside and outside a cell are given in Table 8.2.

An ion channel is a narrow, water-filled tube, permeable to the few ions and molecules small enough to fit through the tube (approximately 10 Å in diameter); their walls are made of proteins, most of which are negatively charged. They have wide entrances and a cross section that varies in the transport, or streamwise, direction (Figure 1.23). Ion channels are responsible for electrical signaling in

Table 8.2. Concentrations of ions inside and outside the cell

Ion	Concentration	Concentration
	Inside cell (mM)	Outside cell (mM)
Na^+	5–15	145
K^+	140	5
Mg^{2+}	0.5	1–2
Ca^{2+}	10^{-7}	1–2
H^+	7×10^{-5}	4×10^{-5}
Cl^-	5–15	110
Fixed anions	high	0

The fixed anions are those molecules that cannot leave the cell. The divalent cation concentrations are for the free ions in the cytosol; large parts of both Ca^{2+} and Mg^{2+} are bound to proteins and other molecules. From Alberts *et al.* (1998).

nerves and muscles.[3] The ions responsible for the majority of nervous system signaling are Ca^{2+}, Cl^-, K^+, and Na^+, and these four ions are transported by most ion channels. Ion channels are responsive to different stimuli: a membrane potential change or a *ligand*, a molecule that attaches itself to the channel, causing it to open.

Ion channels respond to the stimuli in a process called *gating*, the opening or closing of the pore. The open pore has a selective permeability to ions, only allowing certain ions to flow through their electrochemical potential gradients. The ions flow at a very high rate, greater than 10^6 ions per second, with a single ion taking on the order of 1 ms to traverse the channel. This rate of transport is much greater than that of a *carrier* protein, a protein to which an ion is bound, which performs a similar function – see Table 8.3 (Alberts *et al.*, 1998). This rush of ions through the channel changes the electrical potential across the membrane, the *transmembrane potential*, which in turn changes the electrical body force driving the ions through the channel. Thus the operation of an ion channel is much like the electrokinetic phenomenon of *electro-osmotic flow*, which will be discussed in Chapter 9.

As mentioned earlier, the two main characteristics of ion channels are that they are *selective* and *gated*. Selectivity depends on the relative size of the ion channel and the ion that it is supposed to pass. Large ions will not move through small channels owing to both steric (size) and electrostatic effects. As has been mentioned, in general, ions are hydrated by water molecules, which makes the ions appear bigger than they really are. For the highly selective ion channels, it is generally assumed that these hydrated water molecules must be released from the ions so that they pass through the ion channel in single file.

[3] In this text, we have used the term *channel* for a passage that is rectangular in cross section. Ion channels are approximately circular; nevertheless, we use the accepted term *ion channel* rather than the clumsy *ion tube*.

Table 8.3. Gating mechanism and location of the four most common ion channels

Channel	Gate	Location
Na^+	voltage	plasma membrane nerve cell axon
K^+	voltage	plasma membrane nerve cell axon
Ca^{2+}	voltage	plasma membrane nerve cell axon
Cl^-	ligand	plasma membrane neurons

From Alberts *et al.* (1998).

Ion channels are *gated*, as mentioned earlier. Voltage gated channels are generally thought to operate based on a potential drop across the channel, the *transmembrane potential*, allowing the ion concentrations upstream and downstream, according to the Donnan potential discussed in the previous chapter. Thus ion channels open and close when needed, and it is generally thought that the opening or closing of the channel is random. The transmembrane potential affects the probability of an ion channel being open or closed. According to the Donnan potential, the transmembrane potential is related to the concentrations inside and outside the cell by

$$\Delta\phi = \frac{RT}{zF} \ln \frac{C_o}{C_i} \tag{8.9}$$

where C_o and C_i are the concentrations outside and inside the cell and z is the ion valence. Given the concentrations in Table 8.2, the appropriate transmembrane potential can be determined. From the Nernst equation (8.9), the transmembrane potential for ion channels is of the order of $\Delta\phi = \frac{RT}{F}$. Note that the Donnan potential assumes electrical neutrality everywhere in the system.

The state of an ion channel, that is, determining whether it is open or closed, is determined by measuring the *electrical current* through the channel. Today, this can be done using what is called a *patch clamp* device, described in detail by Alberts *et al.* (1998) and in many other places. The patch clamp itself is just a glass tube that is attached to the cell enclosing the target ion channel. A constant voltage source is passed through the solution that houses the cell, with the electrical circuit closed via the insertion of a metal wire into the solution in the glass tube.

The transmembrane potential can be held at a desired value ("clamped") with the current measured at this given transmembrane potential. The transmembrane potential can be varied to assess its influence on the gating process. Thus the current–voltage relationship is a useful measure of the behavior of ion channels. Generally, the results of the patch clamp recording is a time trace of the current, on the order of pA, showing sudden changes in the current as the channel opens or closes.

Much more detail about ion channels is available in a number of texts, including Alberts *et al.* (1994), Alberts *et al.* (1998), Friedman (2008), and Hille (2001). It should also be mentioned that ion channels can be modeled as perm-selective channels (Section 7.18), and the analysis is much the same as in that section (see

Gillespie, 1999; Gillespie & Eisenberg, 2001; Hollerbach *et al.*, 2001; Barcilon *et al.*, 1997; Barcilon *et al.*, 1992). In most cases, ion channels do not seem to operate near the limiting current.

8.8 Chapter summary

The objective of this chapter is to introduce the reader to the fundamentals of molecular and cell biology that are important for understanding biomolecular transport. Many nanoscale devices are designed to transport proteins, DNA, and polysaccharides. Thus knowing the basic structure and properties of these molecules will aid in understanding how these devices should be designed.

In particular, it has been seen that the two most important characteristics of a protein are its structure and function. Because of the inherent complexity of proteins, biologists have defined four levels of structure for analysis, beginning with the primary structure, the amino acid sequence specific to the protein. Determining a protein's conformation is an important exercise, and this is often done either by experiment or increasingly by molecular simulation.

Different proteins perform different tasks, and in general, there are seven distinct types of proteins, though authors may differ on the terminology used. Each of these seven types of proteins has been discussed in this chapter, and examples have been presented for each type of protein.

One of the consequences of a protein's structure and function is its ability to bind to other molecules. The nature of protein binding is discussed in this chapter, specifically antigen–antibody binding, which is so important in fighting disease. Protein binding is viewed as a biochemical reaction similar to those reactions discussed in the previous chapter.

The last three sections discuss the processes that occur at the cellular and subcellular levels. The living cell is the foundation of life itself and comprises mainly water and other macromolecules. Ions and other small molecules pass into and out of cells through ion channels specific to the passage of a given ion. The walls of an ion channel consist of proteins, most often negatively charged, and so electrostatic and electrodynamic interactions are extremely important in their operation.

Many of the principles presented in this chapter will be revisited when biomedical applications are considered in Chapter 12.

EXERCISES

About these exercises: Several of these problems can be used as writing exercises and some are meant to be open-ended.

8.1 Given that a sodium atom has one electron in its outer shell, and a chlorine atom has seven, sketch the the ionic bond that is formed between Na^+ and Cl^-.

8.2 Discuss qualitatively how the conformation of a protein may affect its transport in a synthetic nanopore membrane. Identify the main dimensional parameters that govern its transport.

8.3 Transcription is the process by which a cell uses DNA to make RNA. Discuss the role of transcription within the context of the entire process of the synthesis of proteins.

8.4 Discuss in detail how a protein is synthesized.

8.5 Plot the yield of a complex [AB] given by the solution of equation (8.6), either in concentration or number of molecules, as a function of the equilibrium constant K.

8.6 Solve the previous problem in dimensionless form, that is, nondimensionalize equation (8.6), and show that each term in the equation is of the same order of magnitude.

8.7 Choosing a single common protein, such as albumin, using the Web or another appropriate source, map the charge on the surface, and calculate the total surface charge. What does the valence of a protein mean?

8.8 A globular protein of diameter 3 nm that is negatively charged approaches a nanopore membrane having 10 nm pores. The membrane is negatively charged. Discuss the ability of the globular protein to pass through the membrane.

8.9 An ion channel is gated, that is, it opens and closes. Suggest possibilities for the time scale of the opening and closing procedure. What is the time scale during the period that the ion channel is opened? For the calcium channel, use the Web or another appropriate source to determine the flow rate of calcium through the channel.

8.10 Cystic fibrosis is a disease caused by a malfunctioning of ion channels. Discuss the specifics of what goes wrong and the current treatment, if any, for the disease.

8.11 The term *lipid raft* has recently been used to describe the structure of the cell membrane. Precisely what is a lipid raft?

9 Electrokinetic Phenomena

9.1 Introduction

It was shown in Chapter 2 that as the length scale of microchannels approaches the nanoscale, pressure-driven flow becomes increasingly difficult to achieve. The precise cutoff depends on the desired flow rate, but for a volume flow rate of $Q = 1$ μL/min, electro-osmotic flow (EOF) becomes significantly more efficient around an individual channel height of $h = 40$ nm. Using the material studied in Chapters 4–8, we can now investigate flows of ionic and biomolecular species in micro- and nanochannels.

Electrokinetic phenomena are generally grouped into four classes:

1. Electro-osmosis (EOF): the bulk motion of a fluid caused by an electric field
2. Electrophoresis: the motion of a charged particle in an otherwise motionless fluid or the motion of a charged particle relative to a bulk motion
3. Streaming potential or streaming current: the potential induced by a pressure gradient at zero current flow of an electrolyte mixture
4. Sedimentation potential: the electric field induced when charged particles move relative to a liquid under a gravitational or centrifugal or other force field

Note that electrophoresis and sedimentation potential refer to particles and electro-osmosis and streaming potential refer to electrolyte solutions. The first three of these are by far the most important for our applications, and in this chapter, we discuss the phenomenon of electro-osmosis first, followed by a discussion of electrophoresis and streaming potential.

In this chapter, electro-osmosis and electrophoresis will receive the most attention. It is important to note that for channels with charged walls, the electrical double layer (EDL) will be present independent of whether there is electro-osmotic flow. Elements of electrochemistry will permeate through this chapter, along with the discussion of electrokinetic phenomena. This is especially true of the asymptotic analysis of mixtures of arbitrary valence that could have been discussed in Chapter 7. In the end, I thought that that section goes better in this chapter.

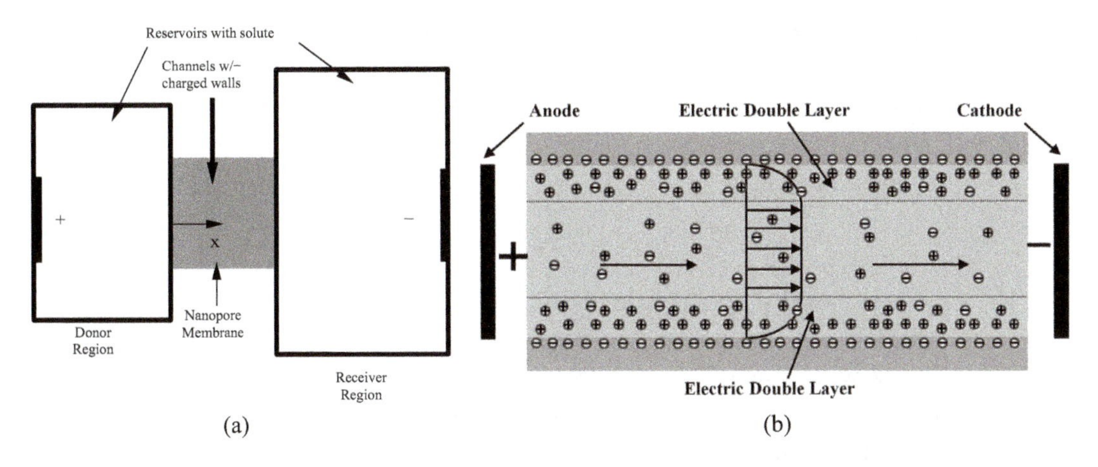

Figure 9.1 (a) Geometry of a typical nanopore membrane, showing electrodes in the donor and receiver regions (not to scale). The walls in the channels are usually negatively charged, but positively charged walls have also been used. (b) Side view of a typical channel in the nanopore membrane, showing a sketch of the EDL. The channel is usually long and wide, as in Figure 1.4.

We first discuss several aspects of electro-osmotic flow in a wide channel, the most common geometry in the field of micro- and nanofluidics, and show that asymptotic solutions may be obtained for species of arbitrary valence in a binary solution. We discuss walls having dissimilar wall charge or ζ potential and follow that with a study of species transport and the channel current and current density. These sections are followed by analysis of electro-osmosis in several canonical geometries as was done for viscous flow in Chapter 4; these canonical flows include the annulus and nozzles and diffusers; nozzles are often used in *electroporation*, the opening of a cell for injection of DNA or genes (Chen & Conlisk, 2008; Wang *et al.*, 2008). The section on electro-osmosis concludes with a discussion of *dispersion*, the spreading of a solute in channel flows.

Next electrophoresis, here defined as the relative motion of particles in a fluid, is described; the models range from a simple force balance for thin and thick EDLs, yielding a simple algebraic expression for the electrophoretic velocity, to a full numercial solution valid for the entire range of EDL thicknesses. The streaming potential, the potential that is induced by an electrically conducting fluid in a pressure-driven flow, is presented next. The streaming potential is often used to measure the average ζ potential in a channel, and it is shown how this can be done. The chapter concludes with a brief description of the sedimentation potential and Joule heating.

9.2 Electro-osmosis

9.2.1 The relationship between velocity and potential

A biomedical application device, described previously, is illustrated in Figure 9.1a. The electro-osmotic flow is driven by electrodes placed in upstream and downstream reservoirs. This geometry may be useful as a drug delivery vehicle. Fluid flows from the upstream reservoir into the mouth of the channel(s) and very quickly reaches a fully developed condition. The electrical potential is composed of two parts corresponding to the presence of the EDL and the externally imposed potential

$$\psi = \phi + \phi_i \tag{9.1}$$

where the subscript i denotes the imposed field. Here we assume that the imposed field is characterized by a constant electric field E_0. In this section, we will work with dimensionless variables almost exclusively. The interested reader is encouraged to determine the appropriate length and velocity scales that are used to obtain the dimensionless equations and solutions for the potential and velocity.

The dimensionless form of the streamwise momentum equation in the fully developed flow region is thus given by (Chapter 3)

$$\epsilon^2 \frac{\partial^2 u}{\partial y^2} = \epsilon^2 \frac{\partial p}{\partial x} - \beta \sum_i z_i X_i \tag{9.2}$$

and the Poisson equation for the potential is

$$\epsilon^2 \frac{\partial^2 \phi}{\partial y^2} = -\beta \sum_i z_i X_i \tag{9.3}$$

where the partial derivatives in this one-dimensional fully developed analysis are really total derivatives $\epsilon = \lambda/h$ and $\beta = c/I$, as in Chapter 3. Here X_i is the mole fraction, but as mentioned previously, we could scale the concentrations on the ionic strength, in which case, $\beta = 1$. Equating the right-hand side of equation (9.3) in equation (9.2) gives

$$\frac{\partial^2 u}{\partial y^2} = \frac{\partial p}{\partial x} + \frac{\partial^2 \phi}{\partial y^2} \tag{9.4}$$

Note that the pressure is scaled on the dimensional electro-osmotic velocity $U_0 = \epsilon_e \phi_0 E_0/\mu$, where E_0 is the imposed electric field in the streamwise direction and the dimensional potential is scaled on $\phi_0 = RT/F$. This equation may be integrated twice, and for a potential specified at each wall, $y = 0, 1$,

$$u(y) = \phi(y) + \frac{1}{2}\frac{dp}{dx}(y^2 - y) + (\zeta_0 - \zeta_1)y - \zeta_0 \tag{9.5}$$

where here ζ_0 and ζ_1 are the dimensionless (scaled on RT/F) ζ potentials at each wall $y = 0, 1$. Note that if both walls have the same ζ potential and there is no axial pressure gradient, then $u(y) = \phi(y) - \zeta_0 = \phi(y) - \zeta$, and thus the velocity is equivalent to the potential perturbation from the ζ potential, as discussed in Chapter 3. The potential $\phi = \phi(y)$ is the dimensionless potential due only to the presence of the EDLs. Thus, if the walls of the channel are not charged, there is no EDL, and even with the presence of the electrodes, there is no bulk electro-osmotic flow.

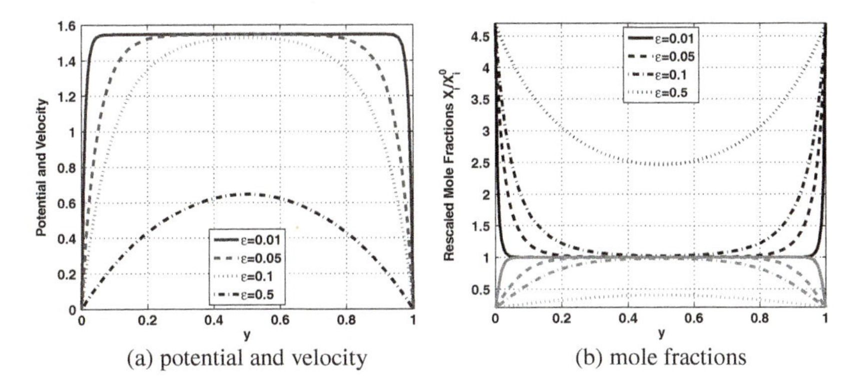

(a) potential and velocity

(b) mole fractions

Figure 9.2 Potential, velocity, and rescaled mole fractions for a 1:1 electrolyte for various values of ϵ. Here the dimensional potential on both walls is $\zeta^* = -40$ mV. In (b), the mole fractions are rescaled based on the upstream reservoir mole fractions as X_i/X_i^0 (note that the values are $O(1)$ – always a good thing). The cations are plotted in black lines, and the anions are plotted in gray lines.

Any of the potential distributions described in Chapter 7 may be used to describe the velocity. For a pair of monovalent ions in channels for which the double layers are overlapping, under the Debye–Hückel approximation,

$$\phi(y) = \zeta \frac{\cosh \frac{y-1/2}{\epsilon}}{\cosh \frac{1}{2\epsilon}} \tag{9.6}$$

In the general nonlinear case, the solution must be calculated numerically. For a thin double layer $\epsilon \to 0$, the cosh solution reduces to

$$\phi = \zeta e^{-\frac{y}{\epsilon}} \tag{9.7}$$

as has been indicated before. Note that in this case, as $y \to \infty$, $u \to -\zeta$.

For the case of a thin double layer, $\epsilon = \lambda/h \ll 1$, recall that the solution to the nonlinear equation for the potential for a pair of symmetric electrolytes of valence z which satisfies $\phi = \zeta$ at $y = 0$ and $\phi \to 0$ as $y \to \infty$ (the dimensionless y coordinate is scaled on ϵ) is

$$\phi = \frac{4}{z} \tanh^{-1} \left(e^{-z\frac{y}{\epsilon}} \tanh \frac{z\zeta}{4} \right) \tag{9.8}$$

The electro-osmotic part of the velocity field near the wall at $y = 0$ is given by

$$u(y) = \phi(y) - \zeta = \frac{4}{z} \tanh^{-1} \left(e^{-z\frac{y}{\epsilon}} \tanh \frac{z\zeta}{4} \right) - \zeta \tag{9.9}$$

where ζ is the dimensionless ζ potential at $y = 0$. For negatively charged walls, the ζ potential will be negative, and thus the velocity in the outer region will be positive. This means that the cathode is located downstream and the anode is located upstream. A similar expression for the velocity holds near $y = 1$.

Figure 9.2 depicts the solution for the velocity and potential and mole fractions for several values of ϵ in the absence of a pressure gradient. Note that both overlapped and thin EDL cases are shown. As $\epsilon \to 0$, the core of the channel becomes electrically neutral, whereas for $\epsilon = 0.5$, it is not. Moreover, for a thin

EDL, the velocity is constant, which is advantageous for reducing dispersion. Note that the mole fractions are rescaled on the reservoir mole fraction for each species X_i^0, which makes the value of the rescaled concentration $O(1)$ – a good thing.

It is useful to discuss the scaling of a typical concentration. Equations (9.2) and (9.3) contain the dimensionless parameter $\beta = c/I$, where c is the total concentration, including the solvent, and I is the reference ionic strength. This is a direct result of using mole fractions as the dependent variable, a scaling that is typical in engineering and chemistry. Because the total concentration includes the solvent, the mole fractions of the electrolytes are usually small, with the combined parameter $\beta X_i^0 = c_i/I = O(1)$, where X_i^0 is the cation mole fraction in the upstream reservoir.

The dimensionless volume flow rate through the channel is given by

$$Q = \int_0^1 u dy \tag{9.10}$$

where here $y = 0$ at the bottom wall of the channel and $y = 1$ at the top wall of the channel. For the cosh potential profile, the overlapped double layer situation, the dimensionless flow rate is given by

$$Q = \zeta \left(\epsilon \tanh \frac{1}{2\epsilon} - 1 \right) - \frac{1}{12} \frac{dp}{dx} \tag{9.11}$$

Even though, strictly speaking, $\epsilon = O(1)$ if the double layers overlap, one can take the limit $\epsilon \rightarrow 0$, as we have seen in equation (9.6), to show that

$$Q = -\zeta - \frac{1}{12} \frac{dp}{dx} \tag{9.12}$$

The effect of a pressure gradient is depicted in Figure 9.3. For a negatively charged wall, a favorable pressure gradient $dp/dx < 0$ increases the flow rate, while an adverse pressure gradient $dp/dx > 0$ reduces the flow rate. Note that eventually, the flow will reverse beginning in the center of the channel, where the Poiseuille velocity is a maximum.

Because the boundary conditions for the cosh potential distribution are constant potential at the walls, the surface charge density must be calculated from the potential solution. In dimensionless form,

$$\sigma = -\frac{\partial \phi}{\partial y} = \zeta \tanh \frac{1}{2\epsilon} \tag{9.13}$$

at $y = 0$ with a similar expression at $y = 1$. Here the dimensionless surface charge density is scaled on $\epsilon_e \phi_0 / h$.

Notation alert: Here come the asterisks again! We need to use Q^* to distinguish it from its dimensionless counterpart Q.

Looking only at the electro-osmotic part of the flow, the *electro-osmotic pumping*, for thin EDLs, the velocity in the core of the channel far from the wall(s) is

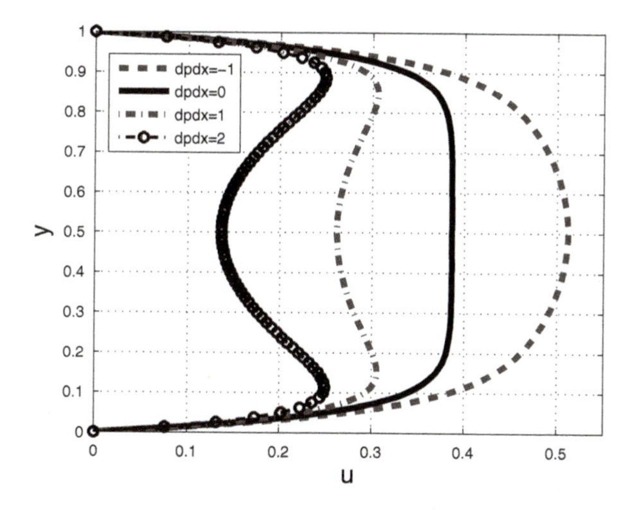

Figure 9.3 Dimensionless velocity distribution for both electro-osmotic and pressue gradient effects. Here the electric field corresponds to 0.06 V over a channel of length $L = 3.5$ μm; the channel height $h = 20$ nm and the electrolyte solution concentration is 0.1 M in the reservoir. The ζ potential is −10 mV on both walls. The dimensionless pressure gradient $\partial p/\partial x$ is −1, 0, 1, and 2, respectively.

constant, and $U_\infty = -\zeta$. In dimensional form, this equation is given by

$$U_\infty^* = -\frac{\epsilon_e \phi_0 E_x}{\mu}\frac{\zeta^*}{\phi_0} = -\frac{\epsilon_e E_x \zeta^*}{\mu} \tag{9.14}$$

where $\zeta^* = \zeta RT/F$ is the dimensional ζ potential; this is the *Smolukowski formula* for the velocity in the bulk of the channel. The dimensional flow rate is thus given by

$$Q^* = -\rho A \frac{\epsilon_e E_x \zeta^*}{\mu} \tag{9.15}$$

where A is the cross-sectional area. The *electro-osmotic mobility* is defined by

$$\mu_{eo} = \frac{U_\infty^*}{E_x} = \frac{-\zeta^* \epsilon_e}{\mu} \tag{9.16}$$

For constant potential boundary conditions, the velocity and the potential perturbation from the ζ potential are equivalent; as in classical boundary layer theory, the width of the EDL can be defined as that location where the velocity and potential reach 99 percent of their free stream value. In the Debye–Hückel approximation for $\epsilon \ll 1$, the dimensionless width is given by $\delta_e = 4.61\epsilon$, where δ_e is nondimensional, scaled on the channel height. Thus the dimensional width of the EDL is given by $\delta_e^* = 4.61\lambda$, a result that was discussed in Chapter 7. A fully numerical result in this case also results in $\delta_e^* = 4.6\lambda$. Note that the Debye length is not a good measure of the asymptotic extent of the EDL and that the constant 4.6 is similar to the result obtained from the Blasius boundary layer in fluid mechanics, for which the boundary layer thickness is given by $\delta_{BL} = 5\text{Re}^{-1/2}$ for large Reynolds number.

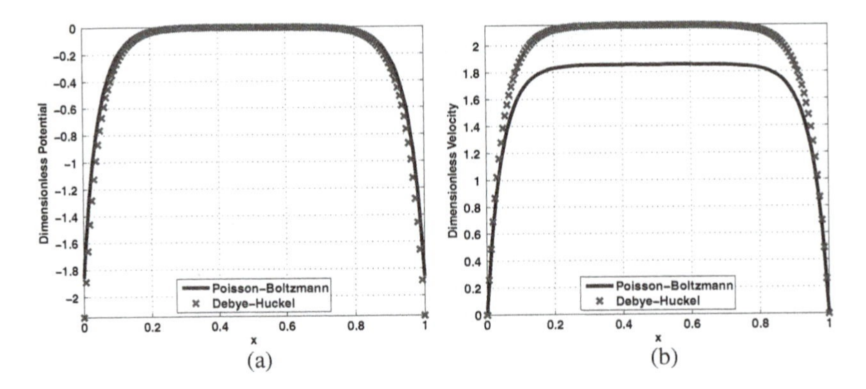

Figure 9.4 Comparison of the results for the dimensionless potential and velocity calculated from the Poisson–Boltzmann equations and the Debye–Hückel approximation, using the fixed surface charge boundary condition. The height of the channel is $h = 20$ nm, and the electrolyte solution is 0.1 M Na^+ and 0.1 M Cl^-. (a) Dimensionless potential for a fixed surface charge density $\sigma = -0.04$ C/m^2. The corresponding dimensionless wall potential is $\zeta = -1.86$ (-48 mV in dimensional form) based on the Poisson–Boltzmann equations and -2.15 (-55.6 mV in dimensional form) based on the Debye–Hückel approximation. (b) Dimensionless velocity.

9.2.2 The Debye–Hückel approximation reviewed

The Debye–Hückel (DH) approximation is a powerful tool enabling an analytical solution not only for the potential but for the velocity as well. It is thus of interest to inquire as to what the error is in making this approximation; thus we compare the full nonlinear solution for the potential and velocity with the DH approximation. In Figure 9.4 are results comparing the nonlinear solution for a 1:1 electrolyte with the DH approximation, using the fixed surface charge condition. The calculations from Figure 9.4 assumes concentrations of $c_{A0} = 0.1$ M and $c_{B0} = 0.1$ M in the bulk. Note in Figure 9.4a, that the ζ potential on the wall is different for the DH approximation. This results in a DH approximation for the velocity that is *larger* than the full nonlinear solution.

The results presented on Figure 9.4 indicate that for a given surface charge density, the Debye Huckel approximation overestimates the magnitude of the zeta potential, and hence the flow rate in a channel with thin EDLs. This is because the dimensionless Poisson-Boltzmann (PB) estimate $\zeta_{PB} = 2\sinh^{-1}\sigma/2$ is always smaller than that of the Debye–Hückel estimate $\zeta_{DHA} = \sigma$ for any given value of (positive/negative) σ (see for example, Figure 9.5). On the other hand, if the constant potential boundary condition is used, the DH approximation will under-predict the PB result since the approximation is only valid for small potential.

9.2.3 Another similarity revealed

Using the similarity of the velocity and potential, the skin friction coefficient, c_f, can be related to the surface charge density. At low Reynolds number, the skin friction can be defined in the same way as for high Reynolds numbers, and since $u = \phi - \zeta$,

$$c_f = \frac{2}{Re}\frac{\partial u}{\partial y} = \frac{2}{Re}\frac{\partial \phi}{\partial y} \tag{9.17}$$

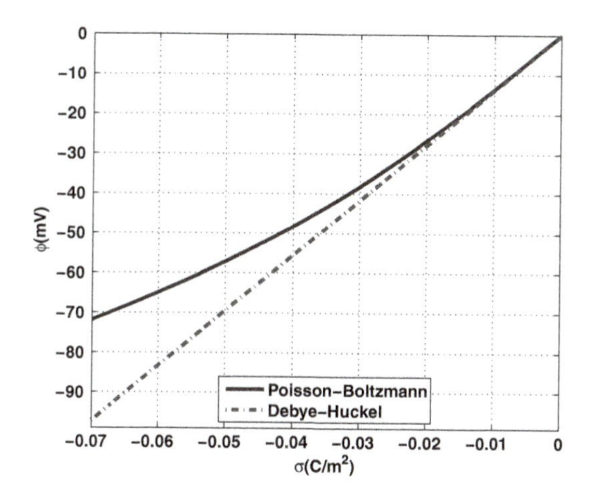

Figure 9.5 Comparison of ζ potential for different surface charge density calculated from Poisson–Boltzmann and Debye–Hückel approximations. The concentration in the bulk is 0.1 M Na^+ and 0.1 M Cl^-.

Now, in this dimensionless form, if the dimensional surface charge density is scaled on $\epsilon_e \phi_0/\lambda$, then

$$\sigma = -\epsilon \frac{\partial \phi}{\partial y} \tag{9.18}$$

if the outward unit normal is in the negative coordinate direction, for example, at $y = 0$. Thus, in terms of surface charge density,

$$c_f = -\frac{2\sigma}{\epsilon Re} = -\frac{2\zeta}{\epsilon Re} \tag{9.19}$$

for a negatively charged wall. Note that the first equality holds independent of the DH approximation and that the second equality holds in the limit of a thin double layer. Both relationships suggest that experimental measurements of the shear stress can lead to a significant amount of information about the ζ potential and surface charge density.

It is not the convention to do so, but for very small Reynolds numbers, conceptually, it is more reasonable to scale the shear stress on the viscous scale $\mu U_0/h$, where U_0 is the electro-osmotic velocity scale, and in this case, $c_f = \partial u/\partial y = \partial \phi/\partial y$.

9.2.4 Asymptotic solution for binary electrolytes of arbitrary valence

By far the most common situation in microchannel flows is the case where the EDL is thin, as depicted in Figure 9.6. In Figure 9.2 it is seen that for this case, the velocity is a constant in the core of the channel $u = -\zeta$ for a pair of monovalent ionic species; in this section, we show how this constant can be obtained directly by the method of *matched asymptotic expansions* without the need for an analytical solution for the potential and for arbitrary values of the valence of the species.

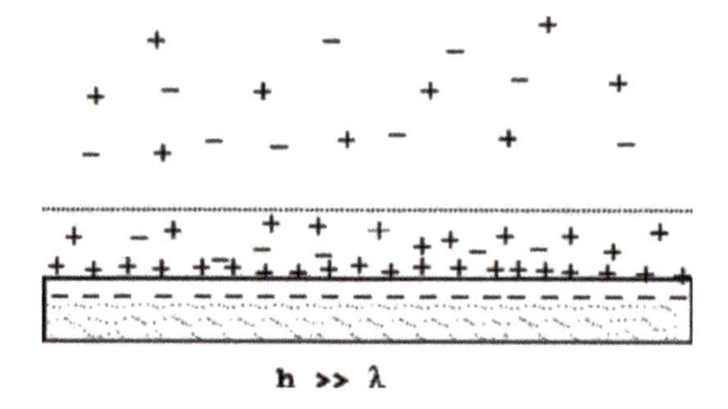

Figure 9.6 If the EDL is thin, the other channel wall is not visible. The flow direction is left to right.

We use here terminology that is introduced in Appendix A. For $\epsilon \ll 1$ and thus as $\epsilon \to 0$, the highest order differentiated term in the Poisson equation is small, except in the region immediately next to the wall. This means that away from the wall, the boundary condition at the wall cannot be satisfied. This is what is known as a *singular perturbation problem* (VanDyke, 1975). For many problems in fluid mechanics and heat and mass transfer, these methods can greatly simplify the solution process. Of course, the Blasius boundary layer problem of fluid mechanics is an example of a singular perturbation problem.

Consider the case of two species of arbitrary valence, with the cation (we assume a negatively charged wall, although this assumption is not necessary) $X_A = g$ and the anion $X_B = f$. It is easier to start in the region near the walls. For example, near $y = 0$, as we have done before, let $Y = y/\epsilon$, and this transformation blows up the EDL so that we can see inside it. Then, because of the no-flux condition at the walls,

$$\frac{\partial g}{\partial Y} + z_g g \frac{\partial \phi}{\partial Y} = 0 \tag{9.20}$$

This is simply the one dimensional Boltzmann equation, which has the well-known solution

$$g = g_R e^{-z_g \phi} \tag{9.21}$$

where g_R is the cation mole fraction in the upstream reservoir, where the potential due to the EDLs on the sidewalls of the membrane is taken to be zero. The solution for f follows in the same way:

$$f = f_R e^{-z_f \phi} \tag{9.22}$$

For thin EDLs, the quantities f_R and g_R are equivalent to the mole fractions in the bulk.

In the method of matched asymptotic expansions, this *inner* solution for the mole fractions of the species (if that is taken as the scale) must be matched with an *outer* solution valid away from the wall.[1] To obtain the matching conditions,

[1]The reader may want to refer to Appendix A for some simple examples of the appropriate matching procedure.

the outer solution for the electrolyte of positive valence, g_o (o for "outer"), must be

$$g_o = \lim_{Y\to\infty} g = g_R e^{-z_g \phi_o} \tag{9.23}$$

Similarly,

$$f_o = \lim_{Y\to\infty} f = f_R e^{-z_f \phi_o} \tag{9.24}$$

Since the core region must be electrically neutral with the volume charge density being zero,

$$f_o = -\frac{z_g}{z_f} g_o \tag{9.25}$$

Thus, from the limit of the inner solution,

$$z_g g_R e^{-z_g \phi_o} = -z_f f_R e^{-z_f \phi_o}, \tag{9.26}$$

and solving for the outer solution ϕ_o,

$$\phi_o = \frac{1}{z_g - z_f} \ln \frac{-z_g g_R}{z_f f_R} \tag{9.27}$$

and $u_o = \phi_o - \zeta$. Note that for $z_g = -z_f$, $\phi_o = 0$; that is, the potential has the same value as in the reservoir. The anion mole fraction in the core of the channel is thus

$$f_o = \sqrt{-\frac{z_g}{z_f} g_R f_R e^{-(z_g+z_f)\phi_o}} \tag{9.28}$$

with a similar solution for g_o. For a $(1, -1)$ electrolyte,

$$f_o = g_o = g_R \tag{9.29}$$

Evaluating the Boltzmann distributions at the wall and dividing the two equations, the dimensionless ζ potential is given by

$$\zeta = \frac{1}{z_f - z_g} \ln \frac{g_w}{f_w} \tag{9.30}$$

so that the dimensionless flow rate $u_o = Q = -\zeta$, as in the previous section. The dimensional volume flow rate is thus given by

$$Q^* = \rho A \frac{\epsilon_e \phi_0 E_x}{\mu} \left(\frac{1}{z_g - z_f} \ln \frac{g_w}{f_w} \right) \tag{9.31}$$

where A is the cross-sectional area of the channel and $\phi_0 = RT/F$. For a negatively charged wall, $g_w > f_w$.

Note that since the EDL is so thin, the flow rate to leading order does not require the solution in the inner region near the wall. This is a powerful result that is independent of the potential distribution within the EDL.

9.2.5 Walls with different ζ potentials

Fabrication procedures for microchannels often involve injection molding (Devasenathipathy *et al.*, 1998), laser ablation (Bianchi *et al.*, 1993), and etching of an open groove on a substrate followed by sealing of the groove by a coverslide of another material. For example, Bianchi *et al.* (1993) use a molded Polydimethylsiloxane (PDMS) channel sealed by a cover made of glass in their experiments. In such situations, the ζ potential on the sealing wall (say, top) differs substantially from that on the side and bottom walls.

When the ζ potentials of the bottom and top wall are not the same, the axial electro-osmotic velocity varies linearly across the channel and the familiar plug profile of EOF no longer occurs. One serious consequence of this nonuniform distribution is increased Taylor dispersion of samples that reduce the efficiency of microfluidic separations such as capillary electrophoresis. However, walls having different ζ potentials are used in what is termed an *asymmetric clamping cell* to calculate the average ζ potential of the two walls. The procedure is described in Masliyah and Bhattacharjee (2006).

Consider the case of monovalent electrolytes. For the case where the walls have differing ζ potentials and hence differing wall mole fractions, the dimensionless velocity in the outer region satisfies

$$u_o = Ay + B \tag{9.32}$$

where E and F are constants. As in the previous section, $\phi_o = 0$. Because of the no-flux condition, the outer solution for the mole fractions is a constant

$$f_o = g_o = f_R = g_R \tag{9.33}$$

In the inner region near $y = 1$, for example,

$$u_i = \phi_i + D \tag{9.34}$$

and near $y = 0$, we have

$$u_i = \phi_i + E \tag{9.35}$$

Clearly to satisfy the no-slip condition, $D = -\zeta_1$ and $E = -\zeta_0$. The constants A and B are found by the matching condition

$$\lim_{Y\to\infty} u_i = \lim_{y\to 0} u_o \tag{9.36}$$

where again $Y = y/\epsilon$ and thus $B = -\zeta_0$ and $A = \zeta_0 - \zeta_1$. This procedure can easily be extended to multivalent species.

Integrating the solution for the velocity in the outer region, the dimensionless flow rate is given by

$$Q = -\frac{1}{2}(\zeta_0 + \zeta_1) + O(\epsilon) \tag{9.37}$$

and the flow rate is proportional to the average ζ potential.

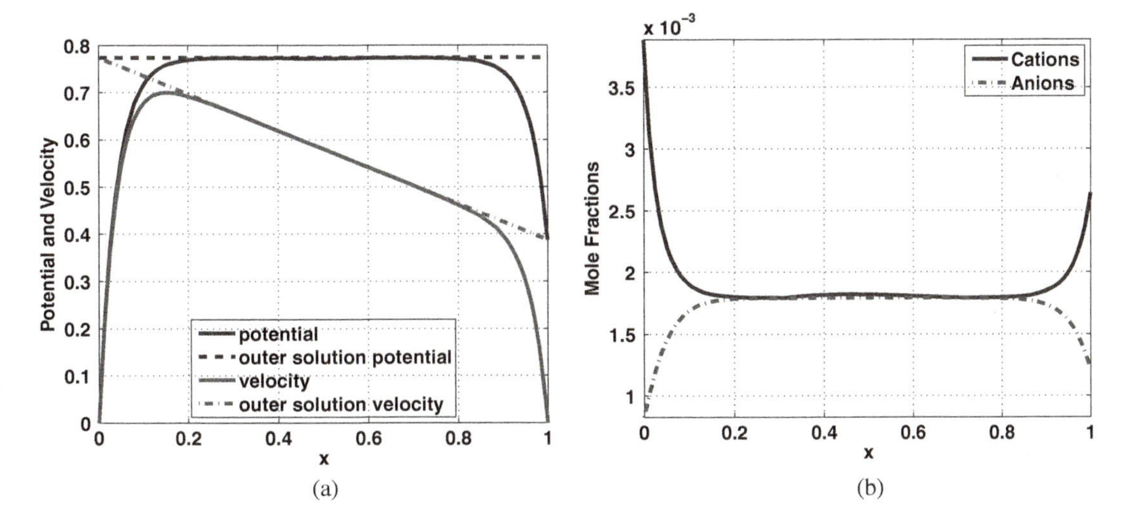

Figure 9.7 Results for the dimensionless velocity and potential along with mole fractions for the asymmetric case of a 1:1 electrolyte mixture. Here the electric field corresponds to 6 v over a channel of length $L = 3.5$ μm; the channel height $h = 25$ nm. (a) Velocity and potential. The outer solutions of the singular perturbation analysis are also shown. (b) Mole fractions. Here the zeta potential is -20 mV at $y = 0$ and -10 mV at $y = 1$. The Boltzman distribution is used for the mole fractions, and the concentrations of each electrolyte are 0.1 M in the upstream reservoir. Here the potential is plotted as $\phi - \zeta$.

Two examples of walls having different ζ potentials are depicted in Figures 9.7 and 9.8, both for a 1:1 electrolyte. In Figure 9.7 $\zeta^* = -20$ mV at $y = 0$ and $\zeta^* = -10$ mV at $y = 1$ so that at $y = 0$, $\zeta = -20/26 = 0.77$, and at $y = 1$, $\zeta = 0.38$. Numerical solutions have been calculated for both the inner and outer regions, and the outer solution for the velocity and potential obtained asymptotically is

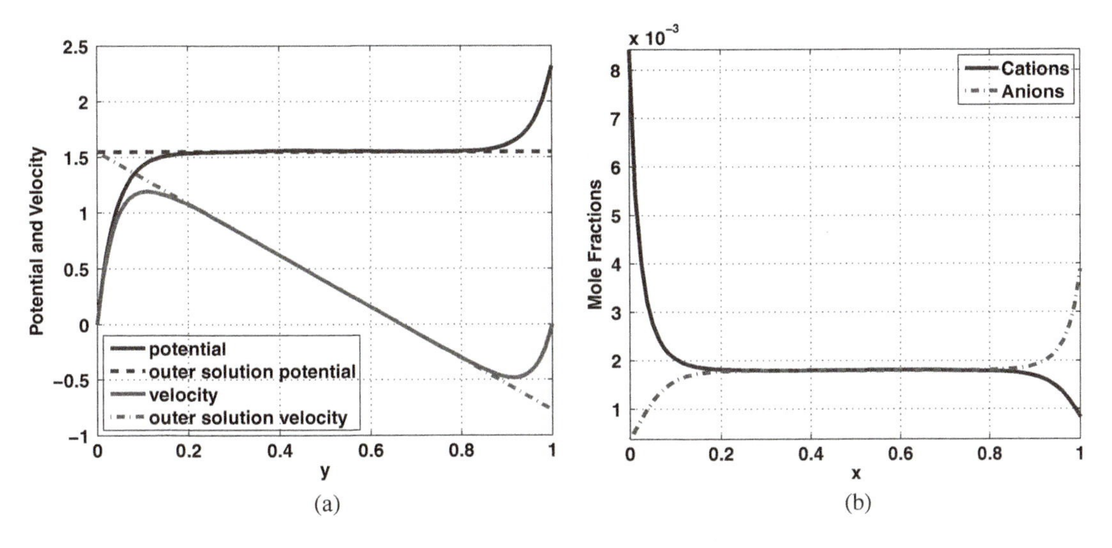

Figure 9.8 Results for the dimensionless velocity and potential (a) along with mole fractions (b) for the asymmetric case of a 1:1 electrolyte mixture. Here the electric field corresponds to 0.06 V over a channel of length $L = 3.5$ μm; the channel height $h = 25$ nm and the NaCl solution concentration is 0.1 M in the reservoir. The ζ potential is -40 mV at $y = 0$ and 20 mV at $y = 1$, and the potential is plotted as $\phi - \zeta$.

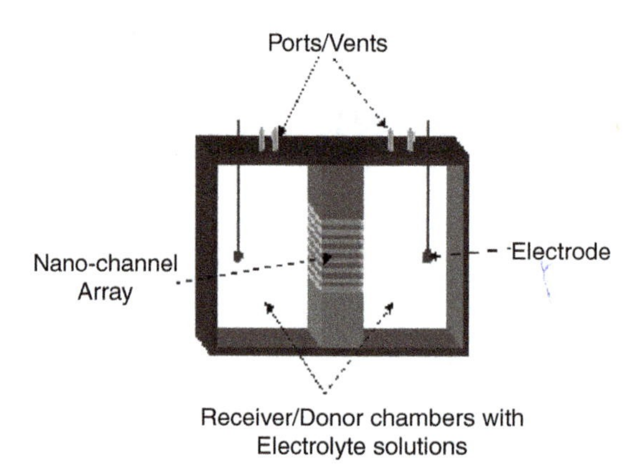

Figure 9.9 An electro-osmotic pumping system with a nanopore membrane; flow is from left to right, with the donor region on the left and the receiver region on the right. Typically, each channel is a slit pore, and typical dimensions in the direction of flow $L \sim 1$ μm, spanwise $W \sim 40$ μm, with the height $h \sim 10$–50 nm.

indicated by the dashed lines in Figures 9.7a and 9.8a. The potential curve is shifted up by ζ to better compare the potential and velocity. Note the linear character of the velocity in the outer region, while the potential remains constant in the core; as mentioned earlier, the variation of velocity will lead to increased dispersion.

In Figure 9.9, the walls are actually of opposite-sign ζ potential leading to reversed flow near $y = 1$. The dimensional flow rate using only the outer solution is $Q^* = 1.37 \times 10^{-18}$ m^3/sec for the parameters of Figure 9.7 and $Q^* = 2.76 \times 10^{-18}$ m^3/sec for the parameters of Figure 9.8.

9.2.6 Species velocities in electro-osmotic flow: Electromigration

Notation alert: For the most part, in this section, the equations will be in dimensional form.

As has already been noted in Chapter 3, the flux of a given species in an electrolyte solution is due in general to Fickian diffusion, electrical migration, and bulk fluid motion. Thus most biomolecular transport studies begin with the expression for the dimensional flux of species A in the form

$$\vec{N}_A = -cD_A \nabla X_A + cm_A z_A X_A \vec{E} + cX_A \vec{V} \tag{9.38}$$

where D_A is the diffusion coefficient, $m_A z_A = z_A\,(FD_A/RT)$ is often termed the *ionic mobility*, F is Faraday's constant, c is the total concentration, X_A is the mole fraction, $\vec{u}$ is the dimensional velocity, and $\vec{E}$ is the dimensional electric field. Also, R is the universal gas constant, and T is temperature.

In particular, equation (9.38) in the streamwise, x direction of Figure 9.1 is given by

$$N_{Ax} = -cD_A \frac{\partial X_A}{\partial x} + cm_A z_A X_A E_x + cX_A u \tag{9.39}$$

The quantity N_{Ax} is a flux and has units of mole/m^{2}sec, and the quantity

$$\frac{N_{Ax}}{c} = u_A \tag{9.40}$$

where c is the total concentration, including solvent, is thus the velocity of the transported species A in the x direction, and so the speed of the molecule at any point is given by

$$u_A = \frac{N_{Ax}}{c} = -D_A \frac{\partial X_A}{\partial x} + m_A z_A X_A E_x + X_A u \tag{9.41}$$

Then the position of any ion or biomolecule assumed to be residing at a point in the channel is obtained by solving

$$\frac{dx_A}{dt} = u_A \tag{9.42}$$

subject to the appropriate initial conditions.

To estimate the order of magnitude of each term in this equation, consider a typical channel in the nanopore membrane in Figure 9.9 (Chen & Conlisk, 2008). First, the diffusion term for a length $L = 10^{-6}$ m $= 1$ μm and $D_A = 10^{-9}$ m^2/sec is

$$D_A \frac{\partial X_A}{\partial x} \sim 10^{-9} \frac{\text{m}^2}{\text{sec}} \Delta X_A/(10^{-6}\text{ m}) \sim 10^{-3} \Delta X_A \frac{\text{m}}{\text{sec}} \tag{9.43}$$

for a typical ionic solute having a change in concentration ΔX_A over the length L. The term proportional to the electric field, for ionic species, is termed *electromigration*, and its magnitude is

$$m_A F z_A X_A E_x \sim 4 \times 10^{-14} \frac{\text{mole}}{\text{J}} \frac{\text{m}^2}{\text{sec}} \times 96{,}500 \frac{\text{C}}{\text{mole}} \times 0.03V/10^{-6}\text{ m} \tag{9.44}$$

$$\sim 10^{-3} z_A X_A \tag{9.45}$$

for $E_x = 3 \times 10^4$ V/m. The species convective term is of the order

$$X_A U \sim \frac{\epsilon_e E_x RT}{F\mu} X_A \sim 5 \times 10^{-4} X_A \tag{9.46}$$

where the viscosity of water is used for μ and the velocity scale $U = \epsilon_e E_x \phi_0/\mu$. Note that the convective term nominally is of the same order of magnitude as the migration term and that the diffusion term is much smaller if $\Delta X_A \ll X_A$. Also, the electromigration term for a cation is greater than zero, whereas that for an anion is less than zero. Thus, for a channel having negatively charged walls, a cation or other biomolecule of positive charge (assuming that the biomolecule can be represented as a point charge) will be transported faster than an anion in an electro-osmotic flow. This is the basis for electrokinetic separation of molecules, particularly ions and biomolecules.

Equation (9.39) shows that a cation will move in a direction opposite to the anion in an electrolyte solution in the absence of an electro-osmotic flow (EOF). The cation will move to the cathode and the anion to the anode. This is depicted in

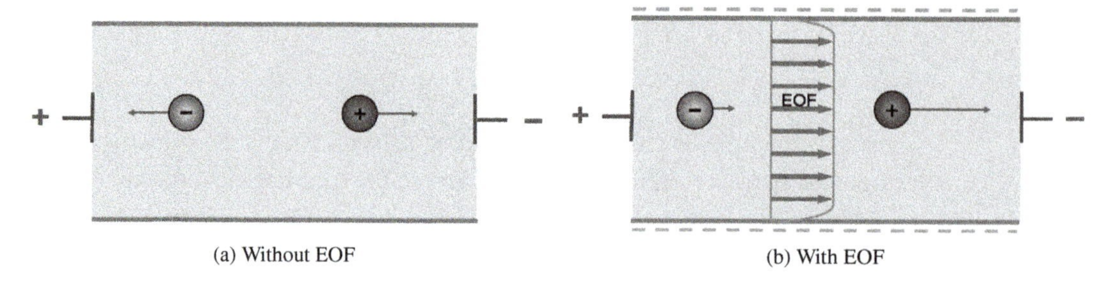

Figure 9.10 Sketch of the motion of ions in an electrolyte solution in a channel having negatively charged walls.

Figure 9.10a. In the presence of electro-osmotic flow between negatively charged walls, the cation and anion will move in the same direction for large enough electro-osmotic flow velocity. The cation will travel faster than the anion; this is depicted in Figure 9.10b.

The simplest model for the motion of a biomolecule in a nanopore membrane is to consider it to be a single species within the multicomponent mixture (say, Phosphate Buffered Saline (PBS)). Consider the transport of an uncharged biomolecule such as glucose. Being neutral, glucose is transported solely by the bulk fluid motion if the concentration is uniform because its valence is zero. Thus an uncharged molecule will always move to the receiver region in Figure 9.10.

The motion of an anion or a negatively charged biomolecule will be controled by the magnitude of the electric field. As noted, in the fully developed region of the channel, the electromigration term is less than zero and the bulk convection term is greater than zero for an anion, and so depending on the magnitude of each term, the anion could move to the receiver or stay in the donor. It is this electrodynamic interaction that determines the transport properties of the membrane. However, because such a biomolecule may adsorb to the surface, not all of the analyte that originated in the donor may reach the receiver. If the polarity is reversed and the negative electrode is on the donor side, the flow will be reversed.

The discussion here assumes steady flow, which means that there is an infinite amount of solute in the donor region. In practice, this is often not the case, and solute may be injected periodically as a plug rather than in a continuous manner. This situation will require an analysis similar to Wilson (1948) for charged solutes; this transient problem for uncharged solutes is discussed in Chapter 5.

9.2.7 Current and current density in electro-osmotic flow

In many flow-through biochemical sensing devices, changes in electrical current are often used to identify an analyte (Braha *et al.*, 2000). The definition of the *electrical current density* in a flowing electrolyte solution having concentration gradients is

$$\vec{J} = F \sum_i z_i \vec{N}_i \tag{9.47}$$

where

$$\vec{N}_i = -D_i \nabla c_i + \frac{D_i}{RT} z_i F c_i \vec{E} + c_i \vec{u} \tag{9.48}$$

The units of current density are amp/m^2. In Section 7.18, we analyzed a permselective membrane having an imposed current, and in this section, we extend the calculation of the current in a stagnant medium having no gradients presented in Section 6.8 to include both electro-osmotic flow and thus velocity in addition to potential and concentration gradients. In this section, the current is generated by flow and the presence of the EDLs and is not imposed.

Notation alert: In this section, $\vec{J}$ is the current density and not the molar flux of species A relative to the bulk velocity field $\vec{J}_A$, as in Chapter 3. Also, here there should be no confusing the current I for ionic strength.

It is convenient to take the origin of the coordinate system at the centerline of the channel. Integrating the streamwise component of the current density across the channel, the total current per unit width of the channel is given by

$$I = \int_{-h}^{h} J_x dy \tag{9.49}$$

Substituting for the streamwise flux, and neglecting streamwise gradients of concentration, as is common (see the previous section), the current per unit width is given by

$$I = E_x \frac{F^2}{RT} \sum_i z_i^2 D_i \int_{-h}^{h} c_i dy + \int_{-h}^{h} \sum_i F z_i c_i u dy \tag{9.50}$$

Taking the dimensional velocity field to be

$$u(y) = E_x \frac{\epsilon_e}{\mu}(\phi - \zeta) \tag{9.51}$$

then

$$I = 2h\langle\sigma_e\rangle E_x - E_x \frac{\epsilon_e^2}{\mu} \int_{-h}^{h} (\phi - \zeta) \frac{\partial^2 \phi}{\partial y^2} dy \tag{9.52}$$

where $\langle\sigma_e\rangle$ is the average electrical conductivity of the solution defined by

$$\langle\sigma_e\rangle = \frac{F^2}{RT} \frac{1}{2h} \sum_i z_i^2 D_i \int_{-h}^{h} c_i dy \tag{9.53}$$

with

$$\sigma_e = \frac{F^2}{RT} \sum_i z_i^2 D_i c_i \tag{9.54}$$

being the electrical conductivity. Note that we have replaced the volume charge density with the second derivative of the potential from Poisson's equation. Since $\phi = \phi(y)$, the partial derivative $\partial^2\phi/\partial y^2 = d^2\phi/dy^2$ is really a total derivative.

Integrating by parts, the total current is given by

$$I = 2h\langle\sigma_e\rangle E_x + E_x \frac{\epsilon_e^2}{\mu} \int_{-h}^{h} \left(\frac{d\phi}{dy}\right)^2 dy \tag{9.55}$$

It then remains to specify a solution for the potential. If the EDLs are overlapped, and the potential is symmetric about the centerline, recall that the potential is given by

$$\phi = \frac{\zeta \cosh \kappa y}{\cosh \kappa h} \tag{9.56}$$

where κ is the inverse Debye length. Performing the integration,

$$I = E_x \left(2h\langle\sigma_e\rangle + \frac{\epsilon_e^2 \zeta^2}{\mu h} \frac{\left(\frac{\epsilon}{2}\sinh(2/\epsilon) - 1\right)}{\epsilon^2 \cosh^2(1/\epsilon)} \right) \tag{9.57}$$

Note that the current–voltage relationship is linear because $E_x = -\Delta\Phi/L$, where $\Delta\Phi$ is the voltage drop across the channel. Equation (9.57) is the analogue of the equation $V = IR$ for lumped parameter static systems; here the V is really a voltage drop, which we have written as $\Delta\Phi$.

The equation for the current can be written in dimensionless form. From equation (9.57), the quantity

$$I_0 = \frac{E_x \epsilon_e^2 \phi_0^2}{\mu h} \tag{9.58}$$

where $\phi_0 = RT/F$ has units of C/m sec = amp/m and so can be used for the current scale. Thus, in dimensionless form,

$$I^* = \frac{I}{I_0} = \Sigma + (\zeta F/RT)^2 \frac{\frac{\epsilon}{2}\sinh(2/\epsilon) - 1}{\cosh^2(1/\epsilon)} \tag{9.59}$$

where $\epsilon = 1/\kappa h = O(1)$ and $\Sigma = 2hE_x\langle\sigma_e\rangle/I_0$.

Thus, in general, the current in the channel consists of two parts: the molar flux due to the migration of ions induced by the electric field, called the *conduction current* (the first term in equation (9.59)), and the flux due to the bulk electro-osmotic flow of the electrolyte, the *convective current*. The convective current is displayed in Figure 9.11. Note that if the EDLs are thin ($\epsilon \to 0$) (recall we can do that because the cosh solution approaches the correct solution as $\epsilon \to 0$), the current due to the convective term is nonzero only within the EDLs on the walls and thus contributes little to the current density. That is why water, while a better conductor than air, is still a dielectric.

9.2.8 Electro-osmotic flow in an annulus

As in classical viscous flow, there are a number of canonical geometries for which the electro-osmotic flow field may be obtained, analytically and numerically. Electro-osmotic flow in a cylinder is covered in the exercises at the end of this chapter, and in this section, electro-osmotic flow through an annulus having fixed

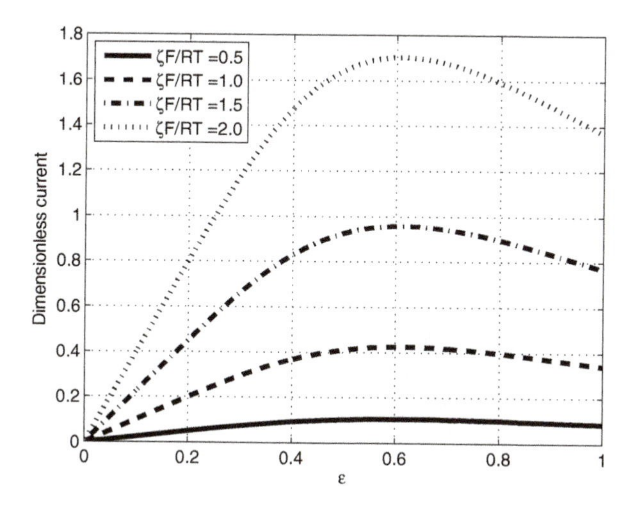

Figure 9.11 The convective current as a function of ϵ for several values of the dimensionless ζ potential.

walls is considered. If the inner cylinder is moving, this problem is a simplified model for the electrophoretic transport of a DNA molecule through a nanotube, discussed in detail in Chapter 12.

Notation alert: As you might have guessed by now, all the electro-osmotic flow problems discussed so far vary in a single dimension. Thus all the partial derivatives could be written as total derivatives, and as such, I have made no attempt to be consistent in which notation to use. Either set at notation works for me.

The dimensionless governing equations for the velocity and potential in fully developed electro-osmotic flow of a binary monovalent electrolyte, with cation g and anion f in a long annulus, shown in Figure 9.12, are

$$\epsilon^2 \left(\frac{\partial^2 \phi}{\partial r^2} + \frac{1}{r}\frac{\partial \phi}{\partial r} \right) = -\beta\,(g - f) \tag{9.60}$$

$$\epsilon^2 \left(\frac{\partial^2 u}{\partial r^2} + \frac{1}{r}\frac{\partial u}{\partial r} \right) = -\beta\,(g - f) \tag{9.61}$$

along with the boundary conditions that at $r = \alpha, 1$, $\phi = \zeta$, $u = 0$. Here g and f are mole fractions, and as before, $\beta = c/I$. These equations can be solved numerically in the nonlinear, high-potential case, and an analytical solution may

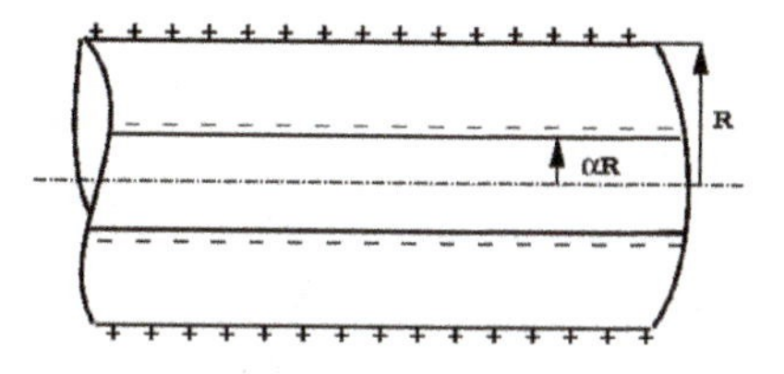

Figure 9.12 Geometry for flow through an annulus.

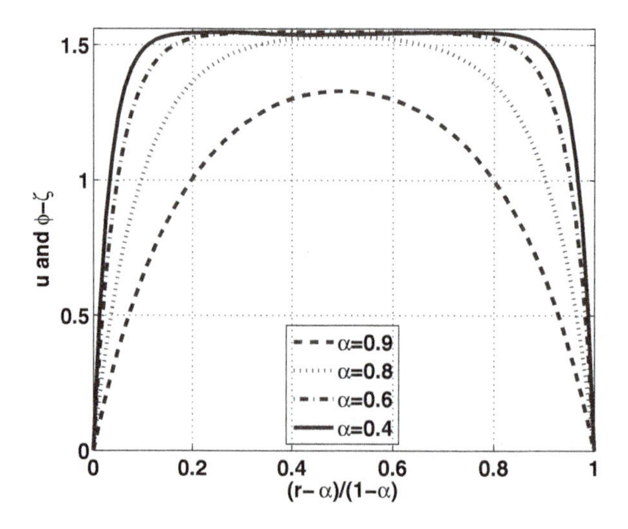

Figure 9.13 Results for the electro-osmotic flow in an annulus; here the potential and the velocity are equivalent. The different curves correspond to different values of $\alpha = a/b$, the ratio of the radii, and the constant surface potential boundary condition is used. Here $b = 50$ nm and $\Delta V/L = 0.6$ V/3.5 μm, and the ionic strength of each species is 0.1 M in the upstream reservoir.

be obtained in the DH approximation. Note that again, in dimensionless form, $u = \phi - \zeta$.

In DH limit, the governing equation for the potential is

$$\frac{1}{r}\frac{\partial}{\partial r}\left(r\frac{\partial \phi}{\partial r}\right) = \frac{\phi}{\epsilon^2} \tag{9.62}$$

This is a modified Bessel equation, and applying the boundary conditions at $r = \alpha$ and $r = 1$, the solution for the potential is

$$\phi(r) = \frac{\zeta I_0\left(\frac{r}{\epsilon}\right)\left[K_0\left(\frac{1}{\epsilon}\right) - K_0\left(\frac{\alpha}{\epsilon}\right)\right] + \zeta K_0\left(\frac{r}{\epsilon}\right)\left[-I_0\left(\frac{1}{\epsilon}\right) + I_0\left(\frac{\alpha}{\epsilon}\right)\right]}{I_0\left(\frac{\alpha}{\epsilon}\right)K_0\left(\frac{1}{\epsilon}\right) - I_0\left(\frac{1}{\epsilon}\right)K_0\left(\frac{\alpha}{\epsilon}\right)} \tag{9.63}$$

where I_0 and K_0 are modified Bessel functions of order 0 and $\epsilon = \lambda/b$.

The results for the potential and velocity are shown in Figure 9.13. The curves are similar to those seen in a channel; note that the EDLs appear to thin as $\alpha \to 0$. This is really not the case and is a result of normalizing the horizontal axis to be on the domain [0, 1]. If the inner cylinder is moving and the outer cylinder is fixed (the Couette flow problem), the velocity induced by the inner cylinder moving can be added linearly to the preceding velocity.

9.2.9 Electro-osmotic flow in nozzles and diffusers

Electro-osmotic flow in micro- and nanonozzles is important in many applications, for example, in electrical measurements on living cells (Lehnert *et al.*, 2002) and for the injection and manipulation of DNA fragments in a mass spectrometer (Luginbuhl *et al.*, 2000). Nozzles are also sometimes used for single molecule detection, analysis, and separation of biomolecules.

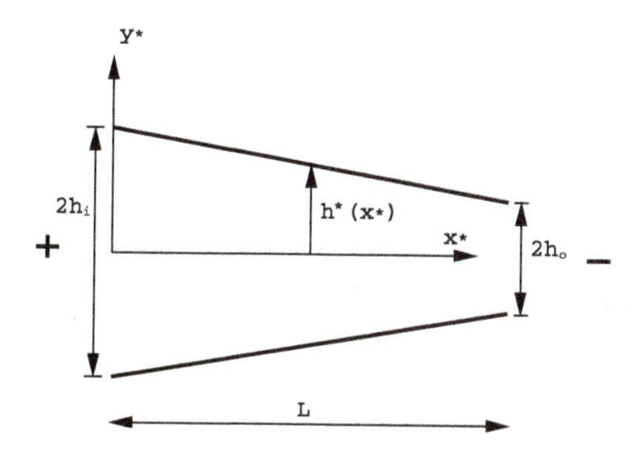

Figure 9.14 The geometry of a typical micro-nano nozzle; the cross section is assumed rectangular.

A sketch of a typical nozzle geometry is depicted in Figure 9.14. The nozzle (or diffuser) is assumed to have a rectangular cross section for simplicity, and the spanwise width of the channel is assumed to be much larger than the height of the channel, so the problem can be considered two-dimensional. The length of the channel is L, and the channel height is assumed to be linear; also,

$$h^*(x) = h_i + \frac{h_o - h_i}{L} x^* \tag{9.64}$$

where h_i, h_o are the half heights of inlet and outlet of the channel, and in dimensionless form,

$$h(x) = 1 + \frac{h_o - h_i}{h_i} x \tag{9.65}$$

where $h = h^*/h_i$ and $x = x^*/L$. The walls of the channel are defined by $y = \pm h(x)$. The imposed electric field $\vec{E}^*$ satisfies Maxwell's equations in the form $\nabla \bullet \vec{E}^* = 0$ so that if i denotes the imposed electric field, $E^*_{xi}(x)A(x) = E^*_{xi}(0)A(0)$, where A is the cross-sectional area of the channel. The total dimensionless electric field may be written

$$E_{Tx} = \frac{h_i E_0 E_x}{\phi_0} - \epsilon_1 \frac{\partial \phi}{\partial x} \tag{9.66}$$

$$E_{Ty} = \frac{h_i E_0 E_y}{\phi_0} - \frac{\partial \phi}{\partial y} \tag{9.67}$$

where $\epsilon_1 = h_i/L$. If the imposed electric field at the inlet $E_0 = E^*_i(0)$ is taken as the scale of the imposed electric field, the dimensionless imposed electric field in the x direction is $E_{xi} = 1/h(x)$.

Solutions for the potential and velocity are discussed under the DH approximation and the lubrication approximation for which $(h_i - h_o/L)\mathrm{Re} \ll 1$ in the case of a nozzle, with a corresponding restriction for a diffuser. The Reynolds number is based on the inlet height h_i; see Section 4.9 for the derivation of this restriction.

THIN EDLS: $\epsilon \ll 1$

For thin EDLs, the electro-osmotic velocity is similar to that in a straight channel, and

$$u_{\text{eof}} = \frac{\zeta}{h}\left(e^{-\left(\frac{h-y}{\epsilon}\right)} - e^{-\left(\frac{h+y}{\epsilon}\right)} - 1\right) \tag{9.68}$$

where ζ is the dimensionless potential on both walls; here, for negatively charged walls, $\zeta < 0$, and so $u_{\text{eof}} > 0$. Superimposing a pressure driven velocity, if necessary, the two-dimensional velocity field is given by

$$u = \frac{1}{2}\frac{dp}{dx}\left(y^2 - h^2\right) + \frac{\zeta}{h}\left(e^{-\left(\frac{h-y}{\epsilon}\right)} - e^{-\left(\frac{h+y}{\epsilon}\right)} - 1\right) \tag{9.69}$$

Because the cross-sectional area of the channel varies with x, there is a transverse velocity that is obtained by integrating the continuity equation

$$\begin{aligned} v = -\epsilon_1 \Bigg[&\frac{1}{2}\frac{d^2p}{dx^2}\left(\frac{y^3}{3} - h^2 y - \frac{2}{3}h^3\right) + \frac{dp}{dx}hh'(y+h) - \frac{\epsilon_1 \phi_w h'}{h^2}(y+h) \\ &+ \left(\frac{\epsilon \phi_w h'}{h^2} + \frac{\phi_w h'}{h}\right)\left(e^{-\frac{h-y}{\epsilon}} - e^{-\frac{h+y}{\epsilon}} - e^{-\frac{2h}{\epsilon}} + 1\right)\Bigg] \end{aligned} \tag{9.70}$$

The pressure is an unknown, as noted in Section 4.9, and is obtained by requiring that the continuity equation is satisfied; the result is a differential equation for the pressure, which is given by

$$\frac{1}{3}h^3\frac{d^2p}{dx^2} + h'h^2\frac{dp}{dx} - \frac{\zeta h'}{h}e^{-\frac{2h}{\epsilon}} + \frac{\zeta h'\epsilon}{h^2}\left(1 - e^{-\frac{2h}{\epsilon}}\right) = 0 \tag{9.71}$$

Assuming that the nozzle is connected to reservoirs, the boundary conditions for the pressure are taken to be

$$p = p_i, \quad x = 0 \quad \text{and} \quad p = p_o, \quad x = 1 \tag{9.72}$$

It should be noted that there is a rather extensive literature on the appropriate boundary conditions to be used in lubrication problems, and these are the simplest that can be considered (Pinkus & Sternlicht, 1961).

Integrating equation (9.71) once between the inlet of the channel and a given position x,

$$h^3\frac{dp}{dx} = 3\zeta\left(\frac{e^{-\frac{2h}{\epsilon}}}{h} - \frac{\epsilon}{h} - \bar{E}i\left(1, \frac{2h}{\epsilon}\right)\right) + C \tag{9.73}$$

where C is an integration constant and $\bar{E}i$ is the exponential integral given by

$$\bar{E}i(x) = -\int_{-x}^{\infty}\frac{e^{-t}}{t}dt \tag{9.74}$$

Equation (9.73) can be integrated for $\epsilon \ll 1$, neglecting the first term in equation (9.73) because it is exponentially small and using the asymptotic expansion

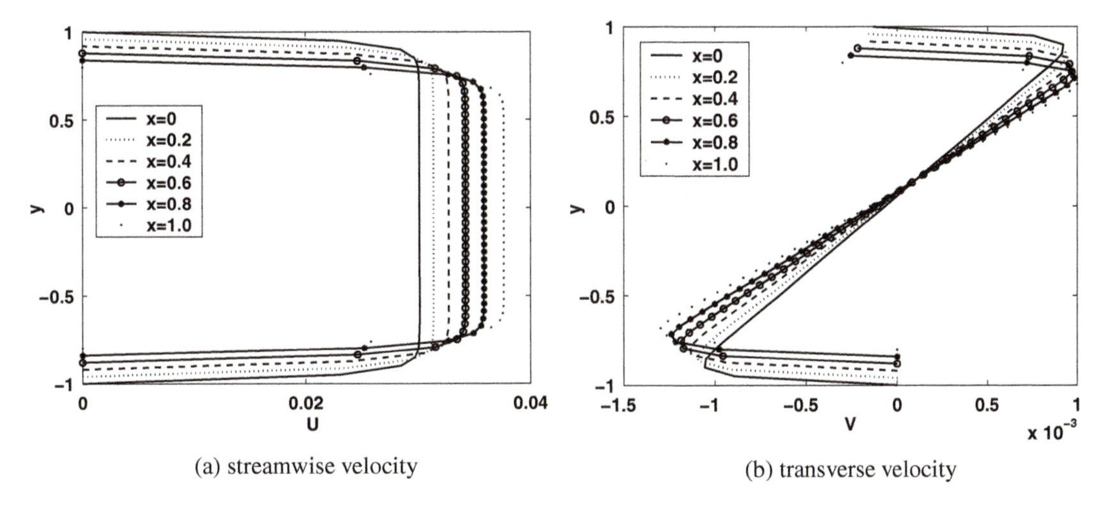

(a) streamwise velocity

(b) transverse velocity

Figure 9.15 The streamwise and transverse velocity of EOF in a microdiffuser. The height of the diffuser is 20 μm at the inlet and 130 μm at the outlet, and the length of the diffuser is 650 μm; $\epsilon = 5 \times 10^{-5}$, and the EDLs are thin compared to the diffuser. The imposed electric field is 8000 V/m, and the ζ potential is −15 mV. The pressure is zero both at the inlet and the outlet. The concentration of each electrolyte component in the upstream reservoir is 0.1 M.

for the exponential integral; applying the boundary conditions,

$$p(x) = \frac{p_o h_o^2(h^2 - h_i^2) + p_i h_i^2(h_o^2 - h^2)}{h^2(h_o^2 - h_i^2)} + \epsilon\left(\frac{\zeta}{h'h^3} - \frac{\zeta\,(h_i^2 + h_i h_o + h_o^2)}{h'h^2 h_i h_o(h_i + h_o)} + \frac{\zeta}{h' h_i h_o (h_i + h_o)}\right) \tag{9.75}$$

It is seen that the pressure distribution comprises two parts – that induced by the inlet and outlet pressure difference as well as the presence of the EDLs – and for thin EDLs, this correction is small. If there is no pressure difference ($p_i = p_o$), then the pressure field is due entirely to the presence of the electro-osmotic flow; the pressure is small, and $p = O(\epsilon)$.

Results for the electro-osmotic flow in a microdiffuser are shown in Figure 9.15. In this example, the height of the diffuser is 20 μm at the inlet and 130 μm at the outlet, and the length of the diffuser is 650 μm; the parameter $\epsilon^2 = 0.04 \ll 1$. Each electrolyte concentration is 0.1 M in the reservoir, and the ζ potential of the PMMA walls is −15 mV (Kirby and Hasselbrink, 2004c). The Debye length is 0.7 nm ($\epsilon = 5 \times 10^{-5}$), and the EDLs are very thin compared to the height of the diffuser. As required by continuity, the velocity decreases in the streamwise direction.

OVERLAPPED EDLs: $\epsilon \sim 1$

The solution for the potential in this case is formally the same as for the straight channel to leading order in ϵ_1, and

$$\phi = \zeta \frac{\cosh \frac{y}{\epsilon}}{\cosh \frac{h}{\epsilon}} \tag{9.76}$$

where y is measured from the centerline of the channel; the velocity is

$$u_{eof} = \frac{\phi}{h} = \frac{\zeta}{h}\left(\frac{\cosh \frac{y}{\epsilon}}{\cosh \frac{h}{\epsilon}} - 1\right) \tag{9.77}$$

The v velocity is obtained from the continuity equation, as discussed earlier, and

$$v = \epsilon_1 \left[-\frac{1}{2}\frac{d^2p}{dx^2}\left(\frac{y^3}{3} - h^2 y - \frac{2}{3}h^3\right) + \frac{dp}{dx}hh'(y+h) + \frac{\zeta h' \sinh\left(\frac{h}{\epsilon}\right)\sinh\left(\frac{y}{\epsilon}\right)}{h\cosh^2\left(\frac{h}{\epsilon}\right)} \right. \\ \left. + \frac{\epsilon\zeta h' \sinh\left(\frac{y}{\epsilon}\right)}{h^2\cosh\left(\frac{h}{\epsilon}\right)} - \frac{\zeta y h'}{h^2} + \frac{\zeta h' \epsilon_1 \tanh^2\left(\frac{h}{\epsilon}\right)}{h} + \frac{\zeta\epsilon_1\epsilon h' \tanh\left(\frac{h}{\epsilon}\right)}{h^2} - \frac{\zeta h'}{h} \right] \tag{9.78}$$

Again, using the continuity equation, the equation for the pressure is given by

$$\frac{1}{3}h^3\frac{d^2p}{dx^2} + h'h^2\frac{dp}{dx} = -\frac{\zeta h'}{h}\tanh^2\left(\frac{h}{\epsilon}\right) - \frac{\zeta h'\epsilon}{h^2}\tanh\left(\frac{h}{\epsilon}\right) + \frac{\zeta h'}{h} \tag{9.79}$$

At this point, further integration of this equation is difficult, and the solution for pressure has to be found numerically.

Because of the complexity of the pressure equation, a numerical calculation is used to obtain the pressure distribution within the nozzle. Second-order finite difference methods are used to discretize the pressure equation (9.79), as discussed in Chapter 10, and the Thomas algorithm is used to solve the resulting set of difference equations.

Figure 9.16 shows the electro-osmotic flow in a nanonozzle. The height of the nanonozzle at the inlet is 13 nm, and the length of the nozzle is 65 nm; the height at the outlet is 8 nm, and so the parameter $\epsilon_1^2 = 0.04 \ll 1$. The Debye length is 2.5 nm ($\epsilon = 0.19$), and the EDLs are overlapped, as shown in the parabolic potential and velocity profiles in Figure 9.16. The imposed electric field is 8000 V/m. The streamwise velocity is parabolic, and the transverse velocity is $O(\epsilon_1)$ and small compared to the streamwise velocity.

In Chapter 12, we will return to this problem because this device has applications in delivering DNA and other biomolecules to cells. An opening is created in the cell membrane, and material is often delivered to the cell through a nozzle of the type described here. The process of creating an opening in a living cell is called *electroporation*.

9.2.10 Dispersion in electro-osmotic flow

As was seen in the case of Poiseuille flow in Section 5.7, dispersion of a solute is characterized by a dispersion coefficient, which is indicative of how much a solute can spread, over and above molecular diffusion, as it travels down a channel. Dispersion of a solute is caused by the local variation of the fluid velocity.

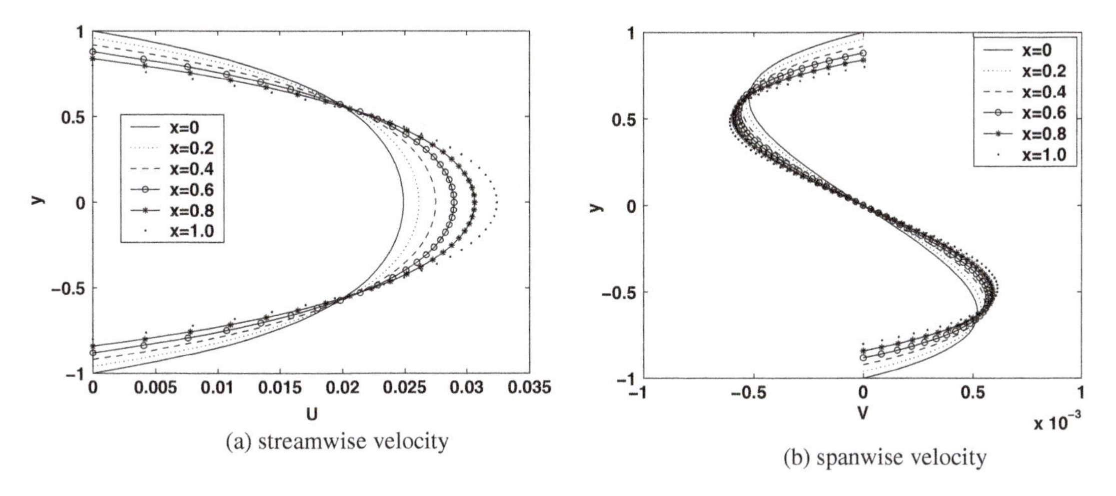

(a) streamwise velocity

(b) spanwise velocity

Figure 9.16 Results for electro-osmotic flow in the converging nanonozzle. The height of the nozzle is 13 nm at the inlet and 8 nm at the outlet, and the length of the nozzle is 65 nm; $\epsilon = 0.19$, and the EDLs are overlapped. The imposed electric field is assumed to be 8000 V/m, and the ζ potential of the walls is -5 mV. The pressure is zero both at inlet and outlet. The concentration of each electrolyte solution in the upstream reservoir is 0.1 M.

In the exercises of Chapter 5, you showed that for a very wide channel, the dispersion coefficient for an uncharged solute in a Poiseuille flow is

$$K = D_A \left(1 + \frac{2h^{*2} \langle u \rangle^2}{105 D_A^2} \right) \tag{9.80}$$

Dispersion also occurs in electro-osmotic flow when the EDLs are overlapped.

Notation alert: In this section, it is convenient to place the origin of the y coordinate at the centerline of the channel. In this case, the electro-osmotic velocity is slightly different from that when the coordinate origin is at the bottom wall, as in the Poiseuille flow in Chapter 4.

As with the current calculation, it is convenient to place the origin of the y coordinate at the centerline of the channel. Assuming that the DH approximation holds, the expression for the dimensional EOF velocity in a channel with the origin at the centerline of the channel having overlapped double layers is given by

$$u = \frac{\epsilon_e E_x \zeta}{\mu} \left(\frac{\cosh y}{\cosh h} - 1 \right) \tag{9.81}$$

where $y = y^*/\lambda$ and $h = h^*/\lambda = \kappa h^*$. The average velocity for this distribution is given by

$$\langle u \rangle = \frac{\epsilon_e E_x \zeta}{\mu} \left(\frac{\tanh h}{h} - 1 \right) \tag{9.82}$$

Griffiths and Nilson (1999) obtained the expression for the dispersion coefficient in the form

$$K = D_A \left(1 + \frac{1}{3} \left(\frac{\langle u \rangle \lambda}{D_A} \right)^2 f(h) \right) \tag{9.83}$$

where f is the function

$$f(h) = \frac{(6+h^2)\sinh^2 h - \frac{9}{2}h \sinh h \cosh h - \frac{3}{2}h^2}{(h\cosh h - \sinh h)^2} \tag{9.84}$$

Substituting for the average velocity, the dispersion coefficient becomes

$$K = D_A \left(1 + \frac{1}{3}\left(\frac{U\lambda}{D_A}\right)^2 \left[\frac{\tanh h}{h} - 1\right]^2 f(h)\right) \tag{9.85}$$

where $U = \epsilon_e E_x \zeta/\mu$ is a velocity scale based on the ζ potential. Note that the quantity $Pe_\lambda = U\lambda/D_A$ is a Peclet number based on the Debye length λ. In the limit $h/\lambda \to \infty$, that is, for very thin EDLs, the dispersion coefficient approaches the value

$$K \sim D_A \left(1 + \frac{1}{3} Pe_\lambda^2\right) \tag{9.86}$$

and as the Peclet number $Pe_\lambda \to 0$, there is no dispersion at all: $K = D_A$.

While we have calculated the dispersion coefficient for the case of overlapped double layers, dispersion can occur due to other effects. We have already mentioned Joule heating, but dispersion may also occur due to geometrical irregularities, such as curvature and roughness, and due to variation of channel properties, such as changes in ζ potential.

Field flow fractionation (FFF) (Giddings *et al.*, 1976; Griffiths & Nilson, 2006) is the term given to a variety of methods to separate and identify analytes based on size and charge. FFF generally uses a pressure-driven flow in the main streamwise direction and an applied field, such as electric and magnetic fields in the transverse direction, which is perpendicular to the primary flow direction. The result is that species will travel at different speeds, depending on their position relative to the walls of the channel, their size, and their charge. This leads to the formation of bands of particle motion similar to what occurs with ions in electro-osmotic flow (see Figure 9.11).

A field that requires a fundamental understanding of dispersion is *chromatography*, which is the term given in the chemistry community to the separation of mixture components flowing through a tube or channel. The carrier fluid can either be a liquid or a gas, and the solid medium is usually a porous material. In chromatography, adsorption of an analyte on the walls of a device is a central feature of the separation process.

The solutes separate into bands of sample constituents due to dispersion, each migrating with a speed characteristic of its relative interaction with the stationary and mobile phases. Thus an analyte with a strong interaction with the stationary phase moves more slowly than one with a weak interaction with the stationary phase.

Chromatographic techniques can be classified according to the state of matter of the mobile phase. In gas chromatography (GC) (Lee *et al.*, 1984), the mobile phase is a gas (the sample constituents have to be vaporizable), and in liquid chromatography (LC) (Snyder & Kirkland, 1979), the mobile phase is a liquid. In supercritical fluid chromatography (SFC) (Anton *et al.*, 1998), the mobile phase is a supercritical fluid.

9.3 Electrophoresis: Single particles

9.3.1 Introduction

The subject of electrophoresis, also known in the earlier literature as cataphoresis, is a vast one that has exploded in recent years with the development of lab-on-a-chip devices for separations and rapid molecular analysis. As discussed in Abramson (1931), electrophoresis seems to have first been described by F. Reuss in a paper he published in 1809. Reuss observed the motion of clay particles relative to the electro-osmosis of water through a quartz-sand medium. Smoluchowski (1918) and Henry (1931) made significant theoretical contributions to the study of electrophoresis. Today there are a number of books on the subject, Shaw (1969) for one, and there are a number of journals, including one appropriately named *Electrophoresis*.

Applications of electrophoresis permeate colloid science. Even in the absence of dispersion, charged molecules such as proteins, amino acids, DNA, RNA, polymers, and other molecules can be separated based on charge, shape, and size using electrophoresis. These separations can be performed in channels or cylindrical pores having uncharged walls (no electro-osmosis) or charged walls. Clearly uncharged molecules can also be separated from charged molecules.

Electrophoretic separation of colloidal particles is achieved by passing a sample in a solvent through that conduit that can be a channel or tube. Electrophoretic separations can occur in the absence or presence of flow. Of course, if flow is added, say, the walls are charged, separation will still occur in the manner described in Section 9.2.6.

There are different experimental variants of electrophoresis; *capillary electrophoresis* or *capillary zone electrophoresis*, using a tube to separate species based on different size and charge, is perhaps the most common. The solvent is often an aqueous solution, but higher-viscosity solvents, such as gels, may also be used, a process known as *gel electrophoresis*. Recall that a gel is a liquid made up of high-molecular-weight material, long-chain polymers; these polymers could consist of proteins, carbohydrates, and polysaccharides.

It should be noted that electrophoresis and electromigration are similar processes. The difference between the two is that in electrophoresis, the size of the molecule or particle is explicitly accounted for, usually based on the Stokes drag law, and the influence of the particle EDL is often included, as will be seen. Conversely, though the size of the particle in electromigration is accounted for in the Stokes–Einstein expression for the diffusion coefficient, the influence of the particle EDL is not included. Thus the term *electromigration* is usually reserved for ions and very small molecules dissolved in solution.

In this section, the fundamental theory of electrophoresis is presented, and the work of Smolukowski and Henry is covered in detail.

Electrokinetic Soil Remediation

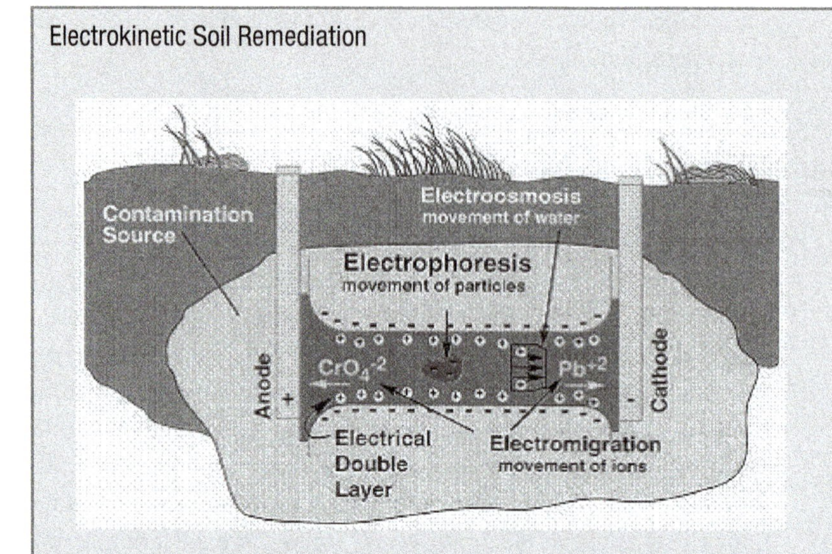

The electrokinetic remediation process. Image courtesy of Sandia National Laboratory.

Electro-osmotic flow can be used in soil remediation. Electrodes are placed in the treatment area as shown in the figure above and the flow through the porous soil media drives the cations toward the cathode and the anion toward the anode. Using this procedure, harmful negatively charged pollutants, such as heavy metals (e.g. chromium ions, CrO_4^{-2}) and polar organics in mud sludge, can safely be removed. In the figure the lead Pb^{+2} moves to the negatively charged cathode and the chromium moves to the positively charged anode.

Note that the electrokinetic remediation process in soil takes place in a porous media so that the EOF is through channels of arbitrary shape and tortuosity. Nevertheless, the basic physical principles involved in electrokinetic remediation are the same as that in a channel of rather uniform dimension.

9.3.2 Electrophoretic mobility

Consider the motion of a single charged particle in a neutral solvent in a stagnant fluid. The insertion of the particle will create an EDL of opposite charge to the surface of the particle to keep the system electrically neutral. Because of the presence of the EDL, there will be locally a net body force of (dimensional) magnitude $\rho_e E_x$, where E_x is the dimensionally applied electric field. If the particle is negatively charged, it will move in a direction opposite to the direction of the electric field, and the solvent will only be affected by the motion of the particle; that is, far from the particle, the solvent velocity vanishes.

Suppose that the Debye length is much larger than the particle radius, as depicted in Figure 9.17a. Then, equating the electrical force with the viscous drag using Stokes's drag law results in

$$qE_x = 6\pi\mu U_e a \tag{9.87}$$

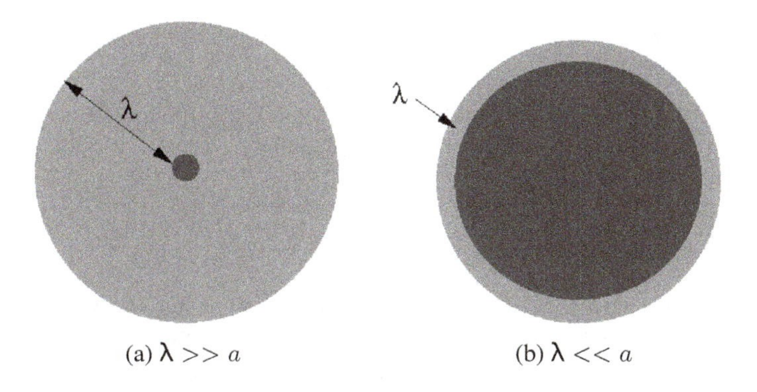

Figure 9.17 A particle moving in an electric field. (a) Debye length much greater than the particle radius. (b) Debye length much smaller than the particle radius, *a*.

where q is the net charge on the particle. Thus the *electrophoretic velocity* is given by

$$U_e = \frac{qE_x}{6\pi \mu a} \tag{9.88}$$

As with electro-osmotic flow, the *electrophoretic mobility* is defined as

$$\mu_e = \frac{U_e}{E_x} \tag{9.89}$$

Recall that the potential due to a charged sphere in the DH limit is given by

$$\phi(r) = \frac{a}{r}\zeta e^{(a-r)/\lambda} \tag{9.90}$$

The total charge on the sphere was calculated in Chapter 7, and the result is

$$q = 4\pi \epsilon_e a^2 \frac{d\phi}{dr}|_a^\infty = 4\pi \epsilon_e a^2 \zeta \left(\frac{1}{\lambda} + \frac{1}{a}\right) \tag{9.91}$$

Substituting for q into equation (9.88),

$$U_e = \frac{2}{3}\frac{\zeta \epsilon_e E_x}{\mu} \quad \text{for } \frac{a}{\lambda} \ll 1 \tag{9.92}$$

This is the so-called Debye-Hückel (DH) expression for the electrophoretic velocity (Figure 9.22a).

In the limit of very large a/λ, larger particles, and a thin double layer, curvature effects may be neglected. In this limit, the electrophoretic velocity is equivalent to the electro-osmotic velocity far from a flat surface, and we have already shown this velocity to be given by

$$U_e = \frac{\zeta \epsilon_e E_x}{\mu} \quad \text{for } \frac{a}{\lambda} \gg 1 \tag{9.93}$$

This is the Helmholtz–Smolukowsky expression (Figure 9.22b).

Now consider the case in which the particle is immersed in an electro-osmotic flow having a velocity U. Then the mixture will experience a net electrical force and will move in the direction of the electric field. Thus the balance of electrical and drag forces becomes

$$qE_x = 6\pi \mu(U_e - U) \tag{9.94}$$

and the particle can go either in the same direction as the electric field or in the opposite direction. The velocity U corresponds to the electro-osmotic velocity far from the particle. For thin double layers on the bounding surfaces of the channel, the electro-osmotic velocity U will be constant.

9.3.3 Henry's solution

In the previous section, several assumptions were made in the calculation of the electrophoretic velocity of a single colloidal particle. First, it was assumed that the local electric field were unaffected by the presence of the particle. This is unlikely to be the case in practice. Second, it was assumed that the dielectric constant of the particle and the fluid were the same. And finally, the electrophoretic velocity of the particle was calculated only for $\epsilon = \lambda/a \ll 1$ and $\epsilon \gg 1$. Henry (1931) removed these assumptions; his work is the subject of this section.

Notation alert: In this section, we will use the notation in the chemistry community for $\lambda = 1/\kappa$. As has been mentioned before, chemists like to refer to the ratio of the particle radius to the Debye length as κa.

Henry was concerned with the dependence of the electrophoretic velocity on the shape of the particle and considered both spheres and cylinders in his analysis. It is assumed that the Reynolds number is small and thus that the Stokes's flow assumption holds. The solution for the applied electric field is simply added to the solution for the EDL, as we have done previously. The potentials due to the applied field satisfy

$$\nabla^2 \psi = \nabla^2 \psi' = 0 \tag{9.95}$$

where

$$\nabla^2 = \frac{1}{r^2}\frac{\partial}{\partial r}\left(r^2 \frac{\partial}{\partial r}\right) + \frac{1}{r^2 \sin\theta}\frac{\partial}{\partial \theta}\left(\sin\theta \frac{\partial}{\partial \theta}\right) \tag{9.96}$$

and the potential has been assumed to be symmetric about the transverse angle, where the prime indicates the potential inside the particle. The boundary conditions are continuity of surface charge at the particle surface:

$$\epsilon_e \frac{\partial \psi}{\partial r} = \epsilon_e' \frac{\partial \psi'}{\partial r} \quad \text{at } r = a \tag{9.97}$$

Continuity of the potential requires

$$\psi = \psi' \quad \text{at } r = a \tag{9.98}$$

$$\psi = -E_x r \cos\theta \quad \text{as } r \to \infty \tag{9.99}$$

The solutions for the potentials are given by

$$\psi = -E_x \left(r + \frac{\xi a^3}{r^2}\right) \cos\theta \tag{9.100}$$

outside the particle and

$$\psi' = -E_x(1+\xi)r\cos\theta \tag{9.101}$$

inside the particle, where $\xi = (\epsilon_e - \epsilon_e')/(2\epsilon_e - \epsilon_e')$.

Turning to the momentum equation, the velocity field satisfies

$$\mu\nabla^2\vec{u} - \nabla p = \rho_e\nabla(\psi + \phi) \tag{9.102}$$

$$\nabla\cdot\vec{u} = 0 \tag{9.103}$$

where ϕ is the potential due to the EDL on the sphere. The sphere is assumed to be held fixed in the flow, and thus the boundary conditions in a coordinate system fixed to the sphere are

$$u_r = -U\cos\theta \quad \text{as } r\to\infty \tag{9.104}$$

$$u_\theta = U\sin\theta \quad \text{as } r\to\infty \tag{9.105}$$

$$u_r = u_\theta = 0 \quad \text{as } r = a \tag{9.106}$$

Henry obtained the solution for the velocity field in a rather elegant but somewhat lengthy manner. The most important result is the form of the force balance on the particle; this is given by

$$-6\pi\mu U_e a + \epsilon_e E_x\int_\infty^a \Gamma dr + \epsilon_e E_x a^2\frac{\partial\phi}{\partial r}|_{r=a} + qE_x = 0 \tag{9.107}$$

By Gauss's law, the last two terms cancel, and thus

$$U_e = \frac{\epsilon_e E_x}{6\pi\mu}\int_\infty^a \Gamma dr \tag{9.108}$$

where Γ is defined by

$$\Gamma = \frac{\partial\phi}{\partial r} + \xi a^3 r\int_\infty^r \frac{\nabla^2\phi}{r^4}dr \tag{9.109}$$

In general, a value for Γ can be computed from a numerical solution for the potential. Avoiding such a situation (remember that Blasius (1908) had computed a numerical solution to the boundary layer equations in 1908), Henry suggested that for a first approximation, the DH approximation could be used (equation (9.90)). In this case, performing the integrations and expanding for large $\kappa a > 20$,

$$U_e = \frac{\epsilon_e\zeta E_x}{\mu}\left(1 - 3\frac{1}{\kappa a} + 25\left(\frac{1}{\kappa a}\right)^2 - 220\left(\frac{1}{\kappa a}\right)^3 + \cdots\right) \tag{9.110}$$

and for $\kappa a < 5$,

$$U_e = \frac{2\epsilon_e\zeta E_x}{3\mu}\left(1 + \frac{\kappa a}{16} - \frac{5(\kappa a)^3}{48} - \frac{(\kappa a)^4}{96} + \frac{(\kappa a)^5}{96}\right.$$
$$\left. + \left(\frac{(\kappa a)^4}{8} - \frac{(\kappa a)^6}{96}\right)e^{\kappa a}\int_\infty^{\kappa a}\frac{e^{-t}}{t}dt\right) \tag{9.111}$$

It is seen that there is a gap in the validity of the two formulas between $\kappa a = 5$ and $\kappa a = 20$. Of course, these formulas are only valid for potentials less than 26 mV, although we saw previously in Chapter 7 that the DH approximation for

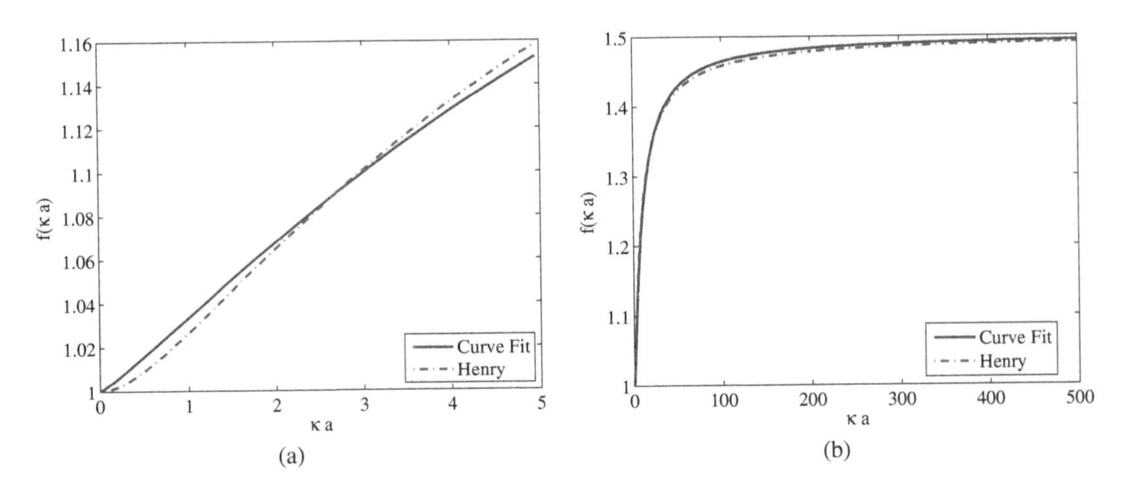

Figure 9.18 Plot of Henry's solution and the curve fit for different ranges of κa.

the potential is accurate at much higher potentials. Note that there is an error in Henry's original paper that is corrected by Dukhin and Derjaguin (1974). It is important to note that this solution neglects the induced asymmetry of the ionic cloud surrounding the particle as it moves through the fluid.

There is a useful curve-fit to Henry's formula, and this formula was developed by Masliyah and Bhattacharjee (2006). If Henry's formula is written in the form

$$\mu_e = \frac{U_e}{E_x} = \frac{2\epsilon_e \zeta}{3\mu} f(\kappa a) \tag{9.112}$$

then a simple but very accurate curve-fit is given by

$$f(\kappa a) = \frac{3}{2} - \frac{1}{2\left(1 + A\epsilon^{-B}\right)} \tag{9.113}$$

where $A = 0.072$ and $B = 1.13$; this function is plotted in Figure 9.18, along with Henry's solution for the entire range of κa.

It is seen that Henry's model for the electrophoretic velocity does not appear to depend on size for $\kappa a \gg 1$. This fact was actually demonstrated by Abramson (1931). He showed in a series of papers that shape does not affect the electrophoretic mobility of quartz, glass, and clay particles of irregular forms and of size varying from 3 to 15 μm. Moreover, Abramson (1931) also found that the electrophoretic mobility of leucocytes, blood platelets and red blood cells are the same as their irregularly shaped aggregates. This, he suggested, is because the particles get covered with an adsorbed layer of protein and so behave as if they have the same surface characteristics.

9.3.4 The full nonlinear problem

While Henry's solution is elegant and gives good results at low potentials, at high surface charge and potential of the particle, Henry's solution neglects some important effects. His theory does not account for the distortion of the counterion cloud surrounding the particle, as depicted in Figure 9.19, and cannot describe the high potentials that occur in the range $5 < \kappa a < 20$. Overbeek (1943)

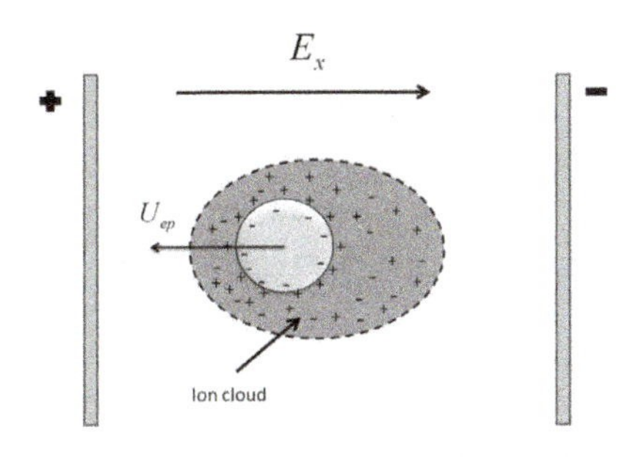

Figure 9.19 Electrophoretic motion of a colloidal particle at high potential, showing asymmetry of the counterion cloud.

and Booth (1950) both sought to remove this restriction, but their results are expressed as a power series in the ζ potential and thus can only describe the result for moderate values of the potential.

Wiersma *et al.* (1966) produced numerical solutions, but because of convergence problems, they could not produce results for a ζ potential greater than $\zeta = 125$ mV or a value of $1/\epsilon = \kappa a = 2.765$. O'Brien and White (1978) calculated the solution to the problem, avoiding the convergence issues encountered by Wiersema *et al.* For higher values of the potential, where the Henry and Wiersema *et al.* solutions are not valid, they found that a maximum occurs in the mobility for $\kappa a > 6$ and near a value of $F\zeta/RT \sim 5$. O'Brien and White's results are depicted in Figure 9.20. Recall that the results for $\kappa a \to \infty$ correspond to the DH expression for the electrophoretic velocity, while the case $\kappa a \to 0$ corresponds to the Helmholtz–Smolukowsky expression. The works of Wiersema *et al.* and O'Brien and White are for a positively charged particle; however, the results apply by simple sign changes to the case of a negatively charged particle.

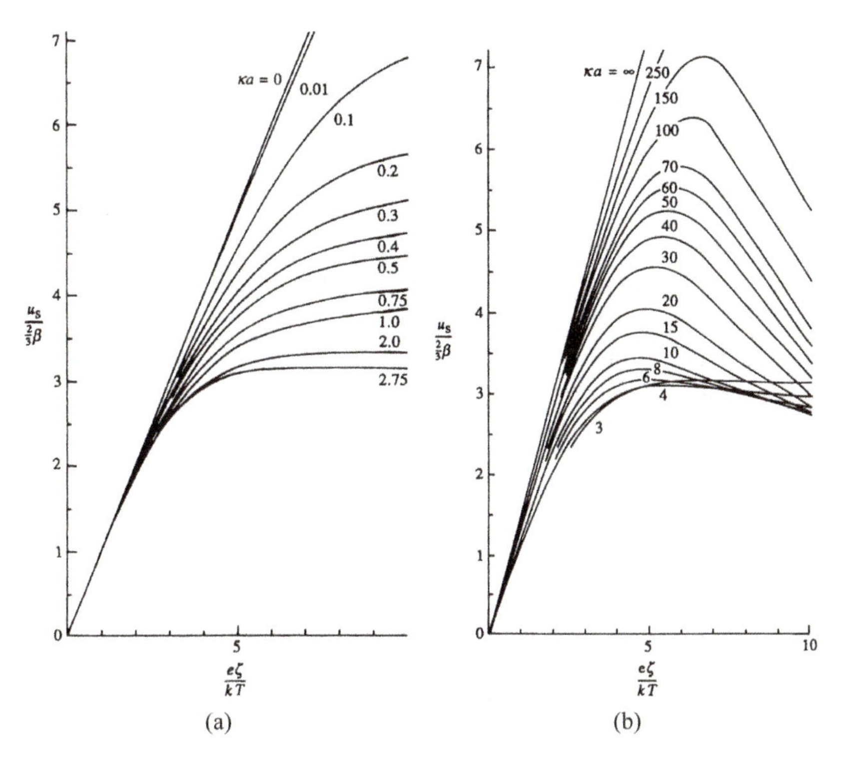

Figure 9.20 Electrophoretic mobility of a colloid particle from O'Brien and White (1978): (a) $\kappa a \leq 2.75$; (b) $\kappa a \geq 3$. Here $y = F\zeta/RT$, and E is the dimensional electric field.

The effect of multivalent ionic species was also considered by O'Brien and White. They found that the effect of counterion valence is substantial, with the mobility significantly decreasing as the counterion valence increases. The increase in counterion valence from 1 to 3 decreases the electrophoretic mobility by a little over a factor of 3; that is, the decrease is linear in valence.

9.4 Streaming potential

We now proceed to the third electrokinetic phenomenon, the streaming potential. When an electrically conducting fluid is passed through a channel or tube under an imposed pressure drop, there is an electrical potential difference set up as a result because the net current in the channel must vanish (i.e., there are no upstream and downstream electrodes). From the governing equations in the direction of flow in the fully developed region, the streamwise momentum equation is given by

$$\mu \frac{d^2u}{dy^2} = \frac{dp}{dx} - \rho_e E_x \tag{9.114}$$

where ρ_e is the volume charge density and E_x is the streamwise component of the induced electric field. Note that as we have discussed previously, the velocity is composed of a pressure-driven component and an electrically driven component. The *streaming potential* corresponds to the potential associated with E_x in equation (9.114). In this section, the manipulations will be performed in dimensional form, and it is left to the exercises to perform the analysis in dimensionless form.

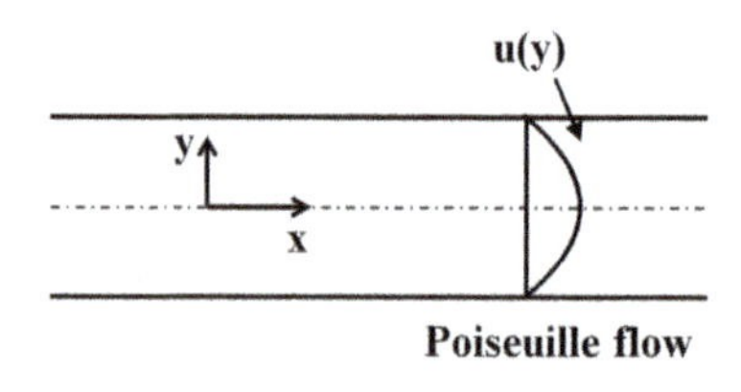

Figure 9.21 Geometry for the calculation of the streaming potential.

The geometry of interest is a channel that is wide and long compared to its height. The geometry is depicted in Figure 9.21. Note that the origin is taken at the center of the channel so that the height is $2h$ and the flow is assumed to be fully developed. This coordinate system is chosen so that a comparison with the result for the cylindrical tube given by Newman (1972) can be made easily.

Recall that the current density is defined by

$$\vec{J} = F \sum_i z_i \vec{N}_i \tag{9.115}$$

where $\vec{N}_i$ is the dimensional flux, given by

$$\vec{N}_i = -D_i \nabla c_i + \frac{D_i}{RT} z_i F c_i \vec{E} + c_i \vec{u} \tag{9.116}$$

Integating the streamwise component of the current density across the channel to obtain the total current, the result is

$$I = \int_{-h}^{h} J_x dy \tag{9.117}$$

Substituting for J_x, at zero current, the result for the induced electric field is given by

$$E_x \frac{F^2}{RT} \sum_i z_i^2 D_i \int_{-h}^{h} c_i dy + \int_{-h}^{h} \sum_i F z_i c_i u dy = 0 \tag{9.118}$$

where we have neglected the streamwise gradients of the concentration.

Further progress may be made by specifying the form of the velocity field. We know that the electrically driven component is proportional to the electrical potential, and so the velocity may be written as

$$u(y) = E_x \frac{\epsilon_e}{\mu}(\phi - \zeta) + \frac{1}{2\mu} \frac{dp}{dx}(y^2 - h^2) \tag{9.119}$$

Substituting into equation (9.118) and noting from the equation for electrical potential that $\sum_i F z_i c_i = -\epsilon_e \, d^2\phi/dy^2$, it follows that[2]

$$2h\langle\sigma_e\rangle E_x - E_x \frac{\epsilon_e^2}{\mu} \int_{-h}^{h} (\phi - \zeta) \frac{d^2\phi}{dy^2} dy - \frac{\epsilon_e}{2\mu} \frac{dp}{dx} \int_{-h}^{h} (y^2 - h^2) \frac{d^2\phi}{dy^2} dy = 0 \tag{9.120}$$

where $\langle\sigma_e\rangle$ is the average electrical conductivity of the solution defined in Section 9.2.7.

Finally, integrating the two integrals by parts, we obtain

$$2h\langle\sigma_e\rangle E_x + E_x \frac{\epsilon_e^2}{\mu} \int_{-h}^{h} \left(\frac{d^2\phi}{dy^2}\right)^2 dy + \frac{\epsilon_e}{\mu} \frac{dp}{dx} \int_{-h}^{h} y \frac{d\phi}{dy} dy = 0 \tag{9.121}$$

so that the induced streaming potential electric field is given by

$$E_x = -\frac{\partial \phi_{sp}}{\partial x} = -\frac{\frac{\epsilon_e}{\mu}\frac{dp}{dx} \int_{-h}^{h} y \frac{d\phi}{dy} dy}{2h\langle\sigma_e\rangle + \frac{\epsilon_e^2}{\mu} \int_{-h}^{h} \left(\frac{d\phi}{dy}\right)^2 dy} \tag{9.122}$$

Equations (9.121) and (9.122) are the analogues of the equation for a cylindrical tube given by Newman (1972), p. 194, equations (63-10).

As a final step in the analysis, note that

$$\int_{-h}^{h} y \frac{d\phi}{dy} dy = (y\phi)\,|_{h}^{h} - \int_{-h}^{h} \phi dy \tag{9.123}$$

$$= 2h\zeta - 2h\langle\phi\rangle \tag{9.124}$$

so that the electric field per width of the channel is given by

$$E_x = -\frac{\partial \phi_{sp}}{\partial x} = -2h \frac{\frac{\epsilon_e}{\mu}\frac{dp}{dx}(\zeta - \langle\phi\rangle)}{2h\langle\sigma_e\rangle + \frac{\epsilon_e^2}{\mu} \int_{-h}^{h} \left(\frac{d\phi}{dy}\right)^2 dy} \tag{9.125}$$

For thin EDLs, the convective portion of the total current is very small and of $O(\epsilon)$, where $\epsilon = \lambda/h$ in dimensionless terms because over much of the channel,

[2] Because fully developed flow is assumed, $\partial\phi/\partial y = d\phi/dy$, though we need not make this sharp distinction, as discussed previously.

the concentrations will be constant and $\sum_i z_i c_i = 0$ to preserve electroneutrality. Thus, for thin double layers, as with the electrical current discussed in Section 9.2.7, the streaming potential difference is very small.

The ζ potential can be measured by using the streaming potential. In a microchannel having thin EDLs, the potential will be a constant at a given cross section we can choose to be zero, and in this case, equation (9.125) becomes

$$E_x = \frac{\zeta \epsilon_e}{\langle \sigma_e \rangle \mu} \frac{dp}{dx} \tag{9.126}$$

With the definition $dp/dx = \Delta p/L$, solving for the ζ potential,

$$\zeta = \frac{\langle \sigma_e \rangle \mu \Delta \phi_{sp}}{\epsilon_e \Delta p} \tag{9.127}$$

Measurement of $\Delta \phi_{sp} = \phi_{\text{upstream}} - \phi_{\text{downstream}}$ and $\Delta p = p_{\text{upstream}} - p_{\text{downstream}}$ then yields the average ζ potential in the channel.

As an example, the transmembrane pressure drop across a nanopore membrane simulating the function of the kidney (Conlisk *et al.*, 2009) is about $\Delta p = 2$ psi. In a single slit pore making up the membrane of $2h = 8$ nm, in 0.1 M NaCl in plasma, a typical average flow speed for the pressure-driven component is $U = 1.83 \times 10^{-5}$ m/sec. For a surface charge density of -3×10^{-3} C/m^2, the ratio

$$\frac{|U_{sp}|}{U} = \frac{\epsilon_e RT |E_{sp}|}{\mu F U} \sim 0.0017 \tag{9.128}$$

so that the streaming potential is small. Note that the streaming potential velocity opposes the pressure-driven velocity.

Analogous to the streaming potential is the streaming current, for which the electric field vanishes rather than the current; a sketch of a streaming current setup appears in Figure 9.22. Consider the situation in which two electrodes inserted near the channel inlet and exit are connected outside the channel through resistanceless electrical connectors. In this situation, both electrodes are at the same potential, and no electric field can exist within the channel ($E_x = 0$ in equation (9.116)). The current I, originating from the net pressure-driven motion of charges within the channel, can be measured in the electrical circuit outside the channel. This is known as the *streaming current method* of calculating the ζ potential on the channel wall.

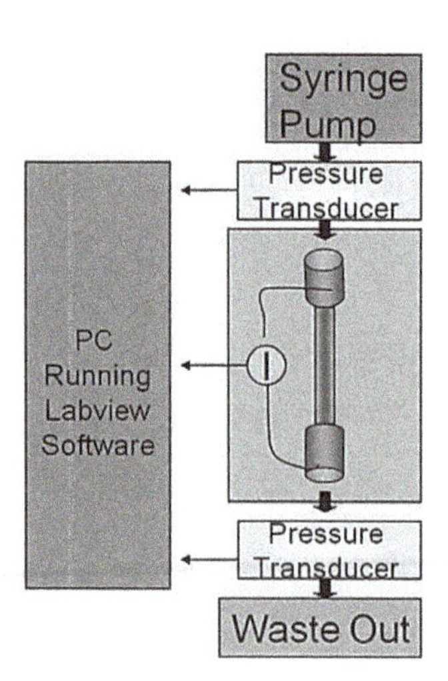

Figure 9.22 An experimental setup for a streaming current measurement of the ζ potential. Contributed by Professor Susan Olesik.

Through simplification of the expression for I obtained earlier in a manner similar to that in the streaming potential method for $\lambda \ll h$, the current per unit width of the channel is given by

$$I = \frac{2h\epsilon_e \zeta \Delta p}{\mu L} \tag{9.129}$$

Note that, unlike equation (9.127), the current predicted by the preceding equation has no dependence on the conductivity σ of the electrolyte. Thus no estimate of electrolyte conductivity is necessary for the streaming current method, unlike for the streaming potential method. However, more sensitive instrumentation may be necessary to measure the typically 1 nA/kPa of current generated at room temperature in the streaming current method than the typically 10–100 mV/kPa of voltage generated in the streaming potential method (Erickson & Li, 2001).

9.5 Sedimentation potential

The *sedimentation potential*, also called the *Dorn effect* (Masliyah and Bhattacharjee, 2006), is the potential induced by the fall of a charged particle under an external force field at zero current (Figure 9.23). It is analogous to the streaming potential in the sense that a local electric field is induced as a result of particle motion under the action of gravity or centrifugal fields. Here we discuss the case of a particle falling under the action of gravity. Just as in electrophoresis, the fall of a particle induces a streaming current caused by the distortion of the EDL around the particle.

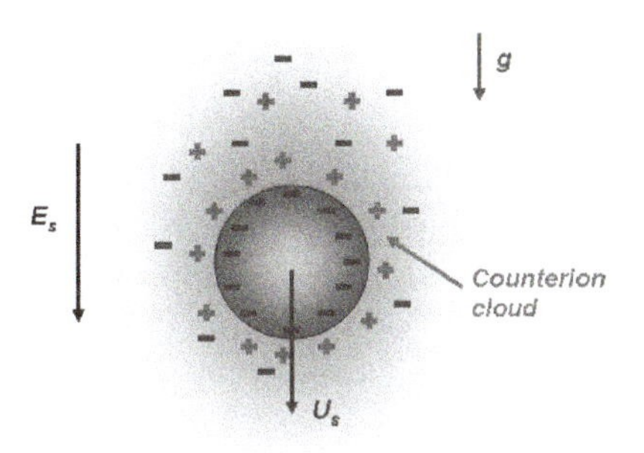

Figure 9.23 A charged particle falling under gravity, illustrating the sedimentation potential.

The governing equations for the sedimentation potential are the same as for electrophoresis of a particle, with the addition of the driving gravitational force term. Thus the methods of determining the sedimentation velocity of a charged particle are similar to those already described for electrophoresis. Moreover, the local velocity field is entirely analogous to the Stokes velocity profile discussed in Chapter 4 for an uncharged particle.

Recall that the sedimentation velocity of a single uncharged particle is given by

$$U_S = \frac{2}{9}\frac{a^2 g}{\nu}\left(\frac{\rho_p}{\rho} - 1\right) \tag{9.130}$$

If $\rho_p/\rho > 1$, then the particle motion is downward in the direction of gravity. Analogous to electrophoresis, a force balance on a charged particle leads to the sedimentation potential, given by

$$E_S = \frac{\epsilon_e \zeta (\rho_p - \rho) g}{\mu \sigma_\infty} \tag{9.131}$$

where σ_∞ is the electrical conductivity of the fluid far from the particle. In a dilute suspension of particle volume fraction α, the sedimentation potential is

$$E_S = \frac{\alpha \epsilon_e \zeta (\rho_p - \rho) g}{\mu \sigma_\infty} \tag{9.132}$$

Both formulas for the sedimentation potential are valid for $\kappa a \gg 1$ and for $\alpha \ll 1$.

As with electrophoresis, the effect of κa is not incorporated in the preceding simple formulas. The sedimentation of a charged particle has been analyzed in detail by Stigter (1980) numerically and by Ohshima *et al.* (1984). Ohshima *et al.* show, in particular, that the sedimentation velocity for a charged sphere is the same as that for an uncharged sphere in the limit of a thin EDL, as noted in Newman (1972) (i.e., $\lambda/a \ll 1$). In particular, the electric field is found to be given by

$$E_S = -\frac{\alpha \epsilon_e \zeta (\rho_p - \rho) g}{\mu \sigma_\infty} f(\kappa a) + \cdots \tag{9.133}$$

where α is the volume density of the particles and ζ is the ζ potential of the particles. Here $f(\kappa a)$ is a function similar to that derived for electrophoresis, and Ohshima *et al.* find that for $F\zeta/RT < 2$,

$$f(\kappa a) = 1 + 2e^{\kappa a} E_5(\kappa a) - 5e^{\kappa a} E_7(\kappa a) \tag{9.134}$$

where E_n are the exponential integrals of order n, defined by

$$E_n(x) = x^{n-1} \int_x^\infty e^{-t} \frac{dt}{t^n} \tag{9.135}$$

Note that equation (9.133) indicates that the electric field acts upward for positively charged particles ($\zeta > 0$) and downward for negatively charged particles ($\zeta < 0$), assuming that gravity acts downward in the positive coordinate direction.

9.6 Joule heating

As already mentioned, at small channel heights, electric fields are more efficient than pressure gradients in driving fluid flow. However, a by-product of applying an external electric field is that it acts as a source of energy, thereby causing unwanted temperature gradients within the channel; that is, Joule heating is the process by which electrical energy is converted to thermal energy because of a resistance to electrical current flow.

Variable temperature affects a number of properties, including viscosity, electrical conductivity, electrical permittivity, and thermal conductivity (Xuan *et al.*, 2004). Viscosity decreases with rise in temperature, and this change in viscosity may lead to increased dispersion.

The current flowing through a microfluidic device generates heat. The electrical power input per unit volume when current passes through a straight microfluidic channel of constant cross section carrying a flowing electrolyte is given by $P_{in} = J_x E_x$, where J_x is the current density from equation (9.48). The axial gradient of the concentration of each ion is significant only a few channel widths from the ends of the channel, and so the diffusional component of current can be neglected. Thus, the electrical power, is,

$$P_{in} = \sigma_e E_x^2 + \rho_e E_x u \tag{9.136}$$

where $\sigma_e = (F^2/RT)\sum_i z_i^2 c_i D_i$ is the electrical conductivity. The last term in equation (9.136) is mechanical work that makes no contribution to the change in thermal energy of the fluid.

Neglecting spanwise variations, the governing equation for the temperature is thus

$$\frac{d}{dy}\left(k(T)\frac{dT}{dy}\right) = \sigma_e(T)E_x^2 \tag{9.137}$$

Note that the thermal conductivity and the electrical conductivity are functions of the local temperature. The temperature dependence of the thermal conductiviy has already been discussed, and it is seen that

$$k(T) = A + BT + CT^2 \tag{9.138}$$

where A, B, and C are constants. For water, $A = -0.383$, $B = 5.254 \times 10^{-3}$, and $C = -6.369 \times 10^{-6}$. The variation of the electrical conductivity with temperature is linear and increases with increasing temperature. Usually, the dependence is of the form (Masliyah and Bhattacharjee, 2006)

$$\sigma_e(T) = \sigma_{e1}(1 + K(T - T_1)) \tag{9.139}$$

where K is a constant and $K = 0.02^\circ\,\mathrm{C}^{-1}$ for freshwater. Equation (9.137) can be solved numerically for the temperature given boundary conditions at the two walls.

Joule heating is a major factor in decreasing the efficiency of microdevices but does not seem to be a major factor in nanodevices. To see this, if the walls of channel dissipate the volumetric generation, then a heat balance on a control volume in the channel results in

$$2q_y A = \sigma_e E_x A h \tag{9.140}$$

from which it is seen that

$$q_y = \frac{1}{2}\sigma_e E_x h \tag{9.141}$$

That is, the heat flux required to be dissipated is much less for a nanochannel than for a microchanel. Indeed, Xuan *et al.* (2004) have shown that in microchannels, Joule heating affects the velocity distribution significantly, changing the classical concave EOF velocity distribution to a convex one. Steep temperature drops were also observed near the ends of the channel. Erickson *et al.* (2003) have investigated the Joule heating effect in PDMS systems and found significant increases in temperature because of the relatively low value of the PDMS thermal conductivity.

9.7 Chapter summary

This book is all about flows of electrically conducting fluids in small channels. As the smallest dimension approaches the nanoscale, it has been shown that pressure driven flow cannot be employed to move fluids.

Thus moving fluids around electrokinetically becomes much more effective. In this chapter, the four main electrokinetic phenomena have been described:

1. Electro-osmosis (electro-osmotic flow): the bulk motion of a fluid caused by an electric field
2. Electrophoresis: the motion of a charged particle in an otherwise motionless fluid or the motion of a particle relative to a bulk motion
3. Streaming potential: the potential induced by a pressure gradient at zero current flow; streaming current: the current induced by a pressure gradient at zero electric field
4. Sedimentation potential: the electric field induced when charged particles move relative to a liquid under a gravitational, centrifugal, or other force field.

Electrophoresis and sedimentation potential refer to particles, and electro-osmosis and streaming potential refer to electrolyte solutions.

By far the greatest emphasis in this chapter is on electro-osmosis and electrophoresis. We have derived the solution for the velocity and potential in a wide rectangular channel (most common in micro- and nanofluidics) for both thin and thick EDLs. The method of matched asymptotic expansions permits the derivation of the solutions for arbitrary values of the ionic valences for different values of the ζ potential on the walls, a situation made tractable by the development of modern fabrication procedures such as injection molding and laser ablation. The electrical current passed by such nanochannels in electro-osmotic flow has also been calculated. For thin EDLs, the current is dominated by the conduction current. We have discussed in particular how ions of different valence can move at different speeds in both micro- and nanochannels, a process called *electromigration*. Ions also move at different speeds due to dispersion, a process we discussed for uncharged molecules in Chapter 5.

Electrophoresis is discussed in this chapter from the point of view of a simple force balance for thin and thick EDLs to full numerical solutions of the governing equations. Of particular value is a curve-fit to the numerical results of Henry (1931), an analysis that spans all but a small range of the parameter κa developed by Masliyah and Bhattacharjee (2006).

We ended this chapter with short discussions of the sedimentation potential and Joule heating, the latter of which can lead to significant dispersion.

The electrical properties of small ions and small to large biomolecules can be analyzed, identified, mixed, separated, and manipulated in nanopore membranes in a variety of applications. Molecules can be separated based on their size and charge due to dispersion and due to their different electrophoretic mobilities; small ions can be separated based on their different electro-osmotic mobilities. These separation properties can be used to estimate a given unknown molecule's size and charge. The field of chromatography, the separation of molecules, particularly biomolecules, relies on basic knowledge of electro-osmotic flow and electrophoresis. It is the electrically conducting nature of the underlying fluid and the imbedded particles that gives rise to the applications described in Chapter 12.

EXERCISES

9.1 What is the best mechanism for achieving a flow rate of 10^{-6} μL/min in a 20 nm channel? Pressure driven or voltage driven? Assume a 1:1 electrolyte in an aqueous solution.

9.2 Calculate the velocity profile under the DH approximation for electro-osmotic flow through a cylinder of radius a. Assume a z:z electrolyte in an aqueous solution and that $\phi = \zeta$ at $r = a$. What is the primary effect of a larger valence?

9.3 An electro-osmotic flow is present between two plates. At $y = 0$, the potential is $\phi = \zeta_0$, and at $y = 1$, it is $\phi = \zeta_1$. If the pressure gradient is zero and the DH approximation is valid, find the relationship between ζ_0 and ζ_1 and other parameters such that the dimensionless volume flow rate $Q = 0$. Assume that the EDLs are thin, and calculate the flow rate through $O(\epsilon)$, where $\epsilon = \lambda/h$. Write this relationship in dimensional form as well as using the potential scale $\phi_0 = RT/F$.

9.4 A flow field induced by a combined pressure-driven and electro-osmotic flow is present between two parallel plates. The ζ potential is the same on the two plates. Under the DH approximation, find the relationship between the pressure gradient and the ζ potential such that the volume flow rate vanishes. Reproduce Figure 9.3 by solving for the electro-osmotic component numerically for h =20 nm for a reservoir concentration of 0.1 M = 0.05 + 0.05 M for electrolytes with valence $z = \pm 1$ and compare with the result in the DH limit.

9.5 Suppose an electrolyte mixture consists of three species (g, f, r) of valence $z_r = 2, z_g = 1, z_f = -1$. Show that in this case, the electrical neutrality condition in the core of the channel leads to

$$x^3 - \frac{g^0}{f^0}x - 2\frac{r^0}{f^0} = 0$$

where $x = e^{\psi_o} = u_o$, the outer solution for the potential and velocity, and the superscript zero denotes the mole fraction at the wall. Compare this result with the solution without species r. It has been observed in experiments that the electro-osmotic velocity can be significantly reduced with the addition of a multivalent cation in a negatively charged channel. If the ionic strength is kept constant, under what conditions will the velocity be significantly reduced.

9.6 A biomolecule of valence $z = -15$ and diffusion coefficient $D = 5 \times 10^{-11}$ m^2/sec are being transported in an aqueous solution in a rectangular channel in a sodium–chloride solution. The concentrations in the reservoir upstream are [Na] = [Cl] = 0.15 M and $[B] = 6.7 \times 10^{-4}$ M. Estimate the convection velocity of the biomolecule with respect to the elecro-osmotic velocity if the pressure gradient is negligible.

9.7 In dimensionless form, show that the electrical current density is conserved: $\nabla \bullet \vec{J} = 0$.

9.8 Using the singular perturbation techniques discussed in Appendix A, develop an expression for the outer solution for the velocity in a cylindrical tube.

9.9 Determine the electro-osmotic velocity profile in a fluid film on a plate that is being dragged out of an electrolyte solution at a velocity U_0. Assume that the shear stress vanishes at the free surface. Write the equation for the velocity in terms of the electrical potential. (Can this be done?) Assume that the surface charge density is fixed at the plate. What is the boundary condition on the potential at the free surface if the fluid outside the film is air?

9.10 An analytical solution for the flow in an annulus has been obtained analytically under the DH approximation. Solve the governing equations numerically for a pair of ions of valence $z = \pm 1$ and $\zeta = 100$ mV on each wall. Compare the result for the velocity with the DH approximation result.

9.11 Polystyrene beads are often used as simulants for biomolecules. Assume that the beads have a radius of $a = 40$ nm, the surrounding mixture is a NaCl solution having a total electrolyte concentration of 1 mM far from the particle, and a surface charge density of $\sigma = -0.01$ C/m^2. Assuming that the electric potential outside the particle is

$$\phi(r) = a/r\zeta e^{(a-r)/\lambda}$$

estimate the electrophoretic velocity of the particle. Compare this value with the value obtained from an electromigration calculation of 1 mm/sec.

9.12 A mixture contains two species of colloidal particles that are negatively charged and spherical. The concentration of the 1:1 aqueous electrolyte is 0.03 M = 0.015 + 0.015 M, and the one species of colloidal is 100 nm in diameter and has a ζ potential of $\zeta = -40$ mV. The other species has a diameter of 20 nm and has $\zeta = -20$ mV. Estimate the time it takes to travel a distance $L = 100$ μm in a channel of height $h = 5$ μm.

9.13 Show that the equation for the streaming potential in a cylindrical capillary of radius a is given by

$$E_x = -\frac{\partial \phi_{sp}}{\partial x} = -\frac{\epsilon_e a^2}{\mu}\frac{dp}{dx}\frac{\zeta - \phi_{av}}{a^2\langle\sigma_e\rangle + \frac{\epsilon_e^2}{\mu}\int_0^a \left(\frac{d\phi}{dr}\right)^2 2r\,dr}$$

Write the equation in dimensionless form as

$$\frac{U_{sp}}{U_p} = \frac{\zeta - \langle\phi\rangle}{\frac{\mu\langle\sigma_e\rangle}{\epsilon_e^2\phi_0^2} + \int_0^a \left(\frac{d\phi}{dr}\right)^2 r dr}$$

where U_p is the velocity scale for the pressure-driven flow and U_{sp} is the streaming potential velocity based on E_x.

9.14 Derive the equation for the streaming potential in a wide channel, including slip at the wall. Where does the slip length appear in the equation? Does it increase or decrease the streaming potential?

9.15 Consider a pressure difference $\Delta P = 2$ psi across the length $L = 4$ μm of a nanochannel with slit-shaped pores of height $2h = 10$ nm and width $W = 45$ μm connected to reservoirs containing 0.143 M sodium–chloride solutions. The surface charge density on each nanochannel wall is $\sigma = 2.8 \times 10^{-3}$ C/m^2. Find the streaming potential drop $\Delta\phi$, assuming that the EDLs do not overlap and the flow rate through the nanochannel in picoliters per second (pico $\equiv 10^{-12}$). What is the the percentage reduction in the flow rate due to the streaming potential effect?

9.16 As mentioned in the text, another way to measure the ζ potential is by the streaming current method. Show explicitly that the current is given by

$$I = \frac{2h\epsilon_e\zeta\Delta p}{\mu L}$$

and thus the ζ is determined by the properties of the electrolyte and the measured current.

9.17 Compare the dispersion coefficient for pressure-driven flow and for electro-osmotic flow. Calculate the ratio

$$\gamma = \frac{\left(\frac{K}{D_A} - 1\right)_{\text{PR}}}{\left(\frac{K}{D_A} - 1\right)_{\text{EOF}}}$$

and show that its value is

$$\gamma = \frac{105 f(\kappa h)}{24(\kappa h)^2}$$

where f is the function defined by equation (9.84). Find the numerical value of both the EOF and the dispersion coefficients for $\kappa h = 1$ and $\kappa h = 100$. Which type of flow is better for achieving minimum dispersion?

9.18 This is an open-ended exercise. Write a research report on nonlinear electrokinetic phenomena. Pay special attention to electro-osmosis of the second kind and induced charge electro-osmosis. How are they similar and/or different?

10 Essential Numerical Methods

10.1 Introduction

Mechanical engineers design new products for consumer use: engines for automobiles, airplanes, and other devices; cars; air-conditioners; heat pumps; compressors; fans; hair dryers; and all sorts of other products. Increasingly, mechanical and chemical engineers are involved in the design of biomedical devices for drug delivery systems, biochemical sensing, and rapid molecular analysis. In this chapter, the basic numerical techniques used to provide design and performance criteria of these devices are described. A simplified view of a general design process is depicted in Figure 10.1.

In this chapter, we shift gears a bit and discuss some basic concepts associated with numerical methods.[1] These methods are required when no simple analytical solution is possible. What is meant by the term *analytical* is that no solution can be found in terms of simple functional forms such as the polynomial, trigonometric, exponential, logarithmic, or hyperbolic functions. In the following, much attention is focused on the basic methods required to solve a nonlinear ordinary differential equation. This requires several different capabilities:

- Numerical differentiation
- Solving sets of linear(ized) equations
- Numerical integration

There are many situations for which numerical methods are required in micro- and nanofluidics. For example, determining the dependence of the ζ potential on pH in Chapter 7 requires a numerical zero-finding technique. The non-linear Poisson equation for the electrical potential discussed in the preceding chapter requires a numerical solution for the potential. In general, the solution of the potential equation for multicomponent and multivalent mistures requires a numerical solution. Indeed, the same is true of the momentum equation in these cases if the analogy between the velocity and potential cannot be invoked.

[1] This chapter is dedicated to Professor James David Allan Walker, who taught me all I know about numerical analysis.

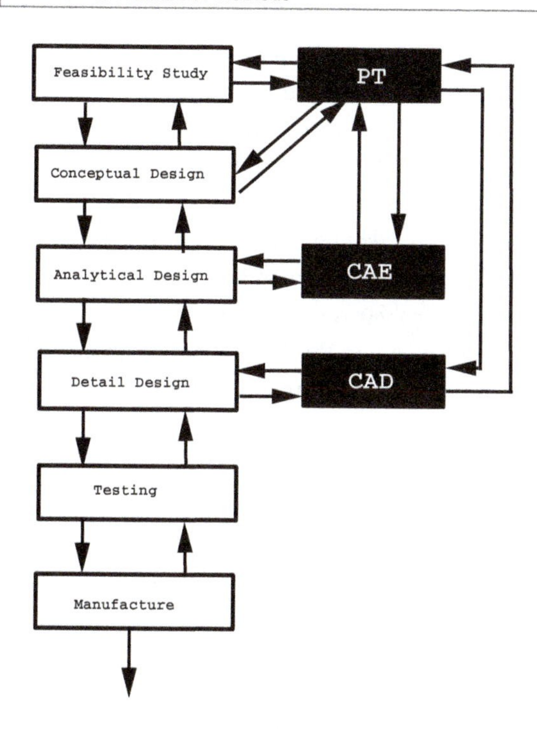

Figure 10.1 A simplified view of a typical device design process. The abbreviation PT means "programming tool," which could be a self-contained toolbox such as Matlab or a self-written computer program such as FORTRAN or C++.

I am sure that the astute reader will note that experimental data acquired in micro- and nanofluidics will require processing using curve-fitting techniques, just like at the macroscale.

The foundation of numerical analysis is the concept of function approximation embodied in the Taylor series. Any continuous and differentiable function has a Taylor series valid locally around a given point, and this concept is discussed first. This is followed by the discussion of zero finding and interpolation and the most common means of comparing with sets of experimental data: curve fitting.

Numerical differentiation and integration formulas are then derived from the Taylor series approximation to a given function. Next, the various methods of solution of linear systems of equations, which arise in the discretization of ordinary differential equations of the boundary value type, are presented. Next is a discussion of the solution of initial value problems so common in problems involving chemical and biochemical kinetics and in the solution of problems governed by a constant Hamiltonian (i.e., constant energy). An example is the numerical solution of the many-body problem that is the core of molecular dynamics simulations; these problems often require the energy to remain constant, a constraint that is satisfied by what are called symplectic integrators. The concept of symplectic integrators is seldom, if ever, discussed in numerical methods courses taught in engineering.

Following this section, the numerical solution of the Poisson–Nernst–Planck system is described as an example. Finally, the chapter ends with a short discussion of partial differential equations and a section on verification and validation of numerical solutions, essential tools in determining the accuracy of a model for describing complex physical systems.

In presenting the material in this chapter, some judgment had to be made as to which methods to include. The basic methods in each area described here work for the vast majority of problems that the student is likely to encounter, and these are small enough to fit nicely in the Matlab environment. The only area of numerical analysis that is not covered here is eigenvalue methods, primarily for lack of space.

There are many books on this subject, and good introductions to these topics, including eigenvalue problems, are given in the books by Fausett (2008) and Gilat and Subramaniam (2008), written for undergraduate engineers. There are many more advanced-level books that include an introduction to computational fluid dynamics, and these texts are far beyond the scope of this book. A Google search for "numerical analysis" will yield thousands of hits, including archival papers discussing special numerical methods for more complicated problems. These types of methods are not discussed here.

For many students, this chapter may be a review and can be skipped. Conversely, because numerical analysis permeates much of the material in this book, this chapter is still useful as a reference for those students whose memory has faded a bit.

Throughout this chapter, example Matlab files will be introduced, which the students can use to begin writing their own "m-files." In addition, for most of the subjects addressed in this chapter, Matlab has its own built-in modules. These are discussed where appropriate as well.

10.2 Types of errors

Before getting into Taylor series and numerical differentiation, we note that there are several types of errors associated with numerical computation. The two types of fundamental errors are the following:

1. Round-off error: this is the error incurred when a computer formulates a floating point number (2.36540) with a fixed word length. On most computers, each number is represented by a 64-bit word length.
2. Truncation error: this error is associated with the particular mathematical formula chosen, a formula usually based on truncation of a Taylor series. This type of error can and should be minimized, within reason.

The first type of error is associated with how a computer represents a number. Numbers on a computer are represented in base 2, and thus all of the numbers in this representation are zeros and ones. For example, the number 6.25 is represented in base 2 as

```
2^4  2^3  2^2  2^1 . 2^{-1}  2^{-2} 2^{-3}
 0    0    1    1  . 0        1      0
 0  + 0  + 4 +  2   + 0   +  .25  +  0
```

Actually, there are many more spaces for digits, and in modern computers, 64 bits are actually available for number representation. Of these 64 bits, 52 spaces are used for the number, 11 for the exponent, and 1 for the sign of the number (Gilat and Subramaniam, 2008; Fausett, 2008). Very large numbers and very small numbers are difficult to represent because of the finite number of digits (52) that can represent the number on a 64-bit machine. A large number may need to be chopped or rounded; for example, 2/3 can be written as 0.6666 or 0.6667. In the first case, the number is chopped, and in the second case, the number is rounded up. The error associated with these procedures is termed *round-off error*. Round-off errors do not often arise in the type of computations described in this book.

Conversely, truncation errors are associated with the order of numerical approximation to a given function or functional operation such as differentiation. For example, truncation error occurs in representing a function by a three-term Taylor series because a Taylor series contains an infinite number of terms. Truncation errors will be discussed extensively in this chapter.

In addition, there are other, more subtle types of errors that involve some decisions on the part of the practitioner. These errors include the following:

- Experimental error that shows up in the original data set: this is most often associated with the uncertainty in experimental data (for which there should be error bars) that may need to be curve fit
- Blunders or programming errors
- Propagated error or error that builds up as the calculation progresses: this is the case in solving first-order ordinary differential equations (ODEs) and parabolic partial differential equations (PDEs).

In this chapter, we will be most concerned with truncation errors, blunders, and propagated errors. Round off error is usually not a factor, but we need to know that it exists.

10.3 Taylor series

So much of numerical analysis is based on Taylor series that it is useful to review it now. Taylor series play a crucial role in zero finding, interpolation, differentiation, and integration.[2]

[2]For many students, this section may be skipped.

Table 10.1. Taylor series of $f(x) = e^x$

x	e^x	1 term	2 term	3 term
0.3	1.3499	1.0000	1.3000	1.3450
0.2	1.2214	1.0000	1.2000	1.2200
0.1	1.1052	1.0000	1.1000	1.1050

The Taylor series of a function $f(x)$ is a polynomial representation such that the coefficients are specified in terms of the derivatives of the function $f(x)$ in a given range of the independent variable x. Suppose that the behavior of a function, the Taylor series of a function about the point $x = c$, is required. Then the Taylor series of f is defined as

$$f(x) = f(c) + f'(c)(x - c) + f''(c)\frac{(x - c)^2}{2!} + f'''(c)\frac{(x - c)^3}{3!} + \cdots \quad (10.1)$$

This is an infinite series; obviously, an infinite number of terms cannot be calculated, and the series will be truncated at the particular number of terms that will give the accuracy required. This truncation process leads to the so-called truncation error. Thus, for a series of $N + 1$ terms,

$$f(x) \cong \sum_{k=0}^{N} \frac{f^{(k)}(c)}{k!}(x - c)^k \quad (10.2)$$

where $f^{(0)}(x) = f(x)$ and N is an integer. Very often, if c is close to x, then N can be as small as 2 or 3.

For example, consider the exponential function $f(x) = e^x$. Here take $c = 0$, and the approximating series is known as the *MacLaurin series*. Here $f(0) = 1$; $f'(0) = 1$ and $f''(0) = 1$, and so

$$e^x = 1 + x + \frac{x^2}{2} + \cdots \quad (10.3)$$

Now suppose it is necessary to find the value of $e^{.3}$ using the Taylor series. This is in anticipation that the values of much more complicated functions may be required. Table 10.1 shows the result of a calculation. Note that the closer the evaluation point is to zero – the expansion point – the better the approximation.

The Taylor series of a function is really a way to approximate the function in a particular region. To be accurate, the Taylor series must converge, and so the point of evaluation, that is, x, must be somewhat near the expansion point c. We should not calculate e^1 using the preceding MacLaurin series around $x = 0$ unless we are preppared to take a large number of terms or take c closer to $x = 1$. Indeed, a Taylor series converges rapidly near the point of expansion, but it may converge slowly or not at all far from that point.

When $f(x)$ is given as a numerical function, values of $f(x)$ between the given points may be calculated using Taylor series. Table 10.2 shows a set of data that is the output of a set of experiments to measure temperature at several points in a

Table 10.2. Temperature as a function of position in a solid

$x(m)$	0	0.5	1.0	1.5	2.0
$T(^oC)$	15.00	12.70	10.00	7.32	5.00

solid. The heat flux q or the heat transfer rate per unit area is proportional to the derivative of the temperature, with the proportionality constant being the thermal conductivity. To calculate the heat flux at $x = 1$ m, the Taylor series is calculated about $x = 1$ m; that is, $c = 1$ m. Then

$$T(x) = T(1) + T'(1)(x - 1) + \cdots \tag{10.4}$$

What is x in this equation? Well, we can take x to be any one of the values in Table 10.2, but it is better that it be close to $x = c = 1$ m. So let us take $x = 1.5$ m, and then the only unknown in the equation is $T'(1)$, which is proportional to the heat flux there. Thus, after truncating the Taylor series after two terms,

$$T'(1) \sim \frac{T(1.5) - T(1)}{1.5 - 1} = \frac{7.32 - 10}{0.5} = -5.36\,\frac{\text{C}}{\text{m}} \tag{10.5}$$

This is only an approximation, but it will get better as the $\Delta x = 0.5$ gets smaller. As will be seen later, this is the *forward difference approximation* to the derivative.

10.4 Zeros of functions

10.4.1 Numerical methods

Consider the following question: for a given function $f(x)$, for what values is the equation $f(x) = 0$? There are a wide range of applications for which this is necessary. In most applications, by preliminary analysis of the problem, the approximate location of the zero(s) will be known, and this feature will help in finding the solution. An example of a function with two zeros is depicted in Figure 10.2. In this section, methods to find the real zeros of functions are discussed.

As an example of the need for finding a zero of a given function, consider the *Redlich–Kwong equation of state* as an extension of the ideal gas law (Moran & Shapiro, 2007), which is

$$p = \frac{RT}{\bar{V} - b} - \frac{a}{\bar{V}T^{1/2}(\bar{V} + b)} \tag{10.6}$$

where a and b are constants that depend on the gas considered. Here $\bar{V}$ is the molar volume in m^3/kmole. The constant b is intended to account for the finite volume of the molecules neglected in the perfect gas model, and the constant a is meant to account for intermolecular forces. This equation is substantially empirical in nature, with a and b determined by curve fits to experimental data,

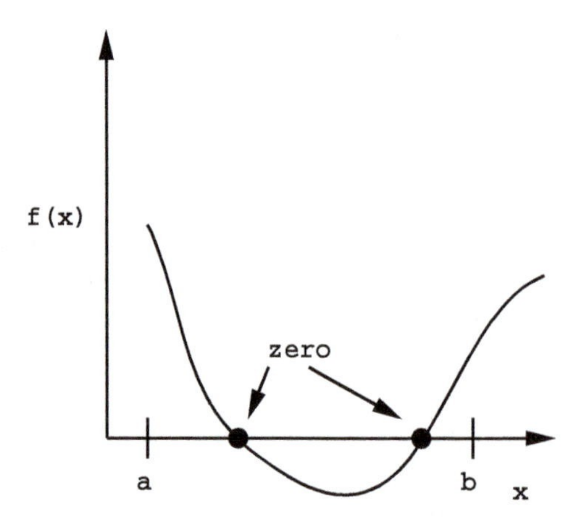

Figure 10.2 Sketch of a function with two zeros.

and so, empirically, it can also be used for liquids. For example, in metric units for water,

$$a = 142.59 \text{ bar} \left(\frac{\text{m}^3}{\text{kmole}} \right) \text{K}^{1/2} \quad \text{and} \quad b = .021 \frac{\text{m}^3}{\text{kmole}} \tag{10.7}$$

Note that given the pressure and temperature, the molar volume $\bar{V}$ cannot be determined analytically. After multiplication by the two denominators, equation (10.6) becomes a polynomial of order 3. For this case, there are explicit formulas for the roots of a third-order polynomial of the form

$$x^3 + Ax^2 + Bx + C = 0 \tag{10.8}$$

and these are given in a table in Appendix B.3 of the book by Murray (2001). The form of the roots can get complicated as the order of the equation increases, and in many cases, it is easier to calculate the roots numerically.

Three methods of numerically finding zeros are described here:

- Bisection
- Secant method
- Newton's method

Before implementing any of these methods, it is useful to graph the function because in any of these methods, a good initial guess is crucial to rapid convergence. Only the case of simple zeros is discussed here, in which

$$f(x) = (x - x_0)F(x), \quad F(x_0) \neq 0 \tag{10.9}$$

BISECTION

The idea in this method is to search for an interval in which $f(a) \times f(c) < 0$, in which case, for any continuous function f, there must be a zero between a and c. This situation is depicted in Figure 10.3. Assuming that a zero has been bounded

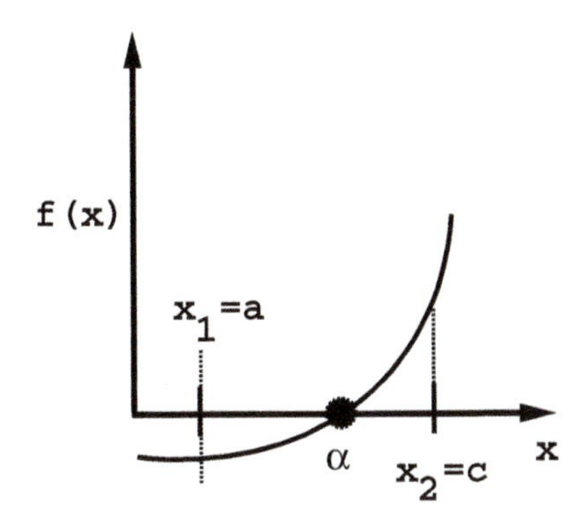

Figure 10.3 Sketch of a function with a single zero. Here α denotes the location of the zero.

by graphing, for example, a pseudo-Matlab script might go something like this:

```
Set x_{1}=a and x_{2}=c
Set x_{3}=\frac{(x_{1}+x_{2})}{2}
Calculate f(x_{3})
Is f(x_{3}) < \epsilon  where \epsilon is a pre-assigned
tolerance?
If yes, stop and the zero, f(\alpha)=0, \alpha=x_{3}
If no, is f(x_{1}) \times f(x_{3}) <0?
If yes, x_{2}=x_{3} go back to line 2
If no, x_{1}=x_{3} go back to line 2
```

The advantage of bisection is that it is simple; the disadvantage is that convergence is slow.

Notation alert: In this and other sections in this chapter, the symbol ϵ is used for the numerical preassigned tolerance. A good starting value is $\epsilon = 10^{-4}$.

SECANT METHOD

In this method, the calculation of x_3 (Figure 10.4) is different. Passing a straight line through $x_1 = a$ and $x_2 = c$,

$$\frac{f(x) - f(x_2)}{x - x_2} = \frac{f(x_2) - f(x_1)}{x_2 - x_1} \tag{10.10}$$

and evaluating at $x = x_3$, where $f(x_3) = 0$,

$$x_3 - x_2 = -\frac{(x_2 - x_1)f(x_2)}{f(x_2) - f(x_1)} \tag{10.11}$$

Note that the zero need not be bounded for the secant method to work. The advantage of this method is that it is faster than bisection, and the only calculation that is different from the bisection procedure is the way x_3 is calculated.

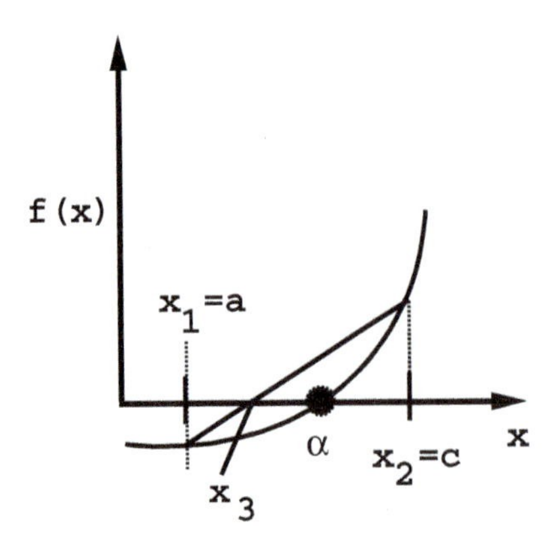

Figure 10.4 The secant method.

NEWTON'S METHOD

Here again assume that a zero exists at x_3 and that the guess, x, is close to x_3. In Newton's method, the zero need not be bounded and we take the equation of the straight line tangent to the curve at x to generate an approximation for x_3 (Figure 10.5). Then, expanding the function f in a Taylor series and chopping after one term, the approximation to the root is

$$f(x_3) = f(x) + f'(x)(x_3 - x) \tag{10.12}$$

Because $f(x_3) = 0$, the estimate of the zero is

$$x_3 = x - f(x)/f'(x) \tag{10.13}$$

Putting this equation in iterative form, the $n + 1$st iterate is given by

$$x_{n+1} = x_n - f(x_n)/f'(x_n) \tag{10.14}$$

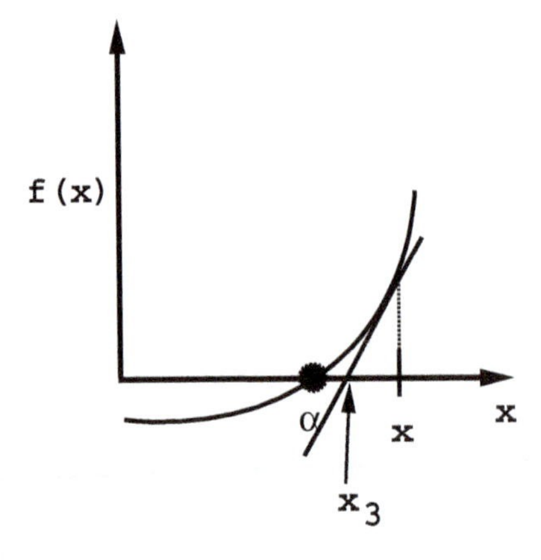

Figure 10.5 Newton's method for finding zeros.

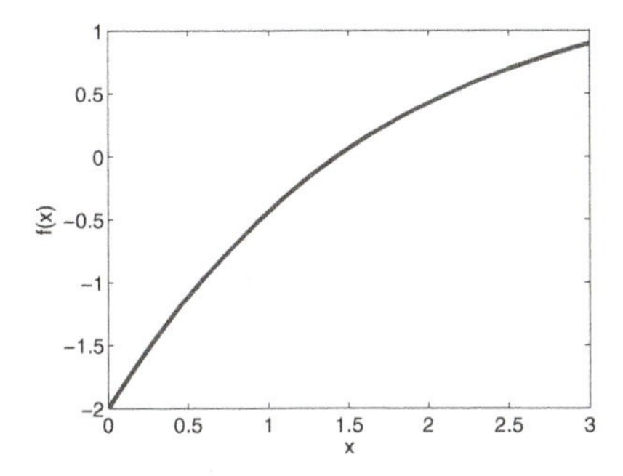

Figure 10.6 Sketch of $f(x) = \log_{10}(1 + x^2) - 2e^{-x}$.

The advantage of this method is that it is faster than the secant method; a disadvantage is that $f'(x)$ must be known, and the scheme may not converge if the guess is not close enough. Newton's method is ideal if $f'(x)$ can be calculated analytically.

MATLAB FUNCTION: *x = fzero('function', guess, tol)*

This Matlab function uses a combination of bisection and the secant method along with inverse interpolation. Consider the function

$$f(x) = \log_{10}(1 + x^2) - 2e^{-x} \tag{10.15}$$

A Matlab script to plot the function may look like the following:

```
% Zero finding example
dx=0.2
xend=3
x0=0
x=x0:dx:xend
y0=log10(1+x.*x) -2.*exp(-x)
plot(x,y0)
```

Figure 10.6 indicates a zero near $x = 1.5$. An m-file defining the function will look as follows:

```
function y=zeroex1(x)
y=log10(1+x.*x)-2.*exp(-x)
```

and this file is called zeroex1.m, created in the Matlab edit window. Now use the Matlab command window to find the zero:

```
>> fzero('zeroex1',1.5)
```

Here the tolerance has not been specified, and Matlab provides the value intrinsically. The program iterates, and the output is

```
Zero found in the interval: [1.4151, 1.56]

ans=

    1.4244

>>
```

Working in the command window in Matlab is the quickest way to get an answer to this simple problem. However, once Matlab is shut down, the result is lost. Writing a Matlab program called an m-file, which can be saved, is the most efficient way to deal with this issue.

As an example, returning to the Redlich–Kwong equation of state, for carbon monoxide gas at $T = 215$ K, $a = 17.26$ bar m^3 K$^{1/2}$/mole^2, $b = 0.02743$ m^3/mole, and $p = 69.2$ bar. The gas constant is 0.08314 KJ/Kgmole° K. Using the secant method with $(a, c) = (0, 0.5)$ with a tolerance of $\epsilon = 0.001$, $\bar{V} = 0.2268$ m^3/kg. There are no other roots in the interval $(a, c) = (0, 10)$. If other real roots occur, then one has to choose the correct root on physical grounds. This example could also have been solved using the methods for polynomials discussed next.

These methods are not confined to analytically defined functions. For example, the bisection method may easily be applied to a numerical function. Newton's method can be as well, except f' would be computed numerically.

10.4.2 Polynomials

A polynomial has the generalized form

$$P_N(x) = \sum_{n=1}^{N+1} a_n x^{N+1-n} = a_1 x^N + \cdots + a_N x + a_{N+1} \tag{10.16}$$

The following statements may be made about polynomials:

1. $P_N(x)$ has N zeros; they may be real and distinct, real and repeating, or complex. If the coefficients a_n are real, the complex roots occur in conjugate pairs.
2. One can estimate the location of all real roots of $P_N(x) = 0$ by plotting $P_N(x)$ for $-1 \le x \le 1$ and $P_N(z)$ for $-1 \le z \le 1$ with $z = \frac{1}{x}$.
3. The largest real root of $P_N(x)$ is approximated by the largest root of

$$a_1 x + a_2 = 0 \quad \text{and} \quad a_1 x^2 + a_2 x + a_3 = 0 \tag{10.17}$$

4. The smallest root is approximated by the smallest root of

$$a_N x + a_{N+1} = 0 \quad \text{and} \quad a_{N-1} x^2 + a_N x + a_{N+1} = 0 \tag{10.18}$$

5. The sum of all the roots is $-a_2/a_1$

Any of the methods described here may be used to find the zeros of a polynomial. In the case of multiple real roots, the value of the zero must be isolated based on the initial guess of the local zero. This information may be obtained by plotting the function.

MATLAB FUNCTION: *r = roots(p)*

Suppose the root of $f(x) = x^3 + x^2 - 3x - 3$ is desired. Then an m-file in the edit window to find the roots may look as follows:

```
>> p=[1 1 -3 -3]
>> r=roots(p)
```

The vector or array p comprises the coefficients of the polynomial. In the command window, execute the m-file by invoking its name "> name1," and in the command window, it is found that

```
p =

     1     1    -3    -3

r=

   -1.7321
    1.7321
   -1.0000
```

As mentioned previously, the Redlich–Kwong equation of state can be put in polynomial form as

$$p\bar{V}^3 - RT\bar{V}^2 - \left(b^2 + RTb + \frac{a}{T^{1/2}}\right)\bar{V} + \frac{a}{bT^{1/2}} = 0 \tag{10.19}$$

This equation, given the constants a, b and the temperature T, can be solved using the roots program. In this case, the roots routine can be called from a program that calculates the constants in the polynomial, unlike the preceding example, for which the constants are given. In these types of problems, care should be taken that all the terms in the equation have the same units.

10.5 Interpolation

Interpolation is the process of evaluating a function at a point where the function is not specifically known. The need arises when measurements are made at discrete points or when solutions to problems are obtained numerically. Clearly measurements of a specific quantity or a computation of a numerical solution at an infinite number of points cannot be obtained. The function may be approximated by a polynomial or by some other set of functions such as cubic splines. In this

Table 10.3. A function f tabulated at a discrete number of points

x_0	f_0
x_1	f_1
x_2	f_2
x_3	f_3
x_4	f_4

section, polynomial interpolation is described and then used to develop the basic numerical integration formulas.

Notation alert: The symbol h in this chapter is generally used for the spacing of the mesh points at which a function is tabulated, as is usually the convention in numerical analysis. Recall that the symbol h has been heretofore used for the height or half-height of a micro- or nanochannel.

10.5.1 Linear interpolation

Suppose a function f is tabulated at a discrete number of points, as in Table 10.3, and it is necessary to find the value of the function at some point x, $x_0 < x < x_1$. Such a value of x can be expressed as

$$x = x_0 + s(x_1 - x_0) = x_0 + sh \quad \text{for } 0 < s < 1 \tag{10.20}$$

where it is assumed that the data are tabulated at equal intervals, defined as $h = x_i - x_{i-1}$. The simplest function that can be used to evaluate the function f between any two points is a straight line, and for $x_0 < x < x_1$, the straight line is defined by

$$p(x) = \frac{x_1 - x}{x_1 - x_0} f(x_0) + \frac{x - x_0}{x_1 - x_0} f(x_1) \tag{10.21}$$

and this case is depicted in Figure 10.7a. This equation can be generalized for any of the intervals $[x_{i-1}, x_i]$, and the result is

$$p(x) = \frac{x_i - x}{x_i - x_{i-1}} f(x_{i-1}) + \frac{x - x_{i-1}}{x_i - x_{i-1}} f(x_i) \tag{10.22}$$

Note that by design, the function p agrees with the function f at x_{i-1} and x_i.

If the points are tabulated at equal intervals, equation (10.21) is simplified to

$$p(x) = p(x_0 + sh) = f_0 + sh\left(\frac{f_1 - f_0}{h}\right) \tag{10.23}$$

and now this equation looks like the first two terms of a Taylor series of f about x_0. Here s is determined by the value x and is a known quantity. To make this equation accurate, sh must be small; that is, the interval h should be as small as possible.

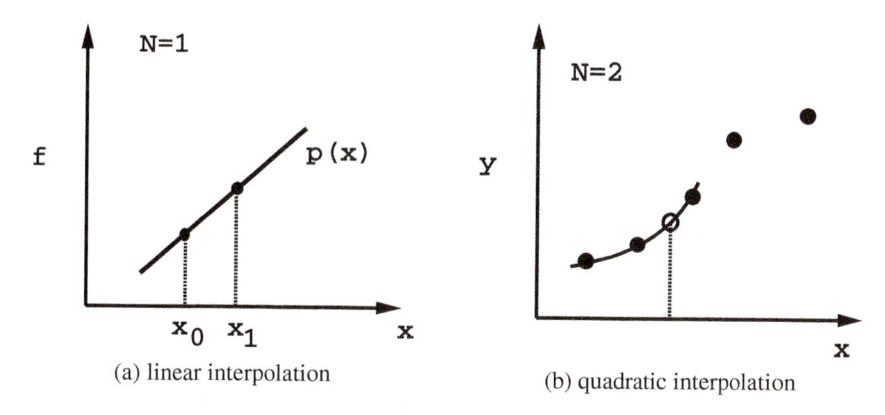

Figure 10.7 Sketch of interpolating polynomials: (a) linear interpolation; (b) quadratic interpolation. The open symbol in (b) denotes a functional value at a point in between the tabulated points.

Let us find $f(1)$ and $f(2.6)$ for the data set in Table 10.4 using linear interpolation. Note that the data are not tabulated at equal intervals. Furthermore, it is clear that $f(1) = 9$ from the linear interpolation formula; that is, $f(1)$ is just the average of $f(0)$ and $f(2)$. Now, for $x = 2.6$, $x_0 = 3$, and $s = -0.4$,

$$p(2.6) = p(x_0 + sh) = f_0 + sh\left(\frac{f_1 - f_0}{h}\right)$$
$$= 28 - 0.4(1)(28 - 11)/1 = 28 - 6.8 = 21.2 \qquad (10.24)$$

10.5.2 The difference table

Suppose a function is tabulated at equal intervals in the form shown in Table 10.5. Then a *difference table* can be formed by calculating successive differences. Each column is composed of a difference between two previous differences. Let us calculate the difference table for $f(x) = x^2$. The results are depicted in Table 10.6.

Note that the third differences are zero; in general, for a polynomial of degree N, the $N+1$st differences are all zero. The difference table is especially useful for assessing the order of polynomial by which a set of data may be approximated. As in Table 10.6, the successive differences must decrease for a well-behaved function. Clearly the difference table can also be used to assess the accuracy and consistency of an experimental data set. The difference table may also be used for *extrapolation* or trying to predict the value of a function outside the range of the tabulated data. This is risky, however, and should be avoided. The difference table can even be used to assess the accuracy of a discretization of an ordinary or partial differential equation, as will be seen.

Table 10.4. A sample set of tabular data

x	0	2	3	4
f	7	11	28	63

Table 10.5. The difference table

x_0	f_0	Δ	Δ^2	Δ^3	Δ^4
x_1	f_1	$f_1 - f_0$			
x_2	f_2	$f_2 - f_1$	$f_2 - fy_1 + f_0$		
x_3	f_3	$f_3 - f_2$	$f_2 - 2f_1 + f_0$	.	
x_4	f_4	$f_4 - f_3$	$f_2 - 2f_1 + f_0$	.	.

10.5.3 Lagrangian polynomial interpolation

In general, if a function is defined by $N + 1$ points, the function can be approximated by a polynomial of degree N whose values agree with the numerical function at the nodal points. If the data set in Table 10.3 is considered, a quadratic function is passed through any three points. Then three equations in three unknowns can be written to find each of the coefficients a_i in the polynomial:

$$p(x) = a_3 + a_2 x + +a_1 x^2 \tag{10.25}$$

The coefficients a_i can be calculated directly, and the result is in analogy with equation (10.22):

$$p(x) = \frac{(x - x_1)(x - x_2)}{(x_0 - x_1)(x_0 - x_2)} f_0 + \frac{(x - x_0)(x - x_2)}{(x_1 - x_0)(x_1 - x_2)} f_1 + \frac{(x - x_0)(x - x_1)}{(x_2 - x_0)(x_2 - x_1)} f_2 \tag{10.26}$$

Note that the points need not be equally spaced.

The general form of an interpolating polynomial of order N is

$$p_N(x) = \sum_{i=0}^{N} l_i(x) f(x_i) \tag{10.27}$$

where

$$l_i(x) = \left(\frac{x - x_0}{x_i - x_0}\right) \cdots \left(\frac{x - x_n}{x_i - x_n}\right) \tag{10.28}$$

Note that this expression is a product of N factors, and so l_i is a polynomial of degree N. The denominators of the l_i are just numbers, and $p_N(x_j) = f(x_j)$, that is, the interpolating polynomial matches the function at each of the grid points. Suppose there are 10 points running from $x = 1$ to $x = 10$. Then, theoretically,

Table 10.6. The difference table for $f(x) = x^2$

x	$f(x)$	Δ	Δ^2	Δ^3	Δ^4
0	0				
1	1	1			
2	4	3	2		
3	9	5	2	0	
4	16	7	2	0	0

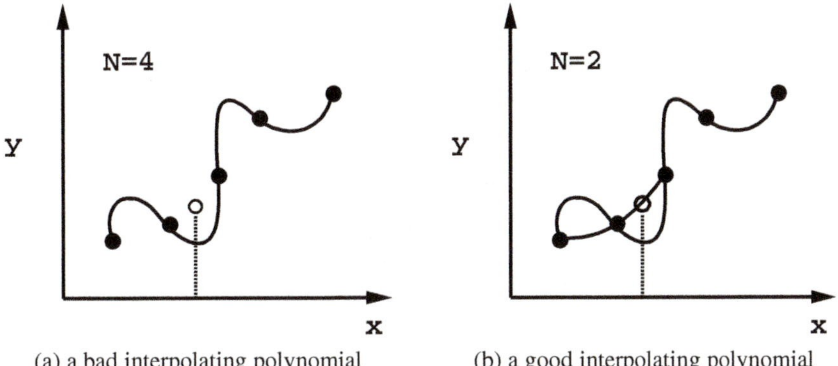

(a) a bad interpolating polynomial (b) a good interpolating polynomial

Figure 10.8 Sketch of interpolating polynomials. As the order of the polynomial increases, the deviation between the functional points and the interpolating polynomial may be large; *N* is the number of intervals. (a) In this case, the interpolating polynomial is not accurate. (b) A quadratic polynomial locally fitted between the first three points is much better.

a polynomial of degree 11 could be fit through all of the points. However, this is dangerous, and if the value of f at $x = 2.5$ of Table 10.4 is desired, for example, the best solution is to locally put a quadratic polynomial through three nearby points. If the interpolating polynomial is of too large a degree, it may oscillate, as depicted qualitatively in Figure 10.8.

The general form of the polynomial in equation (10.27) includes the case of unequal intervals. In this case, a *divided difference table* may be formed. Given a function tabulated at two points $b > a$, the divided difference is given by $\Delta_d = f_b - f_a/b - a$. Each column proceeds as in the difference table calculation; one computes Δ_d^2, Δ_d^3, and so on.

10.5.4 Newton interpolation formulas

The Lagrangian formulas given earlier can be simplified, and consider the quadratic interpolation formula. In this section, assume that the numerical function is tabulated with a uniform spacing. Given three points, say, x_0, x_1, and x_2, the task is to find the value of the function at a point $x = x_0 + sh$, where h is constant. Then, substituting into equation (10.26) and after some algebra, the interpolating polynomial $p(x)$ is given solely in terms of s by

$$p(x) = f_0 + s(f_1 - f_0) + \frac{s(s-1)}{2}(f_0 - 2f_1 + f_2) \tag{10.29}$$

where $0 < s < 2$. This equation can be written in terms of the entries in the difference table as

$$p(x) = f_0 + s\Delta f_0 + \frac{s(s-1)}{2}\Delta^2 f_0 \tag{10.30}$$

where, for example, $\Delta f_0 = f_1 - f_0$.

Table 10.7. Error associated with the Newton interpolation formulas

Formula	Error
Forward	$\frac{h^3}{3!}s(s-1)(s-2)y^{(3)}(\xi)$
Backward	$\frac{h^3}{3!}s(s+1)(s+2)y^{(3)}(\xi)$
Central	$\frac{h^3}{3!}s(s^2-1)y^{(3)}(\xi)$

The variable ξ is a point between x_0 and x_2 whose value is unknown.

Note that one term has been added to the linear interpolation formula. Equation (10.30) is called a *forward formula* because the formula uses points ahead of x_0. Recall that sh should be as small as possible. If, for example, the required interpolation point is near x_2, this formula would not be appropriate because s is too large. Near x_2, the interpolating point is given by $x = x_2 + sh$ and now $s < 0$ so that

$$p(x) = f_2 + s(f_2 - f_1) + \frac{s(s+1)}{2}(f_0 - 2f_1 + f_2) \tag{10.31}$$

where $-2 < s < 0$. This formula is called a *backward formula* because the points behind x_2 are used as a way of keeping s small.

There is also a *central formula*, this is

$$p(x) = f_1 + \frac{s}{2}(f_2 - f_0) + \frac{s^2}{2}(f_0 - 2f_1 + f_2) \tag{10.32}$$

where now $-1 < s < 1$ and $x = x_1 + sh$.

Which of the forward, backward, or central formulas is used depends on where the function is to be evaluated in a given table of values. Near the beginning of the difference table, use the forward formula; near the end, use the backward formula; and away from either end, use the central formula. The central formula should always be used, if possible. There are *truncation errors* associated with both the Newton and Lagrangian formulas; the error associated with the Newton formulas is depicted in Table 10.7.

As an example, Yoda and coworkers (Sadr *et al.*, 2006) have measured the fluid velocity near the wall for electro-osmotic flow in a micron-sized channel with the results depicted in Table 10.8. The ionic strength in the upstream reservoir is 37 μM, and the surface charge density of the channel is $\sigma = 8.63 \times 10^{-4}$ C/m^2. The Debye length is estimated at $\lambda = 18$ nm. There are nine points here, and theoretically, a 10th-order polynomial could be fit to these nine points.

Table 10.8. Experimental data for the velocity near the wall in electro-osmotic flow in a microchannel (Sadr *et al.*, 2006)

y(nm)	0	30	60	90	120	150	180	210	240
$u(\frac{\text{m}}{\text{sec}}) \times 10^5$	0	0.33	2.01	2.33	2.51	2.61	2.65	2.69	2.69

To illustrate the use of Newton's formula, suppose that the value of the velocity at $y = 100$ nm for Yoda's data in Table 10.8 is required. First consider the forward formula with $(x_0, x_1, x_2) = (90, 120, 150)$. Then $s = (100 - 90)/30 = 0.33$, and the linear formula gives

$$p(100) = 2.33 + 0.33(2.51 - 2.33) = 2.39 \times 10^{-5} \frac{\text{m}}{\text{sec}} \tag{10.33}$$

For the quadratic forward formula, the result is

$$\begin{aligned} p(100) &= 2.33 + 0.33(2.51 - 2.33) + \frac{0.33(0.33 - 1)}{2}(2.33 - 2 \times 2.51 + 2.61) \\ &= 2.40 \times 10^{-5} \frac{\text{m}}{\text{sec}} \end{aligned} \tag{10.34}$$

Finally, for the central formula $x_0 = 60$, $x_1 = 90$, and $x_2 = 120$, and again for $s = 0.33$, we have

$$\begin{aligned} p(100) &= 2.33 + \frac{.33}{2}(2.51 - 2.01) + \frac{.33^2}{2}(2.01 - 2 \times 2.33 + 2.51) \\ &= 2.40 \times 10^{-5} \frac{\text{m}}{\text{sec}} \end{aligned} \tag{10.35}$$

Note that the differences between the linear and the quadratic formulas are small and that the forward and central formulas give the same result. Thus this data set is well behaved. Note that only two digits have been kept in the solution because the data are only accurate to two significant digits. This error in the data thus produces error in the estimate of the velocity at $y = 100$ nm.

10.5.5 Matlab interpolation functions

Matlab has two functions that can be used for interpolation. One is *polyfit*, called in the form $p = polyfit(x, f, m)$, where p is the output list of coefficients for the best fit to the function and m is the input order of the interpolating polynomial. The function $polyval(p, xrange)$ then evaluates the polynomial at any point x. As with the zero-finding routines, it can be used in the command line of Matlab or in a Matlab m-file.

The function *interp1* calculates the interpolated value of a function at a single point. The format is $fi = interp1(x, f, xi, 'method')$, where x and f have the same meaning as in this section, xi is the value of x corresponding to the value of fi, and *'method'* is the method of interpolation, which can be one of several methods, such as quadratic interpolation. If no method is specified, then linear interpolation is used.

A data set that contains oscillations is not a well-behaved data set. To illustrate the problem associated with oscillations, consider the data set in Table 10.9. The Matlab program looks as follows:

```
n=5;
x=[0 2 4 5 6 7];
```

Table 10.9. Table of numerical data to illustrate the problem with oscillations in a data set

x	0	2	4	5	6	7
y	0	0.58	0.34	0.33	−0.28	−0.53

```
y=[0 .58 .34 .33 -.28 -.53];
x0=0;
dx=0.001;
xend=7.0;
xrange=x0:dx:xend;
p=polyfit(x,y,n5);
f=polyval(p,xrange);
plot(x,y,'o',xrange,f);
xlabel('x');ylabel('y');
title('Plot of data points and functions');
legend('Actual Data','Polyfit');grid;
```

Note that there are five intervals, and thus a fifth order polynomial ($n = 5$) can theoretically be fit through the five points. The result of this is shown in Figure 10.9. Note the significant oscillation in the first, second, and last intervals. Such a curve is clearly not accurate for $0 < x < 4$. This is clearly not a very good result, but the situation is improved by the use of cubic splines, which is considered next.

10.5.6 Cubic spline interpolation

Straightforward polynomial interpolation of two types has been discussed. First was polynomial interpolation over $N + 1$ points. For N sufficiently large, it has been shown that polynomial interpolation of the function over the entire range can

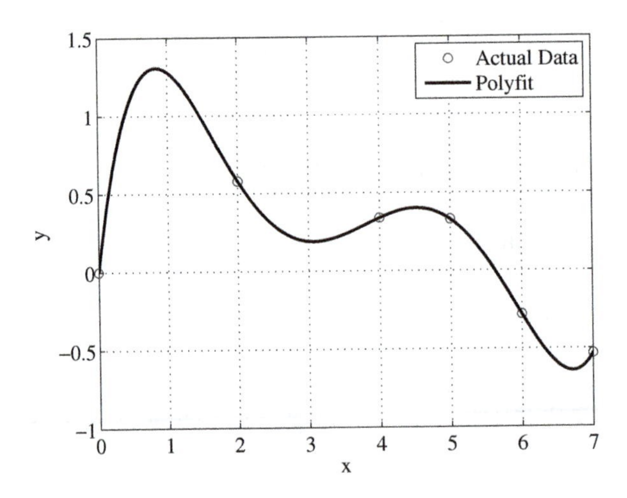

Figure 10.9 Using a single fifth-order polynomial to interpolate a data set is not always best.

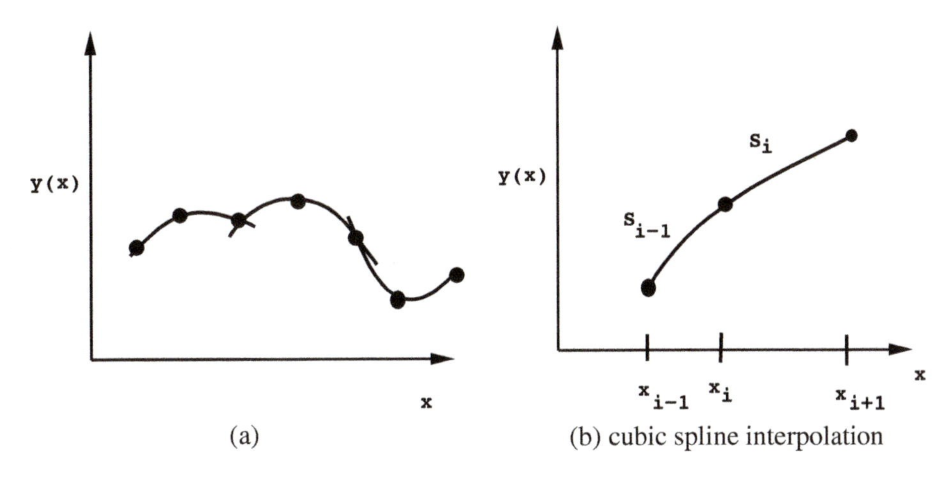

Figure 10.10 (a) Sketch of a series of polynomial approximations. (b) Two successive intervals for the cubic spline approximation.

lead to oscillatory errors. Second, piecewise polynomial interpolation has been discussed and is endorsed. The trouble with piecewise polynomial interpolation is that the interpolating polynomials are not differentiable at the end points of each interval, as shown in the schematic of Figure 10.10a.

There is an alternative to this type of interpolation: *cubic splines*. A spline was a draftman's device used to draw smooth curves. A smooth cubic polynomial is used in each interval. For technical reasons, odd powers of polynomials seem to work better than even powers. For example, in quadratic spline interpolation, the second derivative is still discontinuous at the grid points or "knots." There are many other kinds of splines, but we confine ourselves to the cubic spline.

The cubic spline interpolation procedure begins with two successive intervals $[x_{i-1}, x_i]$, and $[x_i, x_{i+1}]$, $i = 1, N$ where N is the number of intervals (Figure 10.10b); then the cubic spline interpolating function is

$$S_i(x) = a_i(x - x_i)^3 + b_i(x - x_i)^2 + c_i(x - x_i) + d_i \tag{10.36}$$

valid for $i = 1, \ldots, N$. The coefficients are determined by the boundary conditions at the end of each interval.

Now, in each interval, $[x_i, x_{i+1}]$, let $z_i = S''(x_i)$, the second derivative of the spline function evaluated at x_i. Note that S'' is a linear function that takes the values z_i at $x = x_i$ and z_{i+1} at $x = x_{i+1}$ Clearly S_i'' can be written

$$S_i''(x) = \frac{z_{i+1}}{h_i}(x - x_i) + \frac{z_i}{h_i}(x_{i+1} - x) \tag{10.37}$$

and integrating twice,

$$S_i(x) = \frac{z_{i+1}}{6h_i}(x - x_i)^3 + \frac{z_i}{6h_i}(x_{i+1} - x)^3 + C(x - x_i) + D(x_{i+1} - x) \tag{10.38}$$

To find the constants C and D, note that at the grid points, the spline must agree with the functional values of y and

$$S_i(x_i) = y_i \tag{10.39}$$

$$S_i(x_{i+1}) = y_{i+1} \tag{10.40}$$

which gives two equations in two unknowns for C and D. The result is

$$S_i(x) = \frac{z_{i+1}}{6h_i}(x - x_i)^3 + \frac{z_i}{6h_i}(x_{i+1} - x)^3 + \left(\frac{y_{i+1}}{h_i} - \frac{h_i z_{i+1}}{6}\right)(x - x_i) \tag{10.41}$$

$$+ \left(\frac{y_{i+1}}{h_i} - \frac{h_i z_i}{6}\right)(x_{i+1} - x) \tag{10.42}$$

This is the result for the cubic spline; however, the z_i, z_{i+1} are unknown, and we need to be able to calculate them. To do this, differentiating equation (10.42),

$$S_i'(x) = \frac{z_{i+1}}{2h_i}(x - x_i)^2 + \frac{z_i}{2h_i}(x_{i+1} - x)^2 + \frac{y_{i+1}}{h_i} - \frac{h_i z_{i+1}}{6} - \frac{y_i}{h_i} + \frac{h_i z_i}{6} \tag{10.43}$$

and evaluating at $x = x_i$,

$$S_i'(x_i) = -\frac{z_i h_i}{3} - \frac{z_{i+1} h_i}{6} + \frac{y_{i+1} - y_i}{h_i} \tag{10.44}$$

Similarly,

$$S_{i-1}'(x_i) = \frac{z_{i-1} h_i}{6} + \frac{z_i h_{i-1}}{3} + \frac{y_i - y_{i-1}}{h_{i-1}} \tag{10.45}$$

The advantage of splines over direct polynomial interpolation is that the interpolating function derivative can be made continuous at the mesh points or knots x_i by requiring

$$S_i'(x_i) = S_{i-1}'(x_i) \tag{10.46}$$

and so

$$h_{i-1} z_{i-1} + 2(h_{i-1} + h_i) z_i + h_i z_{i+1} = 6(b_i - b_{i-1}) \tag{10.47}$$

where

$$b_i = \frac{y_{i+1} - y_i}{h_i} \tag{10.48}$$

Equations (10.47) are a system of *tridiagonal linear equations* and are valid for $i = 2, N - 1$; the points $i = 1$ and $i = N$ are the end points. We will be solving systems of this type later in this chapter. At the end points, because there are no continuity conditions, we must specify something about the z_i. The most common cubic spline conditions are that $z_1 = 0, z_N = 0$. Other conditions can also be used: $z_1 = z_2, z_N = z_{N-1}$, which is a first order approximation to a zero derivative at the end points. The results are usually not affected greatly by use of either of these conditions.

Let us interpolate the data set in the following Matlab program to find $y(.7)$, $y(1.7)$, $y(4.7)$ using cubic splines using the Matlab spline function. The Matlab program looks as follows:

```
x = [0.0 0.5 1.0 1.5 2.0 2.5 3.0 3.5 4.0 4.5 5.0];
y = [0.0 0.03060 0.11490 0.23209 0.35283 0.44606 0.48609
0.45863 0.36413 0.21785 0.04657];
y_spline1 = interp1(x,y,1.7,'spline');
y_spline2 = interp1(x,y,0.7,'spline');
y_spline3 = interp1(x,y,4.7,'spline');
disp('Interpolation Estimates');
disp('        x         ');
disp('       1.7        ');
disp('        y         ');
disp(y_spline1);
disp('        x         ');
disp('       0.7        ');
disp('        y         ');
disp(y_spline2);
disp('        x         ');
disp('       4.7        ');
disp('        y         ');
disp(y_spline3);
```

Running this program yields the following:

```
>>  splines1
Interpolation Estimates
        x
       1.7
        y
      0.2817


        x
       0.7
        y
      0.0589


        x
       4.7
        y
      0.1507


>>
```

As another example, consider the data set in the following Matlab program. Let us find and plot the cubic spline interpolation curve. The Matlab file is similar to that above, but now the Matlab *spline* command is used:

```
xdata=[0.0 0.6 1.5 1.7 1.9 2.1 2.3 2.6 2.8 3.0]
ydata=[-0.8 -0.34 0.59 0.59 0.23 0.1 0.28 1.03 1.5 1.44]
x1=0;
dx=0.1;
x2=3.0;
xspline=x1:dx:x2;
yspline=spline(xdata,ydata,xspline)
plot(xdata,ydata,'o',xspline,yspline,'-')
title('Cubic Spline Curve')
xlabel('x-axis'),ylabel('y-axis')
```

The results are shown in Figure 10.11. Note that the curve is smooth and approximates the numerical function reasonably well between the mesh points.

10.6 Curve fitting

Curve fitting is merely another form of approximation of a function. In interpolation, a polynomial was passed through a few (good) or all of the points (bad for too many points). However, as with all experimental methods, these data will have some error owing to the experimental method used. Thus it may be unrealistic to require an approximating polynomial to pass through each data point because the experiment is only repeatable within the experimental error. Thus a lower-order polynomial may not pass through *any* of the points may be appropriate. The question is, what function gives the best fit? Usually, the number of data points

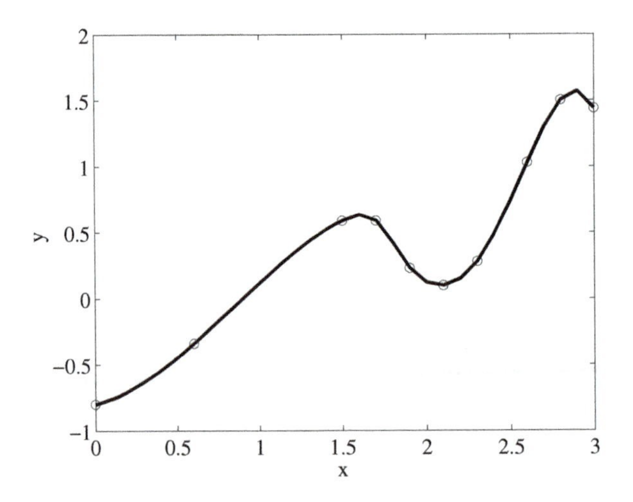

Figure 10.11 Cubic spline interpolation of the data set just discussed. Note that the curve is smooth and differentiable between the mesh points.

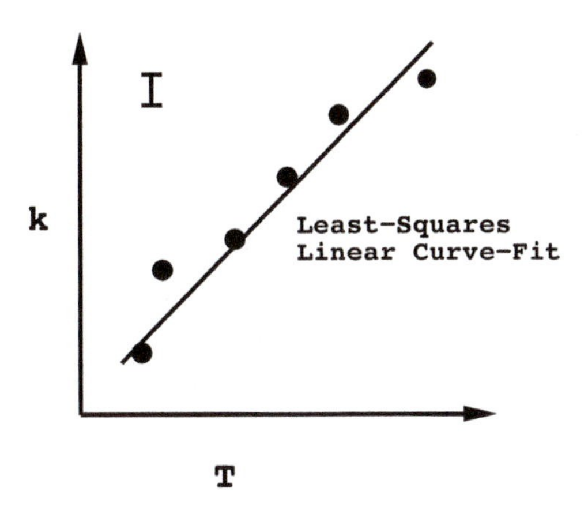

Figure 10.12 Qualitative sketch of the thermal conductivity with temperature and a postulated curve fit.

greatly exceeds the order of the polynomial, so the question of how to determine the coefficients of the polynomial must be addressed.

The results of this section are useful because they will yield one approximating function for an entire set of data, unlike the interlacing polynomials concept. Furthermore, the function should differ only a little from the data, even if additional data points are added.

The thermal conductivity (Figure 10.12) of a material is often approximated by a linear function, as done in Chapter 2:

$$k = k_0(1 + bT) \tag{10.49}$$

with T in °C and k_0 and b constant. Let us write this equation in the form

$$k(T) = \alpha + \beta T \tag{10.50}$$

where α and β are constants fit to experimental data. The question is, how should α and β be chosen so that the difference between the data and the approximating function is, in some sense, a minimum?

Let

$$f_i = \alpha + \beta T \tag{10.51}$$

and define the error as

$$e_i = k_i - f_i \tag{10.52}$$

Define the error E as

$$E = e_1^2 + \cdots + e_N^2 \tag{10.53}$$

The *least squares* criterion requires that E be a minimum, and at a minimum,

$$\frac{\partial E}{\partial \alpha} = -\sum_{i=1}^{N} 2(k_i - (\alpha + \beta T)) = 0 \tag{10.54}$$

$$\frac{\partial E}{\partial \beta} = -\sum_{i=1}^{N} 2T_i(k_i - (\alpha + \beta T)) = 0 \tag{10.55}$$

These are two equations in two unknowns for α and β, which can be solved by the methods to be discussed in Section 10.9.

Curve fitting is the most common way of generating an approximate function to represent a given data set. This is because the experimental data have errors in them, and also, a curve fit usually leads to a single functional form valid over all data points. That is not to say that there may be parameter ranges over which a given set of data requires two or even more functional forms; nevertheless, the curve-fitting procedure is usually optimal for analyzing experimental data.

A set of data may not be well approximated by a linear curve fit. In this case, a similar procedure could be used to approximate a data set with a polynomial curve fit. Moreover, if a given function $f(x) \sim Ax^{\beta}$, then a linear curve fit of the logarithm of the function is possible:

$$\ln f \sim \ln A + \beta \ln x \tag{10.56}$$

Matlab's interpolation functions *polyfit* and *polyval* may also be used for curve-fitting purposes. Matlab recognizes whether the order of the approximating polynomial is less than $N - 1$, where N is the number of points in the data set, and automatically uses the curve-fit procedure.

As an example, Sadr *et al.* (2006) have measured the electro-osmotic mobility $\mu_e = \frac{U}{E_x}$ inside the electrical double layer as a function of reservoir concentration (Figure 10.13). The velocity U is the average velocity over a distance of about 200 nm from the wall. The data appear in the Matlab file:

```
conc=[   0.19 1.9 3.6 18.4 36]; % mM
mob=[ 5.14 2.41 2.02 1.40 1.14]; %mobility x 10^4 cm^2/V/s
conc2=log10(conc)
mob2=log10(mob)
p1=polyfit(conc2,mob2,4)
fit1=polyval(p1,conc2);
p2=polyfit(conc2,mob2,2)
fit2=polyval(p2,conc2);
p3=polyfit(conc2,mob2,1)
fit3=polyval(p3,conc2);
plot(conc2,fit1,'-',conc2,mob2,'o',conc2,fit2,'*',
    conc2,fit3,'-.')
xlabel('log conc')
ylabel('log mobility')
```

Note that the log of both the concentration and the mobility are being curve fit with first a fourth-order polynomial through each point (interpolation), then a quadratic, then a linear fit. Note that all the functions seem to work well as shown by the results in Figure 10.13. The linear curve fit is defined by

$$\log_{10} U = -0.2816 \log_{10} c + 0.4856 \tag{10.57}$$

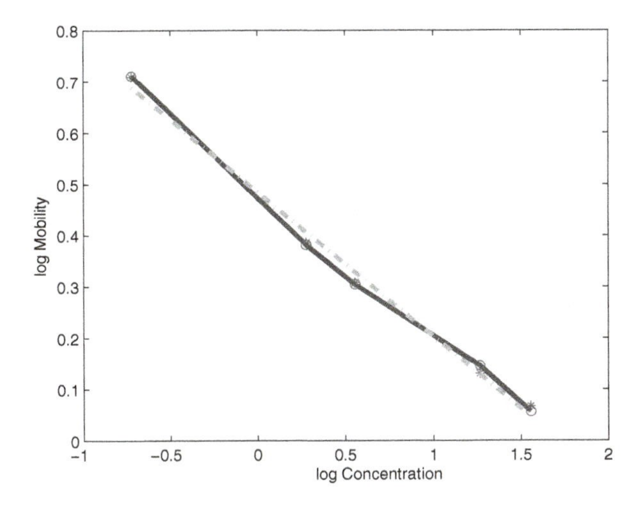

Figure 10.13 Curve fit to the mobility – concentration data taken by Sadr *et al.* (2006). The solid line is a fourth-order polynomial fit, and the dashed line is the least squares linear curve fit.

and note that the fourth order polynomial is very accurate even away from the grid points, indicating that this is a well-behaved data set.

10.7 Numerical differentiation

As has been seen many times, there are many instances where numerical differentiation is required. Often a given physical quantity that has been measured is defined only at a discrete number of points, as was seen in the previous section. Numerical differentiation is also required in the discretization of ordinary and partial differential equations. Moreover, certain physical quantities are defined by derivatives, and thus numerical differentiation is required as part of the numerical solution of ordinary and partial differential equations. For example, the shear stress is defined in terms of the derivative of the velocity, and the surface charge density is defined in terms of the derivative of the electrical potential.

The formulas for the derivative of a given function, analytical or numerical, come from the Taylor series, as with the various methods of function approximation such as interpolation.

10.7.1 Derivatives from Taylor series

Consider the function $f(x)$, and recall that the Taylor series approximation to $f(x)$ near x is given by

$$f(x+h) = f(x) + f'(x)h + f''(x)\frac{h^2}{2!} + f'''(x)\frac{h^3}{3!} + \cdots \tag{10.58}$$

where h is small. Note that this formula is written in a slightly different form from equation (10.1) for reasons that will become obvious.

The *forward difference* approximation to the derivative is obtained by neglecting the second derivative term and

$$f'(x) \cong \frac{f(x+h) - f(x)}{h} \tag{10.59}$$

Evaluating at a discrete value of x, say, x_i,

$$f_i' \cong \frac{f_{i+1} - f_i}{h} \tag{10.60}$$

In a similar way, the *backward difference* approximation is obtained by replacing h by $-h$, and the result is

$$f_i' \cong \frac{f_i - f_{i-1}}{h} \tag{10.61}$$

Because the finite difference formulas are approximations, there is error; for the forward and backward formulas, the error is

$$E \cong f''(\xi)\frac{h}{2} \tag{10.62}$$

where ξ is some point satisfying $x \le \xi \le x + h$. Note that the error gets smaller as the grid size h gets smaller.

A more accurate formula, may be obtained by adding the backward formula and the forward formula, resulting in the *central difference approximation*:

$$f_i' \cong \frac{f_{i+1} - f_{i-1}}{2h} \tag{10.63}$$

The error in the central difference approximation is smaller than the error in the forward or backward difference, being

$$E \cong f'''(\xi)\frac{h^2}{3} \tag{10.64}$$

This *central difference* approximation should be used whenever possible.

The second derivative may be obtained from Taylor series in a similar manner. Adding

$$f(x+h) = f(x) + f'(x)h + f''(x)\frac{h^2}{2!} + f'''(x)\frac{h^3}{3!} + \cdots \tag{10.65}$$

to the corresponding formula for $f(x-h)$ yields the *central difference* approximation to the second derivative:

$$f_i'' \cong \frac{f_{i+1} - 2f_i f_{i-1}}{h^2} + O(h^2) \tag{10.66}$$

It will turn out that derivatives higher than the second will not generally be required.

The expressions for the first and second derivatives in the forward, backward, and central differences appear in the interpolation formulas discussed previously.

10.7.2 A more accurate forward formula for the first derivative

There are times when the more accurate central formula cannot be used. An example is the calculation of the shear stress at a surface or the surface charge density in an electrolyte solution. In these cases, the last point in the fluid is on the boundary, and using a simple forward formula may not be accurate enough, especially when the function is varying rapidly near the surface. However, the Taylor series approximations can be manipulated to derive a more accurate formula.

Consider the forward formula with h replaced by $2h$:

$$f(x+2h) = f(x) + f'(x)2h + f''(x)\frac{4h^2}{2!} + f'''(x)\frac{8h^3}{3!} + \cdots \tag{10.67}$$

and the standard forward formula is

$$f(x+h) = f(x) + f'(x)h + f''(x)\frac{h^2}{2!} + f'''(x)\frac{h^3}{3!} + \cdots \tag{10.68}$$

Multiply the last equation by –4 and add to the first, we find

$$f(x+2h) - 4f(x+h) = -3f(x) - 2f'(x)h + f'''(x)\frac{4h^3}{3!} + \cdots \tag{10.69}$$

Then

$$f'(x) \cong \frac{-f(x+2h) + 4f(x+h) - 3f(x))}{2h} \tag{10.70}$$

or, in discrete form,

$$f_i' \cong \frac{-f_{i+2} + 4f_{i+1} - 3f_i}{2h} \tag{10.71}$$

The error in this formula is still $O(h^2)$, yet all of the points are on one side of the point i. Such formulas are called *sloping difference formulas*.

The temperature distribution through a copper slab with a mean conductivity $k = 83$ W/m/C generates the data in Table 10.10. We might be asked to calculate the heat flux at $y = 0$. Note that the temperature distribution within the slab is nonlinear, and so the heat transfer cannot be from conduction alone. The definition of the heat flux is

$$q = \frac{Q}{A} = -k\frac{dT}{dy} \tag{10.72}$$

If the two point forward formula is used for dT/dy, then

$$\frac{Q}{A} = -k\frac{dT}{dy} = -83\ \frac{\mathrm{W}}{\mathrm{mC}} \times \frac{(28.6-30)\ \mathrm{C}}{.4\quad \mathrm{m}} = 290.5\ \frac{\mathrm{W}}{\mathrm{m}^2} \tag{10.73}$$

This formula is not very accurate and would generally not be used unless absolutely necessary. The more accurate three-point sloping difference formula results

Table 10.10. Results from experimental measurements of the temperature in a slab

$y(m)$	0.0	0.4	0.8	1.2	1.6	2.0
$T(C)$	30.0	28.6	26.2	22.1	20.0	18.3

in

$$\frac{Q}{A} = -k\frac{dT}{dy} = -83\ \frac{\text{W}}{\text{mC}} \times \frac{(-26.2 + 4(28.6) - 3(30))}{.4}\ \frac{\text{C}}{\text{m}} = 186.75\ \frac{\text{W}}{\text{m}^2} \tag{10.74}$$

Note that this answer is considerably different from the two-point forward formula.

At an interior point such as $y = 1.2$, the two-point central difference formula, should be used, and

$$\frac{Q}{A} = -k\frac{dT}{dy} = -83\ \frac{\text{W}}{\text{mC}} \times \frac{(20.0 - 26.2)}{.8}\ \frac{\text{C}}{\text{m}} = 643.25\ \frac{\text{W}}{\text{m}^2} \tag{10.75}$$

For comparison, the result using the backward formula is

$$\frac{Q}{A} = -k\frac{dT}{dy} = -83\ \frac{\text{W}}{\text{mC}} \times \frac{(22.1 - 26.2)}{.4}\ \frac{\text{C}}{\text{m}} = 850.75\ \frac{\text{W}}{\text{m}^2} \tag{10.76}$$

and the forward result is

$$\frac{Q}{A} = -k\frac{dT}{dy} = -83\ \frac{\text{W}}{\text{mC}} \times \frac{(20.0 - 22.1)}{.4}\ \frac{\text{C}}{\text{m}} = 435.75\ \frac{\text{W}}{\text{m}^2} \tag{10.77}$$

Note the considerable difference in the results and that the result using the central difference is just the average of the forward and backward formulas.

As will be seen later in this chapter, using the central difference approximations to the first and second derivatives is the key step in the numerical solution of second-order differential equations of the boundary value type. This will lead to a linear system of equations of tridiagonal form that is quickly and easily solved by several methods. More accurate formulas can be derived for the first and second derivatives; formulas for higher derivatives, such as the third and fourth derivatives, may also be developed. However, these formulas are often difficult to use and lead to considerably more complicated sets of linear equations when solving an ordinary differential equation of the boundary value type. Thus only the first and second derivatives will usually be required. These other formulas are compiled in several places, in particular, Gilat and Subramaniam (2008).

10.8 Numerical integration

Integration and differentiation are complementary in the sense that the integral of the first derivative of a function is the function itself. There are many cases when a function must be integrated numerically; that is,

$$I = \int_a^b f dx \tag{10.78}$$

or

$$I_i = \int_a^x f dy \tag{10.79}$$

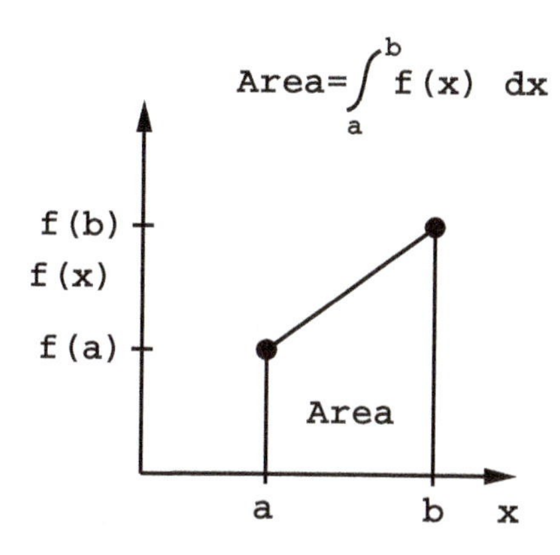

Figure 10.14 The trapezoidal rule is obtained by assuming that the function to be integrated is linear in the integration variable. The integral is simply the area under the curve.

I_i is an *indefinite integral* because x, a variable, appears on one of the integration limits, while I is a *definite integral*. As has been seen, the *volume flow rate* of a fluid flowing in a duct or tube is defined by

$$Q = \int_A u dA \tag{10.80}$$

where u is the fluid velocity and A is the cross-sectional area through which the fluid flows.

As with differentiation, there are just a few formulas that are accurate enough for most problems, although typically, textbooks discuss many more methods than what are described here. Different integration formulas are derived based on how the function is approximated over a given interval; typically, a linear or quadratic approximation of the function is adequate, and only these formulas will be discussed in detail.

10.8.1 The trapezoidal rule

Suppose that the definite integral of a function must be calculated and that the integrand is given by a numerically generated function. In that interval, $[a, b]$ f may be approximated by a linear function, as in Figure 10.14, and

$$f(x) \cong p(x) = A + Bx \tag{10.81}$$

Note that

$$A = f(a) - \frac{a(f(a) - f(b))}{b - a} \tag{10.82}$$

and

$$B = \frac{f(b) - f(a)}{b - a} \tag{10.83}$$

so that $p(x)$ agrees with $f(x)$ at the points a and b. Then, substituting into the integral,

$$\int_a^b f(x)dx \cong \int_a^b p(x)dx \tag{10.84}$$

$$\int_a^b p(x)dx = \int_a^b (A + Bx)dx = A(b-a) + \frac{B}{2}(b^2 - a^2) \tag{10.85}$$

Substituting for A and B,

$$\int_a^b f(x)dx \cong \int_a^b g(x)dx = \frac{h}{2}(f(b) + f(a)) \tag{10.86}$$

where $h = b - a$. This is the *trapezoidal rule*. Note that $1/2(f(a) + f(b))$ is just the average of the function value over the interval $[a, b]$. Thus

$$\int_a^b f(x)dx = \frac{h}{2}(f(b) + f(a)) + error \tag{10.87}$$

and the error is just the error in approximating the function by Taylor series:

$$error \cong -\frac{1}{12}(b-a)^3 \tag{10.88}$$

Equation (10.87) can be extended to more than one interval; consider the case of two intervals. Then, because the integral is additive,

$$\int_a^c f(x)dx = \int_a^b f(x)dx + \int_b^c f(x)dx \tag{10.89}$$

and using the formula for the trapezoidal rule,

$$\int_a^c f(x)dx = \frac{b-a}{2}(f(b) + f(a)) + \frac{c-b}{2}(f(b) + f(c)) + E \tag{10.90}$$

If the grid points are equally spaced, then

$$\int_a^c f(x)dx = \frac{h}{2}f(a) + +hf(b) + \frac{h}{2}f(c) + E \tag{10.91}$$

In general, for N points,

$$\int_a^c f(x)dx = \frac{h}{2}(f_0 + f_{N+1}) + h\sum_{j=1}^{N} f_j + E \tag{10.92}$$

A typical mesh is depicted in Figure 10.15. The error is now

$$E \cong -\frac{1}{12}(c-a)h^2 f''(\xi) \tag{10.93}$$

where $a < \xi < c$, but ξ is unknown. The trapezoidal rule is the simplest of the integration formulas to derive, and it is also accurate enough for most purposes (Figure 10.15).

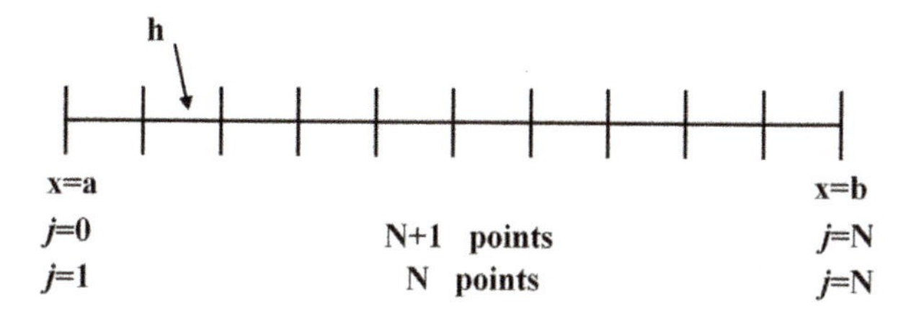

Figure 10.15 A typical mesh for an integration scheme. The first numbering system ($j = 0 - N$) is often used in textbooks. The second numbering system is used in programming. Note that the mesh size $h = (b - a)/N$ in the first case and $h = (b - a)/(N - 1)$ in the second case. Note that the numbering system includes the end points.

Consider the data set in Table 10.11; the Matlab file for the trapezoidal rule is straightforward:

```
%A Program for the Trapezoidal Rule,
x=[0 .5 1.0 1.5 2.0 2.5 3.0 3.5 4.0];
f=[1.90 2.39 2.71 2.98 3.20 3.20 2.98 2.74 2.63];
plot(x,f)
xlabel('x')
ylabel('f')
N=length(x);
h=(x(N)-x(1))/(N-1);
sum=f(1);
for i=1:N-1;
sum=sum+2.*f(i);
end
sum=sum+f(N);
Int=sum.*h./2
```

Note that the counter begins with $i = 1$. If the name of this integration m-file is 'trapint,' then running the m-file produces (without the plot shown)

```
>>trapint

Int =

  12.1825

>>
```

This value is compared with that for Simpson's rule, discussed next. Note that the trapezoidal rule may be used if a function is tabulated at unequal intervals.

Table 10.11. Numerical data set for calculating the definite integral

x	0	.5	1.0	1.5	2.0	2.5	3.0	3.5	4.0
$f(x)$	1.90	2.39	2.71	2.98	3.20	3.20	2.98	2.74	2.63

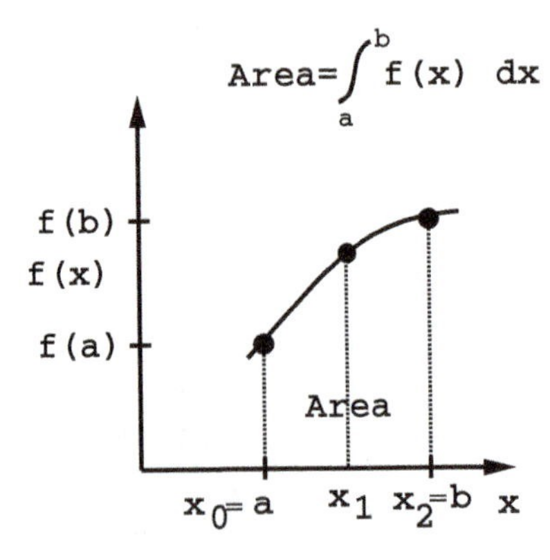

Figure 10.16 The Simpson's 1/3 rule is obtained by assuming that the function to be integrated is quadratic in the integration variable.

10.8.2 Simpson's rules

There are many Simpson's rules, depending on how the function is approximated. The most common rule is the Simpson's 1/3 rule, in which a quadratic function is used to approximate the numerical function. To derive the formula, the Newton forward interpolation formula is used (Figure 10.16):

$$p(x) = f_0 + s\Delta f_0 + \frac{s(s-1)}{2}\Delta^2 f_0 \tag{10.94}$$

where $\Delta f_0 = f_1 - f_0$ is called the *forward difference operator*. Then

$$\int_{x_0}^{x_2} f(x)dx = \int_{x_0}^{x_2} p(x)dx + E \tag{10.95}$$

Now recall that $x = x_0 + sh$, and so

$$\int_{x_0}^{x_2} p(x)dx = h\int_0^2 \left(f_0 + s\Delta f_0 + \frac{s(s-1)}{2}\Delta^2 f_0\right) ds \tag{10.96}$$

$$\int_{x_0}^{x_2} p(x)dx = \frac{h}{3}(f_0 + 4f_1 + f_2) \tag{10.97}$$

and thus

$$\int_{x_0}^{x_2} f(x)dx = \frac{h}{3}(f_0 + 4f_1 + f_2) + E \tag{10.98}$$

and the error E turns out to be

$$E = -\frac{1}{90}h^5 f^{iv}(\xi) \tag{10.99}$$

If the function is approximated by a cubic polynomial over four points, or three intervals, another Simpson's rule, the 3/8 rule, is obtained:

$$\int_{x_0}^{x_3} f(x)dx = \frac{3h}{8}(f_0 + 3f_1 + 3f_2 + f_3) + E \tag{10.100}$$

where now

$$E = -\frac{3}{80}h^5 f^{iv}(\xi) \tag{10.101}$$

Note that the error has not been reduced by adding another point.

If the function is tabulated at $N+1$ points, then the Simpson's 1/3 rule becomes

$$\int_{x_0}^{x_{N+1}} f(x)dx = \frac{h}{3}\left(f_0 + 4\sum_{j odd}^{N-1} f_j + 2\sum_{j even}^{N-2} f_j + f_{N+1}\right) + E \tag{10.102}$$

where N is the total number of intervals and to use Simpson's rule, N must be even. The error is now

$$E = -\frac{(x_N - x_0)}{180}h^4 f^{iv}(\xi) \tag{10.103}$$

a reduction in error of $O(h^2)$ compared to the trapezoidal rule.

Let us compare the value of the definite integral using the trapezoidal rule and Simpson's 1/3 rule. The matlab m-file for Simpson's rule is similar to that for the trapezoidal rule:

```
% function I=simp(f,a,b,n)
%A Program for Simpson's Rule
x=[0 .5 1.0 1.5 2.0 2.5 3.0 3.5 4.0];
f=[1.90 2.39 2.71 2.98 3.20 3.20 2.98 2.74 2.63];
N=length(x);
plot(x,f)
xlabel('x')
ylabel('f')
N=length(x);
h=(x(N)-x(1))/(N-1);
sum=f(1);
for i=2:2:N-1
   sum=sum+4.*f(i);
end
for i=3:2:N-2
   sum=sum+2.*f(i);
end
sum=sum+f(N);
Int=sum.*h./3
```

Running the Matlab file produces

```
>>simpint

Int =

   11.2583

>>
```

which is a bit different than the trapezoidal rule value. The Simpson value is considered more accurate, based on its smaller truncation error.

The question arises of what happens if the function is tabulated over an odd number of intervals. Then Simpson's rule cannot be used near the beginning of the data. One option is to use the trapezoidal rule; however, the trapezoidal rule is much less accurate than Simpson's rule. The solution is to use a high-accuracy forward interpolation formula that is of the same order as Simpson's rule, $O(h^5)$. Such a formula leads to

$$\int_x^{x+h} f(x)dx = \frac{h}{24}(9f_0 + 19f_1 - 5f_2 + f_3) \tag{10.104}$$

for the integration over the first interval, and Simpson's $1/3$ rule may be used to complete the calculation.

10.8.3 Matlab integration functions

Matlab has several built-in functions that can be used for integration. The first is *quad*, which is called as $I = quad('function', a, b, tol)$, where a, b are the integration limits and *tol* is a number related to the error; the smaller the value of *tol*, the smaller the error. This Matlab function uses a form of Simpson's rule that adapts the grid size locally based on the behavior of the function.

The other function of interest here is *trapz*, called by $I = trapz(x, f)$. This Matlab function is designed for numerically given functions like that discussed in the preceding example. It uses the trapezoidal rule, as you might guess from its name.

10.8.4 The indefinite integral

Suppose a function $f(x)$ is perhaps the result of some other numerical computation. Then the indefinite integral is given by

$$I_i(x) = \int_a^x f(x)dx + J(a) \tag{10.105}$$

Note that the indefinite integral $I_i = I_i(x)$. By the trapezoidal rule, then,

$$I_i(x_{j+1}) = J_{j+1} = J_j + \frac{1}{2}h(f_j + f_{j+1}) \quad \text{for} \quad j = 1, \ldots, N \tag{10.106}$$

With Simpson's rule, we need three points, and so we have to start at x_2, for which

$$I_{i2} = \frac{h}{3}(f(a) + 4f_1 + f_2) \tag{10.107}$$

and in the general case,

$$I_i(x_{j+1}) = J_{j+1} = J_{j-1} + \frac{h}{3}(f_{j-1} + 4f_j + f_{j+1}) \quad \text{for} \quad j = 2, \ldots, N \tag{10.108}$$

To obtain $I_i(1)$, a sloping integration formula that is of the same order of accuracy as the Simpson's rule could be used, as discussed earlier, for the definite integral.

For example, the electric field is defined as the gradient of the electrical potential. In one dimension, the electric field is defined to be

$$E = -\frac{d\phi}{dy} \tag{10.109}$$

so that the potential is given as the indefinite integral of the electric field or

$$\phi(y) = -\int_{y_0}^{y} E(y)dy \tag{10.110}$$

10.8.5 Other formulas

It should be pointed out that there are many other formulas such as Newton–Cotes formulas and Romberg integration, often called Richardson extrapolation. The Newton–Cotes formulas result from different approximations to the function over the interval of integration. Romberg integrations utilize solutions for different mesh sizes to enhance the accuracy of the final result. I have found that the trapezoidal rule and the Simpson's 1/3 rule are sufficient for most applications. In fact, I have never used any formulas other than the trapezoidal rule and Simpson's 1/3 rule for a standard integration.

The methods described here should not be used for Fourier integrals of the form

$$I = \int_0^{\pi} f(x)\cos nx dx \tag{10.111}$$

because for large values of n, the integrand becomes highly oscillatory. Two approaches that can be used are the Filon approach (Abramowitz & Stegun, 1972) and the fast Fourier transform (Fausett, 2008).

10.8.6 Grid (mesh) size

The question is, how do we know that the answer for the integral is correct? Well, the answer lies in the number of points taken in the integration: N in the two preceding Matlab files for the trapezoidal and Simpson's rule. For a numerical function such as earlier, N cannot be changed, so the accuracy of the integral is fixed. However, when an analytical function is to be integrated, such as

$f(x) = e^{-x}$, the value of N should be increased systematically until the answer for successive values of N is within a specified tolerance such as 10^{-4}. Obviously, a larger value of N may be required if the function varies rapidly; an example of a function that varies rapidly near $x = 0$ is $f(x) = e^{-Ax}$ with $A \gg 1$. Try numerically integrating this function over the interval [0, 1] with $A = 100$. Of course, the answer may be obtained analytically and is $I = 1/A(1 - e^{-A})$.

This issue is easily fixed by changing variables. Noting that e^{-Ax} is zero for large A, except near $x = 0$, the solution is to change variables to $X = Ax$, and then, for $x = 1$, $X = A \gg 1$, and so now the interval [0, 1] has become $[0, \infty]$ in X in the limit $A \rightarrow \infty$. In this case, the numerical solution may easily be computed.

There is a method in choosing mesh size, equivalent to choosing N. If the integration is over [0, 1], a good starting point is choosing $N = 11$, which gives 10 intervals, and $h_1 = 0.1$. Note that the value of N includes the end points. Once the solution for $N = 11$ is calculated, the procedure is to repeat the calculation for $N = 21$, giving $h_2 = h_1/2 = 0.05$. This process is repeated until the answer for two successive grid sizes or mesh differs by a small specified tolerance. If three-digit accuracy is required (e.g., 0.1235 and 0.1237 are three-digit accurate), then the tolerance is $tol = 10^{-3}$, that is, $I_1 - I_2 < 10^{-3}$.

The Romberg integration process is defined by combining the solution for h_1 and h_2 to achieve $O(h^4)$ accuracy, and the Romberg result is defined by

$$I_R = \frac{4I(h/2) - I(h)}{3} + O(h^4) \tag{10.112}$$

This procedure is equivalent to *Richardson extrapolation* in the solution of ordinary and partial differential equations.

In general, the process of determining the numerical accuracy of a given integral is part of a general process of *verification*, which is discussed later in this chapter.

10.8.7 Singularities

It is important to mention that if there is singular behavior in the region of integration, all these integration methods will fail. For example, the integral

$$I = \int_0^1 \ln x dx \tag{10.113}$$

cannot be performed numerically since $\ln 0 = \infty$ and no numerical procedure will succeed. Conversely, the function $f(x) = x^{-\alpha}$ for $0 < \alpha < 1$ does exist:

$$\int_0^1 x^{-\alpha} dx = \frac{1}{1-\alpha} \tag{10.114}$$

Despite that the integral does exist, none of the integration formulas discussed here will work. Moreover, the function x^α with $\alpha > 0$ has a singularity in the first derivative at $x = 0$, and any Taylor series approximation to the function will thus fail.

There is no general rule on how to deal with integrals having singularities. There are several possible approaches, however, and we discuss these briefly. The first method is the removal of the singular part by performing the singular portion of the integral analytically, and then the remaining regular part can be performed numerically. For example,

$$\int_0^\pi (\ln x + e^{-x^2} \sin x)dx = \pi(\ln \pi - 1) + \int_0^\pi e^{-x^2} \sin x dx \tag{10.115}$$

and the second integral can be performed numerically. Second, a local series expansion can be used to remove the singularity. Consider the evaluation of the integral

$$I = \int_0^1 J_0(x)x^{-1/3}dx \tag{10.116}$$

where J_0 is the Bessel function of order 0. Near $x = 0$, the Bessel function behaves like

$$J_0 = 1 - \frac{x^2}{2^2} + O(x^4) \tag{10.117}$$

Thus, near $x = 0$, where the integrand is singular, the integral can be performed analytically with the rest of the integral being performed numerically. The integral can be evaluated by picking a small number, say, $x = 0.1$, and splitting the integral into two parts:

$$I = \int_0^{0.1} J_0(x)x^{-1/3}dx + \int_{0.1}^1 J_0(x)x^{-1/3}dx \tag{10.118}$$

For the first integral, the behavior of the Bessel function near $x = 0$ is used, say, the first two terms in the series, and the integration is performed analytically. The rest of the integral can be performed numerically.

Finally, sometimes a change in variable will regularize an integral. For example,

$$I = \int_0^b \frac{f(x)dx}{x^{1/2}} \tag{10.119}$$

can be regularized by letting $\eta = x^{1/2}$ so that

$$I = \int_0^b \frac{f(x)dx}{x^{1/2}} = 2\int_0^{\sqrt{b}} \eta f(\eta^2)d\eta \tag{10.120}$$

It should be mentioned that singularities are often the sign of a flaw in a model of a given system. Models of physical systems should not contain singularities unless some physical aspect of the system is not considered. An example is a boundary layer. From an inviscid point of view, the velocity profile is discontinuous near a solid surface. However, if viscous forces are included, the velocity profile is regularized.

10.9 Solution of linear systems

Sets of linear equations occur very often in practice. Consider, for example, the ordinary differential equation

$$y'' + my = e^{-x} \tag{10.121}$$

where m is a constant subject to the boundary conditions $y(a) = 1, y(b) = 0$. To solve this equation numerically, the interval (a, b) is broken up into $N - 1$ smaller intervals each of width $(h = b - a)/(N - 1)$; recall that this was also done in the section on integration, as depicted in Figure 10.15. Let j denote a mesh point; approximating the second derivative by a central difference, then at j,

$$\frac{y_{j+1} - 2y_j + y_{j-1}}{h^2} + my_j = e^{-x_j} \quad \text{for} \quad j = 2, \ldots, N-1 \tag{10.122}$$

or

$$y_{j-1} - (2 - mh^2)y_j + y_{j+1} = h^2 e^{-x_j} \quad \text{for} \quad j = 2, \ldots, N-1 \tag{10.123}$$

These are $N - 2$ equations in $N - 2$ unknowns, and the system is said to be *tridiagonal*. The solution to sets of linear equations is required in curve-fitting problems, as has been seen.

For a small number of equations, say, $N \leq 3$ or 4, you were probably taught to solve them by Cramer's rule. Let us consider the 2 by 2 system

$$a_{11}x_1 + a_{12}x_2 = b_1$$

$$a_{21}x_1 + a_{22}x_2 = b_2$$

with $b_1 \neq 0$. Then Cramer's rule provides the solution as

$$x_1 = \frac{\begin{vmatrix} b_1 & a_{12} \\ b_2 & a_{22} \end{vmatrix}}{\begin{vmatrix} a_{11} & a_{12} \\ a_{21} & a_{22} \end{vmatrix}}, \; x_2 = \frac{\begin{vmatrix} a_{11} & b_1 \\ a_{21} & b_2 \end{vmatrix}}{\begin{vmatrix} a_{11} & a_{12} \\ a_{21} & a_{22} \end{vmatrix}} \tag{10.124}$$

as long as the determinant $a_{11}a_{22} - a_{12}a_{21} \neq 0$. Cramer's rule is easy for $N = 2$ but really gets complicated for N larger. It can be shown, moreover, that the number of multiplications for N equations is $M_c = (N - 1)(N + 1)$. For $N = 20$, $M_c = 97 \times 10^{19}$ multiplications. Clearly a method is required that requires far fewer operations because 50 or 100 or 10,000 or more equations will often be required to solve. Gauss elimination is considered first, which is useful for N up to several hundred, and then iterative methods are considered, which should be used for the really large systems that occur in solving partial differential equations.

Gauss elimination is the main method of solving sets of linear equations, particularly when the matrix of coefficients is not sparse, meaning the matrix is "full." The idea is to reduce one of the equations in the set to one that can be solved immediately for one of the variables and then solve for the rest.

Consider the set of equations

$$
\begin{aligned}
2x_1 + x_2 - 3x_3 &= -1 \\
-x_1 + 3x_2 + 2x_3 &= 12 \\
3x_1 + x_2 - 3x_3 &= 0
\end{aligned}
$$

The matrix of coefficients is given by

$$
\begin{bmatrix} 2 & 1 & -3 \\ -1 & 3 & 2 \\ 3 & 1 & -3 \end{bmatrix} \tag{10.125}
$$

The procedure is to systematically eliminate elements in the matrix of coefficients to get this system in the form of $x_3 = b$, where b is a known constant. To do this, multiply the first equation by $+1/2$ and add to the second equation to obtain

$$
\begin{aligned}
2x_1 + x_2 - 3x_3 &= -1 \\
0 + \frac{7}{2}x_2 + \frac{1}{2}x_3 &= 11\frac{1}{2} \\
3x_1 + x_2 - 3x_3 &= 0
\end{aligned}
$$

Eliminate x_1 in the third equation by multiplying the first equation by $-3/2$ and adding to third, which, after some simplification, yields

$$
\begin{aligned}
2x_1 + x_2 - 3x_3 &= -1 \\
7x_2 + x_3 &= 23 \\
-x_2 + 3x_3 &= 3
\end{aligned}
$$

Now x_2 is eliminated in the third equation by multiplying the second equation by $1/7$ and adding to the third equation:

$$
\begin{aligned}
2x_1 + x_2 - 3x_3 &= -1 \\
7x_2 + x_3 &= 23 \\
3\frac{1}{2}x_3 &= \frac{23}{7} + 3
\end{aligned}
$$

Now we can solve for $x_3 = 2$ and then back substitute to get x_2 and x_1:

$$
\begin{aligned}
x_2 &= \frac{23 - x_3}{7} = 3 \\
x_1 &= \frac{-1 - x_2 + 3x_3}{2} = 1
\end{aligned}
$$

The final form of the set of equations is called *upper triangular form*. For the Gauss elimination procedure, the number of operations can be shown to be $M_G = (1/3)N^3 + N^2 - (1/3)N$, which, for $N = 20$, $M_G = 3060$, much fewer than for Cramer's rule.

This procedure is particularly amenable to programming because the procedure systematically eliminates elements in each column; let us write the system of equations as

$$\begin{aligned} a_{11}x_1 + a_{12}x_2 + \cdots + a_{1N}x_N &= b_1 \\ a_{21}x_1 + a_{22}x_2 + \cdots + a_{2N}x_N &= b_2 \\ &\vdots \\ a_{N1}x_1 + a_{N2}x_2 + \cdots + a_{NN}x_N &= b_N \end{aligned}$$

and the converted system as

$$\begin{aligned} s_{11}x_1 + s_{12}x_2 + \cdots + s_{1N}x_N &= d_1 \\ s_{22}x_2 + \cdots \qquad &= d_2 \\ &\ddots \\ s_{N-1N-1}x_{N-1} + s_{N-1N}x_N &= d_{N-1} \\ s_{NN}x_N &= d_N \end{aligned}$$

Then the back substitution procedure is

$$\begin{aligned} x_N &= \frac{d_N}{s_{NN}} \\ s_{N-1N-1}x_{N-1} + s_{N-1N}x_N &= d_{N-1} \\ &\vdots \end{aligned}$$

and so on down to

$$\sum_{j=1}^{N} s_{1j}x_j = d_1$$

To generate the s_{ij}s, which is the elimination process,

$$\begin{aligned} s_{1j} &= a_{1j} \quad j = 1, N \\ d_1 &= b_1 \end{aligned}$$

That is, the first line is left alone. For the second row,

$$\begin{aligned} s_{2j} &= a_{2j} - \frac{a_{21}}{a_{11}}a_{1j} \quad j = 2, 3, 4, \ldots, N \\ d_2 &= b_2 - \frac{a_{21}}{a_{11}}b_1 \end{aligned}$$

To successively remove the coefficients in the first row, multiply the first equation by a_{21}/a_{11} and subtract from the second equation. This eliminates the x_1 term in the second equation. The general term is

$$s_{ij} = a_{ij} - \frac{a_{i1}}{a_{11}} a_{1j} \quad i \geq 2, j = 2, 3, \ldots, N$$
$$d_i = b_i - \frac{a_{i1}}{a_{11}} b_1$$

In this way, the coefficients multiplying x_1 in every equation are eliminated. Next eliminate the coefficients of x_2 in the third through Nth equations, performing the same operations on the second column, and so on.

In the scheme, it was assumed that $a_{ii} \neq 0$, and so all of the divisions of coefficients can be performed accurately. However, if one of the diagonal elements a_{ii} is small, then significant error may be incurred on division by that element. *Pivoting* prevents the error that would be incurred in a division by a small diagonal coefficient (i.e., if one $a_{ii} \sim 0$); in addition, pivoting can be used to improve the accuracy of the scheme as well. Basically, in pivoting, the order of the equations is altered. For example, consider the following system:

$$.0001x_1 + 1.0000x_2 = 1.0000$$
$$1.0000x_1 + 1.0000x_2 = 2.0000$$

If the computer has no round-off error, then the exact solution is $x_1 = 1.0001$ and $x_2 = 0.9999$. But every computer has round-off error, so let's suppose that the computer rounds to four digits. Then, without pivoting;

$$0.0001x_1 + 1.0000x_2 = 1.0000$$
$$-1.000 \times 10^4 x_2 = -1.000 \times 10^4$$

and to four significant digits,

$$x_2 = 1.0000 \quad x_1 = 0.0000 \tag{10.126}$$

This answer is clearly wrong! To pivot, merely change the order of the equations:

$$1.0000x_1 + 1.0000x_2 = 2.0000$$
$$0.0001x_1 + 1.0000x_2 = 1.0000$$

Now, to four digits, $x_1 = 1.0001$ and $x_2 = 0.9999$.

10.9.1 Solving sets of linear equations in Matlab

There are two ways to solve a linear system in Matlab. If the linear system is defined by $Ax = b$, where A is the matrix of coefficients of the linear system, then left division yields the solution $x = A\ b$. Conversely, the inverse of the matrix A can be calculated; the inverse is defined as that matrix A^{-1} such that $A^{-1}A = I$, where I is the identity matrix. Then the solution x is given by the statement $x = A^{-1} * b$ or by $x = inv(A) * b$, where $inv(A)$ denotes the inverse; check the Matlab help for cautions on the use of the inv command. For example, in the problem just discussed, the answer in Matlab is produced by the script

```
>> A = [0.0001 1
        1    1]

A=

     0.0001 1
        1   1

>>inv(A)

ans=

       -1.0001  1.0001
        1.0001 -0.0001

>> b=[2
       1]

b=

     1
     2

>> x=A\b

x=

      1.0001
      0.9999
```

10.9.2 Iterative solution to linear systems

Gauss elimination is fine for linear systems of less than roughly several hundred equations; for larger numbers of equations, iterative methods are often faster and more efficient. Moreover, these methods can be faster than Gauss elimination when the coefficient matrix is sparse, for example, tridiagonal systems that have elements on the diagonal of the matrix and the two diagonals adjacent.

As an example, consider the following problem:

$$8x_1 + x_2 - x_3 = 8$$

$$2x_1 + x_2 + 9x_3 = 12$$

$$x_1 - 7x_2 + 2x_3 = -4$$

In each equation, we solve for the variable with the largest coefficient; rounding to three digits,

$$x_1 = 1 - .125x_2 + .125x_3$$

$$x_2 = .571 + .143x_1 + .286x_3$$

$$x_3 = 1.333 - .222x_1 + .111x_2$$

We begin with some initial estimate of the solution, say, $x_1 = x_2 = x_3 = 0$, and successively compute values using the previously calculated values. If n denotes iteration, then, say,

$$x_1^{n+1} = 1 - .125x_2^n + .125x_3^n \tag{10.127}$$

This procedure is called *Jacobi iteration*. The following iteration scheme is generated for a zero initial guess:

Iteration	x_1	x_2	x_3
1	1	.571	1.333
2	1.095	1.095	1.048
3	.995	1.026	.969
4	.993	.990	1.000
5	1.002	.998	1.004
6	1.001	1.001	1.001
7	1.000	1.000	1.000

The Jacobi iteration scheme is, in general,

$$x_i^{n+1} = \frac{b_i}{a_{ii}} - \sum_{j=1}^{N} \frac{a_{ij}}{a_{ii}} x_j^{(n)} \quad i \neq j \quad n = 1, \ldots \tag{10.128}$$

Note that all the values of x_j on the right-hand side of the equation are at the previous or (nth) iteration.

We can do much better by using updated x_j as soon as they are generated. For example, in the Jacobi solution of the previous equation,

$$x_2^{(n+1)} = .571 + .143x_1^{(n)} + .286x_3^{(n)} \tag{10.129}$$

Because x_1 has already been updated, it is more efficient to use the latest value:

$$x_2^{(n+1)} = .571 + .143x_1^{(n+1)} + .286x_3^{(n)} \tag{10.130}$$

This is called the *Gauss–Seidel iteration*. The Gauss–Seidel iteration produces the following:

Iteration	x_1	x_2	x_3
1	1	.714	1.032
2	1.041	1.014	.990
3	.997	.996	1.002
4	1.001	1.000	1.000
5	1.000	1.000	1.000

Note that this iteration scheme converges faster than the Jacobi scheme and would always be employed in practice. The Gauss–Seidel iteration is, in general,

$$x_i^{n+1} = \frac{b_i}{a_{ii}} - \sum_{j=1}^{i-1} \frac{a_{ij}}{a_{ii}} x_j^{(n+1)} - \sum_{j=i+1}^{N} \frac{a_{ij}}{a_{ii}} x_i^n \quad n = 1, 2, 3, \ldots \tag{10.131}$$

The *successive overrelaxation method* or (SOR) speeds up Gauss–Seidel further by adding a *relaxation factor* to the solution:

$$x_i^{n+1} = \alpha x_i^{n+1} + (1 - \alpha) x_i^n \tag{10.132}$$

where $1 < \alpha < 2$. Typically $\alpha \sim 1.5$ works very well.

A sufficient condition for convergence of these iteration schemes is that

$$|a_{ii}| > \sum_{j=1}^{N} |a_{ij}| \quad i \neq j \; i = 1, \ldots, N \tag{10.133}$$

That is, the set of equations is *diagonally dominant.* Discretization of ordinary and partial differential equations, if done properly, will result in diagonally dominant systems.

Last, because these are iterative schemes, a test for convergence is required. In general, a relative test should be used; if $xold(i)$ denotes the solution at the nth iteration and $xnew(i)$ is the solution at the $n + 1$st iteration, then the convergence test is

$$\left| \frac{xnew(i) - xold(i)}{xnew(i)} \right| < eps \tag{10.134}$$

where eps is a small number, say, 10^{-4}. A Matlab program using the Gauss–Seidel iteration looks like the following:

```
% Gauss-Seidel solution of a set of linear equations.
function x=gseidel(A,b,max,epsi,n,guess)
% Solution of Ax=b
% xnew is the solution vector
% A is the coefficient matrix
% b is the right side of the linear system
% define new matrix C to be used in solving for xnew
C=-A ;
xold=guess;
xnew=xold;
xnew=xnew';
xold=xold';
for i=1:n
     C(i,i)=0;
   end
```

```
for i=1:n
   C(i,1:n)=C(i,1:n)/A(i,i);
end
for i=1:n
   d(i,1)=b(i)/A(i,i);
end
iter=1;
while(iter <= max)
   xold=xnew;
   for j=1:n
      xnew(j)=C(j,:) * xnew + d(j,1);
   end
   if abs((xnew-xold)./xnew)<epsi
      x=xnew
      disp('Gauss-Seidel iteration scheme converged.')
      disp(iter);
      return;
   end
   %       disp([i   xnew']);
           iter=iter+1;
   end
      disp('Result after max iterations:')
      x=xnew;
```

The colon in the column entry, $C(j, :)$, means to perform the operation for all columns.

The preceding Matlab m-file is self contained. If a separate file is used to define the matrix coefficients, and for the actual Gauss–Seidel iteration, after definition of matrix A and the right-hand side in the calling program, the solution is obtained by writing $x = gseidel(A, b, max, epsi, n, guess)$.

10.9.3 Tridiagonal systems

As has been mentioned, tridiagonal systems arise in the solution of ordinary and partial differential equations. One example is the nonlinear ordinary differential equation for the potential in an electrolyte solution. Consider a system of equations, say, $N = 5$:

$$\begin{aligned}
a_1x_1 + c_1x_2 &= d_1 \\
b_2x_1 + a_2x_2 + c_2x_3 &= d_2 \\
b_3x_2 + a_3x_3 + c_3x_4 &= d_3 \\
b_4x_3 + a_4x_4 + c_4x_5 &= d_4 \\
b_5x_4 + a_5x_5 &= d_5
\end{aligned}$$

Let us work with the one-dimensional arrays a_i, b_i, c_i, d_i, $i = 1, \ldots, N$. Here it is necessary to eliminate only one element from each equation in each column; that is, x_1 from the second equation, x_2 from the third equation, and so on. To do this, multiply the first equation by $-b_2/a_1$ and add to the second equation:

$$a_1 x_1 + c_1 x_2 = d_1$$
$$\left(a_2 - c_1 \frac{b_2}{a_1}\right) x_2 + c_2 x_3 = d_2 - \frac{b_2}{a_1} d_1$$
$$b_3 x_2 + a_3 x_3 + c_3 x_4 = d_3$$
$$\vdots$$

We have eliminated x_1 from equation 2. Now eliminate x_2 from equation 3 by multiplying equation 2 by

$$-\frac{b_3}{a_2 - \frac{c_1 b_2}{a_1}} \tag{10.135}$$

and adding; the last three equations are as follows:

$$\left(a_3 - \frac{b_3 c_2}{a_2 - \frac{c_1 b_2}{a_1}}\right) x_3 + c_3 x_4 = d_3 - \frac{b_3 (d_2 - \frac{b_2}{a_1} d_1)}{a_2 - \frac{c_1 b_2}{a_1}}$$
$$b_4 x_3 + a_4 x_4 + c_4 x_5 = d_4$$
$$b_5 x_4 + a_5 x_5 = d_5$$

We can now see a pattern developing; if a superscript denotes the number of the elimination operation, then

$$a_2^{(1)} = a_2 - \frac{c_1 b_2}{a_1}$$
$$a_3^{(2)} = a_3 - \frac{b_3 c_2}{a_2^{(1)}}$$
$$d_2^{(1)} = d_2 - \frac{b_2 d_1}{a_1}$$
$$d_3^{(2)} = d_3 - \frac{b_3 d_2^{(1)}}{a_2^{(1)}}$$

The general form of these newly formed arrays, say, α and γ, is

$$\alpha_{i+1} = a_{i+1} - \frac{b_{i+1} c_i}{\alpha_i} \quad for\ i = 1, \ldots, N-1 \text{ and } \alpha_1 = a_1 \tag{10.136}$$

$$\gamma_{i+1} = d_{i+1} - \frac{b_{i+1} \gamma_i}{\alpha_i} \quad for\ i = 1, \ldots, N-1 \text{ and } \gamma_1 = d_1 \tag{10.137}$$

The system eventually ends up with all terms below the diagonal eliminated, and this form is

$$a_1x_1 + c_1x_2 = d_1$$
$$\alpha_2x_2 + c_2x_3 = \gamma_2$$
$$\alpha_3x_3 + c_3x_4 = \gamma_3$$
$$\alpha_4x_4 + c_4x_5 = \gamma_4$$
$$\alpha_5x_5 = \gamma_5$$

We now back substitute

$$x_5 = \frac{\gamma_5}{\alpha_5}$$
$$x_4 = \frac{\gamma_4 - c_4x_5}{\alpha_4}$$
$$\vdots$$
$$x_i = \frac{\gamma_i - c_ix_{i+1}}{\alpha_i}$$

for $i = N - 1, \ldots, 1$.

As an example, consider the system

$$x_1 + 2x_2 = 2$$
$$x_1 + 3x_2 + 2x_3 = 7$$
$$x_2 + 4x_3 = 15$$

Computing the arrays α, γ:

$$\alpha_1 = a_1 = 1$$
$$\alpha_2 = a_2 - \frac{b_2c_1}{\alpha_1} = 3 - \frac{1 \times 2}{1} = 1$$
$$\alpha_3 = a_3 - \frac{b_3c_2}{\alpha_2} = 4 - \frac{1 \times 2}{1} = 2$$
$$\gamma_1 = 2$$
$$\gamma_2 = d_2 - \frac{b_2\gamma_1}{\alpha_1} = 7 - \frac{1 \times 2}{1} = 5$$
$$\gamma_3 = 15 - (1)(5) = 10$$
$$x_3 = \frac{\gamma_3}{\alpha_3} = \frac{10}{2} = 5$$
$$x_2 = \frac{5 - (2)(5)}{1} = -5$$
$$x_1 = \frac{2 - (2)(-5)}{1} = 12$$

A Matlab program for this elimination algorithm is

```
% Tridiagonal system solver.
function x=tridiag(a,b,c,d,n)
alpha(1)=a(1);
gamma(1)=d(1);
for i=1:n-1
   alpha(i+1)=a(i+1)-b(i+1)*c(i)/alpha(i);
   gamma(i+1)=d(i+1)-b(i+1)*gamma(i)/alpha(i);
end
x(n)=gamma(n)/alpha(n)
for i=n-1:-1:1
   x(i)=(gamma(i)-c(i)*x(i+1))/alpha(i)
end
```

Different schemes use slightly different elimination procedures. The algorithm that goes by the name of the *Thomas algorithm* defines the two arrays α_i and γ_i as

$$\alpha_1 = \frac{c_1}{a_1}, \gamma_1 = \frac{d_1}{a_1}$$

$$\alpha_i = \frac{c_i}{a_i - b_i c_{i-1}}, \gamma_i = \frac{d_i - b_i d_{i-1}}{a_i - b_i c_{i-1}} \quad \text{for } i = 2, \ldots, N-1$$

with the back substitution step as

$$x_n = \frac{\gamma_n}{\alpha_n}$$

$$x_i = \frac{(\gamma_i - c_i x_{i+1})}{\alpha_i} \quad \text{for } i = N-1, \ldots, 2$$

The Thomas algorithm is extremely popular, but both these elimination schemes work well.

10.9.4 Ill-conditioning and stability

A matrix is said to be *ill-conditioned* if there is a large difference between the largest and the smallest eigenvalues in a system defined as the one-dimensional array of real numbers λ that satisfies $Ax = \lambda x$. The *condition number* is defined as the difference between the largest and smallest eigenvalues. Consider the following system:

$$x_1 + \frac{1}{2}x_2 = \frac{3}{2}$$

$$\frac{1}{2}x_1 + \frac{1}{3}x_2 = \frac{5}{6}$$

for which the *condition number* is 19.281 (Fausett, 2008). The Matlab command to compute the condition number is *cond*, and the script here may look as follows:

```
> A=[1 1/2
     1/2  1/3]
> A=   1.0000         0.5000
       0.5000        0.3333
> cond(A)
> ans =


     19.281

>
```

A condition number that is much greater than 1 indicates a high possibility of ill conditioning.

We have already seen what kinds of errors can occur without pivoting. Difficulties also occur if, say, the coefficients on the right-hand side are not known for certain. For example, the function on the right-hand side of an ordinary differential equation may be the result of experimental data; this often happens. The uncertainty can drive a system into an ill-conditioned state, rendering the calculation of the solution *unstable* in the sense that a small change in the coefficients can lead to large differences in the solution.

Consider, for example, the following system, which is chopped at two significant digits:

$$3.02x_1 - 1.05x_2 + 2.53x_3 = -1.61$$

$$4.33x_1 + .56x_2 - 1.78x_3 = 7.23$$

$$-0.83x_1 - 0.54x_2 + 1.47x_3 = -3.38$$

Solving this system using Gauss elimination with pivoting yields the reduced system

$$4.33x_1 + .56x_2 - 1.78x_3 = 7.23$$

$$0x_1 - 1.44x_2 + 3.77x_3 = -6.65$$

$$0x_1 + 0x_2 - .00362x_3 = .00962$$

keeping three digits, except in the last equation, where we keep five. The solution to this reduced system is

$$x = (0.88, -2.35, -2.66) \tag{10.138}$$

whereas the true solution is

$$x = (1, 2, -1) \tag{10.139}$$

Thus more precision is required here, even though the right-hand side data are only known to, say, two digits. Moreover, if a small change in one of the coefficients is made, the solution to the linear syatem will be much different. Try changing

0.56; in the second equation to 0.50; then

$$x = (0.94, 0.12, -1.72) \qquad (10.140)$$

Very unstable indeed!

In the next two sections, we apply the methods of solving linear systems of equations to the solution of ordinary differential equations.

10.10 Solution of boundary value problems

10.10.1 Introduction

Numerical differentiation and the methods of solution of linear systems provide the means to solve both linear and nonlinear ordinary differential equations. In this section, the focus is on the solution of second-order boundary value problems. Higher-order equations, for example,

$$z^{(iv)} + z' = f(x) \qquad (10.141)$$

which often arise in the field of solid mechanics, can be solved as a system of second-order equations. Recall that the Navier–Stokes equations and the Poisson equation for the potential are second-order equations. For this fourth-order equation, define

$$\begin{aligned} z' &= s \\ w &= s' \quad = z'' \\ w'' + s &= f(x) \end{aligned} \qquad (10.142)$$

Each of these equations can be solved for the variables (z, s, w).

In general, second-order linear equations are of the form

$$y'' + p(x)y' + r(x)y = f(x) \qquad (10.143)$$

where p, r, and f are known functions of x . A second-order nonlinear equation is one, for example, like

$$y'' + yy' = f(x) \qquad (10.144)$$

or, in a more general form,

$$q(x, y, y')y'' + p(x, y, y')y' + r(x)y = f(x) \qquad (10.145)$$

If $f(x) = 0$, the equation is said to be *homogeneous*.

Second-order equations of the boundary value type must satisfy boundary conditions at the end points of the interval. In general, as has been discussed before for specific equations, they are usually of four types:

1. End values specified: $y(a) = A, y(b) = B$
2. End derivatives specified: $y'(a) = dy/dx(a) = C, y'(b) = dy/dx(b) = D$

3. Linear combinations of y, y' specified: $Ey' + Fy = C,\ x = a,\ Hy' + Ry = S, x = b$, where E, F, C, H, R, S are constants
4. Some combination of type 1, 2, or 3 or some other nonlinear conditions: $y'(a) + y^2(b) = 0,\ \ y(b) = B$

The boundary conditions are determined by the physical conditions imposed on the boundary such as the no-slip and slip condition in fluid mechanics, the surface charge density condition for the electrical potential, or the no-flux condition for a chemical species in an electrolyte solution at a solid wall.

At this point, consider the simplest type of equation, the linear equation, subject to specified end values. It should be pointed out that in many cases, closed-form solutions are available such as when p, r in equation (10.143) are constants and $f = 0$.

10.10.2 Linear equations

Often linear equations have an analytical solution in the form of simple functions such as the exponential function or polynomials. However, sometimes the solution to a given equation is so complicated that it is just as easy to solve the problem numerically. For example, evaluating a Bessel function of order zero (Abramowitz & Stegun, 1972) that arises in cylindrical domains and is the solution of the linear ordinary differential equation

$$x^2y'' + xy' + x^2y = 0 \tag{10.146}$$

involves evaluating a complicated integral numerically.

Thus consider the equation

$$y'' + p(x)y' + r(x)y = f \tag{10.147}$$

subject to $y(a) = A,\ \ y(b) = B$. The interval $[a, b]$ is divided into $N - 1$ intervals of equal length (N points), using second-order central differences to approximate the derivatives and evaluating the functions p, r, f at x_j, results in

$$\frac{y_{j+1} - 2y_j + y_{j-1}}{h^2} + p_j\frac{y_{j+1} - y_{j-1}}{2h} + r_jy_j = f_j + E_j \quad \text{for } j = 2, \ldots, N - 1 \tag{10.148}$$

Note that the end points are defined by $j = 1$ and $j = N$ and that E is the error incurred in the differencing and $E = O(h^2)$ for the central difference approximation. Multiplying by h^2 and collecting terms multiplying y_{j+1}, y_j and y_{j-1}, the tridiagonal system is obtained according to

$$y_{j-1}\left[1 - \frac{h}{2}p_j\right] + y_j\left[-2 + h^2r_j\right] + y_{j+1}\left[1 + \frac{h}{2}p_j\right]$$
$$= h^2f_j + h^2E_j \quad \text{for} \quad j = 2, \ldots, N - 1 \tag{10.149}$$

Table 10.12. Numerical solution for the linear ODE

Numerical	Exact
$y_0 = 0$	0
$y_1 = 0.18023$	0.17990
$y_2 = 0.38735$	0.38682
$y_3 = 0.64993$	0.64945
$y_4 = 1.0$	1.0

Provided that the solution y_j is well behaved, the error term, which, on multiplication by h^2, is $E = O(h^4)$, is neglected, and a solution accurate to three or four digits may be obtained by successively increasing the number of grid points, thereby reducing the grid size, that is, by increasing N. The value of N includes the end points, and so there are actually $N - 1$ intervals.

Let us solve the equation $\epsilon y'' - y = x$ with $y(0) = 0$, $y(1) = 1$ using $h = 1/4$ or $N = 5$. Here ϵ is a constant and $p = 0$, $r = -1$, $f = x$, and our equation becomes

$$\epsilon y_{j-1} - \left(2 + \epsilon h^2\right) y_j + \epsilon y_{j+1} = h^2 x_j \text{ with } x_j = jh \; for \; j = 1, \ldots, N-1 \tag{10.150}$$

Writing the tridiagonal system explicitly for $\epsilon = 1$ yields

$$-2.0625 y_2 + y_3 = .015625$$

$$y_2 - 2.0625 y_3 + y_4 = .031250$$

$$y_3 - 2.0625 y_4 = -.953125$$

This is a tridiagonal system of equations, which may be solved by either of our elimination methods. This equation does have an exact solution; for $\epsilon = 1$, it is

$$y = \frac{2}{e^{-1} - e^{1}} \left[e^{-x} - e^{x}\right] - x \tag{10.151}$$

The system of equations yields the solution found in Table 10.12.

Considering that only four intervals have been used ($N = 5$ in this numbering system), the agreement is very good. Comparing the numerical result with an exact solution is one way to verify that the numerical result is correct. However, in most cases, an exact solution will not be available, and as with integration, the value of N is increased systematically until the successive solutions agree to a specified tolerance. To do this, the solution would be calculated for $N = 11$, then $N = 21$, and so on until the desired accuracy is obtained. The reason for these values of N is that it is easy to compare the numbers because half of the points are at the same locations for each increase in N. When solutions for two different mesh sizes agree to within a specified tolerance, the solution is said to be *verified*. Most of the time, programming errors are caught when the grid is refined and the solutions do not agree. A typical Matlab file for this problem is

```
% First example in the linear ODE section.
epsilon=1;
       n=11;
          al=1.;
          dx=al/(n-1);
          n1=n-1;
      y(1)=0;
      y0=y(1);
      y(n)=1;
      yn=y(n);
      x=[0:dx:1];
    for j=1:m
          for i=2:n1
                a(i)=-(2.*epsilon+dx*dx);
                b(i)=epsilon;
                c(i)=epsilon;
                d(i)=x(i)*dx^2;
          end
          d(n-1)=-yn*c(n-1)+d(n-1);
          c(n-1)=0.;
          y=tridiag(n,a,b,c,d,y0,yn);
% yan is the analytical solution for epsilon=O(1) and yan2
% is the analytical solution for epsilon<< 1 near x=1
for i=1:n
       yan(i)=2./(exp(-1)-exp(1))*(exp(-x(i)/sqrt(epsilon))-
exp(x(i)/sqrt(epsilon)))-x(i);
        yan2(i)=2*exp(-(1-x(i))/sqrt(epsilon))-x(i);
end


figure(1);
plot(x,y,'*',x,yan2,'o');
xlabel('x');
ylabel('y');


end
```

Numerical solutions of this equation will be difficult as the value of ϵ decreases. As $\epsilon \to 0$, the second derivative becomes increasingly large near $x = 1$, and the solution approaches the result

$$y = 2e^{-(1-x)/\sqrt{\epsilon}} - x \tag{10.152}$$

The result for various values of ϵ is depicted in Figure 10.17.

As ϵ becomes smaller and approaches zero, we say that this is a *singular perturbation* problem because over much of the domain, $y = -x$, and thus we

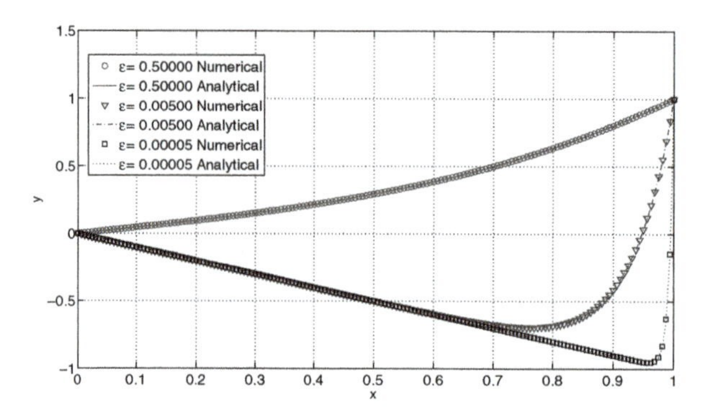

Figure 10.17 Results for several values of ϵ for the linear ordinary differential example. Note that finite difference algorithm is extremely robust and able to resolve the boundary layer, or region of rapid variation, down to $\epsilon = 5 \times 10^{-4}$.

cannot satisfy the boundary condition at $x = 1$. These types of problems were encountered in Chapter 7 and will be seen in subsequent chapters.

Notation alert: Here is where we run into another notation problem. In this chapter, h has denoted mesh size, but in Chapter 7, h denotes a channel height. To avoid confusion in the discretization of the potential equation between h the mesh size and h the channel height, subsequently, we will use H to denote the channel height and $\epsilon = \lambda/H$.

Specifically, recall that the potential in the electrical double layers between two plates satisfies, in the Debye–Hückel limit,

$$\epsilon^2 \frac{d^2\phi}{dx^2} = z^2\phi \tag{10.153}$$

for a solution containing a pair of ionic species of valence z. The resulting difference equations are

$$\epsilon^2\phi_{j-1} - (2\epsilon^2 + z^2h^2)\phi_j + \epsilon^2\phi_{j+1} = 0 \tag{10.154}$$

If the boundary conditions are taken to be $\phi = \zeta$ at $x = 0, 1$, where x is scaled on the channel height, then the first and last equations have a nonzero right-hand side: $\phi_0 = \phi_N = \zeta$. Here $\epsilon = \lambda/H$, where H is the channel height and is equivalent to ϵ as used in Chapter 7.

For $\epsilon \ll 1$, that is, for extremely thin electrical double layers (EDLs), we will need to solve the problem only within the EDL. Near $Y = 0$, we blow up the EDL by defining a *boundary layer variable* $Y = x/\epsilon$ so that the governing equation is

$$\frac{d^2\phi}{dY^2} = z^2\phi \tag{10.155}$$

subject to $\phi = \zeta$ at $Y = 0$. The other boundary condition must be applied at a large value of Y and in the limit $\epsilon \to 0$, $Y \to \infty$ so that if the core is electrically neutral, the boundary condition is $\phi = 0$ as $Y \to \infty$. But numerically, what is

∞? Well, as discussed in Chapter 7, for this problem, $\infty \sim 5$. In the general case, to simulate ∞, the boundary is systematically increased until the solution does not change away from the outer boundary. Of course, there is an analytical solution to this problem, which is

$$\phi(Y) = \zeta e^{-zY} \tag{10.156}$$

and it is not necessary to solve the problem numerically. Note that $Y = x^*/\lambda$, where x^* is dimensional, and so the boundary layer analysis is equivalent to taking the length scale to be the Debye length and not the channel height.

Normally for ordinary differential equations, the direct elimination methods are preferable; it is worth mentioning that the difference equations may also be solved using the iterative methods of Jacobi or Gauss–Seidel or, better, successive overrelaxation. However, the advantage of iterative methods is greatest when the number of equations is very large, say, 1000 or more, and this usually does not happen with ordinary differential equations. To use the iterative methods, the equations must be diagonally dominant, that is,

$$\left|-2 + h^2 r_j\right| \geq \left|1 + \frac{h}{2} p_j\right| + \left|1 - \frac{h}{2} p_j\right| \quad j = 2, \ldots N - 1 \tag{10.157}$$

It turns out that this condition will hold if the mesh size

$$h < \frac{2}{\left|p_j\right|} \quad j = 2, \ldots N - 1 \tag{10.158}$$

and $r_j < 0$. If $r_j > 0$, iterative methods can fail, although even if diagonal dominance is not satisfied, iterative methods may still converge. Difference equations derived from ordinary differential equations are usually diagonally dominant.

Numerical methods are usually required for nonlinear equations, and this subject is considered next.

10.10.3 Nonlinear equations

Notation alert: Just a reminder that the symbol h is used for mesh length in this chapter and not for channel height.

Nonlinear ordinary differential equations cannot normally be solved directly using the linear system solvers discussed earlier. However, the equation can first be linearized by assuming that some portion of the nonlinear term is known. Then the linear system solver can be used iteratively until the process converges. The specific steps are as follows:

1. Linearize the nonlinear terms in the difference equations by guessing their values.
2. Solve the difference equations by a direct method: the Thomas algorithm.
3. Return to step 1, unless two successive iterations differ by less than a prescribed tolerance.

Consider the following equation as an example:

$$y'' + p(x)y^2y' + r(x)y^3 = f(x) \tag{10.159}$$

$$y(a) = A, \quad y(b) = B \tag{10.160}$$

At a typical mesh point, the finite difference approximation to equation (10.159) is

$$y_{j+1} - 2y_j + y_{j-1} + \frac{h}{2}p_j y_j^2 \left[y_{j+1} - y_{j-1}\right] + h^2 r_j y_j^3 = h^2 f_j + O(h^4) \text{ for } j = 2, \ldots, N-1 \tag{10.161}$$

This is a tridiagonal system of equations for the $N - 2$ values of y_j at the interior grid points. The problem is that these equations are nonlinear; here the difference equation should be linearized. To do this, if n denotes the iteration number, for this equation,

$$y_{j+1}^n - 2y_j^n + y_{j-1}^n + \frac{h}{2}p_j \left(y_j^{n-1}\right)^2 \left[y_{j+1}^n - y_{j-1}^n\right] + h^2 r_j \left(y_j^{n-1}\right)^2 y_j^n = h^2 f_j \tag{10.162}$$

and rearranging,

$$y_{j-1}^n \left[1 - \frac{h}{2}p_j \left(y_j^{n-1}\right)^2\right] + y_j^n \left[-2 + h^2 r_j \left(y_j^{n-1}\right)^2\right] + y_{j+1}^n \left[1 + \frac{h}{2}p_j \left(y_j^{n-1}\right)^2\right] = h^2 f_j \tag{10.163}$$

Thus the coefficients of the linear system are

$$a_j = -2 + h^2 r_j \left(y_j^{n-1}\right)^2 \quad j = 2, \ldots, N-1$$

$$b_j = 1 - \frac{h}{2}p_j \left(y_j^{n-1}\right)^2 \quad j = 2, \ldots, N-1$$

$$c_j = 1 + \frac{h}{2}p_j \left(y_j^{n-1}\right)^2 \quad j = 2, \ldots, N-2$$

$$d_j = h^2 f_j \quad j = 2, N-1$$

Note that the boundary conditions in the Matlab program are used for $j = 2$ and $j = N - 1$; for example, if the function is specified at the end point, then

$$d_2 = h^2 f_2 - y_1 \left(1 - \frac{h}{2}p_1 \left(y_1^{n-1}\right)^2\right) \tag{10.164}$$

and depends on the boundary value y_1. The same is true for d_{N-1}.

In general, the solution is guessed initially, and then equations (10.163) are solved successively until

$$\left|1 - \frac{y_j^{n-1}}{y_j^n}\right| < eps \text{ all } j = 2, \ldots N-1 \tag{10.165}$$

where eps is a pre-determined tolerance; usually, $eps = 10^{-4}$ or smaller.

The success or failure of the iterative process can depend on the way the equation is linearized. The best way is to linearize the lowest-order terms in the group of nonlinear terms. For example, suppose the nonlinear term is

$$yy' \tag{10.166}$$

We would not linearize y':

$$yy' \sim yy'^p = y_j^n \frac{\left(y_{j+1}^{n-1} - y_{j-1}^{n-1}\right)}{2h} \tag{10.167}$$

Recall that the nonlinear form of the Poisson equation for the electric potential for binary and symmetric electrolytes is, in the notation of the present section,

$$\epsilon^2 \frac{d^2\phi}{dx^2} = z \sinh z\phi \tag{10.168}$$

In this case, there is no choice but to linearize the right-hand side, and the discretized version is

$$\epsilon^2 \phi_{j-1}^n - 2\epsilon^2 \phi_j^n + \epsilon^2 \phi_{j+1}^n = zh^2 \sinh z\phi_j^{n-1} \tag{10.169}$$

It should be mentioned here that these systems of nonlinear discretized equations can also be solved by using zero-finding techniques. However, I have never found that to be an efficient way to proceed, and those methods are prone to divergence (Gilat and Subramaniam, 2008). Of course, the method described here is also prone to divergence, but in my experience, linearizing the equations and using the direct or iterative solvers is the preferred method.

Often second-order equations are solved as two first-order equations; this approach is called the *shooting method* (Gilat and Subramaniam, 2008). In my opinion, the method is cumbersome to program and involves several estimates of initial values at the point $x = a$ and iteration until the boundary condition at $x = b$ is satisfied. Yet the method is very popular, and the Matlab function *bvp4c* uses this method to solve second-order boundary value problems. Nevertheless, I do not recommend this method.

An example of the solution of a nonlinear equation in the context of a system of two equations is discussed next.

10.10.4 Systems of ordinary differential equations

Consider the case where the system of equations is generated from a higher-order ordinary differential equation; consider the Blasius boundary layer problem

$$\begin{aligned} f''' + ff'' &= 0 \\ f'(0) = f(0) &= 0 \\ f'(\infty) &= 1 \end{aligned} \tag{10.170}$$

where, as noted earlier, the function f' is the velocity. This equation is most easily solved as a system of equations because the third derivative leads to a nontridiagonal system, which is more difficult and time consuming to solve. To convert this equation to a system of equations, write

$$f' = g \quad f = \int_0^x g dx + f_0; \quad f_0 = f(0) = 0 \tag{10.171}$$

$$g'' + fg' = 0 \quad g(0) = 0, \quad g(\infty) = 1 \tag{10.172}$$

To determine ∞, we follow the steps mentioned previously and assume that x_∞ is some "large" number; usually 5–8 is enough, but the precise number depends on the problem. Several values of x_∞ should be checked to make sure the solution does not change much for the smaller values of x. The quantity x_∞ should be as small as possible, obviously. The discretization of equation (10.172) results in

$$\frac{g_{j+1} - 2g_j + g_{j-1}}{h^2} + f_j \frac{(g_{j+1} - g_{j-1})}{2h} = 0 \tag{10.173}$$

and multiplying by h^2,

$$g_{j-1}\left[1 - \frac{h}{2}f_j\right] - 2g_j + g_{j+1}\left[1 + \frac{h}{2}f_j\right] = 0 \quad j = 2, \ldots, N-1 \tag{10.174}$$

Using the trapezoidal rule with the first internal mesh point labeled as $j = 2$,

$$f_j = \frac{h}{2}\left(g_{j-1} + g_{j+1}\right) + f_{j-1} \quad j = 2, \ldots, N \tag{10.175}$$

The problem is that f_j in equation (10.174) is unknown. The final *linearized* equation is thus

$$g^n_{j-1}\left[1 - \frac{h}{2}f^{n-1}_j\right] - 2g^n_j + g^n_{j+1}\left[1 + \frac{h}{2}f^{n-1}_j\right] = 0 \quad j = 2, \ldots, N-1 \tag{10.176}$$

where n denotes the latest iteration level. Note that only $d_{N-1} \neq 0$ because $y(N) = 1$ and thus

$$g^n_{N-2}\left[1 - \frac{h}{2}f^{n-1}_{N-1}\right] - 2g^n_{N-1} = -g^n_N\left[1 + \frac{h}{2}f^{n-1}_{N-1}\right] \tag{10.177}$$

The procedure here would be as follows:

- Specify x_∞, N.
- Guess f' or g, $f' = x$, say, $g = x^2/2$.
- Put the guessed value of f into equation (10.176) .
- Solve using the tridiagonal system solver.
- Integrate to get f.
- Check for convergence.

A Matlab program similar to the linear ODE solver can be written by adding a convergence step, which may look as follows:

```
jc=0
for i=2:n-1;
small(i)=abs((f(i)-fold(i))/f(i));
```

```
if small(i)>eps
jc=1
end
```

Here "jc" is a parameter that signifies whether convergence has been achieved; here "jc = 1" means that the scheme is not converged. This relative test requires all entries of the *small* array to be less than the convergence criteria *eps* "(jc = 0); if only one entry is greater than *eps*, the tridiagonal system must be solved again.

10.10.5 Derivative boundary conditions

As noted in Chapter 3, many heat, mass, and electrochemical problems involve derivative boundary conditions. This entails some adjustments to the tridiagonal solver, and there are two ways to deal with the derivative boundary condition. Consider now the equation

$$y'' + p(x)y' + r(x)y = f(x) \tag{10.178}$$

but now with

$$y_0 = y(a) = A, \;\; y'(b) = B = \text{constant} \tag{10.179}$$

The discretization of equation (10.178) leads to the set of difference equations of the form

$$b_j y_{j-1} + a_j y_j + c_j y_{j+1} = h^2 f_j \;\; j = 1, \ldots, N-1 \tag{10.180}$$

However, note that $y(b)$ (y_N) is unknown, and so the standard back substitution procedure is not possible. There are two ways to deal with this issue.

CENTRAL DIFFERENCES

Here assume that the equation holds at $x_N = b$. Thus, for $j = N$,

$$b_N y_{N-1} + a_N y_N + c_N y_{N+1} = d_N = h^2 f_j \tag{10.181}$$

where y_{N+1} is a fictitious point outside the boundary. To eliminate y_{N+1}, approximate the derivative at $x_N = b$ by central differences:

$$\frac{y_{N+1} - y_{N-1}}{2h} = B$$

$$y_{N+1} = 2hB + y_{N-1} \tag{10.182}$$

Substituting into equation (10.181), the last equation becomes

$$b_N y_{N-1} + a_N y_N + c_N \left(2hB + y_{N-1}\right) = d_N = h^2 f_j \tag{10.183}$$

and on rearranging,

$$(b_N + c_N)\, y_{N-1} + a_N y_N = d_N - 2hBc_N \tag{10.184}$$

The tridiagonal system now has another equation (10.184) added to it, and so the system of equations to be solved is

$$b_j y_{j-1} + a_j y_j + c_j y_{j+1} = d_j \quad j = 2, \ldots, N-1$$
$$(b_N + c_N) y_{N-1} + a_N y_N = d_N - 2hBc_N \tag{10.185}$$

and there are $N-1$ equations in $N-1$ unknowns instead of $N-2$ equations in $N-2$ unknowns.

As a simple example, consider the equation $y'' - y = x$ with $y(0) = 0, y'(1) = 1$, and using $h = 1/4$. Discretization leads to

$$b_j = c_j = 1, \quad a_j = -\left(2 + h^2\right) = -2.0625 \tag{10.186}$$

So the equation at $x = b = x_N$ would read, with $B = 1$,

$$(1+1)y_{N-1} - 2.0625 y_N = .0625 - 2 \times .25 \times 1 \times 1$$
$$2y_{N-1} - 2.0625 y_N = -.4375$$

A weakness of this method is that the error in our difference equations is $O(h^4)$ (i.e., $h^2 \times h^2$) when multiplying through by h^2; however, the Nth equation, because a central difference equation is used, becomes only $h^2 \times h = h^3$, accurate after multiplication by h. This criticism is somewhat overshadowed by the ease of incorporation into the numerical scheme.

SLOPING DIFFERENCES

Instead of using a central difference formula, a sloping difference formula could be used for the derivative:

$$y'(b) = \frac{3y_N - 4y_{N-1} + y_{N-2}}{2h} + O\left(h^2\right) \tag{10.187}$$

Multiplying through by h, it is seen that the equation is only $O\left(h^3\right)$ again. But adding another term to equation (10.187) by suitable manipulation of Taylor series expressions for the first derivative results in

$$y'(b) = \frac{11y_N - 18y_{N-1} + 9y_{N-2} - 2y_{N-3}}{6h} + O\left(h^3\right) \tag{10.188}$$

which, when multiplying through by $6h$, is $O\left(h^4\right)$ accurate. To incorporate equation (10.187) into the numerical scheme, suppose we have used the Thomas algorithm and computed the arrays $[\alpha_i]$, $[\gamma_i]$. Then recall that

$$y_i = \frac{\gamma_i - c_i y_{i+1}}{\alpha_i} \quad i = N-1, \ldots, 1 \tag{10.189}$$

assuming y_0 is the left end point. If $y'(b) = B$, using equation (10.187),

$$3y_N - 4y_{N-1} + y_{N-2} = 2hB \tag{10.190}$$

and using equation (10.189),

$$y_{N-1} = \frac{\gamma_{N-1} + c_{N-1}y_N}{\alpha_{N-1}}$$

$$y_{N-2} = \frac{\gamma_{N-2} + c_{N-2}y_{N-1}}{\alpha_{N-2}}$$

$$y_{N-2} = \frac{\gamma_{N-2} + c_{N-2}\left(\gamma_{N-1} + c_{N-1}y_N\right)/\alpha_{N-1}}{\alpha_{N-2}}$$

Substituting these two values into equation (10.190),

$$3y_N - \frac{4\left(\gamma_{N-1} + c_{N-1}y_N\right)}{\alpha_{N-1}} + \frac{\gamma_{N-2} + \left(c_{N-2}\gamma_{N-1} + c_{N-2}c_{N-1}y_N\right)/\alpha_{N-1}}{\alpha_{N-2}} = 2hB \tag{10.191}$$

Solving for y_N,

$$\begin{aligned}&\left[3\alpha_{N-1}\alpha_{N-2} - 4c_{N-1}\alpha_{N-2} + c_{N-2}c_{N-1}\right]y_N\\&= 2hB\alpha_{N-1}\alpha_{N-2} + 4\alpha_{N-2}\gamma_{N-1} - \alpha_{N-1}\gamma_{N-2} - c_{N-2}\gamma_{N-1}\end{aligned} \tag{10.192}$$

Equation (10.192) determines y_N because everything else is known. The same can be done with equation (10.188); however, the algebra is rather complicated. When equation (10.188) is used, the accuracy problem is solved.

10.10.6 Convergence tests and Richardson extrapolation

As mentioned earlier, the solutions to the nonlinear boundary value problems and other nonlinear problems require the iterative calculations of the solutions to linearized systems of equations. Thus programs must include a test for solution, and if y denotes the solution to the system of equations, this test at the nth iteration is written

$$\left|\frac{y_j^n - y_j^{n-1}}{y_j^n}\right| < eps \text{ all } j \tag{10.193}$$

or in the present notation, all the internal mesh points $2 \le j \le N-1$. It is important to note that this equation must hold for each point in the mesh j. This is a *relative* test that indicates how many significant digits have been retained in the iterative process; for example, if $eps = 10^{-4}$, this formula means that successive iterates must agree to within four significant digits at each internal mesh point. This test is independent of the magnitude of the dependent variable y, in contrast to the absolute test, which omits the denominator in equation (10.193). In that case, if $y \ll O(1)$, the algorithm may converge, even if there are no significant digits that agree in successive iterations.

In any case, common sense should govern the choice of a convergence *criterion*, the value of eps. For very small values of y, the relative test may fail, and in that case, the absolute test can be used locally. In this case, we would usually be satisifed with testing down to a minimum value of y, say, $y \geq 10^{-7}$.

Another test for the accuracy of the algorithm is to use Richardson extrapolation. The idea is that if the formulas used to approximate the derivatives of a boundary value problem are $O(h^2)$ accurate, then at any internal mesh point, the solution is

$$y_j^1 = y_{Tj} + Ah^2 + Bh^4 + \cdots \tag{10.194}$$

where A and B are constants associated with the error in the Taylor series approximation to the derivative and y_{Tj} is the "true" solution. As noted earlier, to determine accuracy, we would reduce the step size h to $h/2$ and repeat the calculation. Thus we have a second solution of the form

$$y_j^2 = y_{Tj} + A\frac{h^2}{4} + B\frac{h^4}{16} + \cdots \tag{10.195}$$

if it is assumed that A and B are approximately constant. The term Ah^2 may be eliminated, and solving for the "true" solution,

$$y_{Tj} = \frac{4y_j^2 - y_j^1}{3} - \frac{Bh^4}{4} + \cdots \tag{10.196}$$

at each mesh point. The key assumption is that A and B are approximately constant in the interval $(x_j - \frac{h}{2^n}, x_j + \frac{h}{2^n})$. This will not normally be exact; however, the approximation will get better as the mesh size decreases. Extrapolation by itself is risky, and we would use this Richardson deferred approach to the limit after halving the interval twice and comparing the extrapolated result with the finest mesh calculation.

10.10.7 Solving boundary value problems with Matlab functions

The Matlab function *bvp4c* will solve boundary value problems of order 2 or greater by converting the ODE to a system of first-order equations. This function solves the resulting first-order system using a finite difference approach. It is called in the form

$$solution = bvp4c('ode', 'odebc', initsol)$$

where ode is a file that contains the system of equations, odebc is the statement of the boundary conditions, and initsol is the initial guess to the solution.

Consider the simple second-order linear equation considered previously: $\epsilon y'' - y = x$ with $y(0) = 0$, $y(1) = 1$ for $\epsilon = 1$. To use bvp4c, this equation is written in the form

$$y' = z \tag{10.197}$$

$$z' = y + x \tag{10.198}$$

The ode file containing the system of first-order equations is

```
function dydx=ode(x,y)
dydx=[y(2)  y(1)+x];
```

and the boundary condition file is

```
function bc=odebc(ya,yb)
y0=0; y1=1;
bc=[ya(1)-y0 yb(1) -y1];
```

The main Matlab program looks as follows:

```
initsol=odeinit(linspace(0,1,11),[1,1])
sol=bvp4c('ode','odebc',initsol)
plot(sol.x,sol.y(1,:))
xlabel('x');xlabel('y')
```

The "linspace" command generates a vector $x = [0, 1]$ using $N = 11$ points.

The Matlab function "bvp4c" is certainly not limited to linear equations, and, for example, the Blasius problem can be solved in this manner.

10.11 Solution of initial value problems

10.11.1 Introduction

By definition, first-order differential equations are of the initial-value type. There are many ways to solve these types of equations using both single-step and multistep methods. To begin to understand these methods, consider a first-order equation of the form

$$\frac{dy}{dx} = f(x, y) \text{ with } y(0) = y_0 \tag{10.199}$$

where $f(x, y)$ is a given function. In some cases, an analytical solution can be found; for $f(x, y) = 1$ and $y(0) = 1$, the solution is clearly

$$y(x) = 1 + \frac{x^2}{2} \tag{10.200}$$

As you might guess, such a simple equation often does not occur in applications. There are several methods for solving such problems, and these methods are discussed next.

There are a wide number of applications for which first-order equations must be solved. As has been seen, the streamlines, pathlines, and streaklines of a fluid flow are governed by first-order equations. First-order equations are particularly common in biochemical reaction kinetics; for example, the action of enzymes is governed by a set of first-order differential equations.

Michaelis and Menten (Murray, 2001) suggested that enzymes react on substrates according to the two-step reaction

$$S + E \underset{k_{-1}}{\overset{k_1}{\rightleftharpoons}} SE \overset{k_2}{\rightarrow} P + E \tag{10.201}$$

with each reaction governed by rate constants k_1 and k_2 and the reverse rate constant k_{-1}. This theory of enzyme action is meant to describe the case of a dilute solution; in concentrated solutions, the rate constants may depend on concentration. This equation states that one molecule of S combines with one molecule of E to form the complex SE, which then produces one molecule of P. The *law of mass action* states that the reaction rate is proportional to the concentration, and letting, for example, $s = [S]$, the concentration of S with corresponding expressions for E, P, SE, there are four equations:

$$\frac{ds}{dt} = -k_1 es + k_{-1}c$$

$$\frac{de}{dt} = -k_1 es + (k_{-1} + k_2)c$$

$$\frac{dc}{dt} = k_1 es - (k_{-1} + k_2)c$$

$$\frac{dp}{dt} = k_2 c \tag{10.202}$$

where $c = [SE]$. The units of the rate constants are $\sec^{-1}$ M^{-1}, where M is molar. The initial conditions are

$$s(0) = s_0, \quad e(0) = e_0, \quad c(0) = p(0) = 0 \tag{10.203}$$

The enzyme E only facilitates the reaction, and so adding the second and third equations and integrating,

$$e(t) + c(t) = e_0 \tag{10.204}$$

Substituting into equations (10.202) (actually, just the first two equations), there are two distinct equations:

$$\frac{ds}{dt} = -k_1 e_0 s + (k_1 s + k_{-1})c \tag{10.205}$$

$$\frac{dc}{dt} = k_1 e_0 s - (k_1 s + k_{-1} + k_2)c \tag{10.206}$$

with the appropriate initial conditions on s, c as earlier. Note that p may be calculated once c is known. The rate constants are usually obtained from experiment, and in the general case, these are two nonlinear first-order differential equations for the concentration s and the complex se.

First-order equations are also common in drug delivery models, called *compartment models* (Saltzman, 2001). *Pharmacokinetics* is the study of the effect of a drug on the body. In its simplest form, pharmacokinetics is the study of the injection of a drug into one or more compartments in the body that is subsequently eliminated in at least one compartment. The major organs of the body are often modeled as two compartment regions in which the drug may be injected, absorbed, or eliminated. For example, if the concentration of a drug is assumed to be constant within a compartment, a well-stirred model, then the concentration within compartment 1 is determined by the equation (Saltzman, 2001)

$$(V_1 + V_2 k(c_2)) \frac{dc_1}{dt} = Q \left(c_0 e^{-\frac{t}{\tau}} - c_1 \right) \tag{10.207}$$

where k is an equilibrium constant but is a function of the concentration in the second compartment c_2, Q is the volume flow rate of blood through the organ, V_1 and V_2 are the volumes of the two compartments, and τ is a time constant. This equation is a simple statement of conservation of mass of the drug. Of course, the concentration of drug in the compartment may not be constant, leading to a partial differential equation.

In this section, several different methods of solving such first-order differential equations are described, beginning first with an analytical method using the Taylor series.

10.11.2 Taylor series method

This method allows the determination of the initial behavior of the solution of a differential equation but is not suited for the development of a longer-term numerical solution. Let us consider the equation

$$\frac{dy}{dx} = e^{-x}(y^3 + 1) \quad \text{with } y(0) = 2 \tag{10.208}$$

Note that the function $f(x, y)$ is nonlinear in y and that an analytical solution, a solution expressible in simple functional form, is not evident. The problem is to compute the solution y for $x > 0$.

To do this, a Taylor series approximation about $t = 0$ is given by

$$y(x) = y(0) + y'(0)x + y''(0)\frac{x^2}{2} + \cdots \tag{10.209}$$

Because $y(0) = 2$, the derivative $dy/dx(0)$ is also known; moreover, the second derivative is also known simply by differentiating equation (10.208). In this way, the first three terms of the Taylor series are

$$y(x) = 2 + 9x + 99\frac{x^2}{2} + \cdots \tag{10.210}$$

Of course, this is not the full solution because there are theoretically an infinite number of terms in the Taylor series. However, equation (10.210) gives a good approximation to the function y for very small values of x. The disadvantage is that the Taylor series will not converge for larger values of x. There is value in this approach because the Taylor series method can be used as a starting formula for a multistep method, discussed later.

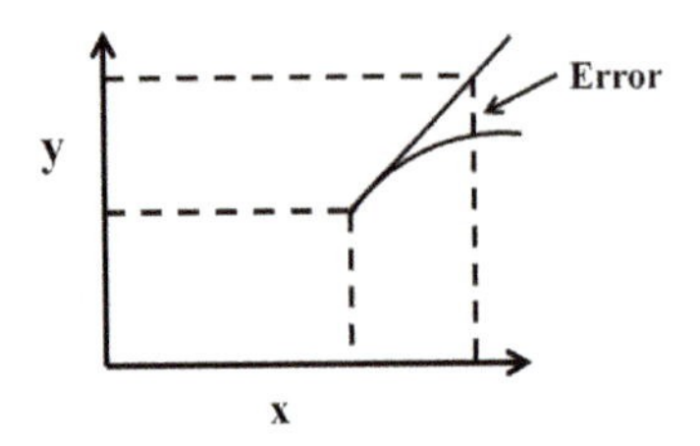

Figure 10.18 A graphical depiction of Euler's method.

10.11.3 Euler methods

Euler methods approximate the derivative dy/dx by a first- or second-order finite difference; the simplest formula is the *Euler forward formula*:

$$y_{j+1} = y_j + hf(x_j, y_j) \quad \text{for } j = 0, 1, 2, \cdots \tag{10.211}$$

where n denotes the number of the time step and $h = \Delta x$ is length of the step in x. This formula approximates the derivative with a forward difference formula. This Euler forward formula is *explicit* in that the solution y_{n+1} follows from the values before. This method is thus easy to program and is fast computationally, and the step $h = \Delta x$ can be changed easily. However, the local error is relatively large, and

$$E_L = h^2 y''(0) + \cdots \tag{10.212}$$

and over a number of time steps of $O(1/h)$, the global or propagated error $E_G \sim O(h)$. In practice, very small steps are required for this method that in general is not used. A graphical or geometric interpretation of the Euler method is depicted in Figure 10.18.

The accuracy can be increased by using a central difference approximation to the derivative, which is equivalent to the equation

$$y_{j+1} = y_j + \frac{h}{2}\left(f(x_j, y_j) + f(x_{j+1}, y_{j+1})\right) \quad j = 0, 1, 2, \ldots \tag{10.213}$$

The only difference is that the function f is evaluated at the midpoint of the interval. The error is now $E \sim O(h^3)$; however, because y_{j+1} is unknown, this equation is *implicit*. The value y_{j+1} can be predicted using the standard Euler formula, equation (10.210). However, error is incurred in just guessing y_{j+1}, and so iteration is necessary. The procedure could go as follows:

Step 1. Predict y_{j+1} using the Euler formula

$$y^P_{j+1} = y_j + hf(x_j, y_j) \tag{10.214}$$

where P indicates the prediction step.

Step 2. Correct this value with the *modified Euler* formula:

$$y^C_{j+1} = y_j + \frac{h}{2}\left(f(x_j, y_j) + f(x_{j+1}, y^P_{j+1})\right) \tag{10.215}$$

where C indicates the correction step.

Step 3. At this point, we have values for y_{j+1}^{P} and y_{j+1}^{C}, and we need to check to see if they are close. We would use a relative test to check to see if

$$\left| \frac{y_{j+1}^{C} - y_{j+1}^{P}}{y_{j+1}^{C}} \right| < eps \tag{10.216}$$

where ϵ is a specified small number that may vary with the problem. A value that often gives good results is $\epsilon = 10^{-4}$.

Step 4. If the test is not satisfied, then repeat step 2 using the formula

$$y_{j+1}^{n+1} = y_j + \frac{h}{2}\left(f(x_j, y_j) + f(x_{j+1}, y_{j+1}^{n})\right) \tag{10.217}$$

where n denotes the number of the iteration.

Generally, only a few iterations are necessary, and it is possible that choosing a small enough $h = \Delta x$ will remove the need to correct or iterate. It can be shown that for this process to converge,

$$\left| h\frac{\partial f}{\partial y} \right| < 2 \tag{10.218}$$

which imposes an upper limit on the mesh length h.

Consider the example problem

$$\frac{dy}{dx} = 1 + y^2, \quad y(0) = 0 \tag{10.219}$$

which has the exact solution $y = \tan x$. Suppose $h = 0.1$ to start, and use two correction steps. Then the solution is displayed in Table 10.13. Note that the modified Euler solution is much better than the Euler solution, although the error even for the modified Euler is growing. In practice, the step size for this problem should be smaller rather than attempting to use more corrector steps. In practice, even the modified Euler method is not used much, and there are better single-step methods, such as Runge–Kutta methods, and these are discussed next.

Table 10.13. Solution to the ordinary differential equation given by equation (10.219)

x	Euler	Mod. Euler	Exact
0	0	0	0
0.1	0.1000	0.1005	0.1003
0.2	0.2010	0.2031	0.2027
0.3	0.3050	0.3100	0.3093
0.4	0.4143	0.4238	0.4227
0.5	0.5315	0.5478	0.5464

10.11.4 Runge-Kutta methods

Runge–Kutta methods are single-step methods that can be made as accurate as necessary by increasing the number of function evaluations. Runge–Kutta methods are of the form

$$y_{j+1} = y_j + a_1 k_1 + a_2 k_2 + a_3 k_3 + \cdots + a_M k_M \tag{10.220}$$

where the a_i are coefficients, depending on the order of the method, and the k_i are values of the function f evaluated at specific points:

$$k_1 = hf(x_j, y_j) \tag{10.221}$$

$$k_2 = hf(x_j + \alpha_1 h, y_j + \beta_1 k_1) \tag{10.222}$$

$$k_3 = hf(x_j + \alpha_2 h, y_j + \beta_2 k_2) \tag{10.223}$$

and so on. The number of terms taken in the expression is the *order* of the method; in equation (10.220), the order of the method is M. The arrays α_i and β_i are chosen from the expression for the Taylor series of the function y about y_j, or

$$y_{j+1} = y_j + hy'_j + \frac{h^2}{2} y''_j + \frac{h^3}{6} y'''_j + \cdots \tag{10.224}$$

where, for example, $y'_j = y'(x_j)$. The equation of interest is

$$y' = f(x, y) \tag{10.225}$$

and consider the second-order Runge–Kutta formula. Note that

$$y''(x) = \frac{df}{dx} \tag{10.226}$$

but $f = f(x, y)$ so that

$$\frac{df}{dx} = \frac{\partial f}{\partial x} + \frac{df}{dx}\frac{\partial y}{\partial x} \tag{10.227}$$

Substituting into equation (10.224),

$$y_{j+1} = y_j + hf_j + \frac{h^2}{2}\left(\frac{\partial f}{\partial x} + \frac{df}{dx}\frac{\partial y}{\partial x}\right)_j \tag{10.228}$$

where $f_j = f(x_j, y_j)$ and $f_j = \frac{\partial y}{\partial x}_j$. Now compare equation (10.228) with the formula

$$y_{j+1} = y_j + a_1 k_1 + a_2 k_2 \tag{10.229}$$

Here it is clear that $k_1 = hf_j$, and $a_1 = 1$ will make the second term in each of the equations the same. From the definition of $k_2 = hf(x_j + \alpha_1 h, y_j + \beta_1 k_1)$, expanding in a Taylor series,

$$k_2 = hf(x_j + \alpha_1 h, y_j + \beta_1 k_1) = hf_j + \alpha_1 h^2 \frac{\partial f}{\partial x}_j + \beta_1 h^2 k_1 f_j \frac{\partial f}{\partial y}_j + \cdots$$

Equations (10.228) and (10.229) match to $O(h^2)$ if the following conditions hold: $a_1 + a_2 = 1$, $a_2\alpha_1 = 1/2$, and $a_2\beta_1 = 1/2$. Note that these conditions have multiple solutions but that the standard symmetrical forms used most often are $a_1 = a_2 = 1/2$ and $\alpha_1 = \beta_1 = 1$. Thus the second-order Runge–Kutta formula is given by

$$y_{j+1} = y_j + \frac{1}{2}(k_1 + k_2) + O(h^3) \tag{10.230}$$

where $k_1 = hf_j$ and $k_2 = hf(x_j + h, y_j + k_1)$.

Higher-order formulas are obtained by taking more terms in the Taylor series, and again, there are multiple solutions for the coefficients a_i, α_i, and β_i. The most common fourth-order formula is

$$y_{j+1} = y_j + \frac{1}{6}(k_1 + 2k_2 + 2k_3 + k_4) + O(h^5) \tag{10.231}$$

where

$$k_1 = hf_j \tag{10.232}$$

$$k_2 = hf(x_j + h/2, y_j + k_1/2) \tag{10.233}$$

$$k_3 = hf(x_j + h/2, y_j + k_2/2) \tag{10.234}$$

$$k_4 = hf(x_j + h, y_j + k_3) \tag{10.235}$$

This is the Runge–Kutta method in most use. The fourth-order Runge–Kutta has several advantages:

- It is very easy to program.
- It is cumulative fourth order accurate after a number of steps of $O(1/h)$.
- Since it is a one-step method, it is self-starting.
- While there are some Runge–Kutta methods that are unstable in some sense, this method is highly stable.

The only real disadvantage of the fourth order Runge–Kutta method is that there are at least four function evaluations at each time step, adding to the expence of the calculation, and all of the information is thrown away as it is calculated. In addition, there is no good estimate of the error incurred. As with boundary value problems, to show that a given solution is accurate, the solution should be calculated using two different grid sizes, say, h and $h/2$. If the solution agrees to a sufficient number of digits, then the solution is said to be *verified.*

To understand a Runge–Kutta calculation, consider the linear ODE

$$\frac{dy}{dx} = 3x + \frac{1}{2}y \tag{10.236}$$

with $y(0) = 1$. Assuming $h = 0.1$, the values of the k_i for the fourth-order scheme at the first nonzero value of x can be calculated as

$$k_1 = hf(0, 1) = 0.1\left(0 + \frac{1}{2}\right) = 0.05 \tag{10.237}$$

$$k_2 = hf(0.05, 1.015) = 0.1(0.15 + 0.5125) = 0.06625 \tag{10.238}$$

$$k_3 = hf(0.05, 1.033125) = 0.06666 \tag{10.239}$$

$$k_4 = 0.08333 \tag{10.240}$$

With these values,

$$y(0.1) = \frac{1}{6}(k_1 + 2k_2 + 2k_3 + k_4) = 1.06652 \tag{10.241}$$

The value at $x = 0.2$ is calculated similarly, beginning with the value $y(0.1)$. The exact solution to this problem may be found as

$$y(x) = 13e^{x/2} - 6x - 12 \tag{10.242}$$

and the value $y(0.1) = 1.06652$, which is identical to the numerical result; actually, the two results agree to the eighth decimal place. For a larger time step, for example, $h = 0.2$, the numerical solution gives $y(0.4) = 1.47820$, while the exact result is $y(0.4) = 1.47823$. Note that now, the two results deviate in the fifth place, and as x increases, the error will get much bigger.

The procedure is easily extendable to systems of equations. Suppose the system of equations is defined by

$$\frac{dy}{dx} = f(x, y, z), \quad y(x_0) = y_0 \tag{10.243}$$

$$\frac{dz}{dx} = g(x, y, z), \quad z(x_0) = z_0 \tag{10.244}$$

Then the k_i can be defined for the first equation in terms of the values of f, and the other coefficients, say, l_i, are defined in terms of the functional values of g; for example, at a given time step n, $k_1 = hf(x_j, y_j, z_j)$ and $l_1 = hg(x_j, y_j, z_j)$. The solutions for (y, z) at the $j + 1$ time step are given by equation (10.224).

The fourth Runge–Kutta method is the most popular method of all the single-step methods; it is easy to program and is self-starting and extremely stable. Conversely, it has four function evaluations, and it is wasteful because it throws away all the information it has calculated. For these reasons, Runge–Kutta methods are not often used in computationally intensive problems such as for molecular dynamics simulations, which are Hamiltonian systems. Nevertheless, it is a good method and can be used in most problems with good success.

10.11.5 Adams–Moulton methods

Multistep methods are those methods derived by Taylor series that span two or more time steps. The most common Adams-Moulton explicit method is given by

$$y_{n+1} = y_n = \frac{h}{24}(55f_n - 59f_{n-1} + 37f_{n-2} - 9f_{n-3}) + O\left(\frac{251}{720}h^5 y_n^{(5)}\right) \tag{10.245}$$

Another formula can be derived that includes the value of the right-hand side f_{n+1}, resulting in the formula

$$y_{n+1} = y_n = \frac{h}{24}(9f_{n+1} + 19f_n + 5f_{n-1} + f_{n-2}) + O\left(\frac{19}{720}h^5 y_n^{(5)}\right) \tag{10.246}$$

Note that the error associated with equation (10.246) is on the order of 10 times less than that associated with equation (10.245). Also it is seen that equation (10.246) is implicit. Thus an iterative scheme similar to a modified Euler scheme can be performed with a predictor, corrector step corresponding to equations (10.245) and (10.246), respectively. It is left as an exercise for the reader to apply the Adams–Moulton method to the simple ODE addressed in the previous section.

This method is the best of the ones discussed so far. Note that it is fast and requires only two function evaluations at each time step, compared to Runge–Kutta, which requires four evaluations. However, because it is a multistep method, a starting formula must be used, and Runge–Kutta could be used for this purpose because the error is of the same magnitude. This method sometimes appears under the name *Adams–Bashforth*. There are many other such methods corresponding to their order and the different departure points for the Taylor series.

10.11.6 Symplectic integrators

It has been mentioned that a second-order initial value problem should usually be solved as a system of first-order equations. For certain problems associated with computing particle trajectories, the highly accurate Runge–Kutta and Adams–Moulton methods may not yield optimal results. This is especially true for systems defined by a *Hamiltonian*, an example being the computation of particle trajectories in classical molecular dynamics simulations discussed in Chapter 11. The methods described in this section are not normally included in presentations of numerical methods in engineering textbooks.

A Hamiltonian system is defined as a system of equations that can be written as

$$\dot{\xi} = \frac{\partial H}{\partial \eta} \tag{10.247}$$

$$\dot{\eta} = \frac{\partial H}{\partial \xi} \tag{10.248}$$

where H is the Hamiltonian and $\dot{\xi} = d\xi/dt$. In classical mechanics, for example, $\xi = y$ is often a particle position and η is its momentum $\eta = mv = m\dot{y}$; m is a particle mass.

For example, the harmonic oscillator is defined by the equation

$$m\frac{d^2y}{dt^2} + ky = 0 \tag{10.249}$$

which may be split into two first order equations for the positions and momentum as

$$\dot{y} = \frac{\partial H}{\partial p} \tag{10.250}$$

$$\dot{p} = -\frac{1}{m}\frac{\partial H}{\partial y} \tag{10.251}$$

where the Hamiltonian is given by

$$H = \frac{1}{2}ky^2 + \frac{p^2}{2m} \tag{10.252}$$

The key feature of systems of equations defined by Hamiltonians is that the total energy in the system, the Hamiltonian, $H = \text{constant} = H_0$, is a constant fixed by its value at the initial time. To show this, note that the work done by a force is given by

$$W_{12} = \int_1^2 F(y)dy \tag{10.253}$$

where the force F may be defined by a potential

$$F(y) = -\frac{d\Phi}{dy} \tag{10.254}$$

and so

$$W_{12} = \Phi_2 - \Phi_1 \tag{10.255}$$

In the same way, integrating the force part of the equation

$$W_{12} = \frac{1}{2}m(v_2^2 - v_1^2) \tag{10.256}$$

and equating the two expressions for the work done, it is seen that $H_1 = H_2$, meaning that $dH/dt = 0$ and

$$H = \text{constant} = H_0 \tag{10.257}$$

Numerical schemes that preserve this feature of the problem are called *symplectic*.

Suppose the harmonic oscillator problem is solved by using the explicit Euler method. Then

$$y_{n+1} = y_n + hp_n \tag{10.258}$$

$$p_{n+1} = p_n - hy_n \tag{10.259}$$

where it has been assumed that the spring constant and the mass have the value of 1, $k = m = 1$. Then, calculating the Hamiltonian directly, it is evident that

$$H_{n+1} = \frac{1}{2}y_{n+1}^2 + \frac{p_{n+1}^2}{2} = H_n(1 + h^2) \tag{10.260}$$

so that the Hamiltonian grows at each time step, violating the condition that the Hamiltonian be constant; that is, the Euler method is not symplectic. In the same way, it can be shown that the classical fourth-order Runge–Kutta method and the Adams–Moulton method are not symplectic, although Runge–Kutta methods can be made symplectic by adjustment of the coefficients (Hairer *et al.*, 2006).

By adjusting the explicit Euler scheme so that one of the equations is implicit, the scheme will be symplectic. Suppose the Euler scheme for the harmonic oscillator is given by

$$y_{n+1} = y_n + hp_{n+1} \tag{10.261}$$

$$p_{n+1} = p_n - hy_n \tag{10.262}$$

Then direct substitution for the Hamiltonian shows that

$$H_{n+1} = \frac{1}{2}y_{n+1}^2 + \frac{p_{n+1}^2}{2} = H_n + h^2(p_{n+1}^2 - y_n^2) \tag{10.263}$$

Because both p and y are harmonic, the error in the Hamiltonian is bounded and does not grow. This algorithm is called the *symplectic Euler method.*

Consider now the general second-order initial value problem defined by

$$\frac{d^2y}{dt^2} = f(y, t) \tag{10.264}$$

subject to

$$y = \frac{dy}{dt} = 0 \ at \ t = t_0 \tag{10.265}$$

Let $v = \dot{y}$; then the second order equation can be written as two first order equations according to

$$\dot{v} = f \tag{10.266}$$

$$\dot{y} = v \tag{10.267}$$

The second-order equation (10.264) may be solved directly by approximating the second derivative at time $t = t_n$ by the central difference approximation, and

$$y_{n+1} - 2y_n + y_{n-1} = h^2 f_n \ \text{ for } n = 1, \cdots \tag{10.268}$$

Note that this method, called a *Verlet* algorithm, is reversible in the sense that exchanging $n + 1$ and $n - 1$ and changing h to $-h$ does not alter the equation. This is an especially important property for systems of equations that describe particle paths (Hairer *et al.*, 2006).

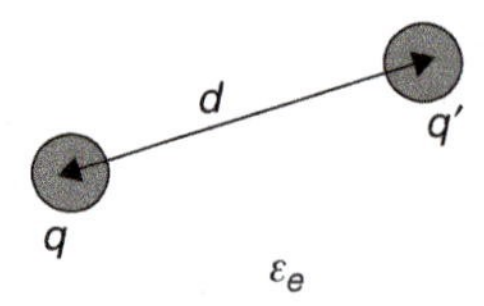

Figure 10.19 Two charged particles in an electrolyte solution.

Equation (10.268) is equivalent to the set of two approximations to the first-order systems

$$v_{n+1/2} = v_n + \frac{h}{2} f_n \quad \text{for } n = 1, \cdots \tag{10.269}$$

$$y_{n+1} = y_n + h v_{n+1/2} \quad \text{for } n = 1, \cdots \tag{10.270}$$

and the use of the central difference approximation

$$v_n = \frac{y_{n+1} - y_{n-1}}{2h} \quad \text{for } n = 1, \cdots \tag{10.271}$$

To complete the scheme, one method of updating the velocity at the point $n + 1/2$ is

$$v_{n+1} = v_n + h f_{n+1/2} \quad \text{for } n = 1, \cdots \tag{10.272}$$

In the molecular dynamics literature, these symplectic methods are called Verlet methods (Verlet, 1967).

The Verlet Algorithm

The term *Verlet algorithm* refers to Loup Verlet who "invented" the algorithm and published the paper in 1967 (Verlet, 1967). According to Hairer *et al.* (2006), this method actually appeared in Newton's *Principia*. The method appears to have been used in astronomy to predict the trajectories of the planets in the solar system by Störmer (1907). Thus, the method is often called the *Störmer-Verlet* algorithm. In the numerical solution of partial differential equations, such as the heat equation, it is often called the *leap-frog* method.

For example, consider the case of two colloidal particles of mass m and total charge q and q' in a fluid of permittivity ϵ_e. For simplicity, the origin can be taken to be the position of one of the particles, and the problem is reduced to finding the trajectory of the second particle relative to the first. The situation is depicted in Figure 10.19. Assume that the only force on the particles is the electrostatic force (omitting the 4π)

$$F = \frac{\epsilon_e q q'}{r^2} \tag{10.273}$$

and that the fluid is at rest. The equations of the two particles can be made dimensionless on the initial distance between the two particles, and then the time scale is given by

$$\tau = \frac{m d^3 \epsilon_e}{| q q' |} \tag{10.274}$$

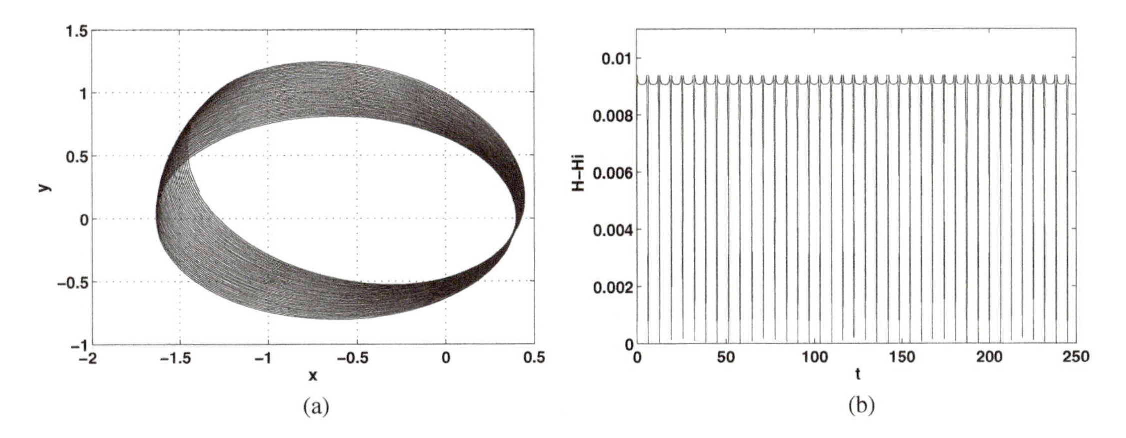

Figure 10.20 The Verlet algorithm integration of the elliptical orbit of two charged particles around each other: (a) the trajectory; (b) time evolution of the deviation of the Hamiltonian from its initial value, $H - H_i$. Here $x_0 = 1 - \beta$, $y_0 = 0$, $v_x(0) = 0$, and $v_y(0) = v_0$, with $v_0 = \sqrt{(1+\beta)/(1-\beta)}$ and $\beta = 0.6$.

where m is the mass of the particle and d is the distance between the two particles. This time scale can be used to define the dimensionless time variable. The dimensionless equations of motion of the particle are thus given by

$$\frac{d^2x}{dt^2} = -\alpha \frac{x}{\left(x^2+y^2\right)^{3/2}} \tag{10.275}$$

$$\frac{d^2y}{dt^2} = -\alpha \frac{y}{\left(x^2+y^2\right)^{3/2}} \tag{10.276}$$

where

$$\alpha = \frac{qq'}{\mid qq' \mid} \tag{10.277}$$

If $\alpha = 1$, the particles are of the same charge, and the dimensionless equations are the same as the classical two-body problem of Kepler in celestial mechanics, for which the force is given by

$$F = \frac{Gm_1m_2}{r^2} \tag{10.278}$$

and the time scale for a particle of mass m is $\tau = \sqrt{h^3/mG}$. For $\alpha = 1$, the particles will revolve around each other in an elliptic orbit as in the two-body problem of celestial mechanics. This is a Hamiltonian system with

$$H = \frac{1}{2}(v_x^2 + v_y^2) + \frac{1}{\sqrt{x^2+y^2}} \tag{10.279}$$

where, in dimensionless form, the momentum is equivalent to the velocity $p_x = v_x$. Thus the Verlet algorithm can be used to integrate the system forward in time.

As a specific example, consider the case where $x(0) = x_0$ and $y(0) = y_0$, with $\dot{x}(0) = 0$ and $\dot{y}(0) = v_0$. The solution for the trajectory using the Verlet algorithm can be compared with that using the simple explicit Euler method. The results are depicted in Figures 10.20 and 10.21. The results for the Verlet algorithm in Figure 10.20 indicate that the trajectory is elliptical and that the Hamiltonian remains

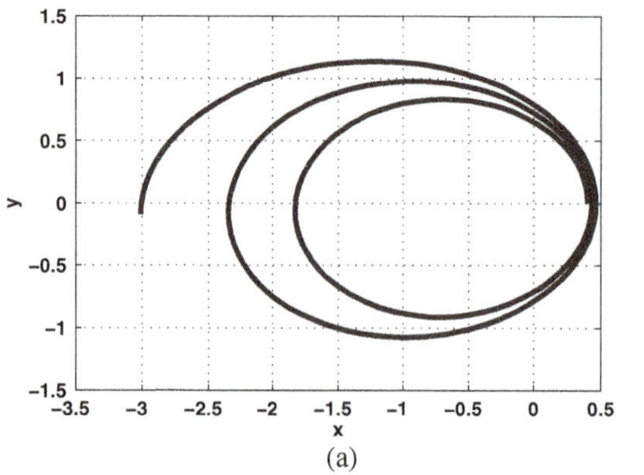

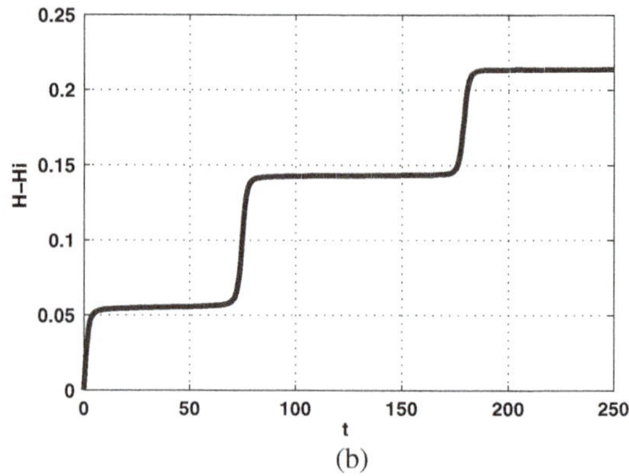

Figure 10.21 Euler integration of the elliptical orbit of two charged particles around each other: (a) the trajectory; (b) time evolution of the deviation of the Hamiltonian from its initial value, $H - H_i$. Here $x_0 = 1 - \beta$, $y_0 = 0$, $v_{x0} = 0$, and $v_{y0} = v_0$, with $v_0 = \sqrt{(1+\beta)/(1-\beta)}$ and $\beta = 0.6$.

approximately constant, with a periodic deviation from the initial value of small amplitude. For the explicit Euler method, the trajectory expands dramatically and the Hamiltonian (the total energy) grows with time, clearly violating the constant energy condition very early in the computation.

It is left as an exercise to solve these equations using a fourth-order Runge–Kutta method or the Adams–Moulton method, neither of which, in its standard form, is symplectic. The interested reader should consult the comprehensive treatment of symplectic algorithms by Hairer *et al.* (2006).

10.11.7 Stiff equations and stability

Stiff differential equations arise when a set of initial value problems contains widely varying time scales. This will occur when the coefficients of a system of initial value problems are large. Consider, for example, the set of equations (Fausett, 2008)

$$\dot{x} = 98x + 198y \tag{10.280}$$

$$\dot{y} = -99x - 199y \tag{10.281}$$

with $x(0) = 1$ and $y(0) = 0$. The solution is given by

$$x(t) = 2e^{-t} - e^{-100t} \tag{10.282}$$

$$y(t) = -e^{-t} + e^{-100t} \tag{10.283}$$

Clearly there are two time scales $\tau_L = 1$ and $\tau_S = 0.01$, a short time scale τ_S, and a long time scale τ_L. If the short time scale must be resolved, then very small time steps are required. Conversely, because the short time scale is short, we may choose not to resolve it.

Stiff equations may be solved by several methods. One way is to use an implicit solver. To see this, consider the equation

$$\dot{y} + \beta y = 0 \tag{10.284}$$

where β is a constant. Euler's method gives

$$y_{n+1} = (1 - \beta h)y_n \tag{10.285}$$

If $1 - \beta h > 1$, the solution grows without bound, and the Euler method is said to be *unstable*. Thus the time step must satisfy $h < 2/\beta$ for stability, and if $\beta \gg 1$, very small time steps are required. Conversely, suppose that the backward Euler method is used, which is defined by

$$y_{n+1} = y_n - \beta h y_{n+1} \tag{10.286}$$

leading to

$$y_{n+1} = \frac{y_n}{1 + \beta h} \tag{10.287}$$

which will be stable, though not necessarily accurate for all h. As can be seen, implicit methods are usually more stable than explicit methods for the same time step.

Another method that can be used is to approximate the right-hand side of the equation at the mid-point of an interval and this equation is

$$y_{n+1} = y_n + 0.5h(f(t_{n+1}, y_{n+1}) + f(t_n, y_n)) \tag{10.288}$$

Applying this equation to the preceding simple equation, we obtain

$$y_{n+1} = y_n \frac{2 - \beta h}{2 + \beta h} \tag{10.289}$$

and so the method is stable. More detail on the nature of the stability of the other numerical methods is given in a variety of texts.

Another way to solve stiff equations is to use a variable time step, the magnitude of which can be based on an estimate of the magnitude of the right hand side of the equation. Other algorithms are described by Fausett (2008) and the references therein.

The stiffness problem may also be addressed using the method of matched asymptotic expansions, as presented in Appendix A for second-order boundary value problems. Stiff equations are common in chemical kinetics problems for which reactions occur on different time scales and lead to *singular perturbation problems* that often may be solved analytically, without recourse to the computer – a powerful tool indeed. Consider the action of enzymes discussed in Section 10.11.1. There the two equations governing the concentration of the substrate material and the complex ES ($= c$) were derived according to (Murray, 2001)

$$\frac{ds}{dt} = -k_1 e_0 s + (k_1 s + k_{-1})c \tag{10.290}$$

$$\frac{dc}{dt} = k_1 e_0 s - (k_1 s + k_{-1} + k_2)c \tag{10.291}$$

subject to

$$s(0) = s_0 \text{ and } c(0) = 0 \tag{10.292}$$

Here e_0 is the concentration of the enzyme; most biochemical reactions require very little of the enzyme. Thus, in most cases of practical interest, $\epsilon = e_0/s_0 \ll 1$.

If the variables are made dimensionless as $t^* = k_1 e_0 t$, $s^* = s/s_0$, and $c^* = c/e_0$, then (after dropping the star)

$$\frac{ds}{dt} = -s + (s + K - \gamma)c \tag{10.293}$$

$$\epsilon \frac{dc}{dt} = s - (s + K)c \tag{10.294}$$

with

$$s(0) = 1 \quad c(0) = 0 \tag{10.295}$$

Here $K = (k_{-1} + k_2)/(k_1 s_0) = K_m/s_0$ and $\gamma = k_2/k_1 s_0$; K_m is called the *Michaelis constant.* Also $\epsilon \ll 1$, and thus there are two time scales, t, the long time scale, and $t/\epsilon = O(1)$, the short time scale. This system of equations is thus stiff. As an example of the order of magnitude of some of these parameters, standard values may be $s_0 = 1$ μM, $e_0 = 0.012$ μM, $k_1 = 0.1$ $(\mu\text{M})^{-1}$ sec^{-1}, $k_{-1} = 0.1$ sec^{-1}, and $k_2 = 0.3$ sec^{-1} (M stands for molar, as usual). Then $K_m = 4$ μM, and $\epsilon = 0.012$.

The leading-order solutions to equations (10.293) and (10.294) are obtained by expanding the variables as $s = S_0 + \epsilon S_1 + \cdots$ and $c = C_0 + \epsilon C_1 + \cdots$. Then S_0 and C_0 satisfy

$$\frac{dS_0}{dt} = -S_0 + (S_0 + K - \gamma)C_0 \tag{10.296}$$

$$S_0 - (S_0 + K)C_0 = 0 \tag{10.297}$$

subject to $S_0(0) = 1$ and $C_0(0) = 0$. Immediately, we see that

$$C_0 = \frac{S_0}{S_0 + K} \tag{10.298}$$

and note that this solution cannot satisfy the initial condition that $C_0 = 0$. This problem will be addressed in the analysis to follow.

Substituting for C_0 into equation (10.296), we find that

$$\frac{dS_0}{dt} = -\frac{\gamma S_0}{S_0 + K} \tag{10.299}$$

which can be integrated to yield

$$S_0 + K \ln S_0 = A - \gamma t \tag{10.300}$$

where A is a constant. The solutions C_0 and S_0 are termed the *outer solutions*, in the language of matched asymptotic expansions.

Setting $\epsilon = 0$ in equation (10.294) means that we have lost the ability to satisfy the initial condition that $c(0) = 0$. The problem defined by equations (10.293) and (10.294) with the initial conditions is thus termed a *singular perturbation* problem. The preceding solutions for S_0 and C_0 are on the time scale $t = O(1)$: the long time scale, as discussed earlier. To complete analysis, the solution on the short time scale $t = O(\epsilon)$ must be calculated. To do this, define a new variable $\hat{t} = \frac{t}{\epsilon}$: the short time scale. Then, on this time scale,

$$\frac{ds}{d\hat{t}} = 0 \tag{10.301}$$

$$\frac{dc}{d\hat{t}} = s - (s + K)c \tag{10.302}$$

In the same way as earlier, we expand the variables as $s = \hat{s}_0 + \epsilon\hat{s}_1 + \cdots$ and $c = \hat{c}_0 + \epsilon\hat{c}_1 + \cdots$, and thus

$$\frac{d\hat{s}_0}{d\hat{t}} = 0 \tag{10.303}$$

so that $\hat{s}_0 = 1$ to satisfy the initial condition. The solution for $\hat{c}_0$ is found to be

$$\hat{c}_0 = \frac{1}{1+K}\left(1 - e^{(1+K)\hat{t}}\right) \tag{10.304}$$

after satisfying the initial condition $\hat{c}_0(0) = 0$. Note that as $\hat{t} \to \infty$, the complex reaches the asymptotic value $\hat{c}_0 = 1/(1 + K)$, which is equal to the limit of the outer solution as $t \to 0$:

$$\lim_{\hat{t}\to\infty} \hat{c}_0 = \lim_{t\to 0} C_0 \tag{10.305}$$

The solutions $\hat{c}_0$ and $\hat{s}_0$ are termed the *inner solutions* in the language of matched asymptotic expansions.

It remains to find the constant A. This is done by matching the inner and outer solutions according to (see Appendix A)

$$\lim_{\hat{t}\to\infty} \hat{s}_0 = \lim_{t\to 0} S_0 \tag{10.306}$$

resulting in $A = 1$, a result that could have been anticipated from the form of the solution.

The inner and outer solutions for s and c can be combined in what is called the *uniformly valid solution*, and for any variable, this is defined by

$$f_{\mathrm{UV}} = f_{\mathrm{inner}} + f_{\mathrm{outer}} - \mathrm{CP} \tag{10.307}$$

where the common part (CP) is the term that appears in both the inner and outer solutions. For s, the common part is $s_{\mathrm{CP}} = 1$, and for c, the common part is $c_{\mathrm{CP}} = \frac{1}{1+K}$. Thus the one term, uniformly valid solution for c is

$$c_{0,\mathrm{UV}} = C_0 + \hat{c}_0 - \frac{1}{1+K} = \frac{S_0}{S_0 + K} - \frac{1}{1+K}e^{-(1+K)\hat{t}} \tag{10.308}$$

$$S_{0,\mathrm{UV}} + K \ln S_{0,\mathrm{UV}} = 1 - \gamma t \tag{10.309}$$

The uniformly valid solutions for the substrate, the complex, and the enzyme are shown in dimensionless form in Figure 10.22. Note the boundary layer behavior near $t = 0$, which is very difficult to resolve in a fully numerical method. Here the parameters given earlier are used, and note that $E + C = 1$ for $\epsilon = 0.1$ for

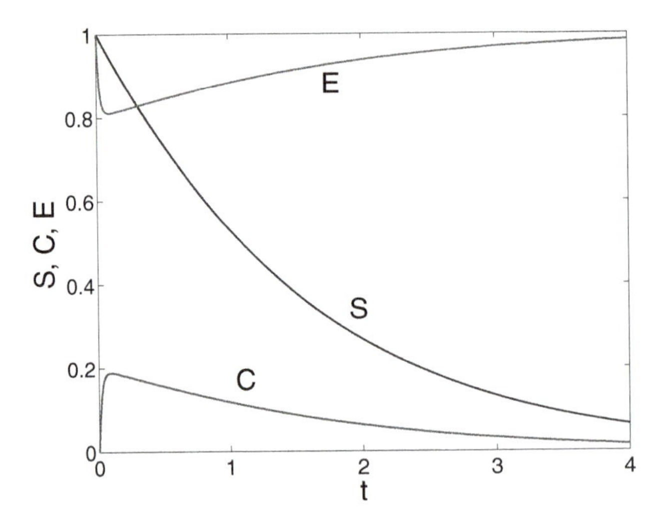

Figure 10.22 Uniformly valid solution for the Michaelis–Menten substrate, enzyme, complex temporal development. Here $E + C = 1$ for $\epsilon = 0.1$ for these values of the parameters.

these values of the parameters. Note also that the substrate (S) variation near $t = 0$ is much slower than that of the complex C; in the kinetics literature, this fact is used to justify solving for the substrate concentration only on the long time scale.

10.11.8 Solving initial value problems using Matlab functions

The Matlab functions `ode23`, `ode45`, and `ode113` may be used to solve initial value problems. These functions are called with a statement of the form

```
[t y] = ode23('f', tspan, y0)
```

where y is the solution, tspan is the interval over which the time integration is to be performed, say, `tspan = [t0 tf]`, and `y0` is the initial condition. For a system of equations, `f` and `y0` are vectors. The function `ode23` uses two second- and third-order Runge–Kutta solvers, whereas `ode45` uses a combination of fourth- and fifth-order methods. The function `ode113` is a variable step size routine that uses a variety of Adams–Moulton methods. There can be additional entries in the calling statements for each of these functions, and the Matlab help resources should be consulted for details. The Matlab functions `ode15s` and `ode23s` are designed specifically to solve stiff initial value problems.

10.12 Numerical solution of the PNP system

As was discussed in Chapter 7, the Poisson–Nernst–Planck (PNP) system is a set of second-order equations for the solutes in the mixture and the electrical potential, and these equations can be solved numerically in one dimension using the methods discussed for boundary value problems. The equations are discretized using the second-order accurate central difference approximations and for all the derivative terms. Recall that the equation for the potential is given in dimensionless form by

$$\epsilon^2 \frac{d^2\phi}{dy^2} = -\rho_e \tag{10.310}$$

where ρ_e is the dimensionless volume charge density and the concentrations are scaled on ionic strength. The finite difference equation for the potential using central differencing is given by

$$\phi_{i-1} - 2\phi_i + \phi_{i+1} = -\frac{h^2}{\epsilon^2}(z_a X_{a,i} + z_b X_{b,i}) \tag{10.311}$$

for $i = 2, \ldots, N-1$ interior grid points, in a mixture having two solute species of valences z_a and z_b.

If the ions are assumed to be point charges, then the equations for the mole fraction have the classical Boltzman distribution. In the solution for the potential, that distribution can be used explicitly. Alternatively, the equation for the mole fraction can be solved in the form

$$\frac{d^2 X_a}{dy^2} - \frac{z_A X_a (z_a X_a + z_b X_b)}{\epsilon^2} + z_A \frac{dX_a}{dy}\frac{dX_a}{dy} = 0 \tag{10.312}$$

Using central differencing, the difference equations are given by

$$\begin{aligned} X_{a,i-1}\left(1 - \frac{1}{4}z_a(\phi_{i+1} - \phi_{i-1})\right) - X_{a,i}\left(2 + \frac{h^2}{\epsilon^2}z_a^2 X_{a,i}\right) \\ + X_{a,i+1}\left(1 + \frac{1}{4}z_a(\phi_{i+1} - \phi_{i-1})\right) = \frac{h^2}{\epsilon^2}z_a z_b X_{a,i} X_{b,i} \end{aligned} \tag{10.313}$$

Note that the quantities inside the parentheses are the linearized coefficients of the tridiagonal system.

The difference equations for the mole fractions are nonlinear, and so iteration is required. Either equation can be solved first, with the result substituted into the potential equation. The set of linearized tridiagonal difference equations can be solved using a direct tridiagonal solver such as the Thomas algorithm or iteratively using the Gauss–Seidel scheme with successive overrelaxation (SOR). The iteration scheme is assumed to converge when

$$\left|\frac{X_{\text{aNew}} - X_{\text{aOld}}}{X_{\text{aNew}}}\right| < eps \tag{10.314}$$

at all grid points, where X_a is the mole fraction of species a and eps is the convergence criterion, usually chosen at first to be $eps = 10^{-4}$.

These equations are subject to boundary conditions that can involve specified values of the potential and mole fractions or derivative boundary conditions. In dimensionless form, the boundary condition for the potential using the surface charge density is

$$\frac{d\phi}{dy} = \frac{-\sigma^0 h}{\epsilon_e \phi_0} \quad \text{at } y = 0 \tag{10.315}$$

for example. In this case, σ^0 is assumed known, and the sign on the right side of the equation depends on the direction of the outward unit normal to the surface. Note that the right side of equation (10.315) is dimensionless. As has been discussed, a central difference approximation for the derivative at the boundary can be written and, after elimination of the fictitious point, yields an additional equation, which is added to the tridiagonal system.

The no-flux condition on the mole fractions corresponds to

$$\frac{dX_a}{dy} + z_a X_a \frac{d\phi}{dy} = 0 \quad \text{at } y = 0 \tag{10.316}$$

A similar procedure is used here, with the $d\phi/dy$ assumed known in the calculation of the solution for X_a.

Of course, because the equations are second-order, another boundary condition is required. This depends on the value of ϵ; for $\epsilon \ll 1$, we would use a specified condition as $y \to \infty$ or, for ϵ not so small, a boundary condition at the other wall could be used.

All the numerical methods used to solve this problem have been discussed in detail, and putting all these modules together to solve the PNP system numerically is left as an exercise.

10.13 Partial differential equations

In Chapter 3, at least in the linear case, second-order partial differential equations are classified into three groups: parabolic, elliptic, and hyperbolic equations. Though the classification procedure is not easily formulated for nonlinear equations; this classification system is usually assumed to apply for nonlinear equations. This classification is based on how information travels in each case; for example, information travels in one direction, forward, in parabolic equations; in two directions for hyperbolic equations; and information is instaneously felt everywhere for elliptic equations. Each of these types of equations must be solved by different methods based on the type of equation. While partial differential equations do not play a large role in this text, it is useful to discuss how to solve them.

As we indicate in Section 3.15, parabolic equations are characterized by the unsteady mass (or heat) transport equation

$$\frac{\partial c_A}{\partial t} = D_A \nabla^2 c_A \tag{10.317}$$

where ∇^2 is the Laplacian operator. Parabolic equations have a single characteristic direction. Parabolic equations are solved by marching the solution forward in time: the characteristic direction. Hyperbolic equations are solved by the method of characteristics along the two directions that information travels: the two characteristic curves. Elliptic equations must be solved at all points simultaneously. The two-dimensional Poisson equation for the electrical potential discussed in the previous section is an example of an elliptic equation.

In this section, the focus is exclusively on elliptic and parabolic equations, which are common in micro- and nanofluidics. Hyperbolic equations are less common in microscale and nanoscale flows.

10.13.1 Elliptic equations

As indicated in Section 3.15, elliptic equations have no characteristic directions. Thus information travels instantaneously in all directions. To illustrate the numerical discretization and solution procedure, let us return to the two-dimensional form of the potential equation in dimensionless form, which is

$$\epsilon^2 \left(\frac{\partial^2 \phi}{\partial y^2} + \frac{\partial^2 \phi}{\partial z^2} \right) = -\rho_e \tag{10.318}$$

where ρ_e is the dimensionless volume charge density. This equation could represent the potential due to the EDLs in a square channel (Bhattacharyya & Conlisk, 2005). Each of the second derivatives is represented by its second-order central derivative expression, and the resulting difference equation is given by

$$\phi_{j-1,k} - 2\phi_{j,k} + \phi_{j+1,k} + \gamma(\phi_{j,k-1} - 2\phi_{j,k} + \phi_{j,k+1}) = -\frac{\Delta y^2}{\epsilon^2} \rho_{ej,k} \tag{10.319}$$

where $\gamma = \Delta y^2 / \Delta z^2$. The labeling of the grid points is depicted in Figure 10.23, and the three-dimensional channel geometry is depicted in Figure 10.24. This is a set of linear or, in the case of nonlinear equations, linearized equations in each of the y, z directions that may be solved directly by sweeping each direction. However, unlike ordinary differential equations, the difference equations are not tridiagonal; thus a Gauss elimination procedure would be used. However, because the matrix of coefficients is relatively sparse, there is much wasted computation.

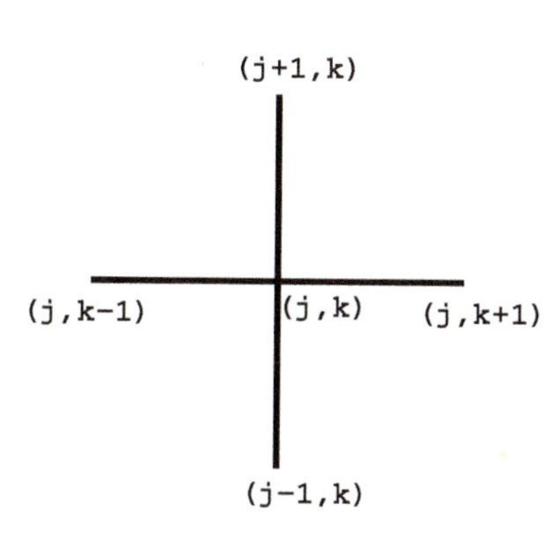

Figure 10.23 Notation for the labeling of the grids surrounding the central point (j, k).

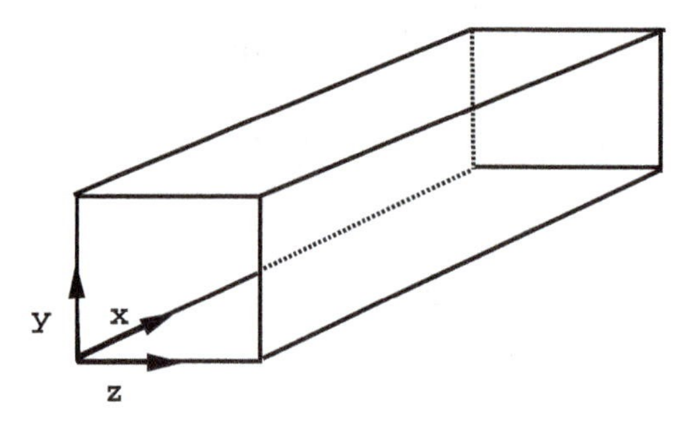

Figure 10.24 Geometry of the channel; u, v, and w are the fluid velocities in the x, y, and z directions, respectively.

Conversely, the equation can be solved iteratively using SOR, and this is the method that is preferred. To do this, solve for $\phi_{k,j}$ and

$$\phi_{j,k} = \frac{1}{2+2\gamma}(\phi_{j-1,k} + \phi_{j+1,k} + \phi_{j,k-1} + \phi_{j,k+1}) + \frac{\Delta y^2}{\epsilon^2}\rho_{ej,k} \tag{10.320}$$

Note that for the volume charge density equal to zero and for $\gamma = 1$, the value of $\phi_{k,j}$ is just the average of the points around it. Both the direct method and the iterative method just discussed will be successful if there is diagonal dominance and the mesh is swept in an orderly fashion. Normally the mesh would be swept horizontally (through the columns) – and then vertically (through the rows) – or the order could be reversed. In the latter case, in Matlab, two "for" loops are required:

```
for i=2:m-1
for j=2:n-1
```

where m is the number of rows and n is the number of columns. Generally, for most problems, the iterative methods converge in far fewer than 100 iterations. Iterative methods also lend themselves to nonlinear problems because iteration is already being performed; that is, a separate iterative loop to account for the nonlinearity is not required.

As with the one-dimensional examples, the SOR method is preferred based on the Gauss–Seidel iteration. It can be shown based on the analysis of the spectral radius of the coefficient matrix that the Gauss–Seidel iterative method converges twice as fast as the Jacobi iteration after a large number of iterations (Smith, 1985).

Specific examples of the solution of elliptic equations, including the Poisson equation for the potential, are left for the exercises.

10.13.2 Parabolic equations

As noted previously, parabolic PDEs have only one set of characteristics, and so information travels in a single direction. This fact has a considerable effect on the numerical method used. Parabolic PDEs can be integrated forward in the timelike variable; the timelike variable need not physically refer to time but can

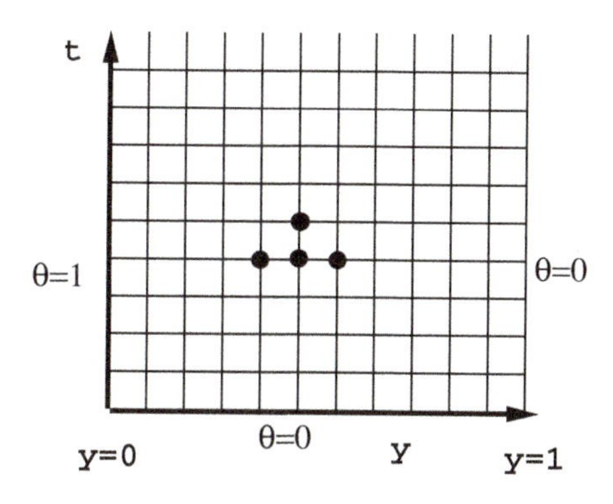

Figure 10.25 The explicit scheme to solve the diffusion equation with specified boundaty conditions; θ is the dimensionless temperature.

be a spatial variable as well. In what follows, however, the timelike variable will be referred to as being time.

In solving parabolic equations, the solution may easily be found explicitly. However, because the error grows in time, very small time steps are required. To remedy this, the equation can be solved implicitly, resulting in a tridiagonal system of equations that can be solved by the direct and iterative methods we have discussed previously. This situation is similar to the case for initial value ordinary differential equations discussed earlier.

EXPLICIT METHODS

Suppose θ represents a dimensionless temperature in the unsteady and one-dimensional heat conduction equation. In an explicit method, the equation is approximated at the previous time step, as shown in Figure 10.25. The result for the temperature θ is

$$\frac{\theta_{n+1,j} - \theta_{n,j}}{\Delta t} = \frac{\theta_{n,j+1} - 2\theta_{n,j} + \theta_{n,j-1}}{\Delta y^2} \tag{10.321}$$

$$\theta_{n+1,j} = \theta_{n,j} + r\left[\theta_{n,j+1} - 2\theta_{n,j} + r\theta_{n,j-1}\right] \tag{10.322}$$

$$\theta_{n+1,j} = r\theta_{n,j+1} + (1-2r)\theta_{n,j} + r\theta_{n,j-1} \tag{10.323}$$

which is an explicit expression for the dimensionless temperature at the new time step. Note that the scheme is dependent on the ratio of the time step to the spatial step $r = \Delta t/\Delta y^2$, and for $r = 1$,

$$\theta_{n+1,j} = \theta_{n,j+1} - \theta_{n,j} + \theta_{n,j-1} \tag{10.324}$$

The advantage of explicit methods is that it is easy to program; however, a disadvantage is that the time step must be very small. In fact, the explicit method is only valid for $r \leq 0.5$; that is, the scheme is convergent and stable only for $r \leq 0.5$ (Smith, 1985).

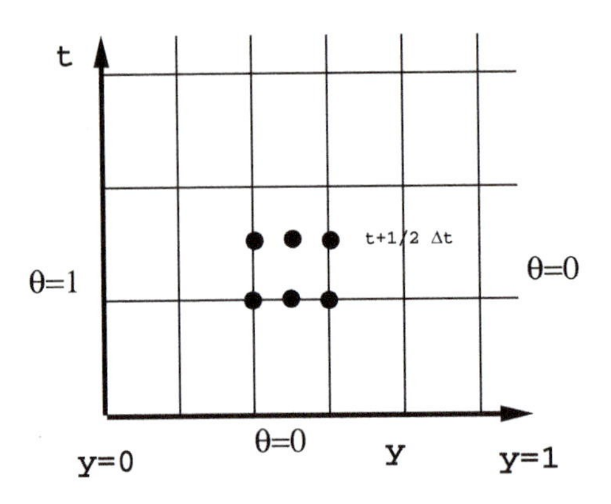

Figure 10.26 The Crank–Nicolson scheme to solve the diffusion equation.

IMPLICIT METHODS: CRANK-NICOLSON

Referring to Figure 10.26, in this case, the equation is approximated at $t_{j+1/2} = \frac{1}{2}(t_{j+1} + t_j) = t + \frac{1}{2}\Delta t$, and the difference equation is

$$\frac{\theta_{n+1,j} - \theta_{n,j}}{k} = \frac{1}{2}\frac{\theta_{n+1,j+1} - 2\theta_{n+1,j} + \theta_{n+1,j-1}}{h^2} + \frac{1}{2}\frac{\theta_{n,j+1} - 2\theta_{n,j} + \theta_{n,j-1}}{h^2} \qquad (10.325)$$

or

$$2\left(\theta_{n+1,j} - \theta_{n,j}\right) = r\left(\theta_{n+1,j+1} - 2\theta_{n+1,j} + \theta_{n+1,j-1}\right) + r\left(\theta_{n,j+1} - 2\theta_{n,j} + \theta_{n,j-1}\right) \qquad (10.326)$$

Collecting terms,

$$\begin{aligned} -r\theta_{n+1,j-1} + (2+2r)\,\theta_{n+1,j} - r\theta_{n+1,j+1} &= 2\theta_{n,j} + r\left[\theta_{n,j+1} - 2\theta_{n,j} + \theta_{n,j-1}\right] \\ &= r\theta_{n,j-1} + (2-2r)\,\theta_{n,j} + r\theta_{n,j+1} \end{aligned} \qquad (10.327)$$

and this is a triadiagonal system of equations to solve for the temperature at each time. Thus the tridiagonal solver or the Thomas algorithm discussed previously can be used to solve the system. Note that r appears again, but here the value of r need not be so small. Indeed, the Crank–Nicolson method is stable for all values of r (Smith, 1985).

There are many other schemes to solve these problems, such as the Keller–Box scheme, in which the second-order equation is converted to two first-order equations. This scheme is very good for nonuniform grids. The Crank–Nicolson scheme is easy to program and is the method of choice in most cases.

NONLINEAR EQUATIONS

Nonlinear partial differential equations are handled in the same way as differential equations. The lowest-order portion of the nonlinear term is linearized. For example, for the equation

$$\rho c \frac{\partial T}{\partial t} = \frac{\partial}{\partial y}\left(K(T)\frac{\partial T}{\partial y}\right) \tag{10.328}$$

with the thermal conductivity given by

$$K(T) = K_0\,(1 + \alpha T)$$

the nonlinearity is in the conductivity term:

$$\frac{\partial}{\partial y}\left(K(T)\frac{\partial T}{\partial y}\right) = K(T)\frac{\partial^2 T}{\partial y^2} + \frac{\partial K}{\partial y}\frac{\partial T}{\partial y} \tag{10.329}$$

In the first term at each iteration, $K(T)$ and $\partial K/\partial y$ are known, and the tridiagonal system is solved in the Crank–Nicolson scheme as discussed previously, in a way similar to the Blasius equation.

10.13.3 The Matlab PDE solver

Matlab has a PDE solver for both parabolic and elliptic equations. The solver `pdepe` solves initial boundary value problems in one space variable and one time variable. However, there must be at least one parabolic equation in the given system of equations. As with Matlab's ODE solver, `pdepe` converts the PDE into an ODE with the ime integration performed with the Matlab function ode15s. The calling syntax is

```
sol=pdepe(m,pdefun,icfun,bcfun,xmesh,tspan)
```

where $m = 0, 1, 2$ denotes rectangular, cylindrical, and spherical coordinates, respectively. The various functions in the calling statement have the same interpretation as for the ODE solver.

10.14 Verification and validation of numerical solutions

There are two main classes of activity for assessing the accuracy and validity of numerical calculations of scientific and engineering problems. *Verification* is the term used to determine whether the discretized problem converges to the continuum problem as the number of mesh point and time steps increases, i.e., whether the given equations have been solved correctly. This procedure is used for both ordinary and partial differential equations (Figure 10.27). One method of assessing the accuracy is to identify the number of digits to which two successive solutions agree. This methodology applies to the finite difference methods described in this chapter and also to finite volume and finite element procedures that are beyond the scope of this book.

Validation is the term given to the process of determining whether the correct physical problem has been solved (Figure 10.27). Validation answers the question,

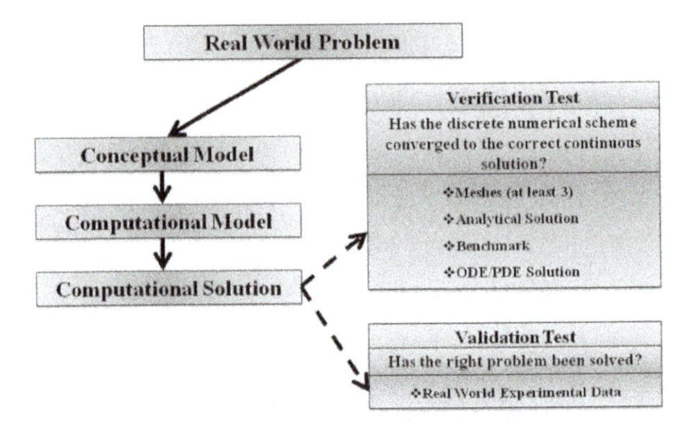

Figure 10.27 Illustration of the process of verification and validation of a numerical solution of an ordinary or partial differential equation. Adapted from Oberkampf *et al.* (1998).

is the computational model an accurate physical representation of the actual real-world problem? Both terms were developed to apply to computational fluid dynamics simulations, but these concepts also apply to the much simpler cases in this book (Oberkampf *et al.*, 1998).

These are serious questions, especially in situations where even the physical properties of a material are unknown. This is one element in the definition of the term *uncertainty*. Moreover, in industry, there are often fiscal constraints that limit the extent to which a given code can be verified and validated. These issues and more are described by Oberkampf *et al.* (1998) in great detail; in this section, we present a short summary of these concepts.

As we have already discussed, the most common way of verifying a computer code is to systematically reduce the grid and compare successive results. Indeed, the main objective of verification is to quantify the effect of the grid on the results. The objective of grid and time step refinement is to estimate the discretization error of the numerical solution and to expose programming errors and blunders. The discretization error should monotonically approach zero as the grid size and time step approach zero. The defining characteristic in this monotonic region is that the order of accuracy of the difference equations being solved is constant as the grid and time step are reduced. When this is true, Richardson's method can be used to extrapolate to an additional accuracy level. At this point, one can say that the numerical scheme is both grid and time step convergent. When a code is accurate to second order, the truncation error is reduced by 4 when the grid size is reduced by 2. This is because the error associated with the discretization is $O(h^2)$, where h is the grid size in any given direction.

An additional complementary means of verifying a code is to compare with an available analytical solution – a solution that can be expressed in "simple" functions such as the circular functions (sine and cosine), exponentials, or combinations of exponentials; the hyperbolic functions; or a Fourier series. A similar procedure could be used to compare with an already verified numerical solution. Both types of solutions are called *benchmark solutions*. One example of a benchmark solution is the Blasius flat plate boundary layer.

Note that both these methods of verification may require simplification of the code to an extent that many of the features of the code used to describe a given physical problem may not be exercised. Thus a systematic reduction of the grid size is the best means of verifying a code. Blunders or coding errors can be identified in this way; if the code is second-order accurate and the results do not exhibit second-order reduction in error, something is wrong. Similar comments apply to time step reduction. Of course, another source of error is lack of convergence of an iterative process, and this aspect of the code must be tested as well. A summary of the verification process is depicted on Figure 10.27.

A relatively recent means of verification of numerical solutions is the method of *manufactured solutions* (Roache, 1998; Roache & Steinberg, 1994; Oberkampf & Blottner, 1998; Roy *et al.*, 2004). In this method, given a partial differential equation, a specific analytical form is chosen and substituted into the equation, generating source terms on the right-hand side. The modified equation, including source terms, is then discretized and solved numerically. A symbolic manipulator such as Mathematica can be used to generate the source terms (Roy *et al.*, 2004). Because the *manufactured solution* generated the right-hand side, the numerical solution of the modified equation should be the manufactured solution itself. A measure of the accuracy is thus the difference between the numerical result and the assumed manufactured solution, or

$$E = f_{\mathrm{Num}} - f_{\mathrm{Man}} \tag{10.330}$$

at each grid point. While manufactured solutions are of greatest value when the grid cannot be decreased, as in coupled problems involving large sets of equations, the method can also be used to determine the accuracy of numerical solutions of ODEs and even integrals.

The most common means of validation is to compare the computational results with experimental data. As is well known, the experimental data contain error, and the owner of the experimental data should always include error bars for a proper comparison. The computational results need not exactly agree with the experimental data. Often in difficult and large systems, the computational model may only predict trends. In the simpler unit problems described here, comparison with the experimental data is likely to be much better. As the complexity of the physical system and its model increases, so, too, does the validation process. A summary of the validation process is also depicted on Figure 10.27.

Calibration, or the process of adjusting certain physical parameters to better agree with experimental data, is *not* validation. However, in very difficult problems, where crucial information is not available, calibration may be a necessary substitute.

Validation at the nanoscale, especially of dynamic phenomena, requires additional scrutiny. Although techniques such as transmission electron microscopy (TEM) and near-field scanning optical microscopy (NSOM) are capable of submicron spatial resolution, most techniques that are nonintrusive and capable of interrogating flow in sealed microchannels are limited to length scales (i.e., spatial resolutions) of at least 1 μm, and hence validation of spatially complex nanoscale computations with experiment is simply not possible. Thus, at length scales below about 1 μm (see Section 3.17), computations in most cases can only be validated using integrated properties such as flow rate or overall force measurements.

Verification and validation activities are much more complicated than is described here (Roache, 1998; Oberkampf *et al.*, 1998). However, for the problems discussed in this book, this section is a good start. Moreover, the concepts of verification and validation can refer to any of the individual numerical procedures such as integration, differentiation, or interpolation.

10.15 Chapter summary

In this chapter, the basic principles of numerical methods are described, from zero finding to the numerical solution of partial differential equations. The chapter is written to be self-contained and is heavily weighted with example Matlab scripts that can be used for all the problems described in this book. The methods presented in this chapter work; indeed, it is not the intent to present all the numerical methods that can be used in a given problem. Thus, for example, in the section on integration, the focus is on the basic trapezoidal rule and Simpson's 1/3 rule. The Matlab intrinsic functions that can be used are also identified. Many, though not all, of the example problems and the problems at the end of this chapter are drawn from topics within the purview of micro- and nanofluidics.

The chapter begins with a discussion of the concept of the Taylor series, which is the most fundamental means of function approximation. After the introduction of the Taylor series, several methods of finding zeros of nonlinear functions are described. Next, the related concepts of interpolation and curve fitting are presented, and the limitations of using higher-order polynomials in the interpolations are identified. In fact, it is shown that sometimes curve fitting is preferred when the data set is noisy.

Numerical differentiation and integration are discussed next; the concept of a central difference that is much more accurate than the forward and backward differences is introduced. Dealing with derivative boundary conditions is also described.

Leading into the discussion of the solution of ordinary and partial differential boundary value problems, the solution of linear systems of equations is presented. Both direct solution methods and iterative methods are presented, and it is noted that iterative solution methods are preferable for large numbers of equations, whereas direct methods are preferable for smaller sets of equations, although the cutoff for the use of each is not distinct.

Solutions of initial value problems are discussed as well. Particularly relevant for micro- and nanofluidics is the concept of symplectic integrators, methods that preserve the condition that the Hamiltonian remain a constant. This property was demonstrated for the classical two-body problem. Symplectic methods are used exclusively in molecular systems governed by a Hamiltonian in molecular dynamics simulations discussed in Chapter 11.

The chapter concludes with the presentation of the methods of solving linear and nonlinear ordinary and partial differential equations. In particular, it is shown how the Poisson–Boltzman equation can be solved numerically for both constant potential and constant surface charge density boundary conditions.

After completion of this chapter, the student should be able to write a Matlab script to solve problems in the areas of numerical analysis described earlier. The tools that Matlab provides are powerful and, in most cases, reasonably straightforward to implement. As can be seen, many of the line drawings presented in this book are drawn using Matlab scripts.

With the knowledge of the material of this chapter, the student should be able to

- Find a zero of a numerical or analytical function such as finding the ζ potential given the surface charge density
- Curve fit or interpolate a given set of data
- Find the volume flow rate through a channel or tube given a numerical function
- Numerically calculate the solution for the electrical potential or velocity for an arbitrary mixture of ionic components of arbitrary valence
- Numerically calculate the solution for the mole fractions to verify the accuracy of the Poisson–Boltzman distribution
- Numerically integrate first-order differential equations in the time domain to find the end products of a biochemical reaction
- Numerically integrate stiff equations using both numerical and asymptotic perturbation methods
- Begin to solve parabolic and elliptic partial differential equations numerically

This is but a small list of the possible applications; there are many more.

EXERCISES

10.1 Find the solution of

$$x = \frac{x^2 + 2 - e^x}{3}$$

accurate to 10^{-5} by reformulating the equation as a zero-finding problem. Use the following methods:

a. From a sketch or several calculated values of the function in the zero-finding formulation, determine the approximate location of the solution.

b. Find the solution by hand calculations using Newton's method. Show $x_1, x_2, f(x_1), f'(x_1)$ and the error at each step.

c. Find the solution with a computer using the secant method.

d. Write an m-file using Newton's method, starting at the same point used in the hand calculation.

10.2 The natural frequencies of vibration of a uniform beam clamped at one end and free at the other are solutions of

$$\cos(\beta b)\cosh(\beta b) = -1$$

where $\beta = \rho\omega^2/EI$, $b = 1.5$m, $\omega =$ frequency, $EI =$ flexural rigidity, and ρ is the density of the beam. Determine the three *smallest* values of β that satisfy this equation to four significant figures. Note that $\beta > 0$

10.3 The solution of partial differential equations required to model phenomena such as surface adsorption and heat transfer arising in microfluidic applications often requires, as an intermediate step, the solution of problems such as

$$x \tan x = p$$

where x needs to be solved for given a numerical value of p, where $p > 0$. Multiple values of x will satisfy this equation.

a. Sketch the graphs of $f(x) = \cot(x)$ and $g(x) = x$ and find the approximate location of the roots of equation (3) when $p = 1$.

b. Calculate the three smallest nonzero roots of equation (3) using an iterative root-finding procedure such as Newton's method for $p = 1$.

c. Use a built-in zero-finding function in Matlab, fzero, to find the three smallest positive roots of x_n of Equation (3) for $p = 0.01, 0.1, 1, 10, 100, 1000$. Plot and tabulate your results.

10.4 For a charged surface immersed in a multispecies electrolyte solution, the surface ζ potential is related to the surface charge density σ by

$$\frac{RT\sigma^2}{2\epsilon_e F^2} = \sum_i c_i^0 \left[\exp\left(\frac{z_i F\zeta}{RT}\right) - 1\right]$$

where F is Faraday's constant, R is the gas constant, and T is the temperature, which can be taken to be 300 K in the calculation. The surface charge density is -0.01 C/m^2, and the concentration of each species in the solution (c_i^0) is listed in the table. Find the surface potential ζ.

Species	c_i^0
Na^+	0.14214 M
Cl^-	0.15834 M
Ca^{2+}	0.00810 M

10.5 Consider the data set of Table 10.9:

a. Draw by hand an interpolating curve by intuition.

b. Use Matlab to verify the results depicted in Figure 10.9.

c. Calculate $y(1)$ using this polynomial.

d. Find an interpolating function using cubic splines.

e. Calculate $y(1)$ using the cubic splines.

All the curves should appear on the same plot. What do you conclude?

10.6 The set of data points in the table is the result of an experiment that yields the velocity distribution in a fluid flow above a plane wall; the data are presented in dimensionless form; $u(y)$ is the velocity in the boundary layer in the x direction. Using a Newtonian interpolation formula, find the velocity at $y = 1.7$, $y = 0.25$, and $y = 2.8$. Use only second-order accurate formulas, then curve ft the data with an appropriate polynomial.

y	0.0	0.5	1.0	1.5	2.0	2.5	3.0
$u(y)$	0.000	0.402	0.682	0.852	0.942	0.981	0.995

10.7 For the data set of the previous problem, do the following:

a. Calculate du/dy at $y = 2$ using the forward formula, the backward formula, and the central formula. Which formula is most accurate?

b. Calculate d^2u/dy^2 at $y = 2$.

c. Calculate the wall shear stress using $\mu = 1.0 \times 10^{-3}$ Nsec/m^2. Assume that y is in millimeters.

10.8 The following problem involves the integration schemes discussed in the text.

a. Write a Matlab function routine to calculate a definite integral using the trapezoidal rule.

b. Write a Matlab function routine to calculate a definite integral using the Simpson 1/3 rule.

c. Test your routine by calculating the integral of the function $f(x) = e^{-x}$ over the range (0, 6) by taking 11, 21, and 41 points. What do you conclude? Compare your answers with the "exact" result.

d. Write an m-file routine to calculate the indefinite integral of the function in b over the range from (0,8). What do you notice about the values of the indefinite integral past 6?

10.9 This problem involves the Simpson rules.

a. Write a Matlab function routine to calculate a definite integral using the Simpson 1/3 rule.

b. Test your routine by calculating the integral of the function $f(x) = e^{-x^2}$ over the range (0, 2) by taking 11, 21, and 41 points. What do you conclude? Compare your answers with the "exact" result of 0.99532.

c. Write an m-file to calculate the integral using QUAD.

d. Write a Matlab function routine to calculate the indefinite integral of the function in b over the range from (0,3). What do you notice about the values of the indefinite integral past 2?

10.10 The total volume flow rate through the boundary layer is defined by

$$Q = \int_0^3 u(y)dy$$

where $u(y)$ is in units of m/sec and Q is in units of m^2/sec. Using the data in the table of problem 10.6, do the following:

a. Write a Matlab script to calculate the volume flow rate Q using Simpson's rule. Compare with the value using the trapezoidal rule.
b. Calculate the indefinite integral $Q(y) = \int_0^y U(\sigma)d\sigma$ from the given data points using the Trapezoidal rule. Assume $Q(0) = 0$. In the solution, include a table of y, $U(y)$ and $Q(y)$, and give the grid spacing. Compare with the Simpson's rule result.

10.11 The data in the table of exercise 10.6, above are from Sadr *et al.*, 2006 in a channel of height $h = 10$ μm and width $W = 25$ μm. Noting that the velocity is constant beyond $y = 210$ nm across the entire channel, assuming symmetry (i.e., that there is another layer with the same velocity as in the table, calculate the volume flow rate through the entire channel. What percentage of the flow rate is confined to the EDL, that is, $y \leq 210$ nm.

10.12 Measurements of some parameters of a physical system give a model of the system represented by the following set of linear equations:

$$2.98x_1 - 1.05x_2 + 2.53x_3 = -1.61$$

$$-1.66x_1 - 1.08x_2 + 2.94x_3 = -6.76$$

$$4.35x_1 + 0.56x_2 - 1.78x_3 = 7.23$$

a. Find the solution to this system of equations *by hand calculations* using Gaussian elimination.
b. Solve the system using Matlab.
c. Repeat b after changing the 0.56 coefficient to 0.50.
d. Calculate the determinant of the two coefficient matrices.

By noting what happens to the solution as we change the coefficient 0.56 to 0.50, what conclusions can be made about the model's susceptibility (or, in engineering terms, *robustness*) to model coefficient errors (and thus measurement errors) in terms of finding the solution?

10.13 Solve the system of equations

$$x_1 + 2x_2 = 2$$

$$x_1 + 3x_2 + 2x_3 = 7$$

$$x_2 + 4x_3 = 15$$

by hand using the Thomas algorithm. Compare the result with the answer in the text for the alternative elimination algorithm.

10.14 Consider the tridiagonal system of equations defined by $a_{ii} = 4 + h$ for $i = 1$ to N, $a_{i+1,i} = 1$ for $i = 1$ to $N - 1$ and $a_{i,i+1} = 1$ for $i = 1$ to $N - 1$. Take the right-hand side $a_{i,N+1} = k^2$. All other entries in the matrix are zero.

a. Solve this system using both the Jacobi and Gauss–Seidel iterative schemes. For simplicity, take $N = 5, h = .05$, and $k = .1$. For the iterative techniques, start with a zero initial guess; the number of iterations required for convergence should be printed out along with the solution.
b. Write a program to solve this system of equations using the two tridiagonal elimination algorithms discussed in class. No two dimensional arrays are to be used in the programs.

10.15 Solve the one-dimensional linearized Poisson equation for the potential

$$\frac{d^2\phi}{dY^2} = z\phi$$

subject to

$$\phi = \zeta \text{ at } Y = 0$$

where ϕ_s is a dimensionless ζ potential and at large distances,

$$\phi \to 0 \text{ as } Y \to \infty$$

numerically for $\phi_s = 1$ and $z = 1, 3$. Compare with the analytical result

$$\phi = \zeta e^{-zY}$$

10.16 Solve the one-dimensional Poisson equation

$$\frac{d^2\phi}{dY^2} = z \sinh z\phi$$

subject to

$$\phi = \zeta \text{ at } Y = 0$$

where ϕ_s is a dimensionless ζ potential and at large distances,

$$\phi \to 0 \text{ as } Y \to \infty$$

numerically using the SOR method for $\phi_s = 1, 4$. Compare with a solution using the direct solver and with the analytical solution given by

$$z\phi = 4 \tanh^{-1}\left(e^{-zY} \tanh \frac{z\zeta}{4}\right)$$

Compare the nonlinear result with the linear result. At what value of the ζ potential does the Debye–Hückel approximation not hold? What is your criterion?

10.17 Repeat the previous problem for the Neumann boundary condition

$$\frac{d\phi}{dY} = -\sigma \text{ at } Y = 0$$

with $\sigma = 1, 4$. What is the dimensional ζ potential at the wall?

10.18 Solve the one-dimensional Poisson equation in cylindrical-polar coordinates:

$$\frac{d^2\phi}{dr^2} + \frac{1}{r}\frac{d\phi}{dr} = z \sinh z\phi$$

subject to

$$\phi = \zeta \text{ at } r = 1$$

$$\frac{d\phi}{dr} = 0 \text{ at } r = 0$$

Note that the value of ϕ at $r = 0$ is unknown, and so an extra equation needs to be added to the linearized system of equations.

10.19 Consider the nonlinear ordinary differential equation

$$\delta\frac{d^2y}{dx^2} - \frac{dy}{dx} - y^2 = 0$$

subject to boundary conditions

$$\delta\frac{dy}{dx}(0) = y(0) - 1$$

$$y(1) = 1$$

a. Solve this equation for $\delta = 1$. Determine the number of points required for three-digit accuracy to the right of the decimal point. Use the Thomas algorithm.

b. Repeat for $\delta = 0.1, 0.01$. What do you notice about the solution near $x = 0$?

c. Plot $y(x)$ for the solutions in (a), (b).

10.20 Using the preceding solutions to the Poisson equation for the potential as a guide, solve the Poisson–Nernst–Planck system for the case of two species of possibly different valences:

$$\frac{d}{dy}\left(\frac{dg}{dy} + z_g g\frac{d\phi}{dy}\right) = 0$$

$$\frac{d}{dy}\left(\frac{df}{dy} + z_f f\frac{d\phi}{dy}\right) = 0$$

$$\epsilon^2\frac{d^2\phi}{dy^2} = -(z_g g + z_f f)$$

with z_f and z_g taking arbitrary values. Assume that g is the cation and f is the anion. The boundary condition on the potential is

$$\phi = \zeta \text{ at } y = 0$$

$$\phi \to 0 \text{ far from the wall}$$

and the no-flux conditions at the wall $y = 0$ for the mole fractions g, f are

$$\frac{dg}{dy} + z_g g \frac{d\phi}{dy} = 0$$
$$\frac{df}{dy} + z_g f \frac{d\phi}{dy} = 0$$

and far from the wall the mixture is electrically neutral and the concentration of each species is 0.1M. Calculate the solution for $z_g = 2$ and $z_f = -1$ and the dimensionless potential at the wall is $\zeta = 0.1$. Is Grahame's solution recovered?

10.21 Extend the previous problem to the channel for the same boundary conditions at $y = 1$ as at $y = 0$.

10.22 Using a method of your choice, solve Airy's equation,

$$\frac{d^2y}{dt^2} - ty = 0$$

subject to $y(0) = 1$ and $\dot{y}(0) = 1$ for $0 \le t \le 5$.

10.23 The spring in a spring-mass damper system often behaves in a nonlinear way. Suppose that a mass is subjected to force that is periodic in time; then Newton's law applied to the vertical deflection of the mass due to the force may be reduced to an equation of the form

$$\frac{dy^2}{dt^2} + y - \frac{1}{6}y^3 = 3\sin(3\omega t)$$

This is called Duffing's equation. In this equation, take $\omega = 0.893$, and the two initial conditions are $y = y' = 0$ at $t = 0$, where $y' = dy/dt$ is the vertical velocity of the mass.

a. Using the fourth-order Runge–Kutta or the Adams–Moulton method, calculate the solution to $t = 20$ sec using a step size of $h = \Delta t = 2$ sec.

b. Repeat for $h = 1, 0.5$ sec to check the accuracy. Is this sufficient for three-digit accuracy?

c. The motion is approximately periodic. Determine the period from a plot of yvst for the appropriate time step. Plot $y' = 0$ on the vertical axis and y on the horizontal axis for a single period. This is the phase plane.

d. Compare your results with the Matlab solver `ode45`.

10.24 For Duffing's equation, find the Hamiltonian and formulate the problem as a set of two first-order equations in Hamiltonian form. Solve the equations using the explicit Euler equation, and compare the result with the symplectic Euler result and your results from the previous problem.

10.25 For a square channel, the Poisson equation for the potential is given by

$$\epsilon^2 \left(\frac{\partial^2 \phi}{\partial y^2} + A^2 \frac{\partial^2 \phi}{\partial z^2} \right) = -(z_g g + z_f f)$$

where A is the aspect ratio of the channel. Solve this equation for a symmetric electrolyte with

$$\phi = \zeta \text{ at } y = 0, 1 \text{ and } z = 0, A$$

for $A = 1, 2$. Assume that the concentration of each ionic species is 0.016M in an upstream reservoir, $z_g = -z_f = 1$, $h = 20$ nm and $\phi_s = -120$ mV. Use the solid wall boundary condition for the mole fractions.

10.26 Write a program to solve the linear diffusion equation

$$\frac{\partial \theta}{\partial t} = \frac{\partial^2 \theta}{\partial y^2}$$

with $\theta(0, t) = 1$, $\theta(1, t) = 0$ and at time $t = 0$ $\theta(y, 0) = 0$.

a. Solve using the explicit scheme for $r = 1/6$ and for $r = 1$ out to a time when the steady state is reached. What do you conclude? Be sure to assess accuracy, and provide a list of values indicating that the solution is three-figure accurate.
b. Plot the solution as a function of time at $y = 0.25, 0.5, 0.75$.
c. Repeat the first two parts using the Crank–Nicolson scheme.

11 Molecular Simulations

11.1 Introduction

In this chapter, the focus is on presenting the fundamentals of molecular simulations, particularly *molecular dynamics* (MD) and its derivatives, particularly *nonequilibrium molecular dynamics* (NEMD), or MD in a system that is exposed to an external field. The number of molecular simulation tools has grown substantially over the last few years, and it would be impossible to mention all the applications that have appeared in the literature. MD simulations have been used in a variety of ways; four of the most important ways in the context of this text that such calculations have been used are as follows:

- Calculating transport properties of new fluids and mixtures for which such properties have not been measured or as a test of the accuracy of such experimental measurements
- Predicting the equilibrium conformation of complex biomolecules and polymers
- Verifying fluid dynamics boundary conditions (Koplik *et al.*, 1989)
- Predicting the flow of pure fluids and complex mixtures in very small channels and circular pores, where the continuum approximation may break down and where direct experimental measurements are not available

Indeed, it is the last class of problems that is of particular interest in this chapter. It is certainly to be noted that while restricted to very short length and time scales, molecular simulations provide information that cannot be found any other way.

As has been seen throughout this book, ionic and biomolecular transport devices are now being used for drug development and delivery, single molecule manipulation, detection and transport, and rapid molecular analysis. Many of these processes are illustrated by natural ion channels a few angstroms in diameter that serve as ion-selective nanoscale conduits in the body, which allow nutrients in and waste products out. Electrokinetic effects, such as electro-osmosis and electrophoresis, can be efficiently utilized to accomplish the desired transport of fluid. As the dimensions of the channels in these new devices shrink from the micro- to the nanoscale, molecular effects may be important for an accurate description of the flow field. However, in this context, continuum theory must still provide a quantitative framework for understanding microfluidics and its interface on nanofluidics.

The complexity of MD simulations is significant. There are literally thousands of potentials from which to choose to describe the molecular interactions between individual atoms and molecules (Sadus, 1997). In this respect, the objective of this chapter is limited. The purpose here is to introduce the reader to the basic concepts of molecular simulations, focusing specifically on MD and how they differ from continuum simulations in both philosophy and scope. The focus is on presenting the equations for both pressure-driven and electro-osmotic flow. Applications to problems involving biomolecular transport are also discussed. There are many assumptions that go into a molecular simulation, and the computational requirements are substantially greater than for continuum calculations.

This book has considered liquid flows exclusively, and this chapter will focus on a particular type of molecular simulation, MD (and NEMD). We discuss only deterministic methods; thus we will not discuss those methods in which some part of the system is not treated explicitly, such as *direct simulation Monte Carlo* (DSMC) techniques (Bird, 1994), which is a statistical method used primarily for gas flows, although it has recently also been used for liquid flows. The same is true for those methods that treat the solvent implicitly such as *Brownian dynamics*. The primary objective of this chapter is to introduce the student and reader to the principles and processes used to develop a MD code and to point the reader to the fundamental literature on the subject, including existing codes, some of which are available free to the public. Indeed, in today's world, the researcher using molecular simulation techniques can access a number of rigorously validated software packages that will be discussed later in the Chapter.

There are several monographs on this subject written at various levels. For good reason, the majority of the books are at the more advanced level (Allen & Tildesley, 1994; Sadus, 1997; Rapaport, 2004; Hinchcliffe, 2003; Frenkel and Smit, 2002; Leach, 1996; Tuckerman, 2010). These books require some knowledge of chemistry, the structure of molecules, intermolecular forces, bonding, and other detailed aspects of computational chemistry. We hope that the essence of MD and NEMD simulations can be presented clearly and effectively with a minimum of such background material although a look back at Chapters 6, 7, and 8 may help set the stage.

There are some lucid review articles relevant to the presentation in this chapter; these articles present the concepts of MD in a way that makes them accessible to the engineering community. These are the reviews by Sagui and Darden (1999), in which the focus is on electrostatic effects, and the review by Allen (2004), which focuses on presenting the fundamentals of MD simulations without a lot of the detail. My belief is that this chapter will more closely follow the philosophies of these reviews.

Generally, the MD process comprises several different steps. These steps include the following:

- Choosing the potential characterizing the nature of the forces between atoms
- Specifying the initial and boundary conditions of the system of atoms
- Specifying how the force field is to be evaluated
- Choosing the numerical method to solve Newton's laws of motion
- If there is an external field imposed on the system, introducing a mechanism to dissipate the energy that is being introduced into the simulated system

The last step is called *thermostating.* We will discuss all these topics. I should mention that in this chapter, it is not necessary to make the distinction between atoms and molecules, and the terms *atom*, *molecule*, *particle*, and *body* will be used interchangeably.

Isaac Newton

The chapter begins with an overview of the nature of molecular simulations, describing the various levels of molecular simulation from computing the electronic wave function of the electron cloud that results in the potential energy between atoms to the approximations inherent in MD calculations. This section is followed by a discussion of the potentials used, followed by a presentation of the details involved in MD and NEMD calculations that require the solution to Newton's laws of motion for many molecules. The chapter ends with a discussion of the types of MD and NEMD computational packages that are available.

Caution: This chapter is *not* for the experienced molecular simulator! It is geared to the late undergraduate or beginning engineering student who is a complete novice when it comes to molecular simulations.

11.2 The molecular world

As we know, molecules, and indeed, all matter, is made up of *atoms*, which are arranged in a specific way to form molecules. The individual atoms are held together by chemical bonds and consists of a positively charged nucleus consisting of protons and neutrons surrounded by a cloud of negatively charged electrons. An atom as a whole must be electrically neutral.

The behavior of a collection of atoms and molecules is completely determined by solving the time-dependent *Schröedinger* equation for the *wave function* Ψ for all the electrons and nuclei in the sample. The wave function contains probabilistic information about the location of each particle as the system evolves in time. For example, let us focus on a single particle (without specifying what the particle actually is) and suppose that particle is constrained to move in the x direction only. Then the wave function $\Psi = \Psi(x,t)$ and quantum mechanics tells us that the probability that the particle will be found between x and $x + dx$ at time t is $|\Psi(x,t)|^2\, dx$. The unit of the wave function is (length)$^{-dN/2}$, where N is the number of particles in the system and d is the dimensionality of the space (Styer, 1996). A good and lucid introduction to the principles of quantum mechanics is the old and short monograph by Gillespie (1970).

At the most basic level, then, a molecular simulation would result in the determination of the wave function for all of the electrons and the nuclei in the system; such a wave function is a function of the electron and nuclei coordinates from which the probabilities that a range of velocities that will occur can be determined, within the limitations imposed by quantum mechanics (Gillespie, 1970).

Of course, this is a daunting task, even for a few particles. Thus some simplifications must be made. One approximation results from the observation that the mass of an electron is much smaller than that of a proton, by over 1800 times. Thus the dynamics of the protons in a system can be decoupled from the dynamics of the electrons and this is called the *Born–Oppenheimer approximation.* In the Born–Oppenheimer aproximation, the electronic wave function is first obtained assuming the nucleus is infinitely massive and stationary. Thus, in this approximation, the wave function provides the electronic energy as a function of nucleus position. This electronic energy becomes the potential energy surface on which nuclear dynamics takes place. Quantum mechanics could also be used to determine nuclear dynamics, but in the overwhelming majority of molecular simulations, the nuclei are assumed sufficiently massive to be described by the classical equations of motion.

The Born–Oppenheimer approximation requires solving for the wave function for many electrons. For example, suppose the wave function is desired for a system of 4100 water molecules, a system that contains 1000 electrons. Then, for the three-dimensional system, the wave function dimension is 3000 – a daunting task indeed!

The process of solving the Schröedinger equation for the wave function is called an *ab initio* calculation, Latin for "from first principles." Molecular dynmics calculations where the electrons are treated explicitly are called *ab initio* molecular dynamics. Even though approximations are made to reduce the computational recourses required, these are still the most expensive MD methods. Thus, often the Born-Oppenheimer electronic energy is assumed to be of a relatively simple form based on physically based information. These are called *empirical potentials* and often come from experimental data or by fitting ab initio results.

The next level of approximation is to perform all of the calculations using some potential energy function that uses empirically determined coefficients; this is what is most commonly termed the *classical MD method.* This method is best illustrated by the *Lennard–Jones potential,* which will be discussed in the next sections, when we present the details of MD calculations. It is this method that is used most often today for molecular calculations. Despite that this empirical method is the most computationally efficient, it still can take days to weeks for a reasonably long computation. The empirical potential method has often been termed *molecular mechanics*. A class of hybrid methods has also been developed in which the electrons are treated explicitly, while empirical potentials are used for the balance of the problem. These methods are known as quantum mechanics–molecular mechanics (QM–MM) methods.

11.3 Ensembles

The state of a thermodynamic system is characterized by a given set of state variables or properties. These properties may include pressure, temperature, composition, density, surface tension, and free energy. Statistical thermodynamics deals with the structure of matter on the molecular level, that is, on the microscopic level. A microscopic state is a specific realization of the system where all velocities and positions of the particles in the system are known. Conversely, a macroscopic state corresponds to averages over all the microscopic states, resulting in macroscopic quantities such as density, pressure, and temperature. All microscopic states that are a realization of a specific macroscopic state belong to it and form what is known as an *ensemble*.

MD simulations are usually performed under well-specified thermodynamic conditions, and thus specific ensembles can be identified. Although any set of thermodynamic variables that specifies a state could be used to determine the ensemble, only a few specific combinations are of practical relevance in MD and NEMD simulations. The most common ensembles are the following:

- Microcanonical ensemble (NVE): the thermodynamic state of a system is defined by its composition N (the number of particles and their type), its volume V, and the total energy E; in this ensemble, the Hamiltonian is constant
- Canonical ensemble (NVT): the thermodynamic state of a system is defined by its composition N, its volume V, and temperature T
- Isothermic-isothermal ensemble (NPT): the thermodynamic state of a system is defined by its composition N, its pressure P, and temperature T

The NEMD calculation of the Poiseuille and electro-osmotic flows in a nanochannel and discussed in Section 11.11 are examples of NVT ensembles. Armed with this background, we can begin to discuss the details of a typical MD calculation, beginning with the potentials.

11.4 The potentials

Although it seems to apply to any molecular level theory, the term *molecular dynamics* is often used to designate the simulation of many-particle systems by solving classical equations of motion. Typical MD simulations involve trajectories of 10^2–10^6 particles evolving for times of 10^2–10^6 ps in simulations cells of 10^1–10^3 nm. Recall that 1 ps $= 10^{-12}$ s.

To perform MD simulations, the important intermolecular forces must be determined; thus the nature of the molecules, whether they are polar or nonpolar, charged or uncharged, and their size must be known. An MD simulation begins with the determination of an intermolecular potential Φ having units of energy from which the force is determined by

$$\vec{F} = -\nabla\Phi \tag{11.1}$$

In the nineteenth century, it was believed that the interaction potential between any two bodies of mass m_1 and m_2 was given by (Israelachvili, 1992)

$$\Phi(r) = -\frac{Cm_1m_2}{r^n} \tag{11.2}$$

where C is a constant. Thus the total potential energy associated with the effect of all of the particles on the other is given by

$$\Phi_{\text{tot}} = \int_a^L N\Phi 4\pi r^2 dr \tag{11.3}$$

where N is the number of particles in the region between r and $r + dr$ and L is some length scale that satisfies $L \gg a$. Substituting for the potential, we find that

$$\Phi_{\text{tot}} = -\frac{4\pi CN}{(n-3)a^{n-3}}\left(1 - \left(\frac{a}{L}\right)^{n-3}\right) \tag{11.4}$$

Thus we see that for $n > 3$, the total energy is independent of the distance L, the size of the box. These are the short range forces. For $n < 3$, the total energy depends on L, and these potentials are associated with a long-range force. Recall in Coulomb's law for the electric field that $n = 1$.

Typical empirical potentials used in a molecular simulation can be classified into two groups: those for bonded interactions and those for nonbonded interactions. A typical potential may be of the form

$$\Phi = \sum_{\text{bonds}} k_{bi}(b_i - b_0) + \sum_{\text{angles}} k_{\theta i}(\theta_i - \theta_0) + \sum_{\text{torsions}} k_\varphi \cos(n\varphi + \delta) + 1)$$
$$+ \sum_i \sum_{j<i} \left[\left(\frac{A_{ij}}{r_{ij}}\right)^{12} - \left(\frac{B_{ij}}{r_{ij}}\right)^6\right] + \sum_i \sum_{j<i} \frac{q_i q_j}{r_{ij}} \tag{11.5}$$

where $r_{ij} = r_i - r_j$ is the distance between two particles. In this expression, the first three terms are bonded potentials and the last two terms are nonbonded potentials. The first term deals with the bond length, the second with the change in bond angles, and the third with changes in the bond rotation. The fourth term in the equation is the Lennard–Jones potential, describing short-range attractive and repulsive forces, and the fourth is the long-range Coulomb interaction. In many empirical MD simulations, the bonded interactions are neglected. This equation could describe the potential for a set of atoms or for small molecules, or even for small sugars or proteins.

An alternative to the repulsive force in the Lennard–Jones potential, the r_{ij}^{-12} term, is the Buckingham potential, defined by

$$\Phi_B = \alpha_{ab} e^{-\beta_{ab} r_{ij}} - \left(\frac{\gamma_{ab}}{r_{ij}}\right)^6 \tag{11.6}$$

and describes the interaction between atoms of type a and b, and α_{ab}, β_{ab} and γ_{ab} are constants. Note that the Buckingham potential behaves very differently from the repulsive Lennard–Jones term as $r_{ij} \rightarrow 0$.

In equation (11.5), it is the long-range electrostatic interactions that are the most troublesome from a computational perspective because they decay only as $1/r_{ij}$. Indeed, the efficiency of MD calculations and the eventual results depend on how these sums are calculated. This and other computational issues are discussed later in the chapter. The constants in equation (11.5), the k_m, A_{ij}, and B_{ij}, are often obtained from an ab initio calculation.

11.5 Using the Lennard–Jones potential

In this section, we will describe the basic computational methodology for an MD simulation using the Lennard–Jones potential, while recognizing that there are many other potentials describing intermolecular potentials in the literature.

The Lennard–Jones potential is perhaps the most popular description of the nonbonded interaction between two particles. The cutoff form of the Lennard–Jones potential is most commonly used in the form

$$\Phi_{LJ}(r_{ij}) = 4\epsilon_W \left[\left(\frac{\sigma}{r_{ij}}\right)^{12} - \left(\frac{\sigma}{r_{ij}}\right)^{6}\right] \quad \text{for } r_{ij} < r_c \tag{11.7}$$

$$= 0 \quad \text{for } r_{ij} \geq r_c \tag{11.8}$$

where ϵ_W is the well depth at $d\phi_{LJ}/dr = 0$, and σ in the Lennard–Jones system is considered to be a particle diameter. Here $r_c \sim 2.5\sigma$ is a cutoff radius and $r_{ij} = |\vec{r}_i - \vec{r}_j|$, the distance between two particles. A sketch of this potential is given in Figure 11.1. The first term in the potential is the repulsive force term, and the second term represents the attractive part of the force field, the van der Waals energy. A Lennard–Jones fluid is a collection of smooth, colliding balls. Note here that the "balls" are not polar, and so strictly speaking, the Lennard–Jones potential cannot describe polar substances like water; nevertheless, a Lennard–Jones calculation can describe some generic properties of liquids, some of which apply to water.

The Lennard–Jones potential has a discontinuity at $r = r_c$ that could adversely affect the numerical results. To avoid this, a *shifted* Lennard–Jones potential is

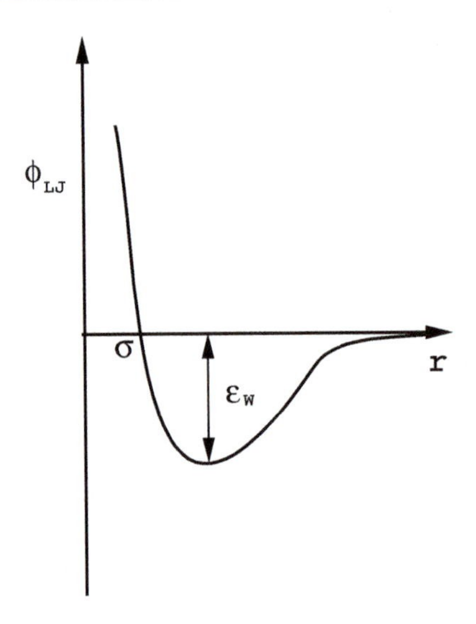

Figure 11.1 Sketch of the Lennard–Jones potential.

defined as

$$\Phi_{LJ,\text{shift}}(r_{ij}) = 4\epsilon_W \left[\left(\frac{\sigma}{r_{ij}}\right)^{12} - \left(\frac{\sigma}{r_{ij}}\right)^{6} \right] - \Phi_{LJ}(r_c) \text{ for } r_{ij} < r_c \tag{11.9}$$

$$= 0 \text{ for } r_{ij} \geq r_c \tag{11.10}$$

In this case, there is not a discontinuity at $r = r_c$, which means that the forces will always be finite.

The force corresponding to the potential is given by the gradient of the potential, and for the shifted potential,

$$\vec{F}_{ij} = -\nabla\Phi_{LJ} \tag{11.11}$$

and performing the differentiation, we find that

$$\vec{F}_{ij} = \left(\frac{48\epsilon_W}{\sigma^2}\right) \left[\left(\frac{\sigma}{r_{ij}}\right)^{14} - \frac{1}{2}\left(\frac{\sigma}{r_{ij}}\right)^{8} \right] \vec{r}_{ij} \tag{11.12}$$

for $r_{ij} < r_c$, and zero otherwise. Note that the potential is continuous at $r_{ij} = r_c$ but the force is discontinuous there; however, experience has shown that the discontinuity does not have a serious adverse affect on the computation if r_c is chosen to be sufficiently large. Then Newton's law is given by

$$m_i \frac{d^2\vec{r}_i}{dt^2} = \vec{F}_i = \sum_{j=1}^{N} \vec{F}_{ij} \tag{11.13}$$

where N is the number of particles (i.e., atoms or molecules) and m_i is the mass of each molecule. The sum in equation (11.13) excludes the value $i = j$.

The problem can also be formulated in terms of the Hamiltonian, which is defined as

$$H = \sum_{i=1}^{N} \frac{p_i^2}{2m} + \sum_{i<j}^{N} \Phi_{LJ}(r_{ij}) \tag{11.14}$$

where p_i is the momentum of particle i and the force associated with the Hamiltonian is given by

$$\vec{F} = -\nabla H = m\vec{a} \tag{11.15}$$

Note that we have described the interaction potential between the atoms or molecules in the fluid; however, if the fluid is confined by walls, the interaction potential between the walls and fluid also must be specified. If the walls are taken to be smooth round spheres, then the interaction potential can also be of the Lennard–Jones type, but with possibly different different values of the Lennard–Jones parameters, σ and ϵ_W, although they are taken to be the same in some cases (Zhu *et al.*, 2005).

Equation (11.13) is usually put in dimensionless form to reduce the number of parameters required to vary to describe the system; chemists often refer to these dimensionless variables as *reduced units*. We define dimensionless variables by chosing σ, m, and ϵ_W to be the units of length, mass, and energy, respectively. Thus, for example, the dimensionless length is

$$r^* = \frac{r}{\sigma} \tag{11.16}$$

The unit of time is $\sqrt{m\sigma^2/\epsilon_W}$, and substituting into equation (11.13), we have (leaving out the asterisk, of course)

$$\frac{d^2\vec{r}_i}{dt^2} = 48 \sum_{j=1}^{N} \left[r_{ij}^{-14} - \frac{1}{2} r_{ij}^{-8} \right] \vec{r}_{ij} + \vec{F}_{\text{exti}} \tag{11.17}$$

where, again, the sum excludes the value $i = j$. Here $\vec{F}_{\text{exti}}$ is an external force field, perhaps the electric field in electro-osmotic flow.

A popular option to obtain the interaction parameters for a Lennard–Jones potential of different species is to use the Lorentz–Berthelot combining or mixing rules (Allen & Tildesley, 1994). The combining rules are empirical expressions that provide cross-interaction parameters (σ, ϵ_W) for molecules from the parameters of the pure atoms. For example, it is assumed that the collision diameter (σ) for interaction between the species a and b is the arithmetic mean of the values for the two pure species:

$$\sigma_{ab} = \frac{1}{2}(\sigma_{aa} + \sigma_{bb}) \tag{11.18}$$

The well depth (ϵ_W) is given as the geometric mean:

$$\epsilon_{Wab} = \sqrt{(\epsilon_{Waa} \times \epsilon_{Wbb})} \tag{11.19}$$

11.6 Molecular models for water

In Chapter 7, we discussed in qualitative terms the structure of water. In this section, we discuss the determination of the Lennard–Jones parameters of water and the specific interactions associated with the water molecule.

The requirements for a water model are that it must be simple, be computationally tractable, and reproduce the basic properties of water. These objectives are not easy to meet and sometimes conflict; thus there is no single water model that is commonly used in MD or NEMD simulations. The computational cost is a crucial factor, and most useful water models include only pairwise potentials with no explicit three-body terms. One of the first MD simulations of water was reported by Rahman and Stillinger (1971), and since then, many other water models have been proposed.

The water monomer (a single water molecule) can be described as a rigid or as a flexible entity, allowing all degrees of freedom for the OH bonds and HOH bond angle, the latter being computationally demanding. In rigid models, the SHAKE algorithm developed by Ryckaert *et al.* (1977) is generally used to constrain the bond lengths.

Molecular simulations of liquids have been confined almost exclusively to the implementation of pairwise rigid water models; three of the currently most used pairwise water models are the following:

- TIP3P water model developed by Jorgensen *et al.* (1983)
- Simple Point Charge (SPC) model of Berendsen *et al.* (1981)
- Simple Point Charge Extended (SPC/E) model of Berendsen *et al.* (1987)

These models use three interaction sites and rigid bond lengths and a rigid angle. The three models include two positive charges corresponding to the hydrogen partial charges and one negative charge corresponding to the oxygen partial charge; moreover, the van der Waals interaction between two water molecules is computed by using a Lennard–Jones potential acting only on the oxygen site. The TIP3P and SPC models are similar, differing only in the values of the angles and bond lengths of the water molecules and in the hydrogen charge; the Lennard–Jones parameters are shown in Table 11.1.

SPC/E is an updated version of SPC including a correction in the potential parameters to include polarization. Detailed studies have been performed to analyze and compare the accuracy and behavior of these classic water models (van der Spoel *et al.*, 1998; Mark & Nilsson, 2001). It should be noted that the assumption of a rigid water molecule is an approximation, and thus results for some properties will be substantially different from the corresponding experimental values. Interactions between water and another species, say, ions, must also undergo a calibration process to obtain the Lennard–Jones parameters from experimental results.

Table 11.1. Original interaction parameters for the SPC, SPC/E, and TIP3P water models

Parameters	SPC	SPC/E	TIP3P
distance (O-H)	0.1 nm	0.1 nm	0.096 nm
angle (H-O-H)	109.47°	109.47°	104.52°
ϵ_W	0.6502 kJ mol^{-1}	0.6502 kJ mol^{-1}	0.63639 kJ mol^{-1}
σ_O	0.3166 nm	0.3166 nm	0.31506 nm
q_O	−0.82e	−0.8472e	−0.834e
q_H	0.41e	0.4238e	0.417e

Because of the computational intensity of MD simulations, boundary conditions are important. In simulations in nanochannels, the boundary conditions can be applied directly to the simulation domain; typically, in a 6 nm channel, there will be about 20 water molecules across the channel (Zhu *et al.*, 2005), and the boundary conditions can be applied directly. Conversely, periodic boundary conditions must be used in the other two directions if the dimensions are larger to reduce the computation time to a reasonable amount – these conditions are discussed next.

11.7 Periodic boundary conditions

Because MD simulations are so time consuming, the region of computation must necessarily be limited. Consider a system of, say, $N = 1000$ atoms or particles in a three-dimensional box. Using walls to specify the boundaries of the system inevitably introduces significant and unwanted perturbations into the calculations. Typically, the influence of the walls in an MD simulation of Poiseuille flow will extend approximately five molecular diameters from the wall. If the walls of a box of particles are of length $n\sigma$, where σ is the molecular diameter (Figure 11.2), the

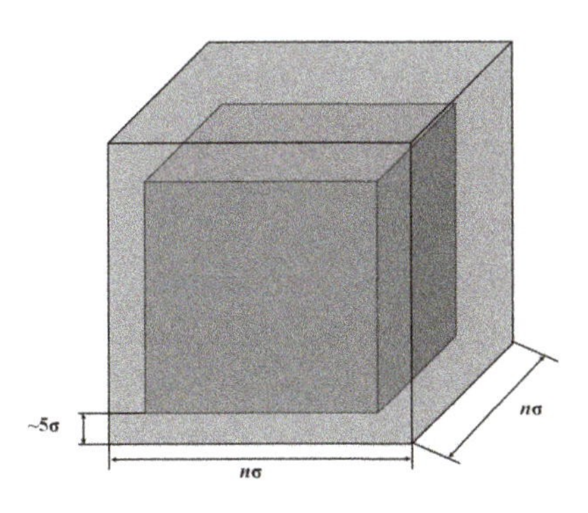

Figure 11.2 A simulation cell showing the need for periodic boundary conditions in molecular dynamics simulations. Figure contributed by Professor Sherwin Singer.

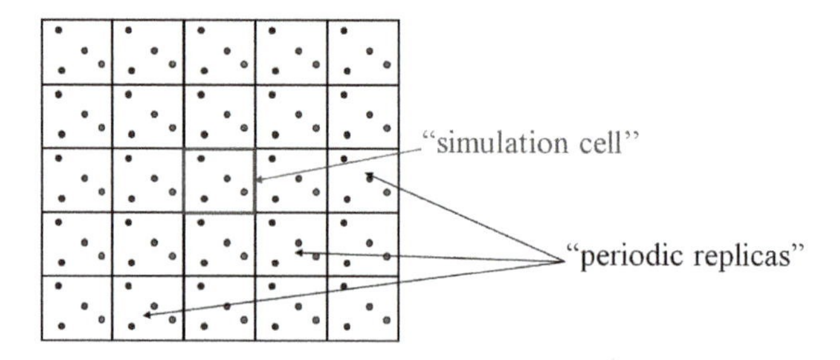

Figure 11.3 Sketch of the primary simulation simulation and its periodic images in two dimensions. Figure contributed by Professor Sherwin Singer.

fraction of the number of particles affected by the walls is given approximately by

$$\text{fraction} = \frac{6(5n^2\sigma^3)}{n^3\sigma^3} = \frac{30}{n} + O(1/n^2) \tag{11.20}$$

Thus, to make sure the walls do not adversely affect the numerical results, we should take $n \gg 30$. For $n = 300$, there will be a total of 27×10^6 particles required, an impossible computation, and still 10 percent of the system would be perturbed by walls. Using periodic boundary conditions offers a more practical way of reducing the computational load while maintaining computational relevance.

The bulk behavior of the fluid using MD calculations can be accomplished by assuming periodic boundary conditions in much the same way that Fourier series are used in continuum periodic problems. The simulation cell and its periodic images are shown in Figure 11.3. Using periodicity allows the simulation of a system that is infinite and without explicit walls. Periodicity does introduce some unphysical features, but these are much less severe than including walls explicitly. The interpretation of the vector $\vec{n}$ is best understood by comparing this expression with a typical Fourier series expansion in three dimensions.

So how many particles are enough? It is generally thought that for most problems, a region bounded by two walls in a given direction should be about 20 particle diameters at a minimum. Consider a molecular dynamics simulation of water. A single water molecule has a diameter of $D = 0.3$ nm and so 20 diameters is $h = 6$ nm. In a three-dimensional box, length $L = 6$ nm, there are about 7200 water molecules so that $30/n = 30/19 > 1$ since $7200^{1/3} \sim 19$. Thus periodic boundary conditions *must* be used. Using periodic boundary conditions reduces the severity of unphysical artifacts caused by the $30/n$ fraction of particles being affected by the wall. Remember, a Lennard–Jones fluid is not water, and typically, the number density of Lennard–Jones particles is taken to be $n = 0.8/\sigma^3$. A simulation using Lennard–Jones balls for the solvent described in the work of Zhu *et al.* (2005) contained 7757 solvent molecules, 31 cations, and 12 anions in their model of electro-osmotic flow. Periodic boundary conditions were used in two of the dimensions.

In using periodic boundary conditions, we write the vector $\vec{r}_i$ as

$$\vec{r}_{i,\vec{n}} = \vec{r}_i + \vec{n}L \tag{11.21}$$

where $\vec{n} = (n_x, n_y, n_z)$ is a vector of integers with $\vec{n} = (0, 0, 0)$ being the primary cell. This is shown in Figure 11.3. Suppose the primary cell is a cube of length L. What is required for the calculation is the potential per cell; we divide the calculation into describing interactions within the same periodic cell and interactions between particles in different cells.

The potential, or energy per cell due to all the interactions between the particles within the same cell, the *intracell* interactions, is given by

$$\Phi_{\text{intra}} = \frac{1}{N_{\text{cell}}} \sum_{\vec{n}}^{N_{\text{cell}}} \sum_{i<j}^{N} \Phi([\vec{r}_i + \vec{n}L] - [\vec{r}_j + \vec{n}L]) \tag{11.22}$$

where N_{cell} is the number of particles within the cell. Strictly speaking, this equation is accurate only in the limit as $N_{\text{cell}} \to \infty$, which we will omit in this discussion. Canceling the $\vec{n}$ terms and then noting that the sum over $\vec{n}$ gives just N_{cell}, we have

$$\Phi_{\text{intra}} = \sum_{i<j}^{N} \Phi(\vec{r}_i - \vec{r}_j) \tag{11.23}$$

That is, we need only use the primary cell ($\vec{n} = 0$) to calculate the intracell interactions.

The *intercell* component of the energy per cell, the interactions between particles in different cells, is given by

$$\Phi_{\text{inter}} = \frac{1}{2N_{\text{cell}}} \sum_{\vec{m}}^{N_{\text{cell}}} \sum_{\vec{n}\neq\vec{m}}^{N_{\text{cell}}} \sum_{i,j}^{N} \Phi([\vec{r}_i + \vec{n}L] - [\vec{r}_j + \vec{m}L]) \tag{11.24}$$

and we note that

$$\Phi([\vec{r}_i + \vec{n}L] - [\vec{r}_j + \vec{m}L] = \Phi([\vec{r}_i - \vec{r}_j + (\vec{n} - \vec{m})L) \tag{11.25}$$

so that all cells with the same relative displacement give the same result. The 2 in the denominator of the intercell potential expression is to avoid counting each i, j pair twice and is an alternative to writing the sum over $i < j$. Here $\vec{m}$ denotes the vector over all of the image cells. The inner sum includes $i = j$, and the $i < j$ is noted to prevent double counting. Now let us look at the sums over $\vec{n}$ and $\vec{m}$; both these sums give an identical contribution, as can be shown by simply taking $N_{\text{cell}} = 2$ and in one dimension, say, $n = 0, 1, 2$. Thus the intercell potential, can be written as

$$\Phi_{\text{inter}} = \frac{1}{2} \sum_{\vec{n}\neq 0}^{N_{\text{cell}}} \sum_{i,j}^{N} \Phi(\vec{r}_i - \vec{r}_j + \vec{n}L) \tag{11.26}$$

Combining the two potentials, the total potential is given by

$$\Phi_{\text{total}} = \sum_{i<j}^{N} \Phi(\vec{r}_i - \vec{r}_j) + \frac{1}{2} \sum_{\vec{n}\neq 0}^{N_{\text{cell}}} \sum_{i,j}^{N} \Phi(\vec{r}_i - \vec{r}_j + \vec{n}L) \tag{11.27}$$

and this is the potential that is used to describe the state of the system when periodic boundary conditions are used. We can combine these two terms, the first having $\vec{n} = 0$ and

$$\Phi_{\text{total}} = \sum_{\vec{n}=0}^{N_{\text{cell}}} \sum_{i<j}^{N\prime} \Phi(\vec{r}_i - \vec{r}_j + \vec{n}L) \tag{11.28}$$

where the prime indicates that $i = j$ is omitted for $\vec{n} = 0$.

One more approximation is made. If a cutoff method is used, the cutoff should not be so large that the motion of a given particle is influenced by its own image. Thus all terms in $\vec{n}$, except the one for which $\mid \vec{r}_i - \vec{r}_j + \vec{n}L \mid$ is the smallest. This approximation is called the *minimum image convention* (Leach, 1996), and the potential is written as

$$\Phi_{\text{total}} = \frac{1}{2} \sum_{i,j}^{N\prime} \text{Min}_{\vec{n}} \Phi(\vec{r}_i - \vec{r}_j + \vec{n}L) \tag{11.29}$$

The minumum image convention is the primary means of truncating the series in $\vec{n}$.

Now that we have the potential, we need to know how to sum the series. The potentials of interest contain both short- and long-range forces; the short-range forces can be handled by the minimum image convention. The long-range electrostatic force fields require a different approach, the *Ewald summation*, and this is discussed next.

11.8 The Ewald sum

In a brilliant paper, Ewald (1921) showed how to sum the long-range electrostatic force contribution to the potential in MD simulations. These interactions decay slower than r^{-3} and thus cannot be truncated like the short-range interactions; moreover, the series is only conditionally convergent, meaning that the end result depends on the order of the summation.

There are several ways to derive the Ewald method; for example, we know that

$$\frac{1}{r} = \frac{\text{erf}(\alpha r)}{r} + \frac{\text{erfc}(\alpha r)}{r} \tag{11.30}$$

where α is some parameter to be defined and erf and erfc are the error and complementary error functions, respectively. The first term is identified as the long-range component of the electrostatic potential since $\text{erf}(\infty) = 1$, and the second term is recognized as the short-range component of the electrostatic potential. The use of the error function in this identity can be motivated by the following arguments, using the potential due to a set of N charges of charge q_i.

The long-range potential at a location $\vec{r}$ due to a point charge placed at $\vec{r}_i$ satisfies Poisson's equation:

$$\nabla^2 \phi_{iL} = -\frac{1}{\epsilon_e} q_i \delta(\vec{r} - \vec{r}_i) \tag{11.31}$$

where δ is Dirac's δ function defined by $\delta(0) = \infty$, and $\delta = 0$ otherwise. The δ function is defined by its action on a function f as

$$\int f(x)\delta(x)dx = f(0) \tag{11.32}$$

If there are many charges, then the sum of all the charges over a volume leads to a volume charge density ρ_e that replaces the individual chages q_i. In this case, the solution to the Poisson equation (11.31) in three dimensions is given by

$$\phi_{iL}(\vec{r}) = \frac{1}{4\pi\epsilon_e} \int_{\mathcal{V}} \frac{\rho_e(\vec{r}_0)}{\mid \vec{r} - \vec{r}_0 \mid} d\mathcal{V}_0 \tag{11.33}$$

Note that this integral is singular at $\vec{r} = \vec{r}_0$, and so its numerical evaluation cannot be done effectively. Knowing that ions and biomolecules are really not point charges, a well-known approximation to the δ function is the Gaussian distribution

$$G_\alpha = (2\alpha^2)^{3/2} e^{-(\alpha r)^2} \tag{11.34}$$

where the standard deviation of this distribution is $1/\sqrt{2}\alpha$. The student can plot the Gaussian distribution for various values of α to show that

$$\lim_{\alpha \to \infty} G_\alpha(r) = \delta(r) \tag{11.35}$$

Let us now write the Poisson equation in a three-dimensional, spherically symmetric domain as

$$\frac{1}{r} \frac{\partial^2}{\partial r^2}(r\Phi_\alpha) = -\frac{G_\alpha}{\epsilon_e} \tag{11.36}$$

Integrating twice, it is evident that

$$r\Phi_\alpha = \frac{1}{2\alpha^2 \epsilon_e} \int_0^r G_\alpha(r)dr \tag{11.37}$$

Using the definition of G_α and noting that

$$\text{erf}(r) = \frac{2}{\pi} \int_0^r e^{-t^2} dt \tag{11.38}$$

we see immediately that

$$\Phi_{\alpha L} = \frac{1}{\epsilon_e r} \text{erf}(\alpha r) \tag{11.39}$$

Thus the long-range contribution to the potential due to a charge q_i having a Gaussian distribution centered at $\vec{r} = \vec{r}_i$ is

$$\Phi_{iL} = \frac{1}{\epsilon_e} \frac{q_i}{\mid \vec{r} - \vec{r}_i \mid} \text{erf}(\alpha \mid \vec{r} - \vec{r}_i \mid) \tag{11.40}$$

A similar argument for the short-range component due to a field of charges given by

$$\rho_{iS} = q_i(\delta(\vec{r} - \vec{r}_i) - G_\alpha(\vec{r} - \vec{r}_i)) \tag{11.41}$$

can be used to show that the short-range potential induced by particle i is

$$\Phi_{iS} = \frac{1}{\epsilon_e} \frac{q_i}{\mid \vec{r} - \vec{r}_i \mid} \text{erfc}(\alpha \mid \vec{r} - \vec{r}_i + \vec{n}L \mid) \tag{11.42}$$

What Ewald really did was to write the overall charge distribution of particle i due to a charge q_j that has short- and long-range components as

$$\rho_{ei,T} = \rho_{ei,L} + \rho_{ei,S} = q_i G_\alpha((\mid \vec{r} - \vec{r}_i \mid) + q_i(\delta(\vec{r} - \vec{r}_i) - G_\alpha(\vec{r} - \vec{r}_i)) \tag{11.43}$$

Using these two results for the long-range and short-range potentials due to all the charges, the total energy can be written, in terms of the simulation cell and all of its periodic neighbors, as

$$\Phi_L = \frac{1}{2\epsilon_e} \sum_n^{N_{\text{cell}}} \sum_{i,j} \frac{q_i q_j}{\mid \vec{r}_i - \vec{r}_j + \vec{n}L \mid} \text{erf}(\alpha \mid \vec{r}_i - \vec{r}_j + \vec{n}L \mid) \tag{11.44}$$

and, for the short-range forces, as

$$\Phi_S = \frac{1}{2\epsilon_e} \sum_n^{N_{\text{cell}}} \sum_{i,j} \frac{q_i q_j}{\mid \vec{r}_i - \vec{r}_j + \vec{n}L \mid} \text{erfc}(\alpha \mid \vec{r}_i - \vec{r}_j \mid) \tag{11.45}$$

where the term $i = j$ for $\vec{n} = 0$ is omitted from the sum. Often the self-energy term is extracted from the long-range interaction; since $\lim_{r \to 0} \text{erfc}(r) = 2/\sqrt{\pi}$, this quantity is given by

$$\Phi_{\text{self}} = \frac{\alpha}{\sqrt{\pi}} \sum_{i=1}^{N} q_i^2 \tag{11.46}$$

and the total potential can be written as

$$\Phi_{\text{Tot}} = \Phi_L + \Phi_S - \Phi_{\text{self}} \tag{11.47}$$

and now we add in the $\vec{n} = 0$ term for $i = j$ in the long-range sum.

The short-range contribution can immediately be summed in real space; however, the long-range sum still needs work because $\text{erf}(\infty) = 1$. So what do we do? We will focus on the one-dimensional situation and then extend the result to three dimensions. Because the system is periodic, we use the finite Fourier transform of the long-range component of the charge density defined by ($i = \sqrt{-1}$ in the exponential of the Fourier transform!)

$$\hat{f}(k) = \int_{-\pi}^{\pi} f(x) e^{-ikx} dx \tag{11.48}$$

which results in

$$\hat{\rho}_L(k) = \sqrt{2\pi\alpha^2} \sum_{j=1}^{N} q_j e^{-ikx_j - \frac{k^2}{4\alpha^2}} \tag{11.49}$$

Taking the Fourier transform of the Poisson equation, the long-range potential is given by

$$\hat{\Phi}_L(k) = \frac{\sqrt{2\pi\alpha^2}}{\epsilon_e} \sum_{j=1}^{N} q_j \frac{e^{-ikx_j - \frac{k^2}{4\alpha^2}}}{k^2} \tag{11.50}$$

Now write the inverse Fourier transform to obtain

$$\Phi_L(x) = \frac{\sqrt{2\pi\alpha^2}}{2\pi\epsilon_e} \sum_{j=1}^{N} q_j \sum_{k=1}^{N} \frac{1}{k^2} e^{-ikx_j - \frac{k^2}{4\alpha^2}} e^{ikx} \tag{11.51}$$

If the system is $2\pi/L$ periodic, then writing

$$k = \frac{2\pi m}{L} \tag{11.52}$$

$$\phi_L(x) = \frac{\alpha^2}{(2\pi)^{5/2}\epsilon_e L} \sum_{j=1}^{N} q_j \sum_{m\neq 0} \left(\frac{m}{L}\right)^{-2} e^{\frac{2\pi i m}{L}(x - x_j) - \frac{\pi^2 m^2}{L\alpha^2}} \tag{11.53}$$

Now, because of the exponential decay factor, the last term, the series is now rapidly convergent, and the electrostatic potential that is long range in physical space becomes short range in transform space, often called *reciprocal space.* It is now a matter of writing the equation in three dimensions, including all the sums, and the final result for the total long-range component of the energy reads

$$\Phi_{\text{Tot},L} = \frac{1}{2\pi L^3} \sum_{i<j} q_i q_j \sum_{\vec{m}\neq 0} \left(\frac{\vec{m}}{L}\right)^{-2} e^{\frac{2\pi i \vec{m}}{L}(\vec{x}_i - \vec{x}_j) - \frac{\pi^2 \vec{m}^2}{L\alpha^2}} \tag{11.54}$$

with again, for $i = j, \vec{m} = 0$ is omitted.

We can now move on to a discussion of the numerical issues faced in MD simulations.

11.9 Numerical issues

11.9.1 Time integration

As discussed in Section 10.11.6, a particularly useful method for MD simulations is to replace the second derivative by a discrete central difference approximation; this approximation may be derived simply from Taylor series approximations, and the approximation at time level k for any position x is defined by

$$\frac{d^2 x_i}{dt^2} = \frac{x_i^{k+1} - 2x_i^k + x_i^{k-1}}{\Delta t^2} + O(\Delta t) \tag{11.55}$$

Such an approximation in the context of MD simulations is called the Verlet (1967) algorithm, and it is symplectic, as noted in Chapter 10, preserving the initial value of the Hamiltonian reasonably well. It is also noted that Newton's equations of motion are reversible in time, as is this Verlet algorithm.

It is well known that particle trajectories computed from satisfying Newton's law are extremely sensitive to the initial conditions. Thus chaos is present in any system whose particle trajectories are calculated from Newton's law. In a potential flow calculation, the author and his colleagues found chaos in the motion of a system of three potential vortices above a plane wall (Conlisk *et al.*, 1989).

As an experienced numerical analyst, I am reading this and saying; so what good are the MD results? Well, I went searching for an explanation and found it in Frenkel and Smit (2002). The objective of an MD simulation is *not* to precisely predict the trajectories of *every* particle in the system. The aim of an MD simulation is to predict the state of a system in a mean sense, the same way that a continuum approach is used to predict the state of a system, perhaps electro-osmotic flow in a channel. This fact is best stated by Frenkel and Smit (2002) themselves, who point out that in a MD calculation,

> We wish to predict the average behavior of a system that was prepared in an initial state about which we know something (e.g. total energy) but by no means everything. In this respect, MD simulations differ fundamentally from numerical schemes for predicting the trajectory of satellites through space: in the latter case we really wish to predict the true trajectory. We cannot afford to to launch an ensemble of satellites and make statistical predictions about their destination. However in MD simulations, statistical predictions are good enough.

This explanation does not rigorously prove that statistical results would be no different if we could compute trajectories with infinite precision, but it can be shown that the results for the particle trajectories are not in too much error; see the discussion at the top of page 73 of Frenkel and Smit (2002).

The most time consuming step in an MD simulation is the calculation of the forces, especially the long-range forces. To do this, at a given particle location, we need to calculate the contribution of all of the other particles on the particle of interest. Thus, with the minimum image convention, there will be $N(N-1)/2$ pairs of interactions that must be evaluated at each time step so that the time required for the force calculation scales as N^2, where N is the number of particles. For a relatively small value of $N = 100$, this means that there are 10^4 evaluations of the force at each time step!

11.9.2 Truncation of interactions

Short-range interactions are truncated by assuming that the potential vanishes beyond a cutoff radius $r_{ij} > r_c$. A typical value for the cutoff radius is $r_c = 2 - 3\sigma$. Alternatively, often a tail contribution of the form is added:

$$\Phi_{\text{tail}} = \int_{r_c}^{\infty} \rho_N \phi(r) 4\pi r^2 dr \tag{11.56}$$

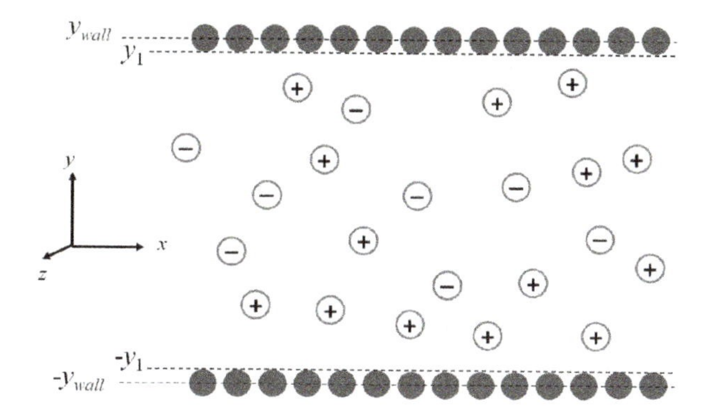

Figure 11.4 Schematic depiction of the simulation cell for electro-osmotic flow in a channel showing the cations and anions. The solvent molecules are not shown. The no-slip condition is applied at $y = \pm y_1 = y_{\text{wall}} - \sigma/2$.

where ρ_N is the average number density. For the Lennard–Jones potential, the result for the tail contribution is

$$\Phi_{\text{tail}} = \frac{8}{3}\pi\rho_N\epsilon\sigma^3\left(\frac{1}{3}\left(\frac{\sigma}{r_c}\right)^9 - \left(\frac{\sigma}{r_c}\right)^3\right) \tag{11.57}$$

where often $\rho_N\sigma^3 = 1$. Note that the tail contribution will diverge unless the potential decays faster than r^{-3}, as discussed earlier. The Coulomb interaction, the long-range electrostatic force, decays as r^{-1} and is evaluated using the Ewald summation procedure discussed in the previous section.

11.9.3 Boundary conditions

Often the wall particle interactions with other wall particles and the fluid particles are assumed to have a Lennard–Jones potential, with perhaps different values of the well depth and σ. Because the particles do not have a unique measure of their spatial extent (atoms are "fuzzy"), the question of the effective channel height is an issue. Consider the situation in Figure 11.4. The stationary wall particles are assumed to be located at $y = \pm y_{\text{wall}}$. An effective height of the channel is thus defined by $y_1 = y_{\text{wall}} - 1/2\sigma$, while in the continuum regime, $y_{\text{wall}} = h$. Thus, in Figure 11.4, $y_1 = h - \sigma/2$. As will be seen, this is an important consideration in the NEMD examples discussed in Section 11.11.

11.10 Postprocessing

So now we have completed the MD (or NEMD) simulation and have thousands of values for the positions and velocities of all the particles. We have run the simulation long enough that the system has reached equilibrium in the sense that the positions and velocities remain relatively constant with time. As we have seen, the purpose of molecular simulations, and MD in particular, is not to precisely

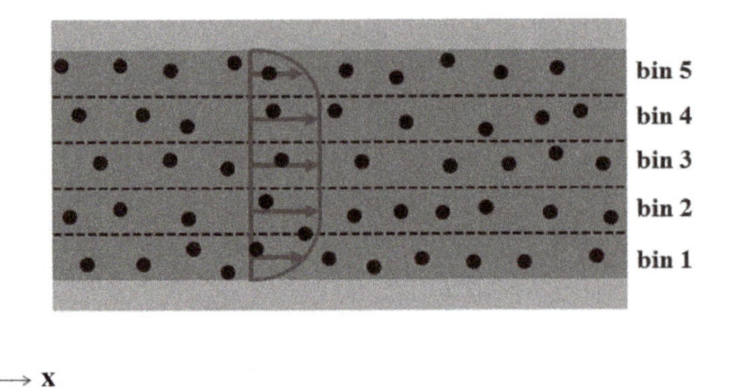

Figure 11.5 Example of "binning" for an MD simulation of electro-osmotic flow. The number of bins is dependent on the problem and $N_b = 5$ is an illustration only.

calculate the trajectories of all the particles but to predict the average behavior of the system.

In a typical MD simulation, positions and velocities for the different species in the system are saved as functions of time. These sets of stored values are used to extract the static and dynamic properties of the system.

To compute properties from the stored particle trajectories, such as density, concentration, and velocity profiles, a method for averaging the noisy data is required. The many-body problem is chaotic in the sense that each particle trajectory is strongly dependent on its original condition. Thus the trajectories can only be calculated statistically or as averages of the trajectories of many partiles. Thus the computational box is divided into *bins*; the binning resolution depends on the property to be measured and the configuration of the simulated system. Consider the calculation of the electro-osmotic flow velocity in a channel, as depicted in Figure 11.5. We have a number of particles in each bin, and the average velocity is defined by

$$\langle u \rangle = \frac{1}{A} \int u dA \tag{11.58}$$

Thus the average in each bin is reduced to a sum (equivalent to using the trapezoidal rule) as

$$\langle u \rangle_{bi} = \frac{1}{N_{bi}} \sum_{j=1}^{N_{bi}} u_j \tag{11.59}$$

where N_{bi} is the number of particles in each bin. The local velocity takes the value $\bar{u}_{bi}$ at the center of each bin and becomes a "velocity at a point" in continuum language. In fact, the velocity $\bar{u}_{bi}$ is actually the average of the center of mass velocities of the molecules in each bin.

Of course, the spatial averaging process takes place only after a suitable number of time steps, and in fact, in a similar way, the results are averaged in temporal bins as well. For example, the time average of the velocity of a particle i after a time t may be

$$\hat{u}_i = \frac{1}{t}\int_0^t u_i(\tau)d\tau \tag{11.60}$$

which can also be turned into a sum. Of course, writing this average after a long time t assumes that the results are independent of the initial condition. Since the many-body problem is chaotic, this will certainly not be the case, and really we should make a number of runs, averaging over a set of initial conditions. Doing this results in an *ensemble average*. It is seen that the postprocessing procedure is extremely complicated, and certainly this presentation is not sufficient to just go out and postprocess numerical data. In fact, an important part of the analysis of the results of MD simulations is to use the methods of *statistical mechanics* (Frenkel and Smit, 2002), the discussion of which is far beyond the scope of this presentation.

So why does statistical mechanics get involved? An MD simulation is based on the numerical solution of a many-body problem. As studied by Conlisk *et al.* (1989), chaos is present in interacting systems that follow Newton's laws. Therefore, in typical MD systems, chaos is expected to be present, although when an equilibrium state is established, the mean properties averaged over a sufficient number of trajectories seem independent of the initial configuration of the system. See the discussion on the objectives of an MD simulation and the quote from Frenkel and Smit (2002) in Section 11.9. Further analysis of this issue can be found in the literature related to chaos in equlibrium and nonequilibrium statistical mechanics (Castiglione *et al.*, 2008; Rice, 2000), and there is more in Frenkel and Smit (2002).

11.11 Nonequilibrium molecular dynamics

11.11.1 Introduction

NEMD refers to the situation where molecules or particles are subjected to an external force field. The use of NEMD simulations to describe the trajectories of molecules in multicomponent mixtures characteristic of problems in biology and chemistry in a bulk fluid flow is in its relative infancy because of the relatively large computation time required. Yet most molecular dynamics simulations are of the NEMD type.

It is important to note that when an external force is applied to a system, a simulation cell will rapidly heat up if heat is not removed. The term *thermostat* is given to the procedure used to remove the excess energy to keep the temperature of the cell constant. Various means have been used to do this such as coupling with a heat bath (Berendsen *et al.*, 1981) and the Nosé–Hoover and Isokinetic thermostats (Heyes, 1998; Rapaport, 2004). In the next section, we discuss the MD simulation of Poiseuille flow.

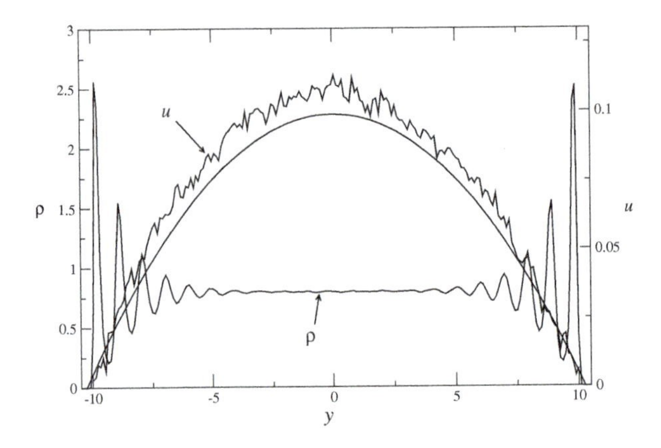

Figure 11.6 Steady state dimensionless density ρ and velocity distributions u established under conditions of planar Poiseuille flow for a pure Lennard–Jones fluid. Flow was induced by a constant force, $f = 0.005$, applied to all particles.

11.11.2 Poiseuille flow

Building on the work of Travis and Gubbins (2000), Zhu *et al.* (2005) has used NEMD to simulate Poiseuille flow in a small channel; the flow is assumed to be in the x direction, as in Figure 11.6. Periodic boundary conditions are used in the x direction, the direction of flow, and the z direction, with dimensionless lengths $L_x = 21.77$ and $L_z = 21.83$, respectively.

It is well known that the Navier–Stokes equations are based on the units of force per unit volume. Indeed, the governing equation for fully developed flow in a channel, repeated from Chapter 4, is given by

$$\frac{d^2u}{dy^2} = \frac{1}{\mu}\frac{dp}{dx} \tag{11.61}$$

for which the solution is given by

$$u(y) = \frac{1}{2\mu}\frac{dp}{dx}(y^2 - h^2) \tag{11.62}$$

where the zero y location is at the center of the channel of height h.

However, in molecular dynamics simulations, it is the force that is specified, and so to compare with the NEMD result, the continuum solution must be modified. The pressure gradient has units of N/m^3 or force per volume. To put the continuum result in a molecular setting, we write the governing equation in the form

$$\frac{d^2u}{dy^2} = -\frac{f_x \rho_N}{\mu} \tag{11.63}$$

where ρ_N is the number density of the particles and f_x is the external imposed force on the system. The continuum result written in terms of these molecular variables is given by

$$u(y) = \frac{f_x \rho_N}{2\mu}\left(y_1^2 - y^2\right) \tag{11.64}$$

where μ is the viscosity, and $y_1 = h - \sigma/2$ (as discussed previously). The pressure gradient is now recognized as

$$\frac{dp}{dx} = f_x \rho_N \tag{11.65}$$

The dimensional distribution of velocity can be made dimensionless by defining a dimensionless number density, viscosity, and force by $\rho_N^* = \rho_N N_A \sigma^3$, $\mu^* = \mu\sigma^2/\sqrt{m\epsilon_W}$, and $f_x^* = f_x\sigma/\epsilon_W$. The dimensionless velocity should then be scaled by $\sqrt{\epsilon_W/m}$ and given in dimensionless variables by (of course, after dropping the asterisk)

$$u(y) = \frac{f_x \rho_N}{2\mu}\left(y_1^2 - y^2\right) \tag{11.66}$$

and the lengths are scaled on σ and not h.

The flow is induced by a constant force, $f_x = 0.005$, applied to all particles. The dimensionless viscosity is given the value of $\mu = 2.13$ (Zhu *et al.*, 2005), and the dimensionless value for the wall height is $h = y_{\text{wall}} = 10.76$, as in Table 11.2. The velocity profile obtained from simulations is compared with the standard result for an incompressible fluid, with overall number density $\rho = 0.8$ and no-slip boundary conditions enforced one particle radius from each wall in the y direction: $y_1 = 10.26$, and y_1 is now dimensionless.

The results of the simulations are shown in Figure 11.6. Note that the continuum expression for the Poiseuille flow in molecular variables slightly underpredicts the NEMD result, but the difference is not large. Also note the molecular layering near the walls, where the dimensionless number density fluctuates, and thus, from the point of view of NEMD, the flow is not incompressible.

We now address a more difficult problem: electro-osmotic flow.

11.11.3 Electro-osmotic flow

In this section, we will assume that the external field is due to an electric field in the form of the Coulombic interaction (Zhu *et al.*, 2005)

$$\Phi_C = \zeta \frac{z_i z_j}{r_i - r_j} \tag{11.67}$$

where the variables are in dimensionless form and

$$\zeta = \frac{e^2}{\epsilon_e \sigma} \tag{11.68}$$

is the dimensionless interaction energy, e is the electron charge, and ϵ_e is the permittivity of the medium.

Here, as in the Poiseuille flow discussed earlier, the ions and wall particles have the same diameter as the solvent, and all particles have the same mass. Strong

Table 11.2. Parameters used in the NEMD simulation of electro-osmotic flow by Zhu *et al.* (2005)

Run	Total	Cations	Anions	Solvent	Wall	y_{wall}	ζ	ϵ_{IS}	M (cation)	M (anion)
A	8638	31	12	7757	836	10.76	5	7	0.2206	0.08538
B	8638	31	12	7757	836	10.76	1	1.4	0.2206	0.08538

ion–solvent attractions are needed to dissolve ions in solution. This means that the dimensionless ion–solvent interaction parameter must be large to keep the ions in solution because the Lennard–Jones spheres modeling the solvent are nonpolar (in contrast to water). The value of the dimensionless ion–solvent well depth, ϵ_{IS}, depends on the Coulomb interaction strength. Explicit values of ϵ_{IS} and other parameters are given in Table 11.2. Similar to the Poiseuille flow, the continuum expression for electro-osmotic flow must be put in molecular variables; the details are given by Zhu *et al.* (2005).

The molarities in Table 11.2 are based on a particle diameter of $\sigma = 2.8818$ Å, which would make the total solution molarity equal to that of water, 55.5 M. The concentrations of ions were chosen to be those estimated for physiological pH salt solutions. Assuming the same value of σ, the wall charge density corresponds to 2.4×10^5 electron charges per μm^2.

The Gaussian isokinetic thermostat (Evans & Morriss, 1990) is used to maintain the system temperature constant. A dimensionless time step of $\Delta t = 0.01$ is used, and the dimensional time is scaled on $\tau = \sqrt{m_{LJ}/\epsilon\sigma}$. As with the Poiseuille flow, only the y and z components of the velocity are thermostatted because there is a nonzero streaming velocity in the x direction.

Velocity profiles for the solvent are exhibited in Figure 11.7a for $\zeta = 5$ and Figure 11.7b for $\zeta = 1$. Note that for the smaller value of the interaction parameter, there is little ion exclusion from the wall, and there is no need to correct for excluded volume effects. For $\zeta = 5$, however, a correction of one ion radius gives excellent agreement with the continuum Poisson-Boltzmann (PB) result.

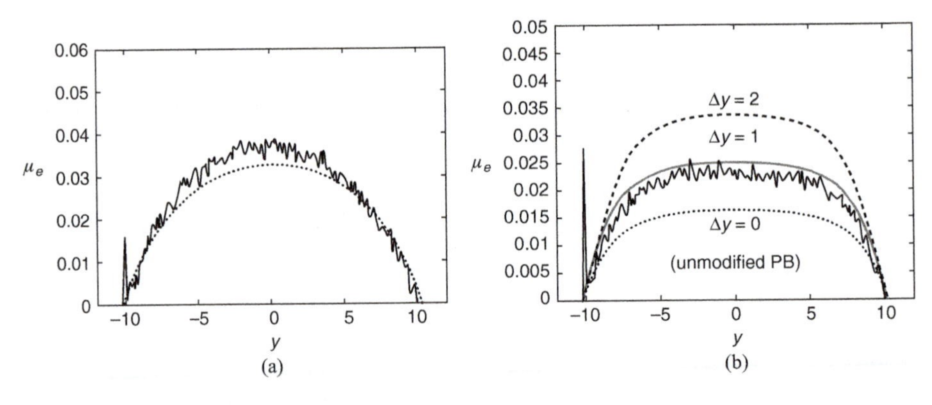

Figure 11.7 Dimensionless electro-osmotic mobility $\mu_e = u/E_x$ across the channel. The dimensionless electric field $E_x = 2$. (a) Interaction parameter $\zeta = 5$ and (b) interaction parameter $\zeta = 1$. For $\zeta = 5$, ion exclusion from the wall is significant. The characteristic velocity is $U_0 = \sqrt{\epsilon_W/m_{LJ}}$ where ϵ is the energy scale in the *LJ*-potential and the electric field is scaled on $E_0 = \epsilon_W/\sigma e$.

Table 11.3. MD/NEMD packages and their availability on the Web

Package	Web site	About
GAUSSIAN	http://www.gaussian.com	ab initio quantum chenistry
LAAMPS	http://laamps.sandia.gov	MD/NEMD (free)
GROMACS	http://www.gromacs.org	MD/NEMD (free)
AMBER	http://ambermd.org	MD force fields (free)
CHARMM	http://www.charmm.org	molecular mechanics
NAMD	http://www.ks.uiuc.edu/Research/namd/	MD/NEMD (free)

GAUSSIAN, an ab initio package, is included because the users of other packages may use force fields derived from GAUSSIAN.

The anion and cation concentrations for $\zeta = 5$ show significant layering near the surface, with significant density fluctuations; however, ions are completely absent from the wall layer. This is due to the strong ion–solvent interaction that prevents the ion from shedding its solvated solvent molecules to any great extent. Conversely, for $\zeta = 1$, the ions can populate the inner wall layer readily. For a discussion of other simulations that suggest that the Newtonian fluid assumption may break down near the interface (Qiao & Aluru, 2003b; Freund, 2002), see Zhu *et al.* (2005). Such behavior is not seen in the present work if the indicated correction of one ion radius is used in specifying the actual width of the channel.

11.12 Molecular dynamics packages

11.12.1 Introduction

The focus in this discussion is on empirical MD simulators as opposed to ab initio packages that calculate atomic interactions based on quantum chemistry. Some of the most popular classical MD software packages are Groningen Machine for Chemical Simulations (GROMACS). Large-Scale Atomic/Molecular Massively Parallel Simulator (LAMMPS), Chemistry at Harvard Molecular Mechanics (CHARMM), and Assisted Model Building with Energy Refinement (AMBER). All these packages are open source and readily available on the Web. The Web site locations are given in Table 11.3.

GROMACS or LAMMPS appear to be the tools of choice among researchers working on nanoscale problems in fluid dynamics. For example, Qiao and Aluru (2003a) have explored electrokinetic problems using GROMACS, and Priezjev *et al.* (2005) have demonstrated the use of LAMMPS to study slip behavior in wetted films.

11.12.2 What MD/NEMD simulators do

What kinds of information do these packages contain?

1. Force fields: These packages contain a number of interaction potentials. Usually in nanofluidics, we are interested in modeling of the electrical double layer, solvation of ions, surface tension, slip, contact lines, interaction of liquids, and polyelectrolyte with carbon nanotubes. For these applications, the Lennard–Jones potential and electrostatic potentials are often sufficient, but more sophisticated models of the interaction between water molecules may be required (e.g., SPC/E; Berendsen *et al.*, 1981). Both GROMACS and LAMMPS have a large number and variety of force fields from which to choose.
2. These packages also allow the user to choose the integrator, include constraints (e.g., constant temperature in NEMD), and provide boundary conditions. To perform MD simulations in ensembles where energy is not fixed and/or in a domain of finite extent in a certain direction, a variety of constraints, such as thermostats, barostats, and periodic boundaries, are needed. Usually, they are built into the MD packages.
3. These codes have been optimized so that the force fields are calculated efficiently. Portability across hardware and platforms is also a consideration.
4. Ease of use and legacy: A close connection between the problem of interest and the sample problems that come with the installation of a package ensures shorter lead times in obtaining meaningful results. LAMMPS comes with examples for atomistic simulation of Couette and Poiseuille flow.

A factor that enhances the versatility of newer packages like NAMD and LAMMPS (both released first in 2002) is that they can use the force fields and input configurations of CHARMM and AMBER. The ease of preprocessing, postprocessing, and visualization should also be considered in choosing a package, for example, the trajectories and initial coordinates of atoms in GROMACS can be visualized without a need for extra software; whereas LAMMPS needs software like VMD or RasMol for visualization.

11.13 Summary

In this chapter, the reader has been introduced to molecular simulations, specifically what is termed molecular dynamics. There are a number of applications, and relative to the subject of this text, these applications include the following:

- Calculating transport properties of new fluids and mixtures for which such properties have not been measured or as test of the accuracy of such measurements
- Predicting the equilibrium conformation of complex biomolecules and polymers
- Verifying the no-slip condition (Koplik *et al.*, 1989)

- Predicting the flow of pure fluids and complex mixtures in very small channels and circular pores, where the continuum theory may break down

A number of steps are involved in the building of a MD code, and these steps have been discussed in this chapter. These steps include the following:

- Choosing the potential characterizing the nature of the forces between atoms
- Specifying the initial and boundary conditions of the system of atoms
- Specifying how the force field is to be evaluated
- Choosing the numerical method to solve Newton's laws of motion
- If there is an external field imposed on the system, rescaling the temperature to insure that it is constant

The last step is called *thermostating*.

In today's world, it is typically not necessary for a researcher to build MD code from scratch. We have discussed several packages that are available to the public, three of them free of charge. That is not to say that these packages are easy to use, however!

EXERCISES

One note: The goal of the exercises following is to help the student understand the fundamentals of why the MD results are processed the way they are. The reason is that all MD simulations exhibit chaotic behavior at some level. This is why the methods of statistical mechanics are often used in postprocessing the MD data.

11.1 Return to the two-body problem discussed in Section 10.11.6 and add a third body. Write the governing equations in dimensionless form for the motion of the three particles. Calculate and analyze the trajectories. What do you find? Be sure to reduce the time step to make sure the trajectories are accurate. Arrange the particles in a triangle initially. Perturb the initial condition of one of the particles by 1 percent. Are the trajectories any different?

11.2 Explicitly write out the governing equations for the motion of the N-body problem for $N = 3$ from equation (11.17). Assume a Lennard–Jones potential.

11.3 Solve this system in dimensionless form numerically using the Verlet algorithm and compare with the same result using Runge–Kutta. Be sure to verify the results by reducing the time step. Calculate the Hamiltonian. What do you find? Assume the particles are in a triangle initially. Plot the trajectories as a function of time. Perturb the initial condition of one of the particles by 1 percent of the distance from one other particle. Do the trajectories change?

11.4 This is an open-ended exercise. Write a report on chaos in fluid mechanics. A good initial source is the paper by Conlisk *et al.* (1989). Chaos theory has also been used as a model for turbulent flow.

11.5 Write out the governing equations as in exercise 11.1, but include the electrostatic term.

11.6 This is an open-ended exercise. Think about the simulations described in Section 11.11. Using the results of your chaos report from exercise 11.4, is the system in exercise 11.1 chaotic? If so, how must you view the results?

11.7 This exercise requires some of the research from your chaos report. Now return to exercise 11.1 and perform a time series analysis of the trajectories. What measures are used to analyze the trajectories?

11.8 Write a report on statistical mechanics and its relationship to molecular dynamics simulations. What is a Green–Kubo relation, and how is it used?

12 Applications

12.1 Introduction

As has been mentioned in Chapter 2, applications of micro- and nanofluidic analyses include drug delivery and its control, DNA manipulation and transport, protein separations, rapid molecular analysis, renal assist devices, biochemical sensing, and cancer treatment. In the present chapter, several of these applications are described, and modeling tools are developed using the ideas developed in the preceding chapters.

Generally, many of the biomedical devices being tested for the preceding aplications are essentially synthetic micro- or nanopore membranes consisting of an array of channels, much as in Figure 1.1. For example, these devices can be used to deliver insulin to a diabetes patient, as needed, in a highly controlled manner. They have also been considered for use as a renal assist device (RAD) whose purpose is to replace some of the functions of the kidney; this device is discussed later. DNA has been shown to be able to navigate through approximately cylindrical nanopores, and this problem is also discussed in this chapter; the in vivo transport of DNA through a nanopore has been demonstrated by Bayley and Cremer (2001).

More general applications of nanotechnology and additional information on some of the applications described may be found in the short monograph edited by Malsch (2005) and in the monograph by Ciofalo *et al.* (1999), respectively. The latter book lists centers and groups that are active in either nanoscale physiological systems or other biological flow problems as of its publication in 1999. Since then, that list has expanded considerably (see Bohn (2009) and the references therein).

Many biomedical applications are described in the four-volume set titled *BioMEMS and Biomedical Nanotechnology*, edited by Ferrari (2006). The titles of each of the four volumes include "Biological and Biomedical Nanotechnology" (Lee & Lee, 2006), "Micro/Nano Technology for Genomics and Proteomics" (Ozkan & Heller, 2006), "Therapeutic Micro/Nano Technology" (Desai & Bhatia, 2006), and "Biomolecular Sensing, Processing and Analysis" (Bashir & Wereley, 2006).

The purpose of this chapter is to investigate several application areas and illustrate the use of the modeling techniques described in the previous chapters. The main application areas discussed here are

- DNA transport
- Transport processes in a RAD
- Biochemical sensing

These three areas were chosen because of the crucial role that micro- and nanofluidics plays in the operation of the devices in each area. These areas are certainly not all inclusive of the myriad applications, and the reader is encouraged to consult the references listed.

In particular, there are four areas that are not covered in this chapter, non-biomedical problems that are of intense interest:

- Electrokinetic remediation
- Batteries
- Fuel cells
- Water desalination and purification

Each of these application areas includes electrochemistry as the primary process driver, and the reader of this text will clearly understand the processes that occur in these subject areas. The reader will be introduced to these areas as exercises at the end of this chapter.

12.2 DNA transport

DNA transport through tubes is used for DNA sequencing, the determination of the order of the individual nucleotide bases that make up the DNA, DNA repair, injection of the repaired DNA into cells (electroporation), and sensing. In the sensing area, Chang *et al.* (2004) and Fan *et al.* (2005) have performed experiments in an effort to understand the dependence of the electrical current through a nanochannel filled with a DNA strand on various flow parameters such as electrolyte concentration, flow geometry, and applied electric fields. These experimental studies are closely related to (and inspired by) efforts to develop a novel DNA sequencing mechanism (Bayley & Cremer, 2001; Kasianowicz *et al.*, 1996; Deamer & Akeson, 2000; Meller *et al.*, 2001). It should be noted that the model described here can easily be generalized to other long-chain, cylindrical polymers.

Both natural and synthetic nanopores have been used to study the transport process, called *translocation* experimentally. Kasianowicz *et al.* (1996), Bayley and Cremer (2001), and Meller *et al.* (2001) experimentally investigated the translocation of single-strand DNA (ssDNA) through natural α hemolysin pores to characterize the DNA translocation properties by analyzing the ionic current signals.

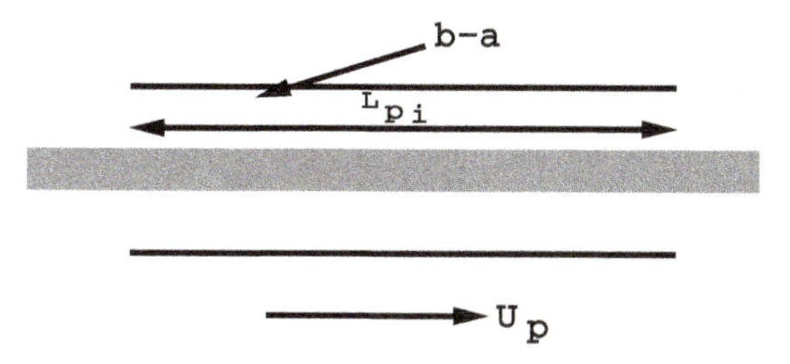

Figure 12.1 Side view of the geometry of the circular tube and circular DNA strand. Here the DNA strand of radius a is moving to the right; the outer radius is denoted by b. The notation L_{pi} denotes the length of the particle inside the tube and is equal to the tube length. The direction of motion of the particle depends on the surface charge densities of the particle and the wall.

Researchers have also investigated DNA transport through synthetic nanopores that are claimed to be more stable, offer more flexible choices in pore shape and surface properties, and are easier to integrate into micro- and nanofluidic devices. Fan *et al.* (2005), Li *et al.* (2003), and Storm *et al.* (2005) investigated the translocation of double-stranded DNA (dsDNA) through a synthetic nanopore experimentally as a function of applied votlage, electrolyte concentration, and other physical parameters. In particular, Li *et al.* (2003) demonstrated the capability of observing individual molecules of dsDNA translocation, and the experimental translocation rate for dsDNA through synthetic nanopores is ~0.01 m/s for 3 kbps (kbps = 1000 base pairs) and 10 kbps dsDNAs. Chen *et al.* (2004) showed that the DNA translocation velocity varies linearly with the applied voltage drop, increasing as the voltage drop is increased. The DNA velocity was demonstrated to be independent of the DNA length based on the experimental data for 3 kbps, 10 kbps, and 48.5 kbps dsDNA. Conversely, the experimental data of Storm *et al.* (2005) on 11.5 kbps and 48.5 kbps dsDNA show that longer DNA strands move more slowly through the nanopore than shorter ones, in conflict with Li *et al.* (2003). This point is thus a matter for further research.

In the next few sections, we describe a continuum model for translocation of DNA through a tube idealized as the moving inner cylinder in the electro-osmotic flow through an annulus discussed in Section 9.2.8.

12.2.1 How does DNA move?

Let us describe in general terms how DNA may move through a tube. The simplest model of a dsDNA molecule is a solid cylinder; it is assumed for simplicity that the dsDNA lies within a cylindrical pore. The radius of the dsDNA is about $a = 2 - 3$ nm, and the synthetic nanopore is usually about $b \sim 10$ nm in radius. DNA, whether single or double stranded, is negatively charged, and so a cathode placed upstream and an anode placed downstream will drive the dsDNA molecule through the tube. Thus the dsDNA is electrophoretically driven through the tube in concert with an electro-osmotic flow. The geometry of the problem is depicted in Figure 12.1.

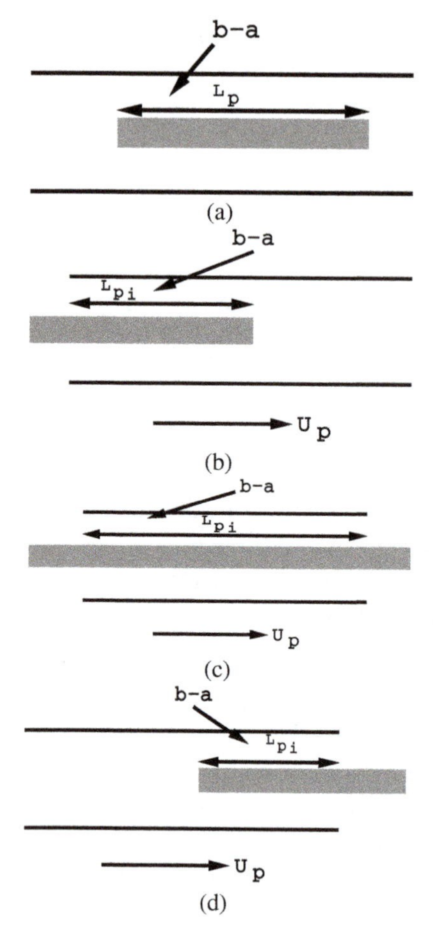

Figure 12.2 Schematic of a cylindrical particle in a cylindrical tube. In all cases, the gap between the particle and the wall of the channel is assumed much smaller than the length of the particle. (a) Length of the particle is less than the length of the tube (Zhao & Bau, 2007). (b) Length of the particle is much longer than the length of the tube; partial entry. (c) Length of the particle is much longer than the length of the tube; particle fills the tube. (d) Length of the particle is much longer than the length of the tube; partial exit. Not all of the particle is shown in (b)-(d).

It is assumed that the molecule and the tube are charged with possibly differing charge densities and that the gap between the tube and the molecule is big enough so that the fluid flow between the tubes can be described by a continuum analysis (Zhu *et al.*, 2005). In addition, it is assumed that electrostatic and other local conditions do not prevent the biomolecule from entering the tube; this is a separate problem and involves possible deformation mechanics, a feature that can be modeled by Brownian dynamics simulations (Hur *et al.*, 2000).

The length of the particle[1] that is inside the tube is denoted by $L_{pi}(t)$, and the various parameter ranges are shown in Figure 12.2.[2] As the particle enters the channel, only a portion of the tube is occupied. During this time, a transient pressure gradient is induced because the channel is only partially filled. As the leading edge of the DNA approaches and reaches the exit of the tube, there is another short transient period in which the pressure gradient relaxes to zero (provided that the trailing edge of the biomolecule has not reached the entrance), resulting in a sudden but not large change of electrophoretic velocity. During the time when the biomolecule fully occupies the channel, the flow is fully developed and steady. As the trailing edge of the biomolecule approaches and enters the tube, another transient period results in the development of a transient-induced pressure gradient, and the electrophoretic velocity suddenly changes again. The current density changes when the DNA occupies the pore, and thus transient current density changes can be expected. These transient periods – their duration, in particular – can be used to determine the length of the DNA strand of interest.

[1] The term *particle* will be used extensively to denote the DNA strand. Indeed, the model described here is generic and will apply to any long-chain charged polymer, natural or synthetic.

[2] The parameter ranges described here have been deduced based on unpublished work by the author and former graduate student Prashanth Ramesh.

The direction of motion of the DNA is determined by the relative magnitude of the surface charge densities of the particle and the tube. The direction of the EOF is always from left to right for a negatively charged tube. For the tube surface charge density $\sigma^1 > \sigma^0$, the particle will move to the right and in the same direction as the EOF. Conversely, for $\sigma^1 < \sigma^0$, the particle will move to the left and against the EOF. This result will be demonstrated later and is an important consideration in designing rapid molecular analysis tool kits.

Though the transient periods and their length are important in DNA analysis systems, these transient periods are short in duration, and the predominant translocation velocity may be determined using a steady flow analysis. This is discussed in the next section.

12.2.2 Mathematical model

The governing equations for the motion of DNA are the same as in the case of electro-osmotic flow in an annulus, which is considered in Section 9.2.8, except for the partial entry and partial exit cases and the fact that the particle (i.e., DNA) velocity is unknown. Assuming that the transient periods are very short on the time scale of the electro-osmotic flow and that the flow is fully developed and steady, the governing equation for the axial electro-osmotic velocity is given in dimensionless form by

$$\epsilon^2\left(\frac{\partial^2 w}{\partial r^2}+\frac{1}{r}\frac{\partial w}{\partial r}\right)=-\epsilon^2\frac{\partial p}{\partial z}-\beta\sum_i z_i X_i \tag{12.1}$$

where the pressure is scaled on the viscosity. Similarly, the equations for the mole fractions and the potential are

$$\frac{\partial^2 X_i}{\partial r^2}+\frac{1}{r}\frac{\partial X_i}{\partial r}+z_i X_i\left(\frac{\partial^2 \phi}{\partial r^2}+\frac{1}{r}\frac{\partial \phi}{\partial r}\right)+z_i\frac{\partial X_i}{\partial r}\frac{\partial \phi}{\partial r}=0 \tag{12.2}$$

$$\epsilon^2\left(\frac{\partial^2 \phi}{\partial r^2}+\frac{1}{r}\frac{\partial \phi}{\partial r}\right)=-\beta\sum_i z_i X_i \tag{12.3}$$

Of course, as mentioned earlier, the concentrations can also be scaled on the ionic strength, for which $\beta = 1$.

It is the boundary conditions that differ from those considered in Section 9.2.8. The speed of the particle must be determined by a force balance. The boundary conditions on the walls are obtained by assuming the no-slip condition at the two walls and using the force balance on the particle surface at $r = \alpha$, with $\alpha = a/b$ to determine the particle velocity. The dimensional force balance at the particle surface is given by

$$\lambda_e E_x + 2\pi\alpha b\mu\frac{\partial w^*}{\partial r^*}=0 \tag{12.4}$$

where λ_e is the line charge density on the particle and μ is the dynamic viscosity; E_x is the constant externally applied electric field. Rearranging and nondimensionalizing this expression using the length scale b and the standard electro-osmotic velocity scale (see Chapter 3), the dimensionless boundary condition for velocity at the inner wall is given by

$$\frac{\partial w}{\partial r} = \frac{-\lambda_e}{2\pi\alpha\epsilon_e\phi_0} \tag{12.5}$$

The boundary conditions for the potential are obtained using the surface charge density boundary condition at the particle surface:

$$\frac{\partial \phi^*}{\partial r^*} = \frac{-\sigma^0}{\epsilon_e} \tag{12.6}$$

where σ^0, the surface charge density of the particle, is approximated as a uniformly distributed charge corresponding to the charge per unit length, λ_e. Thus

$$\sigma^0 = \frac{\lambda_e}{2\pi\alpha b} \tag{12.7}$$

In dimensionless form, equation (12.6) becomes

$$\frac{\partial \phi}{\partial r} = \frac{-\lambda_e}{2\pi\alpha\epsilon_e\phi_0} \tag{12.8}$$

$$= \frac{-\sigma^0 b}{\epsilon_e\phi_0} \tag{12.9}$$

The no-flux boundary condition is used as the boundary condition for mole fractions. Equations (12.2), (12.3), and (12.1) are thus subject to the boundary conditions

$$\frac{\partial w}{\partial r} = \frac{-\sigma^0 b}{\epsilon_e\phi_0}, \quad \frac{\partial \phi}{\partial r} = \frac{-\sigma^0 b}{\epsilon_e\phi_0}, \quad \frac{\partial X_i}{\partial r} + z_i X_i \frac{\partial \phi}{\partial r} = 0 \quad \text{at } r = \alpha \tag{12.10}$$

$$w = 0, \quad \frac{\partial \phi}{\partial r} = \frac{\sigma^1 b}{\epsilon_e\phi_0}, \quad \frac{\partial X_i}{\partial r} + z_i X_i \frac{\partial \phi}{\partial r} = 0 \quad \text{at } r = 1 \tag{12.11}$$

As long as the particle fills the tube, there is no pressure gradient, and in this case, the governing equations for the velocity and potential are seen to be identical, and the boundary conditions differ only at the particle surface. Thus the velocity and the electrical potential differ by a constant

$$w = \phi + C \tag{12.12}$$

where C is a constant. The governing equations can be solved numerically using either a direct tridiagonal solver or an iterative technique, as discussed in Chapter 10. In the next section, we discuss the results of the numerical simulation.

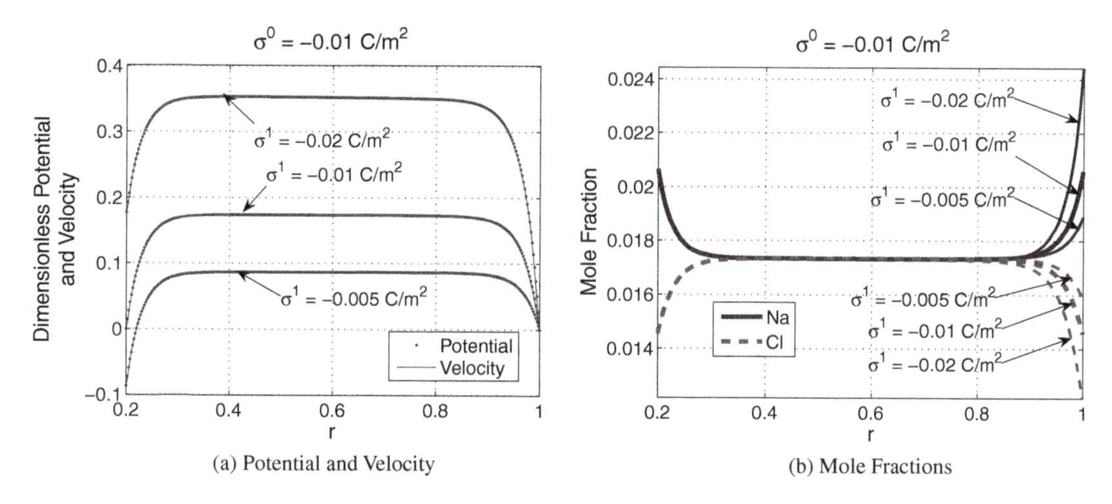

Figure 12.3 The numerical results for (a) potential difference, velocity and (b) mole fractions for the case of a two-component aqueous electrolyte mixture. The surface charge density of the inner cylinder is -0.01 C/m^2, while that of the outer cylinder varies from -0.005 C/m^2 to -0.01 C/m^2 up to -0.02 C/m^2. The electrical current is calculated to be 450.1 pA. The dimension of the annulus is $b = 10$ nm, $\alpha = 0.2$.

12.2.3 Results

Results in this section are for dsDNA, for which the surface charge density is a matter of some debate and quoted values range from -0.01 C/m^2 to -0.8 C/m^2. To illustrate the behavior, computations were carried out for three cases, with the surface charge density on the inner cylinder, the DNA, held constant at -0.01 C/m^2 and the outer wall surface charge density varying from -0.005 C/m^2 up to -0.02 C/m^2, with an applied electric field corresponding to 1.0 V over a channel of length 10 μm and with a fully dissociated electrolyte concentration of 1 M in the upstream reservoir. The outer radius is 10 nm and the inner cylinder radius is 2 nm in the calculations, which is about the radius of dsDNA. Figure 12.3 shows the dimensionless velocity and potential and notes that the double layers are thin.

The variation of the DNA translocation velocity with electrolyte concentration is shown in Figure 12.4. It is evident from this plot that the concentration of the electrolyte taken to be its value in the upstream reservoir plays a strong role in the translocation velocity of the DNA. Although the dsDNA has zero velocity when both surfaces have equal surface charge density (irrespective of the electrolyte concentration), the velocity varies nonlinearly with concentration when the difference of surface charge densities is nonzero. Figures 12.4a and 12.4b show the variation of inner cylinder velocity as a function of the concentration for two representative cases, one with $\sigma^1 = -0.005$ C/m^2 and the other with $\sigma^1 = -0.02$ C/m^2, with the inner cylinder surface charge density held at $\sigma^0 = -0.01$ C/m^2 for both cases. Note that the velocity shown in Figure 12.4a is in the direction opposite to the EOF, whereas in Figure 12.4b, the DNA velocity is in the same direction as the EOF. We are now in a position to calculate

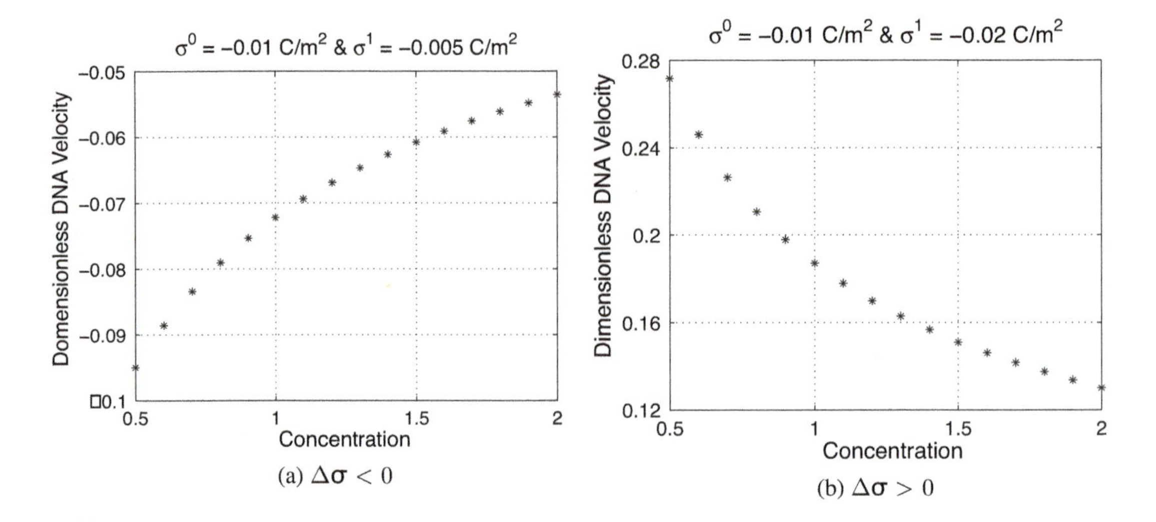

Figure 12.4 Nondimensional velocity as a function of reservoir electrolyte concentration (in M) for two different surface charge densites on the outer wall. Here $\Delta\sigma = \sigma^0 - \sigma^1$.

the total current and its variation with time – the main reason to perform these calculations.

12.2.4 DNA current

The current density and total current in electro-osmotic flow are discussed in Section 9.2.7. In this section, we apply those concepts to the annular model of DNA translocation. The dimensionless current density in the streamwise z direction is given by

$$J_z = \sum_i z_i \left(\frac{\partial X_i}{\partial z} + z_i \Lambda X_i - z_i \frac{\partial \phi}{\partial z} + PeX_i w \right) \tag{12.13}$$

where $\Lambda = E_0 L/\phi_0$, $Pe = ReSc$ is the Peclet number, and the dimensionless current is defined by $J_z = J_z^*/cU_0$. The total current in the channel is seen to comprise four parts: the streamwise concentration gradient, molar flux due to the migration of ions under the applied electric field, the flux due to the perturbation potential gradient, and the flux due to the bulk velocity of the electrolyte. The current through the channel is the integral of the current density across the entire cross section of the channel:

$$I = 2\pi \int_\alpha^1 J_z r dr \tag{12.14}$$

We can now summarize the theoretical picture of the change in current due to the transport of a biomolecule through the tube. Note that the current density contains two terms with axial gradients and two terms that are proportional to the mole fraction of species i. As the biomolecule enters the channel, the current density should be dominated by the two gradient terms. The streamwise current density will jump from its value when the pore is empty to its value when the pore is full. This time period is so short, however, that it will appear as an instantaneous rise or drop. During the time period when the biomolecule is within the pore, the current density will be independent of time and dominated by the terms proportional to the mole fraction, or

$$I = 2\pi \sum_i z_i \int_\alpha^1 (\Lambda z_i X_i + PeX_i w) r dr \tag{12.15}$$

and it is seen that it is crucially dependent on the two dimensionless parameters Λ and Pe.

For thin electric double layers in which the core is electrically neutral and the mole fractions are constant, the streamwise current density reduces to

$$I = \pi \Lambda (1 - \alpha^2) X_{io} \sum_i z_i^2 \tag{12.16}$$

where X_{io} is the (constant) mole fraction of the electrolyte in the core region and also in the upstream reservoir. Thus the current density is solely dependent on the concentration of the electrolytes in the core of the channel. Of course, in the overlapped double-layer case, the integral must be computed numerically, and the bulk flow term will play, perhaps, a significant role.

So how do the values of the particle velocity and the current drops compare with experiment? This is discussed next.

12.2.5 Comparison with experiment

Chen and Conlisk (2010) have incorporated the effect of upstream and downstream reservoirs on the DNA translocation velocity and have compared the computational with experimental data in Table 12.1. The detailed experimental parameters are also listed in the table. The geometry of the nanopore used in the Storm *et al.* (2005, 2003) experiments is similar to the pore shown in Figure 12.5. In this case, the radius of the nanopore at the inlet region is very large compared to the Debye length, and an asymptotic solution is used to calculate the electroosmotic flow velocity (Chen & Conlisk, 2010). In Li *et al.*'s (2003) work, the inlet radius is 50 nm, and the total length is 500 nm; the radius of the cylindrical pore is 5 nm, and the length of the cylindrical pore is 10 nm. The results for the DNA velocity calculated numerically compare well with the experimental data in both cases.

Table 12.1. Comparison on DNA velocity between the numerical results and the experimental data

Source	σ^1 (C/m^2)	L_{DNA} (μm)	V_{DNA} (m/s) Exp	V_{DNA} (m/s) Num
Storm *et al.* (2005)	−0.20	3.91	0.013	0.015
Li *et al.* (2003)	−0.14	0.40	0.010	0.012

Here σ_w is the surface charge density of the nanopore; L_{DNA} is the length of the DNA used in the experiments; $C_{KCl} = 1$ M is the concentration of the KCl solution; V_{DNA} Exp is the experimental data, and V_{DNA} Num is the numerical result.

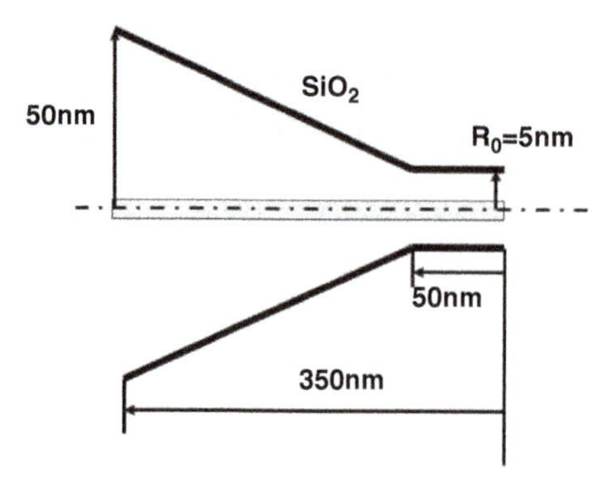

Figure 12.5 The geometry of the synthetic nanopore used for calculation. A 300 nm long conical silica nanopore is connected to a 50 nm long cylindrical pore. The radii of the large and small ends of the conical pore are 50 nm and 5 nm, respectively (Storm *et al.*, 2003).

As shown in Figure 12.6, the current results compare very well with the experimental data from Storm *et al.* (2005) for 11.5 kbps DNA. The baseline current defined as the current through the nanopore without DNA in it is $I_{\text{baseline}} = 7085$ pA, and the amplitude is $\Delta I = 160$ pA. The corresponding experimental data is $I_{\text{baseline}} = 7{,}100$ pA and $\Delta I = 140$ pA. The relative current change $(\Delta I/I)$ is 0.020 for experimental data and 0.022 from the calculation.

12.3 Development of an artificial kidney

12.3.1 Background

End-stage renal disease (ESRD), characterized by a loss of the essential protein *albumin*, among other factors, affects over 375,000 Americans and is increasing

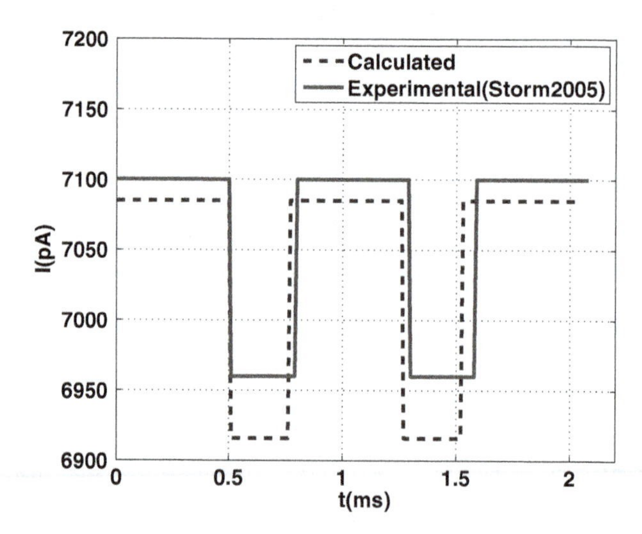

Figure 12.6 The results for current through the nanopore as a function of time (Chen, 2010). For the numerical results, the baseline current is $I_{\text{baseline}} = 7085$ pA, and the amplitude is $\Delta I = 160$ pA. The corresponding experimental data on $I = 7100$ pA and $\Delta I = 140$ pA (Storm *et al.*, 2005).

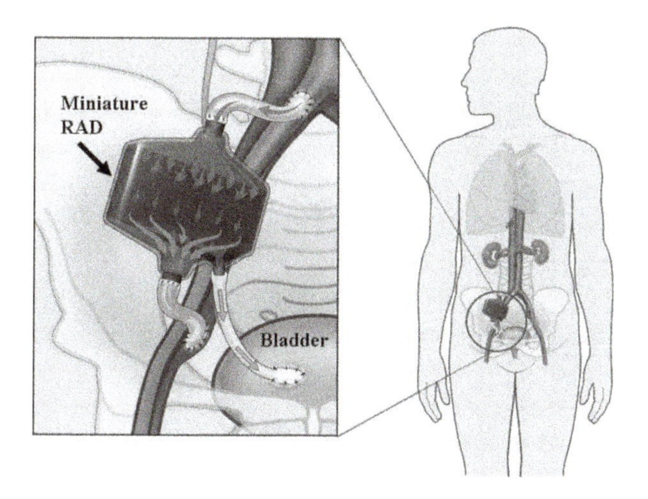

Figure 12.7 Sketch of an implantable renal assist device (RAD). From Conlisk *et al.* (2009).

in prevalence at an annual rate of 8 percent (Fissell & Humes, 2006; Humes *et al.*, 2006).[3] Treatment of ESRD patients by renal transplant is hindered by shortage of donor organs, and dialysis is expensive, inconvenient, and often unsuccessful. Tissue engineering of an artificial kidney is an alternative strategy to dialysis in the treatment of ESRD. The extracorporeal RAD is a bioartificial kidney that combines hemofiltration, the filtration of small ions, and other small molecules from the blood with cell therapy to provide the metabolic, endocrine, and immunological functions of a healthy kidney. A sketch of such a device appears as Figure 12.7.

Dialysis is the primary renal replacement therapy for most ESRD patients. The most common implementation of dialysis is hemodialysis, whereby toxins in a patient's blood are filtered through a dialyzer. The key component of the dialyzer is a semipermeable membrane. Blood is pumped over one side of the membrane at a rate of 200–500 mL/min, while a clean electrolyte solution (dialysate) is pumped at a similar or higher flow rate on the other side. The membrane is engineered to allow for diffusive transport of small molecules, including ions such as sodium and chloride, from the blood, while retaining higher-molecular-weight compounds such as albumin, coagulation proteins, and immunoglobulins in the bloodstream. This configuration results in diffusion of urea, creatinine, and other toxins from blood into the dialysate.

The nanopore membranes described previously in this text can also be used to filter the small ions, such as sodium and chloride, through the membrane, while retaining the (relatively) large proteins such as albumin. The pores in the membrane are on the order of <10 nm in the smallest dimension and are very wide, $W \sim 40$–100 μm of rectangular cross section, and about 4 μm long. These dimensions are mentioned elsewhere in this book. These filtration methods

[3] Shuvo Roy and William Fissell wrote significant portions of this section, and the author is grateful for their permission to use this material.

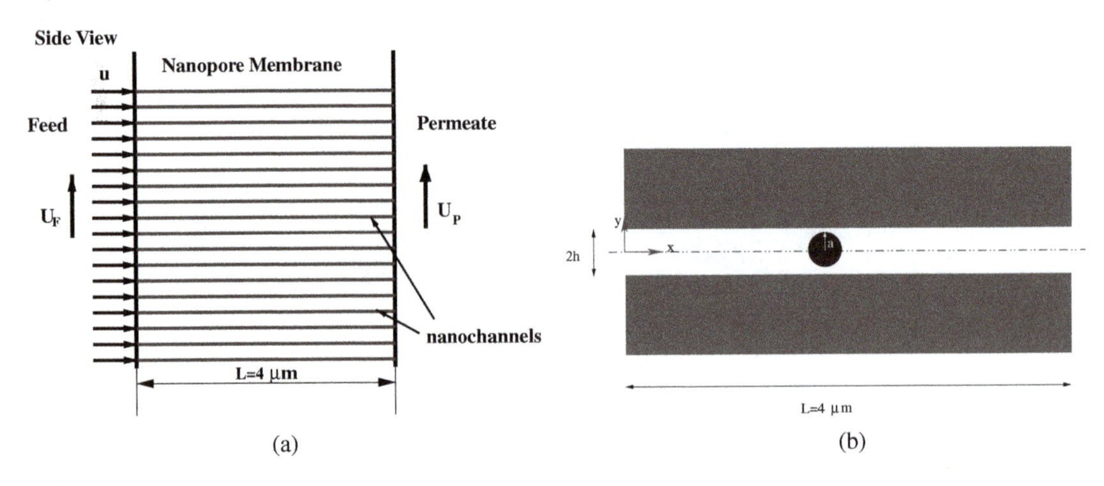

Figure 12.8 (a) A synthetic nanopore membrane for the proposed RAD (Fissell *et al.*, 2007); a side view of a nanopore membrane. (b) Sketch of a spherical particle in a rectangular channel; side view. The dimension into the plane of the paper is about $W = 10$ μm. A scanning electron microscope image of the top view of the membrane appears in Figure 4.1.

mimic filtration mechanisms in nature such as the cell membrane, the respiratory membrane, the vascular endothelium, and the glomerular basement membrane, a key structure in the kidney (Scherrer & Gerhardt, 1971; Deen *et al.*, 2001; Martini, 2001), as well as technology for water treatment, desalination, pharmaceuticals, and food and beverage processing (Zeman & Zydney, 1996). The mathematical modeling of such a membrane is described next.

12.3.2 The nanopore membrane for filtration

A *semipermeable or permselective membrane* allows the passage of certain molecules while restricting the passage of others and is an important component of filtration mechanisms in nature; the cell membrane, respiratory membrane, vascular endothelium, and glomerular basement membrane in the kidney are all semipermeable membranes.

Figure 12.8a is a sketch of a typical membrane used for hemofiltration in the RAD (Fissell *et al.*, 2007). Figure 12.8b shows an individual nanopore of width $2h$ confining a molecule of radius a. The membrane consists of a large number of nanopores ($N \sim 10^4$) and connects an upstream feed microchannel where the average flow speed is U_F to a downstream permeate microchannel where the average flow speed is U_P (Figure 12.8a).

The characteristic dimensions of the feed and permeate channels that serve as the donor and receiver reservoirs in the other applications discussed here are on the scale of millimeters, and the flow rate in the feed channel is of the order of milliliters per second; in typical experiments discussed here, an external pump is used to drive this flow past the membrane (Fissell *et al.*, 2007). The flow speed U_P in the permeate channel is solely due to permeation of the feed solution through the membrane pores and is not imposed externally, as in dialysis or hemodiafiltration. The flow across the nanopore membrane is driven by a transmembrane pressure drop of $\Delta p \sim 1 - 2$ psi, established between the feed and permeate solution (Figure 12.8). This is about the same pressure drop that occurs across the native kidney. Figure 12.8b shows the geometry of an individual nanopore and the coordinate system chosen for the theoretical model.

To simplify the theoretical problem, the pores are assumed to be straight, nonintersecting, and of uniform width. The width $2h$ of each nanopore is $\sim$10 nm; for example, $2h = 8$ nm for the membranes described in Fissell *et al.* (2007). As is evident from Figures 12.8a and 12.8b, the other dimensions of the nanopore are large enough for the pore to be treated as a slit. A hemofiltration chip in the first generation of the RAD consists of an array of several (five to nine) such membranes. The chip is fabricated using a state-of-the-art nanofabrication technology that involves deposition of a sacrificial layer of SiO_2 for the definition of nanopore size and can be controlled precisely enough to result in a nearly monodisperse pore size distribution ($<1\%$ variation across the chip) for 5 nm pores.

The *sieving coefficient* is a measure of the ability of a membrane to retain solutes, in this case, primarily albumin. The sieving coefficient of a solute with concentration c_P in the permeate channel and c_F in the feed channel is defined by $S = c_P/c_F$. A small sieving coefficient $S \sim 10^{-4}$ is desirable, and the inverse of the sieving coefficient $1/S$ can be considered a quantitative measure of its selectivity (Zeman & Zydney, 1996).

12.3.3 Hindered transport

A sketch of the geometry of a nanopore membrane is depicted in Figure 12.8. In this case, the channels are very wide, $W \sim 100$ μm, and so the flow may be assumed to be one-dimensional. Because the pores of the membrane are so small ($h < 10$ nm), the diameter of an albumin molecule ($d = 7$ nm) is comparable to the size of the pore, and transport within the pore will be hindered by the close proximity of the walls. The flux of a biomolecule of valence z in the streamwise direction in the nanopores of interest is given by

$$N_A = -K_d D_A \frac{\partial c_A}{\partial x^*} + z K_d m_A E_x^* c_A + K_c u^* c_A \tag{12.17}$$

where the quantities in this equation are all averaged over the cross section of the pore. For example, u^* is the average velocity, and c_A is the average concentration in the pore. If the ratio of the size of the particle to the pore height is large enough, the particle or biomolecule will tend to stay in the center of the channel. The *hindrance factors* K_c and K_d can be calculated based on this so-called *centerline approximation*. Conversely, Dechadilok and Deen (2006) have shown that this approximation is not necessary, and they obtain

$$K_c = \frac{1-3.02\chi^2+5.776\chi^3-12.3675\chi^4+18.9775\chi^5-15.2185\chi^6+4.8525\chi^7}{1-\chi} \tag{12.18}$$

$$K_d = \frac{1+\frac{9}{16}\chi log(\chi) - 1.19358\chi + 0.4285\chi^3 - 0.3192\chi^4 + 0.08428\chi^5}{1-\chi} \tag{12.19}$$

where $\chi = a/h$ is the ratio of the solute radius (i.e., albumin) to the channel half-height.

Equation (12.17) can be written in dimensionless form as in Conlisk *et al.* (2009), and the flux N_A takes the form

$$N_A = K_c \left[-\frac{1}{Pe_H} \left(\frac{dX}{dx} - zE_x X \right) + uX \right] \tag{12.20}$$

assuming only streamwise mass transport. Here the dimensionless quantities are defined by $x = x^*/L, u = u^*/U_0$, where U_0 is the centerline velocity in Poiseuille flow and $E_x = E_x^* L/\phi_0$ and $N = N^*/(U_0 c X_0)$, where X_0 is the mole fraction of species A, albumin at the entrance to the pore, and c is the total concentration of the mixture. Here, as usual, $\phi_0 = RT/F = 26$ mV. The quantity $Pe_H = K_c U_0 L / K_d D_A$ is the hindered transport Peclet number and can be interpreted as the characteristic ratio of axial convection speed and axial diffusion speed of a solute constrained by the pore.

The condition of a constant flux of the species requires $dN_A/dx = 0$, leading to the governing equation for the species mole fraction X as

$$\frac{d^2X}{dx^2} = Pe_H u(1+s)\frac{dX}{dx} \tag{12.21}$$

This equation is subject to boundary conditions at the entrance and exit of the pore. These boundary conditions require some thought.

As a solute approaches the entrance of a single pore in a nanopore membrane, the concentration or mole fraction of the solute at the pore entrance, say, $x = 0$, changes sharply from its value in the bulk feed solution. This effect can be quantified using a *partition coefficient* $\mathcal{F}$, defined by

$$X(0) = X_F \mathcal{F} \tag{12.22}$$

where $X(0)$ is the solute mole fraction at the slit pore entrance and X_F is the the same quantity in the bulk feed stream. Similarly, near the exit of the pore, we define $X(1) = X_P \mathcal{P}$, where the subscript P denotes *permeate*. The value of the mole fraction X_P depends on the permeate flux and so is unknown.

What about the boundary condition at $x = 1$? The total flux of the solute at the exit of the slit pore should balance the net flux in the permeate stream. The net flow efflux through this region must equal the permeate flux,

$$N_A = uX_P = \frac{X(1)}{\mathcal{P}} \tag{12.23}$$

Inserting N_A from equation (12.20) (evaluated at $x = 1$) into equation (12.23) at

$$\frac{dX}{dx}|_{x=1} = Pe_H u \left(1 + s - \frac{1}{\mathcal{P}K_c}\right) X(1) \tag{12.24}$$

where the dimensionless parameter

$$s = \frac{zDE_x^* FK_d}{RTu^* K_c} = \frac{zE_x^*}{Pe_H u} \tag{12.25}$$

incorporates the effect of any applied or induced electric field (E_x) and $-1 < s < \infty$.

Equation (12.21) can be solved with the boundary conditions (12.22) and (12.24) to give (Conlisk *et al.*, 2009)

$$X(x) = X_F \mathcal{F} \frac{1 + [(1+s)\mathcal{P}K_c - 1]\exp[-Pe_H u(1+s)(1-x)]}{1 + [(1+s)\mathcal{P}K_c - 1]\exp[-Pe_H u(1+s)]} \tag{12.26}$$

The sieving coefficient S defined as $X_P/X_F = X(1)/X_F\mathcal{P}$ calculated from this distribution is

$$S = \frac{(1+s)\mathcal{F}K_c}{1 + [(1+s)\mathcal{P}K_c - 1]\exp[-Pe_H u(1+s)]} \tag{12.27}$$

The asymptotic forms of equation (12.27) for large and small Peclet numbers are

$$S_\infty = \mathcal{F}K_c(1+s) \quad \text{for } Pe_H \gg 1 \tag{12.28}$$

$$S_0 = \frac{\mathcal{F}}{\mathcal{P}} \quad \text{for } Pe_H \ll 1 \tag{12.29}$$

Note that for large Peclet number, $S_\infty \to 0$ as $s \to -1$. Recall that it is desired that the sieving coefficient for albumin be small, specifically, $S \sim 10^{-4}$. This can be achieved by adjusting the electric field E_x.

In the case of a negatively charged species like albumin, a negative value of s implies an applied electric field directed from feed to the permeate side. The case $s = -1$ means that convection effects are balanced by electromigration. In this case, from equation (12.21), the mole fraction is linear, and

$$X = X_0 \mathcal{F}\left(1 - \frac{Pe_H u}{\mathcal{P}K_c + Pe_H u} x\right) \tag{12.30}$$

for which the sieving coefficient is

$$S = \frac{\mathcal{F}\mathcal{P}K_c}{\mathcal{P}K_c + Pe_H u} \tag{12.31}$$

and is independent of K_c since $Pe_H = K_c uL/K_d D$, but the sieving coefficient is dependent on K_d.

Conlisk *et al.* (2009) discuss explicitly the determination of the feed and permeate partition coefficients $\mathcal{F}$ and $\mathcal{P}$ as well as presenting results for the

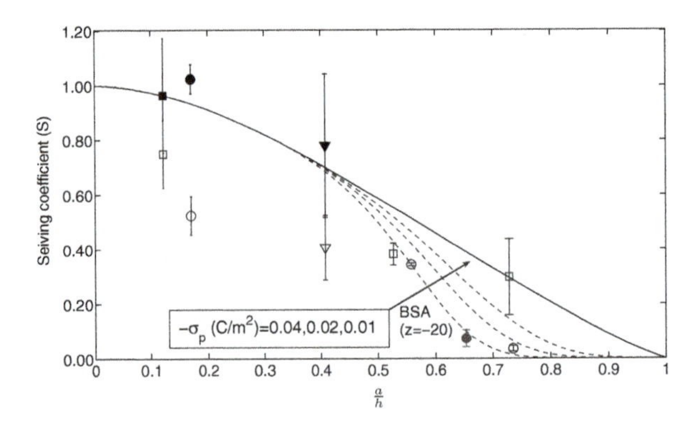

Figure 12.9 Experimental sieving coefficients of bovine serum albumin (BSA), thyroglobulin, and carbonic anhydrase (CA) in $1 \times$ PBS, $10 \times$ PBS (circles colored black), and bovine blood with clotting factors removed as a function of the ratio a/h of protein hydrodynamic radius (a) and pore half-width (h) and comparison with theoretical curves obtained under the assumption of steric partitioning (solid curve) and electrostatic partitioning as applicable to bovine serum albumin (dashed curves). The shapes of the symbols indicate the type of protein: circles, triangles, and squares stand for BSA, thyroglobulin, and CA, respectively. The circle colored black near $a/h = 0.65$ is the only data point in blood. The experiments were conducted in membranes of pore widths $2h = 7, 9.69, 10.9, 12.78, 42$ nm.

mole fraction and sieving coefficient. For example, under the assumption that the solution is dilute and there are no significant electrostatic or other long-range interactions between the solute molecules and the pore wall,

$$\mathcal{F} = \mathcal{P} = 1 - \chi \tag{12.32}$$

where $\chi = a/h$ is the ratio of the solute radius (a) to the pore half-width (h) (Deen, 1987). The expression describes what is termed *steric partitioning*, the partitioning due to the finite size of the particle or biomolecule. Thus the boundary condition at the entrance to the pore is $X(0) = X_F(1 - \chi)$.

We have just seen that an imposed electric field can significantly reduce the sieving coefficient. This is also true if the semipermeable membrane itself is charged. Conlisk *et al.* (2009) compare the results of their analysis for charged pores with experiments conducted at the Cleveland Clinic. The theory for a charged solute in a charged pore is compared with these experiments in Figure 12.9. All data in $1 \times$ PBS feed solution are indicated with hollow symbols, the data in $10 \times$ PBS are indicated with solid symbols colored black, and the only datum in blood (also identifiable uniquely by a horizontal coordinate value of $a/h = 0.65$) is indicated with a solid symbol colored red. The dashed curves are obtained using a charge number of -20 for bovine serum albumin (BSA) and for three different values of surface charge ($\sigma = -0.01, 0.02, 0.04$ C/m^2) density on the pore walls. The experiments were conducted in membranes with the pore widths, $2h = 7, 9.69, 10.9, 12.78$, and 42 nm, and so different experiments could be performed at the same value of χ. The dashed curves indicating electrostatic partitioning give significantly lower sieving coefficients, beginning at $\chi \sim 0.7$, than for steric (size exclusion) effects alone. This value of χ indicates that the nanopore membrane should have $2h \sim 10$ nm or smaller. In addition, it would seem that the use of a charged membrane can reduce the sieving coefficient substantially.

12.4 Biochemical sensing

12.4.1 Introduction

In Section 12.2, we spoke of the fact that the DNA translocation through a nanopore can significantly change the current through the nanopore. How that current change is determined is part of the function of biochemical sensing. In this section, we qualitatively describe the processes involved in biochemical sensing. In this section, we focus primarily on *electrochemical* sensors and *optical* sensors. The Sandia device depicted in Figure 1.2a is an example of an electrochemical sensor.

Chemical and biochemical sensors are used to detect chemical and biochemical species of interest: the analyte. The essential parts of a biochemical sensor are the recognition element or a receptor and a transducer that transforms the event into a signal that represents the effect of the event. Chemical or biochemical sensors that use a biological recognition element, such as antibodies, enzymes, nucleic acids, cells, or a synthetic receptor mimicking that found in nature, are called *biosensors* (McNaught & Wilkinson, 1997; Spichiger-Keller, 1998).

There are many areas of application, including, for example, implantable nanopore arrays for glucose monitoring; because of the size of the market, glucose monitoring has been one of the major applications in this area (Ravindra *et al.*, 2007). The ultimate goal is the continuous monitoring of glucose levels in the blood or urine using an implantable device that will automatically adjust glucose concentration to the proper level. Such continuous monitoring eliminates the need for unpleaseant and painful injections.

Other applications include detection of airborne pathogens, such as nerve and mustard gas, and explosives such as TNT and biological warfare agents such as anthrax and ricin; detection and removal of environmental pollutants (e.g., electrokinetic remediation); and clinical applications such as treatment of cancer, measurement of key biological entities sustaining human health such as folic acid, vitamin B_{12}, biotin, and so on, determining pH level, and detecting pollutants such as botulinum and anthrax in water.

An objective that is now possible in biochemical sensing is the ability to detect and analyze single molecules. Indeed, biochemical sensing and, in particular, single molecule analysis permeates the four-volume work of Ferrari (2006); each of the four volumes contains at least several papers whose basic themes are biochemical sensing, and volume IV is titled *Biomolecular Sensing, Processing and Analysis*.

There are a number of reviews on the subject, including Vo-Dinh (2006), Spichiger-Keller (1998), Rosi and Mirkin (2005), Jianrong *et al.* (2004), D'Orazio (2003); and Scheller *et al.* (2001). The books on biosensors by Eggins (1996) and

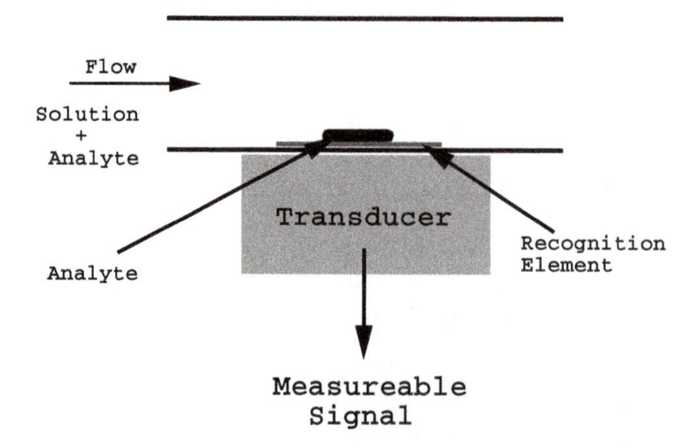

Figure 12.10 Sketch of a biosensor showing the recognition element and the transducer.

Kress-Rogers (1997) are a good source for background on biosensors in general. A partial list of companies in the biosensing business is given by Ravindra *et al.* (2007).

A sensor utilizes a specific means of detecting a given analyte such as the action of an antibody for a specific analyte; this biochemical reaction will then produce a signal (the transducer) such as a color change or a change in electrical potential. Nature has created a particularly large variety of membrane sensors, such as ion channels, that can discriminate between a large number of stimuli (Braha *et al.*, 2000b; Bayley and Cremer, 2001).

The two most important properties of a sensor are its sensitivity and its selectivity. Sensitivity of a sensor is related to its limit of detection, that is, the smallest concentration of an analyte in a sample that will yield a measureable output signal based on a given input signal. If the sensitivity of a sensor is low, false negatives will occur. Selectivity refers to the ability to detect a given target. If a sensor is not selective enough, false positives will occur. A generic sketch of a biosensor is depicted in Figure 12.10.

An extensive discussion of all of the possible applications and the types of sensors available is clearly beyond the scope of this section. The purpose here is to introduce the reader to the components of a sensor, specifically, a biosensor, and how micro- and nanofluidics can be used in their operation. The small size of these systems means that in a biomedical application, the device can be brought to the patient; such devices are called *point-of-care* (POC) systems.

12.4.2 What is a biosensor?

The recommended definition of a biosensor (Thévenot *et al.*, 2001) from the International Union of Physical and Applied Chemistry (IUPAC) is as follows:

> A biosensor is a self-contained integrated device which is capable of providing specific quantitative or semi-quantitative analytical information using a biological

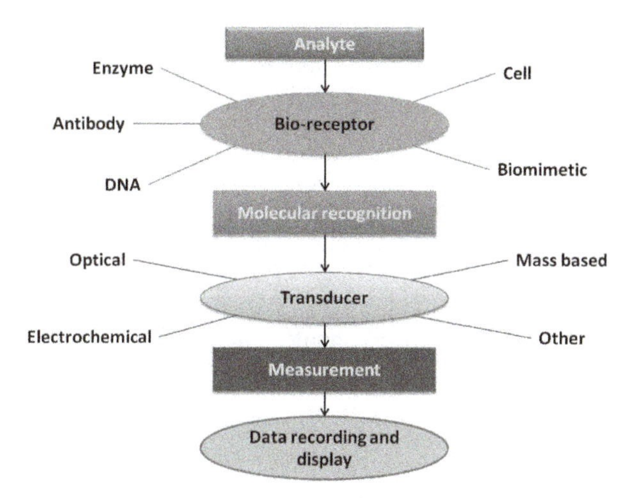

Figure 12.11 Conceptual diagram of the biosensing principle and schematic classification of biosensors according to receptors and transduction mechanism. (Reproduced from Vo-Dinh (2006)).

recognition element (biochemical receptor) which is in direct spatial contact with a transducer element. A biosensor should be clearly distinguished from a bioanalytical system which requires additional processing steps, such as reagent addition. Furthermore, a biosensor should be distinguished from a bioprobe which is either disposable after one measurement, i.e. single use, or unable to continuously monitor the analyte concentration.

Biosensors can be classified according to their receptors as well as their transducers (Vo-Dinh, 2006). Figure 12.11 adapted fromVo-Dinh (2006) provides a conceptual diagram of a biosensor and summarizes these two classification schemes. The first 10 pages of Vo-Dinh (2006) are an excellent overview of biosensors for nonspecialist engineers and scientists. When several biosensors are integrated on a microchip to perform detection of several molecules, the integrated system is called a *biochip* (Vo-Dinh, 2006).

12.4.3 Receptor-based classification of biosensors

The immune system of biological organisms produces antibodies that can target specific foreign proteins (antigens) introduced into the body fluids. This type of interaction is the basis for antibody biosensors that perform *immunoassays*.[4] One example of this antigen–antibody pairing is streptavidin, which is the antigen to the biotin antibody. Biotin is a B-complex vitamin that is soluble in water and promotes healthy cell growth; it is present in eggs and some vegetables. Streptavidin is a protein that is present in the bacterium streptomyces. Nucleic acid biosensors are typically based on *DNA hybridization*, or the study of the

[4] An immunoassay is a test that uses the binding of antibodies to antigens to identify and measure certain substances.

sequence of ATGC pairings. Enzymes can be used as catalysts in various processes, including the processes for detection of an enzyme-tagged molecule; Enzyme-Linked ImmunoSorbent Assay (ELISA) is a well-known immunoassay to detect the presence of antibodies or antigens in a given sample.

12.4.4 Transducer-based classification of biosensors

As shown on Figure 12.11 an alternative to a receptor-based classification system is a transducer-based classification scheme. The most commonly used detection methods are optical, electrochemical, and mass-based systems.

OPTICAL DETECTION METHODS

Optical methods typically use microscopes equipped with sources of radiation such as plane polarized light, laser radiation, X-rays, or UV rays that illuminate the zone to be interrogated in the system of interest. Other common optical system accessories are cameras to record images, prisms, optical fibers, wave guides, and the necessary electronics. Fluoroscopy uses fluorescent-tagged molecules for optical detection. Conventional fluoroscopy uses volumetric illumination. In *total internal reflection fluorescence* (TIRF) microscopy, selective illumination of a surface is possible using evanescent waves; this allows adsorption of analyte to a solid substrate to be detected (Axelrod *et al.*, 1984; Sadr *et al.*, 2004). The intensity of fluorescence (Staubli *et al.*, 2005) as well as a decrease in fluorescence intensity (quenching) may be measured (Humbert *et al.*, 2005).

Another precise optical technique is *surface plasmon resonance*, which uses surface illumination and measures electron charge oscillations instead of fluorescence (Homola *et al.*, 1999). Optical detection systems are very precise and are often capable of giving real-time information (Brecht & Gauglitz, 1997), although they are very expensive.

OTHER DETECTION METHODS

Colorimetric methods use changes in color of a species used as a tag (such as a 4-hydroxyazobenzene-2 carboxylic acid (HABA) tag with streptavidin) to detect a binding event (Green, 1970). In many cases, electrochemical signals, such as electrical conductivity (conductimetry), current (amperometry), or voltage (voltametry), are used for detection (D'Orazio, 2003; Scheller *et al.*, 2001). *Atomic force microscopy* offers a technique based on changes in the force field associated with a binding event (Allen *et al.*, 1996; Rosi & Mirkin, 2005). Some detection techniques depend on the measurement of the change in mass when the analyte attaches to a surface containing a receptor, for example, through measurement of acoustic vibration frequency (Thévenot *et al.*, 2001; Vo-Dinh, 2006). Novel nanofabricated sensors, such as nanowires, colloidal gold particles, and carbon nanotubes, are also being used for biosensing (Rosi & Mirkin, 2005).

12.4.5 Evaluation of biosensor performance

The key parameter affecting the performance of a biosensor is the minimum concentration of analyte the biosensor can detect; this can range from as low as attomolar (Kemery *et al.*, 1998; Gong *et al.*, 2008) to micromolar (Leca-Bouvier & Blum, 2005). The efficiency of a biosensor can be assessed either by determining the ratio of the number of moles of analyte captured on the wall in the solid phase (Figure 12.10) to the maximum capacity of the solid phase or by the ratio of the number of moles captured in the solid phase to the total number of moles supplied from the moving solution. This can be quantified using *binding curves* (a general term common in separations literature) or *dose response curves* (more common in the drug delivery literature) with solution phase analyte moles/concentration as the independent variable and the solid phase captured moles as the dependent variable.

Mixing a solution is often used to enhance binding of an analyte, either to a surface or to another molecule in the bulk. Mixing a sample with a solution can be used to attach the analyte (target) to a complementary tagging molecule. Surface delivery is often required to ensure that the target molecules interact with a detection surface and is often used in conjunction with mixing to enhance the concentration of the analyte at the detection site.

Mott *et al.* (2006), Golden *et al.* (2007), and Mott *et al.* (2009) have developed a toolbox for optimized mixing in a microfluidic channel using complex herringbone-like structures (i.e., grooves) to capture the analyte. The application is the detection of TNT in seawater; the Reynolds number is on the order of $\mathrm{Re} \leq 10$. Several groove configurations are depicted in Figure 12.12a. The channel has a width of $W = 1016\ \mu\mathrm{m}$ and a height $h = 325\ \mu\mathrm{m}$. The effect of the grooves is to steer fluid toward the sidewalls to the center of the channel and toward the floor and the ceiling – the interrogation zones. Using what they call *advection maps*, the authors were able to sample over 500,000 groove configurations for optimizing mixing.

In Figure 12.12b are results for grooves pointing both upstream and downstream. The value of N denotes the total number of grooves on the floor and the ceiling – the sample has passed. The value $N = 20$, for example, indicates that the sample has passed 10 grooves on the floor and 10 grooves on the ceiling. The solid symbols indicate that particles have not yet reached within $0.1h$ of the ceiling or the floor. The hollow cylinders indicate particles that have come within $0.1h$ of the ceiling or the floor. Note that by $N = 50$, almost all of the particles tracked have reached the target threshold. Mott *et al.* (2009) found that the results for the features pointing downstream increased the number of particles reaching the target threshold by almost 3 times compared to the baseline channel with no grooves.

Low Reynolds number vortices may also be generated as a result of an engineered change in surface charge density or ζ potential and used to improve surface delivery of an analyte; the situation is depicted in Figure 12.13. Anderson and Idol (1985) appear to have been the first to investigate this phenomenon (see also

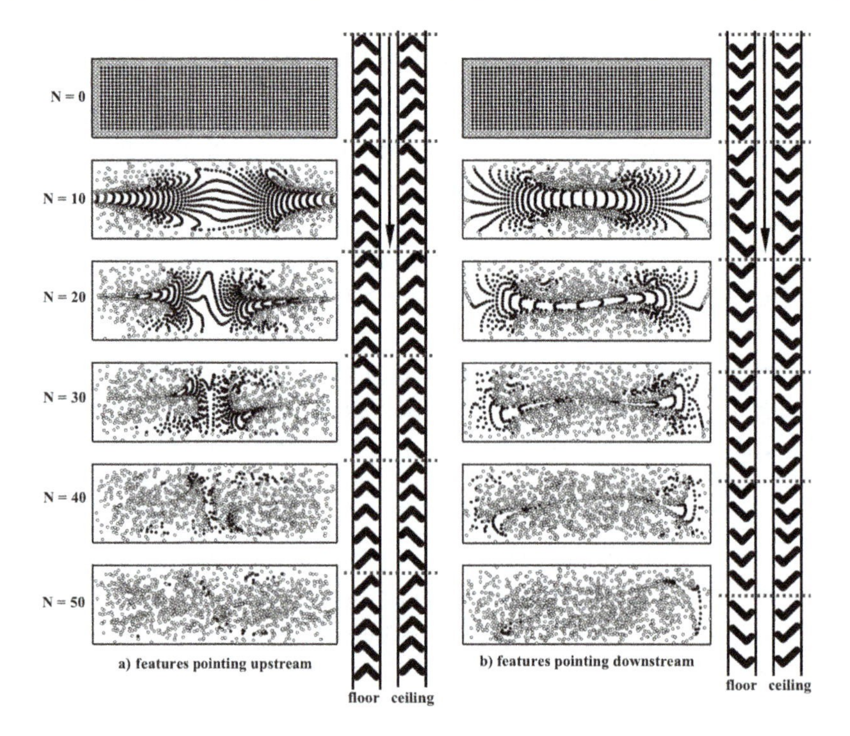

Figure 12.12 (a) Schematic of several features (grooves) placed on the ceiling and floor of a channel to enhance surface delivery. (b) Results for both upstream and downstream facing features; the lighter shade indicates that a sample has come within 10% of the channel height to the surface. Note that nearly all of the sample has reached this threshold. From Mott *et al.* (2009).

Erickson & Li, 2002) analytically, and Stroock *et al.* (2000) have observed this effect experimentally (see also Chen *et al.*, 1997, and the references therein). As has been seen in electro-osmotic flow, the direction of the velocity field depends on the sign of the wall surface charge. An abrupt change in the wall charge or potential of the same sign will cause a "vortex" to form. The results are slightly different from a modeling perspective for the case of a changing wall charge compared with a changing ζ potential, but both result in a vortex that can enhance surface delivery. It should be noted that because of the discontinuous nature of the potential and the finite Peclet number, the solution for the concentrations deviates from the Poisson–Boltzmann distribution. Note that in Figure 12.13, the vortex covers about 25 percent of the channel; it turns out that the size of the vortex decreases with decreasing Peclet number and that the measure of the deviation of the numerical solution from the Boltzmann distribution increases with increasing Peclet number.

12.4.6 Nanopores and nanopore membranes for biochemical sensing

While uncommon just a few years ago, many research groups around the world are exploring the use of nanopores and nanopore membranes for sensing many

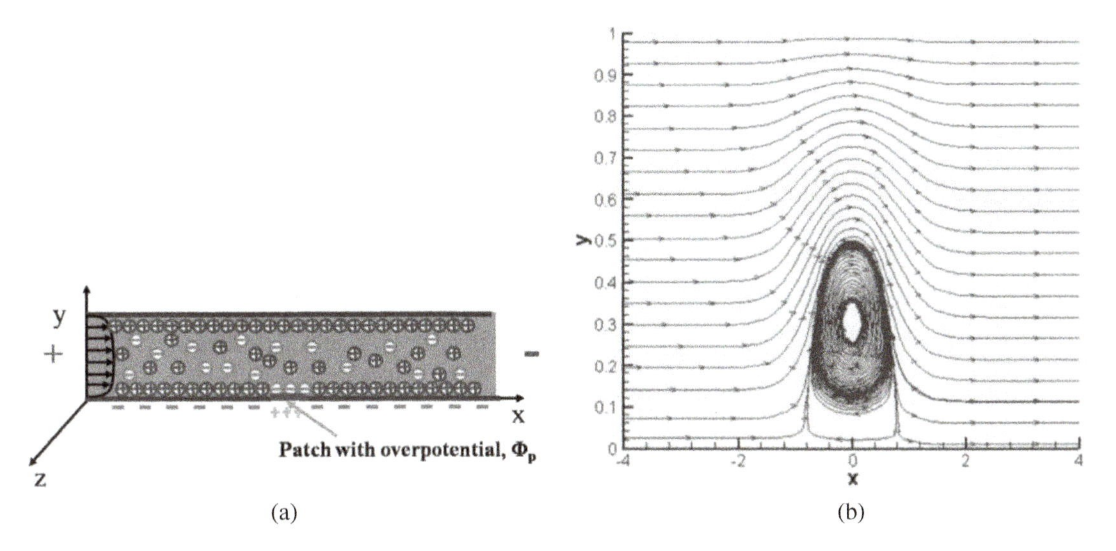

Figure 12.13 Schematic of a channel having a potential discontinuity.

different molecules. Much of the work is still in the research stage, and some of these concepts have been applied to defense-related problems.[5]

It is well known that miniaturization of a given device for detection and rapid molecular analysis has several advantages. First, because the device is small, the amount of analyte and sample required will be small. Also, it is clear that the smaller the flow path, the shorter will be the required analysis times.

Braha *et al.* (2000) showed experimentally that the occupation of a pore by ions could be detected by a modulation of the current passing through the pore. This has been demonstrated theoretically in the preceding analysis of the DNA transport problem. They used the natural α hemolysin (αHL) channel as the recognition element, a pore that is 10 nm long and consists of a water-soluble 293-amino acid polypeptide chain that naturally occurs in biological membranes.

As was mentioned before, Bayley and his co-workers (Bayley & Cremer, 2001) could identify various types of DNA down to single base-pair resolution as they pass through a α hemolysin nanopore. They also showed that electro-osmotic flow could significantly increase the effectiveness of the binding of a neutral molecule (in this case, $\beta-$cyclodextrin) to an αHL pore.

While early micro- and nanofluidic devices used some form of the silica family of surfaces for the channels, many researchers have used PDMS and PMMA for their ease of use and inexpensive fabrication procedures. Saleh and Sohn (2006) have demonstrated that a single nanopore can be used for a wide variety of applications, including immunoassays and DNA sizing. Using the fact that the ratio of the change in current to the total current is proportional to the ratio of the volume of a particle to the volume of the pore (as shown theoretically in Section 12.2.4),

[5] The Defense Advanced Research Projects Agency (DARPA) program Engineered Biomolecular Nanodevices and Systems (MOLDICE) was created to investigate the use of biological sensing elements to detect pathogens.

they were able to measure the radius of a DNA strand, about $a = 2$ nm, a value that was used in the section describing the transport of DNA through a tube.

12.5 Chapter summary

The purpose of this chapter is to introduce the reader to several application areas and illustrate the use of the modeling techniques described in the previous chapters. The main application areas discussed here are as follows:

- DNA transport and electroporation
- Transport processes in a RAD
- Biochemical sensing

These three areas were chosen because of the crucial role that micro- and nanofluidics plays in the operation of the devices in each area. These areas are certainly not all-inclusive of the myriad applications, and the reader is encouraged to consult the references listed in this chapter, especially the four-volume set edited by Ferrari (2006) and the MEMS handbook of Gad-el Hak (2001).

EXERCISES

The exercises here are meant to be open ended, and each exercise contains a writing assignment. Some of these exercises may be appropriately assigned after Chapters 9 and 7 are covered. The writing assignments can be of a length determined by the instructor, depending on the purpose of the exercise. These exercises are designed for the student to make extensive use of the Web.

12.1 Electro-osmotic flow can be used to remove metal contaminants from saturated clay and other soil materials. Using the governing equations for EOF in a porous media in the paper by Shapiro and Probstein (1993), and the parameters defined therein, determine the total amount of the contaminant that can be removed as a function of energy input. Write a paper describing the basic principles of an *electrokinetic remediation* (ER) process and your results. Define the efficiency of an electrokinetic remediation process.

12.2 Carbon easily binds to carbon in a unique way. Carbon nanotubes or buckytubes can be synthesized into a hollow cylinder made up of a single array of carbon atoms; these tubes are called single-wall carbon nanotubes (SWCNT). These tubes have a variety of applications. Write a paper describing the the applications of SWCNTs to biochemical sensing.

12.3 Carbon nanotubes are also used for cancer diagnostics. Write a paper discussing the specific way in which a carbon nanotube is used in cancer diagnostics.

12.4 Another application of nanopore membranes is the purification of water. Write a report on the basics of water purification using *reverse osmosis*, and identify the difference between reverse osmosis membranes and *nanofiltration membranes*. Also explain the use of nanoparticles, such as *carbon nanotubes*, for the purification of water.

12.5 The world depends on a steady supply of fresh water for a variety of activities. Unfortunately, only 1 percent of the total supply of water is fresh water. Nanopore membranes can also be used to desalinate water. Explain how such membranes can be used to desalinate water, and write a paper on your results. Using a given supply of salt water, determine how efficiently the membrane removes the salt from the water.

12.6 Write a report on the electrochemical aspects of the operation of a battery.

12.7 Repeat the previous exercise for a fuel cell.

12.8 Chromatography is the process of separating out chemical species in a porous medium by selective adsorption on a solid surface. Neglecting diffusion, using a simple rectangular control volume in a porous medium of void fraction ϵ, derive first order partial differential equations governing the concentration of the solute in the liquid and that adsorbed on the substrate surface. Consult Rhee *et al.* (1986, page 33) if necessary.

12.9 The artificial kidney depends on a pressure-driven flow for its sieving mechanism. For the parameters of the artificial kidney, write a report discussing the influence of the streaming potential.

12.10 Equation (12.26) determines the influence of an externally imposed electric field on the mole fraction of an analyte. Compare the sieving coefficient for the imposed field with that for a charged membrane.

12.11 Design an implantable drug delivery system for the delivery of insulin to a type I diabetes mellitus patient. Write a paper on your design. Be sure to begin the paper with a definition of diabetes and why the patient needs additional insulin.

12.12 Ion channels supply cells with needed nutrients and are tailored to their specific function. Ion channel walls are made up of charged proteins and are essentially natural nanopores. Write a paper to characterize the specific ion channel of your choice (there are many), and identify the role of ion channels in the disease cystic fibrosis.

12.13 Design a nanoparticle for the treatment of a lung cancer tumor.

12.14 What is a quantum dot?

Appendix A

Matched Asymptotic Expansions

A.1 Introduction

Why study perturbation techniques? The answer lies in the fact that we can simplify the problem significantly if we study the structure of the equation. In some cases, analytical solutions may be obtained, which eliminates any need for computational work. More commonly, the application of these techniques results in a problem that may be solved numerically much easier than the original problem.

Perturbation methods had been known since the days of the ancient astronomers to predict the trajectories of planets subject to small disturbances. The modern concept of *regular and singular perturbation problems* was developed in the 1960s and refined over the 1970s in fluid mechanics to investigate the high Reynolds number flow past a flat plate, the *boundary layer*. The function of a *boundary layer* is to bring the velocity to zero on a solid body. Because the boundary layer is very thin, of $O(Re^{-1/2})$, the velocity will vary rapidly across the layer, making numerical solutions very difficult, if not impossible. In the intervening years, boundary layer–type behavior, the rapid variation of any quantity over a short distance or period of time, has been used in a wide variety of disciplines to describe such behavior.

There are a number of books on this subject, and these appear in the bibliography; they include Kevorkian and Cole (1981, 1996), VanDyke (1975), Nayfeh (1973), Bellman (1964), Bender and Orszag (1999), Holmes (1995), Hinchcliffe (2003), and Howison (2005). The book by VanDyke (1975) is specifically oriented toward fluid mechanics, whereas the others are more general treatments.

A.2 Terminology

Before moving on to some examples, we need to define some terms commonly used in perturbation theory. Ordering of numbers and functions with respect to the small parameter is important. We say, for example, $2 = O(3)$, which is stated in words as "2 is of the order of 3"; similarly, when a given number is "about" 10 times another number, we say $1 \ll 10$, or "1 is much less than 10." Similarly,

$10 \gg 1$. There are no hard boundaries for this terminology, and these ordering symbols should be treated as estimates; for example, we might say $2 \ll 10$, but this estimate is at the boundary of validity. In the case of functions, $f(x) \ll g(x)$ means that "f is much less than g, for all x." The technical definitions of the two symbols $x \ll y$ and $x = O(y)$ may be expressed in terms of the small parameter ϵ as $f(\epsilon) = O(g(\epsilon))$ as $\epsilon \to 0$ if

$$\lim_{\epsilon \to 0} \frac{f(\epsilon)}{g(\epsilon)} < \infty \tag{A.1}$$

and $f(\epsilon) \ll g(\epsilon)$ as $\epsilon \to 0$ if

$$\lim_{\epsilon \to 0} \frac{f(\epsilon)}{g(\epsilon)} = 0 \tag{A.2}$$

It should be noted that $f \ll g$ is sometimes written $f = o(g)$, or "f is little o of g."

As specific examples, we write

$$\sin 2\epsilon = O(\epsilon) \quad \text{as } \epsilon \to 0$$

$$\cos(\epsilon) = O(1) \quad \text{as } \epsilon \to 0$$

We could also write

$$\sin 2\epsilon = o(1) \quad \text{as } \epsilon \to 0$$

Also, we should note that

$$e^{-\frac{1}{\epsilon}} \ll \epsilon^m \quad \text{for all } m.$$

A.3 Asymptotic sequences and expansions

Perturbation methods most always involve a small parameter that often and perhaps usually arises as a result of a nondimensionalization process. Consider the series

$$f = f_0 + \epsilon f_1 + \epsilon^2 f_2 + \cdots \tag{A.3}$$

where $\epsilon \ll 1$. The functions $g_n = \epsilon^n$ are called *gauge functions* and satisfy the condition that $g_{n+1} \ll g_n$ or $g_{n+1} = o(g_n)$ since

$$\frac{g_{n+1}}{g_n} = \epsilon$$

The sequence of functions g_n for $n = 0, 1, 2, \ldots$ is called an *asymptotic sequence*, and the expansion of f is called an *asymptotic expansion*.

Equation (A.3) looks like a Taylor series expansion, and so the question of convergence in the mathematical sense may be posed. However, there are two limiting processes at work here: the limit $\epsilon \to 0$ and the limit $n \to \infty$. For a Taylor series, we take the limit $n \to \infty$ before even thinking about the value of ϵ. In asymptotic and perturbation analysis, the limit $\epsilon \to 0$ is of paramount importance. In some sense, then, in an asymptotic analysis, the error should get smaller as $\epsilon \to 0$. Asymptotic analysis is a highly pragmatic subject, and it is customary to proceed until experience suggests otherwise.

To illustrate the concept, consider the general problem of finding a solution to the equation

$$Lf = a + \epsilon P(f) \tag{A.4}$$

where a and P do not depend on ϵ; for simplicity, we assume that a is a constant and that P is some operation on f. Then, expanding f in an asymtotic series and equating coefficients of ϵ, we find

$$L(f_0 + \epsilon f_1 + \epsilon^2 f_2 + \cdots) = a + \epsilon F(f_0 + \epsilon f_1 + \epsilon^2 f_2 + \cdots) \tag{A.5}$$

$$L(f_0) = a$$

$$L(f_1) = F(f_0)$$

and so on, where we have assumed that $a = O(1)$. For example, suppose $a = 2$ and $L(f) = f$ and $P(f) = f^2$. Then the solution is obviously

$$f = 2 + 4\epsilon + 16\epsilon^2 + E \tag{A.6}$$

where $E = O(\epsilon^3)$ is the error. Note the rising coefficients of the increasing powers of ϵ, and we might ask the question, does the series converge? For *fixed* ϵ as $n \to \infty$, the answer is no because the ratio of the coefficients gets larger and not smaller, as in a convergent Taylor series. But in the limit $\epsilon \to 0$, for *fixed* n, the answer is yes!

A.4 Regular perturbations

A typical example of this type of perturbation is the following problem:

$$y'' - \epsilon y = 0, \quad \epsilon \ll 1 \tag{A.7}$$

subject to

$$y(0) = 1, \quad y(1) = 0 \tag{A.8}$$

The exact solution to this problem is

$$y = \frac{\sin h\, \epsilon^{1/2}(1-x)}{\sin h\, \epsilon^{1/2}} \tag{A.9}$$

Note that the small parameter ϵ does *not* multiply the highest derivative of the equation. This problem is easy to solve, but let us anticipate solving problems of a more difficult nature and try an *asymptotic expansion* of the form

$$y = y_0 + \epsilon y_1(x) + \epsilon^2 y_2(x) + \cdots \tag{A.10}$$

Each term in the asymptotic expansion must combine to satisfy the equation and the boundary conditions that are obtained by substituting directly into the equation; thus

$$n = 0 \ y_0'' = 0 \ y_0(0) = 1, \quad y_0(1) = 0$$

$$n = 1 \ y_1'' = y_0 \ y_1(0) = 0, \quad y_1(1) = 0$$

$$n = 2 \ y_2'' = y_1 \ y_2(0) = 0, \quad y_2(1) = 0$$

for the first three terms of the asymptotic expansion. It should be noted that an asymptotic series need not converge in the mathematical sense, although the solution should approach the true solution as more terms are included.

These equations are easy to solve, and the result is

$$y_0 = 1 - x \tag{A.11}$$

$$y_1 = -\frac{x}{6}(x-1)(x-2) \tag{A.12}$$

and so on. Typical of these methods is that the solution for higher-order terms in the expansion gets more complicated, and for this reason, only one or two terms in the expansion are usually calculated. This is often sufficient for an accurate solution to a physical problem. To show that the perturbation solution agrees with the exact solution for small ϵ, we can expand the exact solution in a Taylor series for small ϵ, and the result is

$$y = 1 - x - \epsilon\frac{x}{6}(x-1)(x-2) + \cdots \tag{A.13}$$

which agrees with the first two terms in the perturbation expansion. For a regular perturbation problem, one expansion is sufficient to describe the solution everywhere in the domain. This is not the case for a singular perturbation.

A.5 Singular perturbations

We should caution here that there are a wide variety of singular perturbation problems, and the singular nature of the problem may appear in a number of ways. The most common way of telling whether the problem is of the singular perturbation type is that the small parameter multiplies the highest-order dervative in the equation. As noted earlier, this is the case when we set $Re = \infty$ in the Navier–Stokes equations.

Let us consider a specific example; we first consider a simple ordinary differential equation. Consider the equation

$$\epsilon y'' + y' + y = 0, \ \ \epsilon \ll 1 \tag{A.14}$$

subject to

$$y(0) = 1, \ \ y(1) = 0 \tag{A.15}$$

The solution is easily calculated as

$$y = Ae^{\alpha_1 x} + Be^{\alpha_2 x} \tag{A.16}$$

where α_1 and α_2 are given by

$$\alpha_1 \sim -1 - \epsilon \quad \text{as} \quad \epsilon \to 0$$
$$\alpha_2 \sim -\frac{1}{\epsilon} + 1 + \epsilon \quad \text{as} \quad \epsilon \to 0$$

Applying the boundary conditions, we find that

$$y(x) = e^{-x(\frac{1}{\epsilon}-1-\epsilon)} - e^{-(\frac{1}{\epsilon}-1-\epsilon)}e^{(\epsilon+1)(1-x)} \tag{A.17}$$

Let us analyze the solution and further simplify the exact solution. Note that for $\epsilon > 0$ and ϵ small, $y \ll 1$, except when $x = O(\epsilon)$. Let us define $\eta = x/\epsilon$; then, to leading order in the small parameter ϵ, the solution may be expressed as

$$y = e^{-\eta} \tag{A.18}$$

Apparently, for $\eta = O(1)$, the solution is given by the preceding equation. This is called the *inner solution*. Note that it satisfies the boundary condition at $\eta = x = 0$. Away from the boundary $\eta \gg 1$, $y = 0$ is the solution. This is called the *outer solution*.

Let us apply the concepts of singular perturbation theory to this problem. To solve this problem, we must determine which terms in the equation are important and where. We note that if $\epsilon = 0$, then

$$y' + y = 0 \tag{A.19}$$

and the solution is

$$y = Ae^{-x} \tag{A.20}$$

This is our outer solution. However, A is an arbitrary constant, and at this point, we do not know how to determine it. We cannot satisfy both boundary conditions because we have only one constant. If we choose to satisfy the boundary condition $y(0) = 1$, then $A = 1$, whereas if we choose to satisfy $y(1) = 0$, then $A = 0$. Which do we choose? Clearly a regular perturbation expansion procedure will not work. (Why?) We must decide how to bring in the first and second derivative terms.

Suppose we decide to balance the two derivative terms and take $\epsilon y'' = O(y')$; then

$$\frac{d}{dx} = O(\epsilon^{-1}) \tag{A.21}$$

and the *boundary layer variable* is

$$\eta = \frac{x}{\epsilon} \tag{A.22}$$

With this scaling, the governing equation becomes

$$\frac{d^2Y}{d\eta^2} + \frac{dY}{d\eta} + \epsilon Y = 0 \tag{A.23}$$

where Y is our *boundary layer–dependent variable*. Thus we expand Y as

$$Y = Y_0 + \epsilon Y_1 + \cdots \tag{A.24}$$

and Y_0 satisfies

$$Y_0'' + Y_0' = 0 \tag{A.25}$$

This equation is easy to solve, and the result is

$$Y_0 = C + De^{-\eta} \tag{A.26}$$

We have assumed that the *boundary layer* is at $x = 0$, and so we must satisfy the boundary condition at $\eta = 0$. Thus

$$C + D = 1 \tag{A.27}$$

The solution is thus

$$Y_0 = 1 + D(e^{-\eta} - 1) \tag{A.28}$$

This is our *inner solution*.

The question is, how do we connect or *match* the inner and outer solutions? In the inner solution, we have A as the arbitrary constant, and in the inner solution, D is the arbitrary constant. A reasonable procedure is to take the limit of the outer solution as the boundary layer is approached and the limit of the inner solution away from the boundary layer. This is called a *matching principle* by Van Dyke (1975), and mathematically, the condition is written as

$$\lim_{x \to 0} y_0 = \lim_{\eta \to \infty} Y_0 \tag{A.29}$$

Clearly, in our problem, this requires

$$A = 1 - D \tag{A.30}$$

Because the outer solution is valid away from $x = 0$, we require it to satisfy the boundary condition at $x = 1$. In this case, $A = 0$, $D = 1$, and we have

$$Y_0 = e^{-\eta} \tag{A.31}$$

$$y_0 = 0 \tag{A.32}$$

The question you may be asking is, why can there not be a boundary layer at $x = 1$? You can show very easily that this assumption leads to an exponentially growing solution in the boundary layer variable:

$$\xi = \frac{1 - x}{\epsilon} \tag{A.33}$$

This is unacceptable, and thus no boundary layer can be present at $x = 1$. You can also show that there cannot be a boundary layer of thickness $\epsilon^{1/2}$; that is, we cannot balance d^2y/dx^2 with y.

As a final note, the solution can be written

$$y = e^{-\eta} + 0 - 0$$

$$y = \text{inner} + \text{outer} - \text{common part}$$

and the solution is now valid everywhere on the interval $0 \leq x \leq 1$. We say that the solution is *uniformly valid*. The common part is that part of the solution that is common to both the inner and the outer solution. Another way of matching the two solutions is to write the outer solution in inner variables and the inner solution in outer variables, equate the two, and take the limit as $\epsilon \to 0$. In this case, this procedure yields

$$Ae^{-\epsilon\eta} = 1 + D(e^{-\frac{x}{\epsilon}} - 1) \tag{A.34}$$

which yields, in the limit $\epsilon \to 0$,

$$A = 1 - D \tag{A.35}$$

as we have already seen.

Appendix B

Vector Operations in Curvilinear Coordinates

If Φ is a scalar function and $\vec{A} = A_1\vec{e}_1 + A_2\vec{e}_2 + A_3\vec{e}_3$ is a vector function of orthogonal curvilinear coordinates u_1, u_2, u_3 ($\vec{e}_1$, $\vec{e}_2$, $\vec{e}_3$ are unit vectors in the direction of increasing u_1, u_2, u_3, respectively), then the following results are valid:

1. $\nabla\Phi = \text{grad}\Phi = \frac{1}{h_1}\frac{\partial\Phi}{\partial u_1}\vec{e}_1 + \frac{1}{h_2}\frac{\partial\Phi}{\partial u_2}\vec{e}_2 + \frac{1}{h_3}\frac{\partial\Phi}{\partial u_3}\vec{e}_3$
2. $\nabla \bullet \vec{A} = \text{div}\vec{A} = \frac{1}{h_1h_2h_3}\left[\frac{\partial}{\partial u_1}(h_2h_3A_1) + \frac{\partial}{\partial u_2}(h_3h_1A_2) + \frac{\partial}{\partial u_3}(h_1h_2A_3)\right]$
3. $\nabla \times \vec{A} = \text{curl}\vec{A} = \frac{1}{h_1h_2h_3}\begin{vmatrix} h_1\vec{e}_1 & h_2\vec{e}_2 & h_3\vec{e}_3 \\ \partial/\partial u_1 & \partial/\partial u_2 & \partial/\partial u_3 \\ h_1A_1 & h_2A_2 & h_3A_3 \end{vmatrix}$
4. $\nabla^2\Phi = \text{Laplacian of }\Phi = \frac{1}{h_1h_2h_3}\left[\frac{\partial}{\partial u_1}\left(\frac{h_2h_3}{h_1}\frac{\partial\Phi}{\partial u_1}\right) + \frac{\partial}{\partial u_2}\left(\frac{h_3h_1}{h_2}\frac{\partial\Phi}{\partial u_2}\right)\right.$
$\left. + \frac{\partial}{\partial u_3}\left(\frac{h_1h_2}{h_3}\frac{\partial\Phi}{\partial u_3}\right)\right]$

where h_1, h_2, h_3 are the *scale factors*:

$$h_1 = \left|\frac{\partial\vec{r}}{\partial u_1}\right|, \quad h_2 = \left|\frac{\partial\vec{r}}{\partial u_2}\right|, \quad h_3 = \left|\frac{\partial\vec{r}}{\partial u_3}\right|$$

B.1 Cylindrical coordinates

$$\begin{array}{lll} u_1 = r & h_1 = 1 & \vec{e}_1 = \vec{e}_r \\ u_2 = \theta & h_2 = r & \vec{e}_2 = \vec{e}_\theta \\ u_3 = z & h_3 = 1 & \vec{e}_3 = \vec{e}_z \end{array}$$

B.2 Spherical coordinates

$$\begin{array}{lll} u_1 = r & h_1 = 1 & \vec{e}_1 = \vec{e}_r \\ u_2 = \theta & h_2 = r & \vec{e}_2 = \vec{e}_\theta \\ u_3 = \phi & h_3 = r\sin\theta & \vec{e}_3 = \vec{e}_\phi \end{array}$$

B.3 Rectangular coordinates

$$\begin{array}{lll} u_1 = x & h_1 = 1 & \vec{e}_1 = \vec{i} \\ u_2 = y & h_2 = 1 & \vec{e}_2 = \vec{j} \\ u_3 = z & h_3 = 1 & \vec{e}_3 = \vec{k} \end{array}$$

Reference: M. R. Spiegel, *Theory and Problem of Vector Analysis: Introduction to Tensor Analysis*, New York, McGraw-Hill.

Appendix C Web Sites

Here are some Web sites that contain much useful information on micro- and nanofluidics. This list is not meant to be complete, and I would welcome additional information on cool Web sites that you may have. In addition, there are individual researchers' Web sites at various universities that I have left out. I have also left out individual company and publisher Web sites. Also, I have grouped these sites into categories; however, understand that in micro- and nanofluidics, this may be dangerous.

C.1 Fluid dynamics and micro- and nanofluidics

1. http://www.efluids.com/
   Very informative Web site that includes many applets and demos covering fluid dynamics in general.
2. http://www.emicronano.com/
   Very informative Web site that includes images, movies, demos, pictures, and solver tools for micro- and nanofluids. Run by the same people as efluids, this Web site concentrates on smaller-scale fluids.
3. http://www.lab-on-a-chip.com/
   This site is an RSS feed site that has many current news headlines and articles from university labs covering microfluidics, microarrays, and labs-on-a-chip. It also has a list of upcoming events regarding microfluidics and other information in the field.
4. http://www.rsc.org/Publishing/Journals/lc/Chips_and_Tips/index.asp
   This is a Q&A Web site dedicated to answering questions and providing information for those interested in labs-on-a-chip.
5. http://www.aip.org/tip/INPHFA/vol-9/iss-4/p14.html
   This is a feature article from *The Industrial Physicist* that explains the explosion of micro- and nanofluidics technologies in recent years. The site describes several technologies benefitting from new discoveries in the microfluidics area and discusses the differences between micro- and nanofluidics.

C.2 General nanotechnology

1. http://nanoHUB.org/
   This site provides online nanocomputation tools in a variety of disciplines such as quantum transport, nanomedicine, and others. Researchers contribute computation tools for use by registered clients. There are currently over 800 contributors. The site also provides a series of courses in mechanics, biology, and chemistry.
2. http://www.nnin.gov/
   This is the site for the National Nanotechnology Infrastructure Network (NNIN), funded through the National Science Foundation. It is an integrated network of facilities for sharing of nanoscale experimental and fabrication tools. The url http://www.nnin.org/nnin_compsim.html is the computational portal of NNIN, which operates in a manner similar to nanoHUB.

C.3 Wikipedia

1. http://en.wikipedia.org/wiki/Microfluidics
   Standard definitions and basic information about microfluidics from Wikipedia. Discusses continuous-flow microfluidics, digital microfluidics, DNA chips, molecular biology, optics, acoustic droplet ejection, and fuel cells.
2. http://en.wikipedia.org/wiki/Nanofluidics
   Standard definitions and basic information about nanofluidics from Wikipedia. Discusses theory and includes several equations, specifically the Poisson–Boltzmann equation. It also has a few paragraphs on nanostructure fabrication and general applications of nanofluidics technologies.
3. http://en.wikipedia.org/wiki/Lab-on-a-chip
   Contains lots of information on labs-on-a-chip, and includes a history of the technology, advantages, disadvantages, fabrications, and relations to global health and examples.
4. http://en.wikipedia.org/wiki/List_of_microfluidics_research_groups
   Contains a large list of microfluidics research groups throughout the world and has links to more information about each group and its research.
5. http://en.wikipedia.org/wiki/Capillary_electrophoresis
   Wikipedia article that gives in-depth information on capillary electrophoresis – a good overview.

Appendix D

A Semester Course Syllabus

Syllabus: Introduction to Micro- and Nanofluidics for Engineers and Physical Scientists (Three hour lecture, 1 hour lab)

Syllabus for a 4 hour (3 hour lecture, 1 hour lab) semester course 15 weeks long, assuming 30 lectures of length 1 hour 12 minutes, two lectures per week, and does not include exam periods. The lab meets once per week, with two lectures on fabrication and experimental methods – the content of lecture 6a. If the class meets three times per week, prorate lectures to 45.

Objective: The objective of this course is to introduce students to the basic physical foundations of incompressible fluid mechanics appropriate to micro- and nanosize conduits. The course will emphasize the fundamental principles involved in the formulation and solution of problems in fluid mechanics and mass transfer for pressure-driven and electrically driven motions of biofluids and individual components such as ions. On completion of the course, the student should be able to extract from a raw physical situation the essential principles from which a useful fluid mechanical model may be developed.

The lab will provide students with an introduction to the design, fabrication, and testing of microfluidic and nanofluidic devices and how practical devices can be analyzed theoretically.

Meeting	Topic	Reading
1	Overview and Historical Perspective	1.1–1.17
	Convergence of Molecular Biology and Engineering	
2	Transport Properties	2.1–2.3
	Viscosity and Diffusion Coefficient	
	Electrical Permittivity	
	Surface Tension and Wettability	
3	Thermodynamics	2.5, 2.6
4	Kinematics/Forces	3.3, 3.4
	Conservation of Mass	3.5
	Navier–Stokes Equations	3.6
5	Mass Transport	3.7

Meeting	Topic	Reading
	Electrostatics: Poisson Equation	3.8
	Energy Transport	3.9
6	Boundary Conditions	3.11
	Dimensional Analysis	3.12
	Fluid and Heat and Mass Transfer Analogies	3.13
6a	Fabrication/Experiments/Theory	3.17
7	Fully Developed Flow in Pipes and Channels	4.1–4.3
	Slip Flow	4.4
8	Fully Developed Suction Flows	4.7
	Developing Suction Flows	4.8
9	Slow Flow Past a Sphere	4.11
	Sedimentation	4.12
	Blood Flow	4.13
10	The Temperature Distribution in Pipe and Channel Flow	5.1–5.4
11	Graetz Problems	5.5
12	Taylor–Aris Dispersion	5.7
	The Stochastic Nature of Diffusion: Brownian Motion	5.9
13	Electric Field	6.1–6.4
14	Electrical Potential	6.5–6.7
	Current and Current Density	6.8
	Maxwell's Equations	6.9
15	Electrochemistry	7.1–7.4
16	Chemical Potential, Electrochemical Potential	7.5–7.7
17	Charged Surfaces	7.9, 7.10
18	Electrical Double Layer	7.11–7.13
19	ζ Potential	7.14
	Surface Charge Density	
20	EDL/Cylinders/Sphere	7.15, 7.16
	Semipermeable Membranes	7.18
21	Molecular Biology	Chapter 8
	DNA and Proteins	
22	Electrosmosis	9.1–9.2.3
23	Continued	9.2.4–9.2.6
24	Continued	9.2.8–9.2.10
25	Electrophoresis	9.3.1, 9.3.2
26	Continued	9.3.3, 9.3.4
27	Streaming Potential	9.4
	Sedimentation Potential	9.5
	Joule Heating	9.6
28	Molecular Simulations	Chapter 11
29	Applications	Chapter 12
30	Continued	Chapter 12

Bibliography

Abramowitz, M., & Stegun, I. A., eds. 1972 *Handbook of Mathematical Functions with Formulas, Graphs and Mathematical Tables*. Washington, DC: National Bureau of Standards.

Abramson, H. A. 1931 The influence of size, shape, and conductivity on cataphoretic mobility, and its biological significance. *J. Phys. Chem.* **35**, 289–308.

Acheson, D. J. 1990 *Elementary Fluid Mechanics*. Oxford: Clarendon Press.

Adrian, R. J. 1991 Particle-imaging techniques for experimental fluid mechanics. *Ann. Rev. Fluid Mech.* **23**, 261–304.

Adrian, R. J., & Yao, C. S. 1985 Pulsed laser technique application to liquid and gaseous flows and the scattering power of seed materials. *Appl. Opt.* **24**, 44–52.

Alberts, B., Bray, D., Lewis, J., Raff, M., Roberts, K., & Watson, J. D. 1994 *Molecular Biology of the Cell*. New York: Garland.

Alberts, B., Bray, D., Hopkin, K., Johnson, A., Lewis, J., Raff, M., Roberts, K., & Walter, P. 1998 *Essential Cell Biology*. New York: Garland.

Allen, M. P. 2004 Introduction to molecular dynamics simulation. In *Computational Soft Matter: From Synthetic Polymers to Proteins,* NIC Series, vol. 23 (ed. Norbert Attig, Kurt Binder, Helmut Grubmuller & Kurt Kremer), pp. 1–28. Julich, Germany.

Allen, M. P., & Tildesley, D. 1994 *Computer Simulation of Liquids*. Oxford: Clarendon Press.

Allen, S., Davies, J., Dawkes, A. C., Davies, M. C., Edwards, J. C., Parker, M. C., Roberts, C. J., Sefton, J., Tendler, S. J. B., & Williams, P. M. 1996 In situ observation of streptavidin-biotin binding on an immunoassay well surface using an atomic force microscope. *FEBS Lett.* **390**(2), 161–164.

Anderson, J. D. 1982 *Modern Compressible Flow: With Historical Perspective*. New York: McGraw-Hill.

Anderson, J. L., & Idol, W. K. 1985 Electroosmosis through pores with nonuniformly charged walls. *Chem. Eng. Commun.* **38**, 93–106.

Anton, K., & Berger, C. 1998 *Supercritical Fluid Chromatography with Packed Columns: Techniques and Applications*. New York: Marcel Dekker.

Archer, D. G., & Wang, P. 1990 The dielectric constant of water and the Debye-Hückel limiting law slopes. *J. Phys. Chem. Ref. Data* **19**, 371–411.

Aris, R. 1956 On the dispersion of a solute in a fluid flowing through a tube. *Proc. R. Soc. A* **235**, 67–77.

Aris, R. 1959 On the dispersion of a solute by diffusion, convection and exchange between phases. *Proc. R. Soc. A* **252**, 538–550.

Atkinson, B., Brocklebank, M. P., Card, C. C. H., & Smith, J. M. 1969 Low Reynolds number developing flows. *AIChE J.* **15**, 548–553.

Axelrod, D., Burghardt, T. P., & Thompson, N. L. 1984 Total internal reflection fluorescence. *Ann. Rev. Biophys. Bioeng.* **13**, 247–268.

Barcilon, V., Chen, D.-P., & Eisenberg, R. S. 1992 Ion flow through narrow membrane channels: Part II. *SIAM J. Appl. Math.* **52**, 1405–1425.

Barcilon, V., Chen, D. P., Eisenberg, R. S., & Jerome, J. W. 1997 Qualitative properties of steady-state Poisson–Nernst–Planck systems: Perturbation and simulation study. *SIAM J. Appl. Math.* **57**, 631–648.

Barrat, J. L., & Bocquet, L. 1999 Large slip effect at a nonwetting fluid-solid interface. *Phys. Rev. Lett.* **82**, 4671–4674.

Bashir, R., & Wereley, S., eds. 2006 *BioMEMS and Biomedical Nanotechnology: Volume IV Biomolecular Sensing, Processing and Analysis*. New York: Springer.

Batchelor, G. K. 1967 *Introduction to Fluid Dynamics*. Cambridge: Cambridge University Press.

Bavier, R., & Ayela, F. 2004 Micromachined strain gauges for the determination of liquid flow friction coefficients in microchannels. *Measure. Sci. Technol.* **15**, 377–383.

Bayley, H., & Cremer, P. S. 2001 Stochastic sensors inspired by biology. *Nature* **413**, 226–230.

Bazant, M. Z., Chu, Kevin T., & Bayly, B. J. 2005 Current-voltage relations for electrochemical thin films. *SIAM J. Appl. Math.* **65**, 1463–1484.

Becker, O. M., & Karplus, M. 2006 *A Guide to Biomolecular Simulations*. Dordrecht, Netherlands: Springer.

Bellman, R. E. 1964 *Perturbation Techniques in Mathematics, Physics, and Engineering*. New York: Holt, Rinehart and Winston.

Bender, Carl M., & Orszag, Steven A. 1999 *Advanced Mathematical Methods for Scientists and Engineers: Asymptotic Methods and Perturbation Theory*. New York: Springer.

Berendsen, H. J. C., Grigera, J. R., & Straatsma, T. P. 1987 The missing term in effective pair potentials. *J. Phys. Chem.* **91**, 6269–6271.

Berendsen, H. J. C., Postma, J. P. M., van Gunsteren, W. F., & Hermans, J. 1981 Interaction models for water in relation to protein hydration. *Intermolecular Forces*, vol. 3 (ed. B. Pullman Reidel, Dordrecht. The Netherlands), pp. 331–342.

Berman, A. S. 1953 Laminar flow in channels with porous walls. *J. Appl. Phys.* **24**, 1232–1235.

Bernoulli, D. 1738 *Hydrodynamics*. Strasbourg: Bernouli. English Translation Dover, New York.

Bhattacharjee, S., & Elimelech, M. 1997 Surface element integration: A novel technique for evaluation of DLVO interaction between a particle and a flat plate. *J. Colloid Interface Sci.* **193**, 273–285.

Bhattacharjee, S., Elimelechi, M., & Borkovec, M. 1998 DLVO interaction between coloidal particles: Beyond Derjaguin's approximation. *Croatica Chem. Acta* **71**, 883–903.

Bhattacharyya, S., & Conlisk, A. T. 2005 Electroosmotic flow in two-dimensional charged micro- and nanochannels. *J. Fluid Mech.* **540**, 247–267.

Bhushan, B., ed. 2007 *Springer Handbook of Nanotechnology*, 2nd ed. New York: Springer.

Bianchi, F., Wagner, F., Hoffmann, P., & Girault, H. H. 1993 Electroosmotic flow in composite microchannels and implications in microcapillary electrophoresis systems. *Science* **261**, 895–897.

Bird, G. A. 1994 *Molecular Gas Dynamics*. Oxford, UK: Clarendon Press.

Bird, R. B., Stewart, W. E., & Lightfoot, E. N. 2002 *Transport Phenomena*, 2nd ed. New York: John Wiley.

Blasius, H. 1908 Grenzschichten in flussigkeiten mit kleiner reibung. *Z. Math. Phys.* **56**, 1–37.

Bockris, J. O'M., & Reddy, A. K. N. 1998 *Modern Electrochemistry, Volume 1 Ionics*, 2nd ed. New York: Plenum Press.

Bohn, P. 2009 Nanoscale control and manipulation of molecular transportin chemical analysis. *Ann. Rev. Anal. Chem.* **2**, 279–296.

Booth, F. 1950 The cataphoresis of spherical, solid non-conducting particles in a symmetrical electrolyte. *Proc. R. Soc. London A* **203**, 514–533.

Braha, O., Gu, Li-Qun, Zhou, Li, Lu, Xiaofeng, Cheley, S., & Bayley, H. 2000a Simultaneous stochastic sensing of divalent metal ions. *Nat. Biotechnol.* **18**, 1005–1007.

Brecht, A., & Gauglitz, G. 1997 Recent developments in optical transducers for chemical or biochemical applications. *Sensors Actuators B Chem.* **38**, 1–7.

Bretherton, F. P. 1961 The motion of long bubbles in tubes. *J. Fluid Mech.* **10**, 166–188.

Breuer, K., ed. 2005 *Microscale Diagnostic Techniques*. Berlin: Springer.

Brown, G. M. 1960 Heat or mass transfer in a fluid in laminar flow in a circular or flat conduit. *A I Che J.* **6**, 179–183.

Bruus, H. 2008 *Theoretical Microfluidics*. New York: Oxford University Press.

Burgeen, D., & Nakache, F. R. 1964 Electrokinetic flow in ultrafine capillary slits. *J. Phys. Chem.* **68**, 1084–1091.

Butler, J. N. 1998 *Ionic Equilibrium: Solubility and pH Calculations*. New York: John Wiley.

Castellan, G. W. 1983 *Physical Chemistry*, 3rd ed. Menlo Park, CA: Benjamin Cummings.

Castiglione, P., Falcioni, M., Lesne, A., & Vulpiani, A. 2008 *Chaos and Coarse Grainning in Statistical Mechanics*. Cambridge, UK: Cambridge University Press.

Chang, H., Kosari, F., Andreadakis, G., Alam, M. A., Vasmatzis, G., & Bashir, R. 2004 DNA-mediated fluctuations in ionic current through silicon oxide nanopore channels. *Nanoletters* **4**, 1551–1556.

Chang, H.-C., & Yeo, L. Y. 2010 *Electrokinetically-Driven Microfluidics and Nanofluidics*. Cambridge: Cambridge University Press.

Chang, R. 2000 *Physical Chemistry for the Chemical and Biological Sciences*. Sausalito, CA: University Science Books.

Chapman, D. L. 1913 A contribution to the theory of electrocapillarity. *Philos. mag.* **25**, 475–481.

Chen, D. P., Lear, J., & Eisenberg, R. 1997 Permeation through an open channel: Poisson–Nernst–Planck theory of a synthetic ion channel. *Biophys. J.* **72**, 97–116.

Chen, L., & Conlisk, A. T. 2008 Electroosmotic flow and particle transport in micro/nano nozzles and diffusers. *Biomed. Microdevices* **10**, 289–298.

Chen, L., & Conlisk, A. T. 2009 Effect of nonuniform surface potential on electroosmotic flow at large applied electric field strength. *Biomedical Microdevices* **11**, 251–258.

Chen, L. & Conlisk, A. T. 2010 DNA translocation phenomena in nanopores, *Biomedical Microdevices*, **12**, 235–245.

Chen, P., Gu, J., Brandin, E., Kim, Y.-R., Wang, Q., & Branton, D. 2004 Probing single DNA molecule transport using fabricated nanopores. *Nano Letters* **4**, 2293–2298.

Chen, R.-Y. 1973 Flow in the entrance region at low Reynolds numbers. *J. Fluids Eng.* **95**, 153–158.

Chu, K. T., & Bazant, M. Z. 2005 Electrochemical thin films at and above the limiting current. *SIAM J. Appl. Math.* **65**, 1485–1505.

Churchill, R. V. 1969 Fourier series and boundary value problems McGraw-Hill, **2**.

Ciofalo, M., Collins, M. W., & Hennessy, T. R. 1999 *Nanoscale Fluid Dynamics in Physiological Process: A Review Study*. Southampton, UK: WIT Press.

Condon, E. U., & Morse, P. M. 1929 *Quantum Mechanics*. New York: McGraw-Hill.

Conlisk, A. T., Guezennec, Y. G., & Elliott, G. S. 1989 Chaotic motion of an array of vortices above a flat wall. *Phys. Fluids A* **1**, 704–717.

Conlisk, A. T., Datta, S., Fissell, W. H., & Roy, S. 2009 Biomolecular transport through hemofiltration membranes. *Ann. Biomed. Eng.* **37**(4), 732–746.

Constant, F. W. 1958 *Theoretical Physics*. Reading, MA: Addison-Wesley.

Conway, B. E. 1981 *Ionic Hydration in Chemistry and Biophysics*. New York: Elsevier.

Crick, F. H. C., & Watson, J. D. 1953 Molecular structure of nucleic acids. *Nature* **171**, 737–738.

Cui, S. T. 2004 Molecular dynamics study of single-stranded DNA in aqueous solution confined in a nanopore. *Molecular Phys.* **102**, 139–146.

Currie, I. G. 2003 *Fundamental Mechanics of Fluids*, 3rd ed. New York: Marcel-Dekker.

Cussler, E. L. 1997 *Diffusion: Mass Transfer in Fluid Systems*, 2nd ed. Cambridge: Cambridge University Press.

Czarske, J., Buttner, L., Razik, T., & Muller, H. 2002 Boundary layer velocity measurements by a laser Doppler profile sensor with micrometre spatial resolution. *Measure Sci. Technol.* **13**, 1979–1989.

Datta, S. & Ghosal, S. 2008 Dispersion due to wall interactions in microfluidic separation systems. *Phys. Fluids* 20, 012103–1–012103–14.

Daune, M. 1993 *Molecular Biophysics: Structures in Motion*. Oxford: Oxford University Press.

Dawson, T. H. 1976 *Theory and Practice of Solid Mechanics*. New York: Plenum Press.

Day, M. A. 1990 The no-slip condition of fluid mechanics. *Erkenntis* **33**, 285–296.

Deamer, D. W., & Akeson, M. 2000 Nanopores and nucleic acids: Prospects for ultrarapid sequencing. *Tibtech* **18**, 147–151.

Debye, P., & Hückel, E. 1923 The interionic attraction theory of deviations from ideal behavior in solution. *Z. Phys.* **24**, 185–206.

Dechadilok, P., & Deen, W. M. 2006 Hindrance factors for diffusion and convection in pores. *Ind. Eng. Chem. Res.* **45**, 6953–6959.

Deen, W. M. 1987 Hindered transport of large molecules in liquid-filled pores. *AIChE J.* **33**, 1409–1425.

Deen, W. M., Lazzara, M. J., & Myers, B. D. 2001 Structural determinants of glomerular permeability. *Am. J. Physiol. Renal Physiol.* **281**, 579–596.

Denbigh, K. 1971 *Principles of Chemical Equilibrium*, 3rd ed. Cambridge: Cambridge University Press.

Derjaguin, B. V. 1934 Friction and adhesion IV: Theory of adhesion of small particles. *Kolloid Z.* **69**, 155–164.

Derjaguin, B. V., & Landau, L. D. 1941 Theory of the stability of strongly charged lyophobic colloids and the adhesion of strongly charged particles in solutions of electrolytes. *Acta Physicochim.* **14**, 633–662.

Desai, T., & Bhatia, S., ed. 2006 *BioMEMS and Biomedical Nanotechnology: Volume III Therapeutic Micro/Nano Technology*. New York: Springer.

Devasenathipathy, S., & Santiago, J. G. 2005 Electrokinetic flow diagnostics. In *Microscale Diagnostic Techniques* (ed. Kenny Breuer), pp. 113–154. Berlin: Springer.

Devasenathipathy, S., Santiago, J. G., & Takehara, K. 1998 Particle tracking techniques for electrokinetic microchannel flows. *Exp. Fluids* **25**, 316–319.

D'Orazio, P. 2003 Biosensors in clinical chemistry. *Clin. Chim. Acta* **334**, 41–69.

Drazin, P. G., & Riley, N. 2006 *The Navier–Stokes Equations: A Classification of Flows and Exact Solutions*. Cambridge: Cambridge University Press.

Dukhin, S. S., & Derjaguin, B. V. 1974 Electrokinetic phenomena. In *Surface and Colloid Science* vol. 7 (ed. E. Matijevic), pp. 1–351. John Wiley.

Eggins, B. R. 1996 *Biosensors: An Introduction*. New York: John Wiley.

Einstein, A. 1905a A new determination of molecular dimensions. PhD thesis, University of Zurich.

Einstein, A. 1905b On the motion of small particles suspended in liquids at rest required by the molecular-kinetic theory of heat. *Ann. Phys.* **17**, 549–560.

Einstein, A. 1956 *Investigation on the Theory of the Brownian Movement*, 4th ed. New York: Dover.

Elimelech, M. 1998 *Particle Deposition and Aggregation: Measurement, Modelling and Simulation*. Burlington, MA: Butterworth-Heinemann.

Erickson, D., & Li, D. 2001 Streaming potential and streaming current methods for characterizing heterogeneous solid surfaces. *J. Colloid Interface Sci.* **237**, 283–289.

Erickson, D., & Li, D. 2002 Influence of surface heterogeneity on electrokinetically driven microfluidic mixing. *Langmuir* **18**, 1883–1892.

Erickson, D., Sinton, D., & Li, D. 2003 Joule heating and heat transfer in poly(dimethylsiloxane) microfluidic systems. *Lab on a Chip* **3**, 141–149.

Ethier, C. R., & Simmons, C. A. 2007 *Introductory Biomechanics: From Cells to Organisms*. Cambridge: Cambridge University Press.

Evans, D. J., & Morriss, G. P. 1990 *Statistical Mechanics of Nonequilibrium Liquids*. London: Academic Press.

Ewald, P. P. 1921 The calculation of optical and electrostatic grid potential. *Ann. Phys. (Leipzig)* **64**, 253–287.

Fan, R., Karnik, R., Yue, M., Li, D., Majumdar, A., & Yang, P. 2005 DNA translocation in inorganic nanotubes. *Nanoletters* **5**, 1633–1637.

Fausett, L. V. 2008 *Applied Numerical Analysis Using Matlab*, 2nd ed. Upper Saddle River, NJ: Prentice Hall.

Fawcett, W. R. 2004 *Liquids, Solutions, and Interfaces: From Classical Macroscopic Descriptions to Modern Microscopic Details*. Oxford: Oxford University Press.

Ferrari, M., ed. 2006 *BioMEMS and Biomedical Nanotechnology*. New York: Springer.

Ferrer, M. L., Duchowicz, R., Carrasco, B., de la Torre, Jose G., & Acuna, A. U. 2001 The conformation of serum albumin in solution: A combined phosphorescence depolarization-hydrodynamic modeling study. *Biophys. J.* **80**, 2422–2430.

Feynman, R. P. 1961 *There's Plenty of Room at the Bottom*. New York: Reinhold.

Fissell, W. H. 2006 Developments towards an artificial kidney. *Expert Rev. Med. Devices* **3**, 155–165.

Fissell, W. H., & Humes, H. D. 2006 Tissue engineering renal replacement therapy. In *Tissue Engineering and Artificial Organs*, Section 5, chap. 60 (ed. J. D. Bronzino), **60**, pp. 1–14. Boca Raton, FL: CRC Press.

Fissell, W. H., Manley, S., Dubnisheva, A., Glass, J., Magistrelli, J., Eldridge, A., Fleischman, A., Zydney, A., & Roy, S. 2007 Ficoll is not a rigid sphere. *Am. J. Physiol. Renal Physiol.* **293**, F1209–F1213.

Franks, F. 1972 *Water: A Comprehensive Treatise*, 7 vols. New York: Plenum Press.

Freifelder, D. 1987 *Molecular Biology*, 2nd edn. Boston: Jones and Bartlett.

Frenkel, D., & Smit, B. 2002 *Understanding Molecular Simulations from Algorithms to Applications*, 2nd ed. San Diego, CA: Academic Press.

Freund, J. B. 2002 Electroosmosis in a nanometer scale channel studied by atomistic simulation. *J. Chem. Phys.* **116**, 2194–2200.

Friedman, M. H. 2008 *Principles and Models of Biological Transport*, 2nd ed. New York: Springer.

Fung, Y. C. 1981 *Biomechanics: Mechanical Properties of Living Tissues*. New York: Springer.

Fuoss, R. M., & Onsager, L. 1955 Conductance of strong electrolytes at finite dilutions. *Proc. Nat. Acad. Sci. U.S.A.* **41**, 274–283.

Gad-el Hak, M. 2001 *The MEMS Handbook*. Boca Raton, FL: CRC Press.

de Gennes, P. G., 2002 On fluid/wall slippage. *Langmuir* **18**, 3413–3414.

Gibbs, J. W. 1961 *The Scientific Papers of J. W. Gibbs*. New York: Dover.

Giddings, J. C., Yang, F. J., & Myers, M. N. 1976 Flow-field-flow fractionation: A versatile new separation method. *Science* **193**, 1244–1245.

Gilat, A., & Subramaniam, V. 2008 *Numerical Methods for Scientists and Engineers*. New York: John Wiley.

Gillespie, D. T. 1970 *A Quantum Mechanics Primer*. Scranton, PA: International Textbook.

Gillespie, D. 1999 A singular perturbation analysis of the Poisson–Nernst–Planck system: Applications to ionic channels. PhD thesis, Rush Medical School, Chicago.

Gillespie, D., & Eisenberg, R. S. 2001 Modified Donnan potentials for ion transport through biological ion channels. *Phys. Rev. E.* **63**, 061902-1–06192-8.

Glazer, A. N., & Nikaido, H. 2007 *Microbial Biotechnology: Fundamentals of Applied Microbiology*, 2nd ed. Cambridge: Cambridge University Press.

Golden, J. P., Floyd-Smith, T. M., Mott, D. R., & Ligler, F. S. 2007 Target delivery in a microfluidic immunosensor. *Biosensors Bioelectr.* **22**, 2763–2767.

Goldstein, S. 1965a *Modern Developments in Fluid Dynamics Volume I*. New York: Dover.

Goldstein, S. 1965b *Modern Developments in Fluid Dynamics Volume II*. New York: Dover.

Gong, M., Kim, B. Y., Flachsbart, B. R., Shannon, M. A., Bohn, P. W., & Sweedler, J. V. 2008 An on-chip fluorogenic enzyme assay using a multilayer microchip interconnected with a nanocapillary array membrane. *IEEE Sensors J.* **8**, 601–607.

Gouy, G. 1910 About the electric charge on the surface of an electrolyte. *J. Phys. A* **9**, 457–468.

Grahame, D. C. 1953 Diffuse double layer theory for electrolytes of unsymmetrical valence types. *J. Chem. Phys.* **21**, 1054–1060.

Granicka, L. H., Kawiak, J., Snochowski, M., Wojcicki, J. M., Sabalinska, S., & Werynski, A. 2003 Polypropylene hollow fiber for cells isolation: Methods for evaluation of diffusive transport and quality of cells encapsulation. *Artificial Cells Blood Substitutes Biotechnol.* **31**, 249–262.

Green, N. M. 1970 Spectrophotometric determination of avidin and biotin. *Methods Enzymol.* **18**, 418–424.

Gribbin, J. 1997 *Richard Feynman: A Life in Science*. New York: Dutton.

Griffiths, S. K., & Nilson, R. H. 1999 Hydrodynamic dispersion of a neutral nonreacting solute in electroosmotic flow. *Anal. Chem.* **71**, 5522–5529.

Griffiths, S. K., & Nilson, R. H. 2006 Charged species transport, separation, and dispersion in nanoscale channels: Autogenous electric field-flow fractionation. *Anal. Chem.* **78**, 8134–8141.

Guo, L. J. 2004 Recent progress in nanoimprint technology and its applications. *J. Appl. Phys. D:* **37**, R123–R141.

Hairer, E., Lubich, C., & Wanner, G. 2006 *Geometric Numerical Integration: Structure Preserving Algorithms for Ordinary Differential Equations*, 2nd ed. Heidelberg, Germany: Springer.

Hardy, R. C., & Cottingham, R. L. 1949 Viscosity of deuterium oxide and water in the range 5°C to 1250°C. *J. Res. Na. Bur. Standards* **42**, 573–578.

Haynes, W. M., ed. 2011–2012 *Handbook of Chemistry and Physics*, 92nd ed. Cleveland, Ohio: CRC Press.

Helmholtz, H. L. F. 1897 Uber den einflu der elektrischengrenzschichten bei galvanischer spannung und der durch wasserstromung erzeugten potentialdiffernz. *Ann. Physik.* **7**, 337–387.

Henry, D. C. 1931 The cataphoresis of suspended particles, Part I. The equation of cataphoresis. *Proc. R. Soc. London A* **133**, 106–129.

Heyes, D. M. 1998 *The Liquid State: Applications of Molecular Simulations*. Chichester, UK: John Wiley.

Hille, B. 2001 *Ion Channels of Excitable Membranes*, 3rd ed. Sunderland, MA: Sinauer Associates.

Hill, T. L. 1963 *Thermodynamics of Small Systems, Part I*. New York: W.A. Benjamin.

Hill, T. L. 1964 *Thermodynamics of Small Systems, Part II*. New York: W.A. Benjamin.

Hinchcliffe, A. 2003 *Molecular Modeling for Beginners*. John Wiley.

Hollerbach, U., Chen, D. P., & Eisenberg, R. 2001 Two- and three-dimensional Poisson–Nernst–Planck simulations of current flow through gramicidin a. *J. Sci. Comput.* **16**, 373–409.

Holmes, M. H. 1995 *Introduction to Perturbation Methods*, 2nd ed. New York: Springer.

Homola, J., Yee, S. S., & Gauglitz, G. 1999 Surface plasmon resonance sensors: Review. *Sensors Actuators BC* **54**, 3–15.

Honig, C. D. F., & Ducker, W. A. 2007 No-slip hydrodynamic boundary condition for hydrophilic particles. *Phys. Rev. Lett.* **98**, 053101.

Howison, S. 2005 *Practical Applied Mathematics*. Cambridge: Cambridge University Press.

Hughes, W. F., & Gaylord, E. W. 1964 *Basic Equations of Engineering Science*. New York: Schaum.

Humbert, N., Zocchi, A., & Ward, T. R. 2005 Electrophoretic behavior of streptavidin complexed to a biotinylated probe: A functional screening assay for biotin-binding proteins. *Electrophoresis* **26**, 47–52.

Humes, H. D., Fissell, W. H., & Tiranathanagul, K. 2006 The future of hemodialysis membranes. *Kidney Int.* **69**, 1115–1119.

Hunter, R. J. 1981 *Zeta Potential in Colloid Science*. London: Academic Press.

Hur, J. S., Shaqfeh, E. S. G., & Larson, R. G. 2000 Brownian dynamics simulations of single DNA molecules in shear flow. *J. Rheol.* **44**, 713–742.

Icenhower, J. P., & Dove, P. M. 2000 Water behavior at silica surfaces. In *Adsorption on Silica Surfaces* (ed. Eugene Papirer), pp. 277–295. New York: Marcel-Dekker.

Iler, R. K. 1979 *The Chemistry of Silica*. New York: John Wiley.

Incropera, F. P., & Dewitt, D. P. 1990 *Fundamentals of Heat and Mass Transfer*, 3rd ed. New York: John Wiley.

Ishido, T., & Mizutani, H. 1981 Experimental and theoretical basis of electrokinetic phenomena in rock-water systems and its application to geophysics. *J. Geophys. Res.* **86**(83), 1763–1775.

Ishijima, A., & Yanagida, T. 2001 Single molecule nanoscience. *Trends Biochem. Sci.* **26**, 438–444.

Israelachvili, J. 1992 *Intermolecular and Surface Forces*, 2nd ed. London: Academic Press.

James, R. O. 1981 Surface ionization and complexation at the colloidl/aqueous electrolyte interface. In *Adsorption of Inorganics at Solid–Liquid Interfaces* (ed. M. A. Anderson & A. J. Rubins), pp. 219–261, chap. 6. Ann Arbor, MI: Ann Arbor Science.

Jianrong, C., Yuqing, M., Nongyue, H., Xiaohua, W., & Sijiao, L. 2004 Nanotechnology and biosensors. *Biotechnol. Adv.* **22**, 505–518.

Jorgensen, P. L. 1990 Structure and molecular mechanism of Na, k-pump. In *Monovalent Cations in Biological Systems* (ed. Charles Alexander Pasternak), pp. 117–154. Boca Raton, FL: CRC Press.

Jorgensen, W. L., Chandrasekhar, J., Madura, J. D., Impey, R. W., & Klein, M. L. 1983 Comparison of simple potential functions for simulating liquid water. *J. Chem. Phys.* **79**, 926–935.

Judy, J., Maynes, D., & Webb, B. W. 2002 Characterization of frictional pressure drop for liquid flows through microchannels. *Int. J. Heat Mass Transfer* **45**, 3477–3489.

Karniadakis, G., Beskok, A., & Aluru, N. 2005 *Microflows and Nanoflows*. New York: Springer.

Kasianowicz, J. J., Brandin, E., Branton, D., & Deamer, D. W. 1996 Characterization of individual polynucleotide molecules using a membrane channel. *Proc. Natl. Acad. Sci. U.S.A.* **93**, 13770–13773.

Kays, W. M., & Crawford, M. E. 1980 *Convective Heat and Mass Transfer*, 2nd ed. New York: Mcgraw-Hill.

Keilland, J. 1937 Individual activity coefficients of ions in aqueous solutions. *J. Am. Chem. Soc.* **59**, 1675–1678.

Kemery, P. J., Steehler, J. K., & Bohn, P. W. 1998 Electric field mediated transport in nanometer diameter channels. *Langmuir* **14**, 2884–2889.

Kestin, J. 1978 Thermal conductivity of water and steam. *Mech. Eng. Mag.* **August**, 47.

Kevorkian, J., & Cole, J. D. 1981 *Perturbation Methods in Applied Mathematics*. New York: Springer.

Kevorkian, J., & Cole, Julian D. 1996 *Multiple Scale and Singular Perturbation Methods*. New York: Springer.

Kirby, B. J. 2010 *Micro- and Nanoscale Fluid Mechanics: Transport in Microfluidic Devices*. Cambridge, UK: Cambridge University Press.

Kirby, B. J., & Hasselbrink, E. F. 2004a Zeta potential of microfluidic substrates: 1. Theory, experimental techniques, and effects on separations. *Electrophoresis* **25**, 187–202.

Kirby, B. J., & Hasselbrink, E. F. 2004b Zeta potential of microfluidic substrates: 2. Data for polymers. *Electrophoresis* **25**, 203–213.

Kirby, B. J., & Hasselbrink, E. F. 2004c Zeta potential of microfluidic substrates: 2. Data for polymers. *Electrophoresis* **25**, 203–213.

Kjelstrup, S., & Bedeaux, D. 2008 *Non-Equilibrium Thermodynamics of Heterogenous systems*. New Jersey: World Scientific.

Knox, J. H., & McCormack, K. A. 1994 Temperature effects in capillary electrophoresis. 1: Internal capillary temperature and effect upon performance. *Chromatographia* **38**, 215–221.

Koltun, W. L. 1965 Precision space-filling atomic models. *Biopolymers* **3**, 665–679.

Koplik, J., Banavar, J. R., & Willemson, J. F. 1989 Molecular dynamics of fluid flow at solid surfaces. *Phys. Fluids A* **1**, 781–794.

Kress-Rogers, E., ed. 1997 *Handbook of Biosensors and Electronic Noses: Medicine, Food and the Environment*. Boca Raton, FL: CRC Press.

Kuo, T.-C., Jr, Cannon, D. M., Jr, Shannon, M. A., Bohn, P. W. & Sweedler, J. V. 2003 Hybrid three-dimensional nanofluidic/microfluidic devices using molecular gates. *Sensors and Actuators A* **102**, 223–233.

Lamb, S. H. 1945 *Hydrodynamics*, 6th ed. New York: Dover.

Landau, L. D., & Levich, B. V. G. 1942 Dragging of a liquid by a moving plate. *Acta Physicochim.* **17**, 42–54.

Landers, J. P., ed. 1994 *Handbook of Capillary Electrophoresis*. Boca Raton, FL: CRC Press.

Langhaar, H. L. 1942 Steady flow in the transition length of a straight tube. *J. Appl. Mech.* **9**, 55–58.

Latini, G., Grifoni, R. C., & Passerini, G. 2006 *Transport Properties of Organic Liquids*. Southhampton, UK: WIT Press.

Lauga, E., Brenner, M. P., & Stone, H. A. 2005 Microfluidics: The no-slip condition. In *Handbook of Experimental Fluid Mechanics* (ed. J. Foss & A. Yarin), pp. 1219–1240. New York: Springer.

Leach, A. R. 1996 *Molecular Modeling: Principles and Applications*. Essex, UK: Longman.

Leal, L. G. 2007 *Advanced Transport Phenomena*. New York: Cambridge University Press.

Leca-Bouvier, B., & Blum, L. J. 2005 Biosensors for protein detection: A review. *Anal. Lett.* **38**, 1491–1517.

Lee, A. P., & Lee, L. James, eds. 2006 *BioMEMS and Biomedical Nanotechnology: Volume I Biological and Biomedical Nanotechnology*. New York: Springer.

Lee, L. J. 2006 Nanoscale polymer fabrication for biomedical applications. In *BioMEMS and Biomedical Nanotechnology: Volume I Biological and Biomedical Nanotechnology* (ed. Abraham P. Lee & L. James Lee), pp. 51–96. New York: Springer.

Lee, M. L., Yang, F. J., & Bartle, K. D. 1984 *Open Tubular Column Gas Chromatography: Theory and Practice*. New York: John Wiley.

Lee, P.-S., & Garimella, S. V. 2006 Thermally developing flow and heat transfer in rectangular microchannels of different aspect ratio. *Int. J. Heat Mass Transfer* **49**, 3060–3067.

Lehnert, T., Gijs, M., Netzer, R., & Bischoff, U. 2002 Realization of hollow SiO_2 micronozzles for electrical measurements on living cells. *Appl. Phys. Lett.* **81**, 5063–5065.

Lempert, W. R., Magee, K., Ronney, P., Gee, K. R., & Haugland, R. P. 1995 Flow tagging velocimetry in incompressible flow using photo-activated nonintrusive tracking of molecular motion. *Exp. Fluids* **18**, 249–257.

Levich, V. G., & Krylov, V. S. 1969 Surface-tension driven phenomena. *Ann. Rev. Fluid Mech.* **1**, 293–316.

Levin, Y., & Flores-Mena, J. E. 2001 Surface tension of strong electrolytes. *Europhys. Lett.* **56**, 187–192.

Li, D. 2004 *Electrokinetics in Microfluidics*. Amsterdam: Elsevier.

Li, J., Gershow, M., Stein, D., Brandin, E., & Golovchenko, J. A. 2003 DNA molecules and configurations in a solidstate nanopore microscope. *Nat. Mater.* **2**, 611–615.

Li, Z., & Liu, B. C.-Y. 2001 A molecular model for representing surface tension for polar liquids. *Chem. Eng. Sci.* **56**, 6977–6987.

Liou, W. K., & Fang, Y. 2006 *Microfluid Mechanics: Principles and Modeling*. New York: McGraw-Hill.

Luginbuhl, P., Indermuhle, P.-F., Gretillat, M.-A., Willemin, F., de Rooij, N. F., Gerber, D., Gervasio, G., Vuilleumier, J. -L., Twerenbold, D., Dugelin, M., Mathys, D., & Guggenheim, R. 2000 Femtoliter injector for DNA mass spectrometry. *Sensors Actuators B* **63**, 167–177.

Malsch, N. H., ed. 2005 *Biomedical Nanotechnology*. Boca Raton, FL: Taylor and Francis.

March, H. W., & Weaver, W. 1928 The diffusion problem for a solid in contact with a stirred liquid. *Phys. Rev.* **31**, 1072–1082.

Mark, P., & Nilsson, L. 2001 Structure and dynamics of the TIP3P, SPC, and SPC/E water models at 298 K. *J. Phys. Chem. A* **105**, 9954–9960.

Martin, F., Walczak, R., Boiarski, A., Cohen, M., West, T., Cosentino, C., & Ferrari, M. 2005 Tailoring width of microfabricated nanochannels to solute size can be used to control diffusion kinetics. *J. Controlled Release* **102**, 123–133.

Martini, F. 2001 *Fundamentals of Anatomy and Physiology*, 5th ed. Prentice Hall.

Masliyah, J. H., & Bhattacharjee, S. 2006 *Electrokinetic and Colloid Transport Phenomena*. Hoboken, NJ: John Wiley.

Maxwell, J. C. 1847 On Faraday's lines of force. *Trans. Cambridge Philos. Soc.* **10**, 27–83.

McCammon, J. A., & Harvey, S. C. 1987 *Dynamics of Proteins ansd Nucleic Acids*. Cambridge: Cambridge University Press.

McNaught, A. D., & Wilkinson, A. 1997 *Compendium of Chemical Terminology (Gold Book)*, Malden: Blackwell.

Meagher, R. J., Light, Y. K., & Singh, A. K. 2008 Rapid, continuous purification of proteins in a microfluidic device using genetically-engineered partition tags. *Lab-on-a-Chip* **8**, 527–532.

Meller, A., Nivon, L., & Branton, D. 2001 Voltage-driven DNA translocations through a nanopore. *Phys. Rev. Lett.* **86**, 3435–3438.

Moran, M. J., & Shapiro, H. N. 2007 *Fundamentals of Engineering Thermodynamics*, 6th ed. New York: John Wiley.

Moran, M. J., Shapiro, H. N., Munson, B. R., & Dewitt, D. P. 2003 *Introduction to Thermal Systems Engineering*. New York: John Wiley.

Mott, D. R., Howell, P. B., Golden, J. P., Kaplan, C. R., Ligler, F. S., & Oran, E. S. 2006 Toolbox for the design of optimized microfluidic components. *Lab-on-a-Chip* **6**, 540–549.

Mott, D. R., Howell, P. B., Obenschain, K. S., & Oran, E. S. 2009 The numerical toolbox: An approach for modeling and optimizing microfluidic components. *Mech. Res. Commun.* **36**, 104–109.

Munson, B. R., Young, D. F., & Okiishi, T. H. 2005 *Fundamentals of Fluid Mechanics*, 2006th ed. New York: John Wiley.

Murray, J. D. 2001 *Mathematical Biology I: An Introduction*, 3rd ed. New York: Springer.

Murray, J. D. 2003 *Mathematical Biology II: Spatial Models and Biological Applications*, 3rd ed. New York: Springer.

Murrell, J. N., & Jenkins, A. D. 1982 *Properties of Liquids and Solutions*, 2nd ed. Chichester, UK: John Wiley.

Nayfeh, A. H. 1973 *Perturbation Methods*. New York: John Wiley.

Newman, J. S. 1972 *Electrochemical Systems*. Englewood Cliffs, NJ: Prentice Hall.

Nguyen, N. T., & Wereley, S. T. 2002 *Fundamentals and Applications of Microfluidics*. Norwood, MA: Artech House.

Oberkampf, W. L., & Blottner, F. G. 1998 Issues in computational fluid dynamics code verification and validation. *AIAA J.* **36**, 687–695.

Oberkampf, W. L., Sindir, M. M., & Conlisk, A. T. 1998 G-077-1998 guide for the verification and validation of computational fluid dynamics simulations. *Tech. Rep.* American Institute of Aeronautics and Astronautics.

O'Brien, R. W., & White, L. R. 1978 Electrophoretic mobility of a spherical colloidal particle. *J. Chem. Soc. Faraday Trans. II* **74**, 1607–1626.

Ohshima, H., Healy, T. W., White, L. R., & O'Brien, R. 1984 Sedimentation velocity and potential in a dilute suspension of charged spherical colloidal particles. *J. Chem. Soc. Faraday Trans. II* **80**, 1299–1317.

Onsager, L., & Samaras, N. N. T. 1934 The surface tension of debye-huckel electrolytes. *J. Chem. Phys.* **2**, 528–536.

Oseen, C. W. 1910 Uber die sstokes'sche formel und uber eine verwandte aufgabein der hydrodynamik. *Ark. Math. Astron. Fys.* **6**.

Overbeek, J. TH. G. 1943 Theory of the relaxation effect in electrophoresis. *Kolloide Beihefte* **54**, 287–364.

Oyanader, M., & Arce, P. 2005 A new and simpler approach for the solution of the electrostatic potential differential equation: Enhanced solution for planar, cylindrical and annular geometries. *J. Colloid Interface Sci.* **284**, 315–322.

Ozkan, M., & Heller, M. J., ed. 2006 *BioMEMS and Biomedical Nanotechnology: Volume II Micro/Nano Technology for Genomics and Proteomics*. New York: Springer.

Papirer, E., ed. 2000 *Adsorption on Silica Surfaces*. New York: Marcel Dekker.

Persello, J. 2000 Surface and interface structure of silica. In *Adsorption on Silica Surfaces* (ed. Eugene Papirer), pp. 297–342. New York: Marcel-Dekker.

Peters, T. 1996 *All About Albumin: Biochemistry, Genetics and Medical Applications*, 3rd ed. San Diego, CA: Academic Press.

Pinkus, O., & Sternlicht, B. 1961 *Theory of Hydrodynamic Lubrication*. New York: McGraw-Hill.

Plawski, J. L. 2001 *Transport Phenomena Fundamentals*. New York: Marcel-Dekker.

Priestley, J. 1767 *The History and Present State of Electricity*. London: Printed for J. Dodsley, J. Johnson and T. Cadell.

Priezjev, N. V., & Troian, S. M. 2006 Influence of periodic wall roughness on slip behavior at liquid/solid interfaces: Molecular scale simulations versus continuum predictions. *J. Fluid Mech.* **554**, 25–48.

Priezjev, N. V., Darhuber, A. A., & Troian, S. M. 2005 Slip behavior in liquid films on surfaces of patterned wettability: Comparison between continuum and molecular dynamics simulations. *Phys. Rev. E* **71**, 41608.

Probstein, R. F. 1989 *Physicochemical Hydrodynamics*. Boston: Butterworths.

Proudman, L., & Pearson, J. R. A. 1957 Expansions at small Reynolds numbers for the flow past a sphere and a circular cylinder. *J. Fluid Mech.* **2**, 237–262.

Qiao, R., & Aluru, N. R. 2003a Atypical dependence of electroosmotic transport on surface charge in a single-wall carbon nanotube. *Nano Lett.* **3**, 1013–1017.

Qiao, R., & Aluru, N. R. 2003b Ion concentrations and velocity profiles in nanochannel electroosmotic flows. *J. Chem. Phys.* **118**, 4692–4701.

Quinke, G. 1859 Ueber eine neue Art elekrischer Ströme. *Prog. Ann.* **107**, 1–47.

Rahman, A., & Stillinger, F. H. 1971 Molecular dynamics study of liquid water. *J. Chem. Phys.* **55**, 3336–3359.

Rapaport, D. C. 2004 *The Art of Molecular Simulation*, 2nd ed. Cambridge: Cambridge University Press.

Ravindra, N. M., Prodan, C., Fnu, S., Padroni, I., & Sikha, S. K. 2007 Advances in the manufacturing, types and applications of biosensors. *JOM* **59**, 37–43.

Raymond, K. W. 2007 *General, Organic and Biological Chemistry: An Integrated Approach*, 2nd ed. New York: John Wiley.

Reid, R. C., Prausnitz, J. M., & Poling, B. E. 1987 *The Properties of Gases and Liquids*, 4th ed. New York: McGraw-Hill.

Revil, A., Pezard, P. A., & Glover, P. W. J. 1999 Streaming potential in porous media 1. Theory of the zeta potential. *J. Geophys. Res.* **104**, 20021–20032.

Rhee, H.-K., Aris, R., & Amundson, N. R. 1986 *First-Order Partial Differential Equations: Volume 1 Theory and Applications of Single Equations*. Englewood Cliffs, NJ: Prentice Hall.

Rhee, H.-K., Aris, R., & Amundson, N. R. 1989 *First-Order Partial Differential Equations: Volume 2 Theory and Applications of Hyperbolic Systems of Quasilinear Equations*. Mineola, NY: Dover.

Rice, S. A. 2000 Active control of molecular dynamics: Coherence versus chaos. *J. Stat. Phys.* **101**, 187–212.

Richardson, S. 1973 On the no-slip boundary condition. *J. Fluid Mech.* **59**, 707–719.

Roache, P. J. 1998 *Verification and Validation in Computational Science and Engineering*. Socorro, NM: Hermosa.

Roache, P. J., & Steinberg, S. 1994 Symbolic manipulation and computational fluid dynamics. *AIAA J.* **22**, 1390–1394.

Robinson, R. A., & Stokes, R. H. 1959 *Electrolyte Solutions*. New York: Academic Press.

Roco, M. 2005 Converging technologies: Nanotechnology and medicine. In *Biomedical Nanotechnology* (ed. Neelina H. Malsch). Boca Raton, FL: Taylor and Francis.

Rosi, N. L., & Mirkin, C. A. 2005 Nanostructures in biodiagnostics. *Chem. Rev.* **105**, 1547–1562.

Roy, C. J., Nelson, C. C., Smith, T. M., & Ober, C. C. 2004 Verification of Euler/Navier–Stokes codes using the method of manufactured solutions. *Int. J. Numer. Methods Fluids* **44**, 599–620.

Russel, W. B., Saville, D. A., & Schowalter, W. R. 1991 *Colloidal Dispersions*. Cambridge: Cambridge University Press.

Ryckaert, J.-P., Ciccotti, G., & Berendsen, H. J. C. 1977 Numerical integration of the Cartesian equations of motion of a system with constraints: Molecular dynamics of n-alkanes. *J. Comput. Phys.* **23**, 327–341.

Sadr, R., Yoda, M., Zheng, Z., & Conlisk, A. T. 2004 An experimental study of electro-osmotic flow in rectangular microchannels. *J. Fluid Mech.* **506**, 357–367.

Sadr, R., Yoda, M., Gnanaprakasam, P., & Conlisk, A. T. 2006 Velocity measurements inside the diffuse electric double layer in electroosmotic flow. *Appl. Phys. Lett.* **89**, 044103-1–044103-3.

Sadr, R., Hohenegger, C., Li, H., Mucha, P. J., & Yoda, M. 2007 Diffusion-induced bias in near-wall velocimetry. *J. Fluid Mech.* **577**, 443–456.

Sadus, R. J. 1997 *Molecular Simulation of Liquids: Theory, Algorithms and Object-Orientation*. Amsterdam: Elsevier.

Sagui, C., & Darden, T. A. 1999 Molecular dynamics simulations of biomolecules: Long range electrostatic effects. *Annu. Rev. Biomolecular Structure*, **28**, 155–179.

Saleh, O. A., & Sohn, L. L. 2006 *An On-Chip Artificial Pore for Molecular Sensing*. New York: Springer.

Saltzman, W. M. 2001 *Drug Delivery: Engineering Principles for Drug Therapy*. Oxford: Oxford University Press.

Saltzman, W. M. 2009 *Biomedical Engineering*. Cambridge, UK: Cambridge University Press.

Scheller, F. W., Wollenberger, U., Warsinke, A., & Lisdat, F. 2001 Research and development in biosensors. *Curr. Opi. Biotechnol.* **12**, 35–40.

Scherrer, R., & Gerhardt, P. 1971 Molecular sieving by the *Bacillus megaterium* cell wall and protoplast. *J. Bacteriol.* **107**, 718–735.

Schnell, E. 1956 Slippage of water over nonwettable surfaces. *J. Appl. Phys.* **27**, 1149–1152.

Sears, F. W., & Zemansky, M. W. 1964 *University Physics*, 3rd ed. Reading, MA: Addison-Wesley.

Shapiro, A. P., & Probstein, R. F. 1993 Removal of contaminants from saturated clay by electroosmosis. *Environ. Sci. Technol.* **27**, 283–291.

Shaw, D. 1969 *Electrophoresis*. London: Academic Press.

Shereshefsky, J. L. 1967 A theory of surface tension of binary solutions I. Binary liquid mixtures of organic compounds. *J. Colloid Interface Sci.* **24**, 317–322.

Sinha, M. K., Roy, D., Gaze, D. C., Collinson, P. O., & Kaski, J.-C. 2004 Role of ischemia modified albumin, a new biochemical marker of myocardial ischemia, in the early diagnosis of acute coronary syndromes. *Emerg. Med. J.* **21**, 29.

Smith, G. D. 1985 *Numerical Solutions of Partial Differential Equations*, 3rd ed. Oxford: Oxford University Press.

Smoluchowski, M. 1918 Versuch einer mathematischen theorie der koagulation kinetic kolloider losungen. *Z. Phys. Chem* **92**, 129–135.

Snyder, L. R., & Kirkland, J. J. 1979 *Introduction to Modern Liquid Chromatography*. New York: John Wiley.

Spichiger-Keller, U. E. 1998 *Chemical Sensors and Biosensors for Medical and Biological Applications*. Wiley-VCH.

Spoel, D., van der, van Maaren, P. J., & Berendsen, H. J. C. 1998 A systematic study of water models for molecular simulation: Derivation of water models optimized for use with a reaction field. *J. Chem. Phys.* **108**, 10220–10230.

Stachel, J., ed. 1998 *Einstein's Miraculous Year*. Princeton, NJ: Princeton University Press.

Staubli, T., Stæckli, T., Knapp, H. F., Alpnach, S., Lausanne, S., de Neuchtel, U., & Neuchtel, S. 2005 Fast immobilization of probe beads by dielectrophoresis-controlled adhesion in a versatile microfluidic platform for affinity assay. *Electrophoresis* **26**, 3697–3705.

Stern, O. 1924 The theory of the electrolytic double layer. *Z. Elektrochem.* **30**, 508–516.

Stigter, D. 1980 Sedimentation of highly charged colloidal spheres. *J. Phys. Chem.* **84**, 2758–2762.

Storm, A. J., Chen, J. H., Zandbergen, H. W., & Dekker, C. 2003 Fabrication of solid-state nanopores with single-nanometre precision. *Nat. Mater.* **2**, 537–540.

Storm, A. J., Chen, J. H., Zandbergen, H. W., & Dekker, C. 2005 Translocation of double-strand DNA through a silicon oxide nanopore. *Phys. Rev. E.* **71**, 051903.

Störmer, C. 1907 Sur les trajectoires des corpuscules electrises. *Arch. Sci. Phys. Nat. Geneve* **24**, 5–18, 113–158, 221–247.

Stroock, A. D., Weck, D. M., Chiu, D. T., Huck, W. T. S., Kenis, P. J. A., Ismagilov, R. F., & Whitesides, G. M. 2000 Patterning electro-osmotic flow with patterned surface charge. *Phys. Rev. Lett.* **84**, 3314–3317.

Styer, D. F. 1996 Common misconceptions regarding quantum mechanics. *Am. J. Phys.* **64**, 31–34.

Tabeling, P. 2005 *Introduction to Microfluidics*. Oxford: Oxford University Press.

Tahery, R., Modarress, H., & Satherly, J. 2005 Surface tension prediction and thermodynamic analysis of the surface for binary solutions. *Chem. Eng. Sci.* **60**, 4935–4952.

Taylor, G. I. 1953 Dispersion of soluble matter in solvent flowing slowly through a tube. *Proc. R. Soc. London A* **219**, 186–203.

Taylor, G. I. 1954 Conditions under which dispersion of a solute in a stream of solvent can be used to measure molecular diffusion. *Proc. R. Soc. London A* **225**, 473–477.

Taylor, R., & Krishna, R. 1993 *Multicomponent Mass Transfer*. New York: John Wiley.

Terrill, R. M. 1964 Laminar flow in a uniformly porous channel. *Aeronaut. Q.* **XV**, 297–299.

Terrill, R. M., & Shrestha, G. M. 1965 Laminar flow through parallel and uniformly porous walls of different permeability. *Z. Angewan. Math. Phys.* **16**, 470–482.

Terrill, R. M., & Thomas, P. W. 1969 On laminar flow through a uniformly porous pipe. *Appl. Sci. Res.* **21**, 37–67.

Thévenot, D. R., Toth, K., Durst, R. A., & Wilson, G. S. 2001 Electrochemical biosensors: Recommended definitions and classification. *Biosensors Bioelectronics* **16**, 121–131.

Thompson, P. A., & Troian, S. M. 1997 A general boundary condition for liquid flow at solid surfaces. *Nature* **389**, 360–362.

Thust, M., Schoning, M. J., Frohnhoff, S., & Arens-Fischer, R. 1996 Porous silicon as a substrate material for potentiometric biosensors. *Measure. Sci. Technol.* **7**, 26–29.

Tokaty, G. A. 1971 *A History and Philosophy of Fluid Mechanics*. New York: Dover.

Travis, K. P., & Gubbins, K. E. 2000 Poiseuille flow of Lennard–Jones fluids in narrow slit pores. *J. Chem. Phys.* **112**, 1984–1994.

Tuckerman, M. 2010 *Statistical Mechanics: Theory and Simulation*. Oxford: Oxford University Press.

Turns, S. R., *Thermal-Fluid Sciences: An Integrated Approach*, Cambridge, UK: Cambridge University Press, 2006.

Tyrrell, H. J. V., & Harris, K. R. 1984 *Diffusion in Liquids: A Theoretical and Experimental Study*. London: Butterworth.

Van Dyke, M. 1975 *Perturbation Methods in Fluid Mechanics*, 2nd ed. Stanford, CA: Parabolic Press.

Venema, P., Hiemstra, T., & van R., Willem H. 1996 Comparison of different site binding models for cation sorption: Description of pH dependency, salt dependency, and cation-proton exchange. *J. Colloid Interface Sci.* **181**, 45–49.

Venturoli, D., & Rippe, B. 2005 Ficoll and dextran vs. globular proteins as probes for testing glomular permselectivity: Effects of molecular size, shape, charge and deformability. *A. J. Physiol. Renal Physiol.* **288**, 605–613.

Verlet, L. 1967 Computer experiments on classical fluids I. thermodynamical properties of Lennard–Jones molecules. *Phys. Rev.* **159**, 98–103.

Verwey, E. J. W., & Overbeek, J. T. G. 1948 *Theory of Stability of Lyophobic Colloids*. Amsterdam: Elsevier.

Vo-Dinh, T. 2006 Biosensors and biochips. In *Biomolecular Sensing, Processing and Analysis* (ed. R. Bashir, Steve Wereley, & Mauro Ferrari), pp. 4–33. New York: Springer.

Volkov, A. G., Paula, S., & Deamer, D. W. 1997 Two mechanisms of permeation of small neutral molecules and hydrated ions across phospholipid bilayers. *Bioelectrochem. Bioenergetics* **42**, 153–160.

Wang, S., Hu, Xin, & Lee, L. J. 2008 Electrokinetics induced asymmetric transport in polymeric nanonozzles. *Lab-on-a-Chip* **8**, 573–581.

Wang, S., Zeng, C., Lai, S., Juang, Y.-J., Yang, Y., & Lee, L. J. 2005 Polymer nanonozzle array fabricated by sacrificial template imprinting. *Adv. Mater.* **17**, 1182–1186.

Wang, Y., Bhushan, B., & Maali, A. 2009 Atomic force microscopy measurement of boundary slip on hydrophilic, hydrophobic and superhydrophobic surfaces. *J. Vac. Sci. Technol. A* **27**, 1–7.

Wehausen, J. V., & Laitone, E. V. 1960 Surface waves. In *Handbuch der Physik* (ed. E. Flugge), vol. IX, pp. 446–758. Berlin: Springer.

Wereley, S. T., & Meinhart, C. D. 2010 Recent advances in micro-particle image velocimetry. In *Ann. Rev. Fluid Mechanics*, vol. 42, pp. 557–576. Palo Alto: Annual Reviews.

White, F. M. 2003 *Fluid Mechanics*, 5th ed. New York: McGraw-Hill.

White, F. M. 2006 *Viscous Fluid Flow*, 3rd ed. New York: McGraw-Hill.

White, F. M., Barfield, B. F., & Goglia, M. J. 1958 Laminar flow in a uniformly porous channel. *J. Appl. Mech.* **25**, 613–617.

Wiersma, P. H., Loeb, A. L., & Overbeek, J. T. G. 1966 Calculation of the electrophoretic mobility of a spherical colloid particle. *J. Colloid Interface Sci.* **22**, 78–99.

Wilson, A. H. 1948 A diffusion problem in which the amount of diffusing substance is finite. *Philos. Maga.* **54**, 48–58.

Wilson, S. D. R. 1982 The drag-out problem in film coating theory. *J. Eng. Math.* **16**, 209–221.

Xuan, Xiangchun, X., Bo, S., David, & Li, D. 2004 Electroosmotic flow with Joule heating effects. *Lab-on-a-Chip* **4**, 230–236.

Yoda, M. 2006 Nano-particle image velocimetry. In *Biomolecular Sensing, Processing and Analysis* (ed. Rashid bashir & Steve Wereley), pp. 331–348. New York: Springer.

Zaltzman, B., & Rubinstein, I. 2007 Electroosmotic slip and electroconvective instability. *J. Fluid Mech.* **579**, 173–226.

Zeman, L. J., & Zydney, A. L. 1996 *Microfiltration and Ultrafiltration: Principles and Applications*. New York: Marcel-Dekker.

Zhao, H., & Bau, H. H. 2007 On the effect of induced electroosmosis on a cylindrical particle next to a wall. *Langmuir* **23**, 4053–4063.

Zhu, W., Singer, S. J., Zheng, Z., & Conlisk, A. T. 2005 Electro-osmotic flow of a model electrolyte. *Phys. Rev. E* **71**, 041501.

Index

anion, 24
aqueous solutions, 14
artificial kidney, 484f
  centerline approximation, 487
  end-stage renal disease, 484
  hindered transport, 487
  influence of electric field, 489
  nanopore membranes as, 485
  sieving coefficient, 487
atomic force microscope, 284

Batchelor, George, 194
batteries, 270
Blasius boundary layer
  similarity analysis, 153
  singular perturbation problem, 153
Blasius boundary layer equations, 150
blood flow, 173f
boundary conditions, 106f
  apparent slip, 108
  electrostatics, 114
  at a free surface, 112
  heat transfer, 116
  heat transfer convection coefficient, 116
  mass transfer, 113
  mass transfer convection coefficient, 113
  no-slip, 109, 111
  slip condition, 110, 147
  slip length, 108, 110, 147
  surface tension effects, 113
  velocity, 107
boundary layer, 17
Brownian motion, 133, 196f
  diffusion coefficient, 200
bubble in a tube, 166f

carbon nanotubes, 494, 498
Casson fluid, 173
cation, 24
cells, 298f
  cell death, 298
  cell doctrine, 298
  cytoskeleton, 298
  cytosol, 299
  endoplasmic reticulum, 299
  eukaryotic cell, 298
  fatty acid, 300
  Golgi apparatus, 299
  hormones, 300
  insulin, 300
  ion channels, 300
  lipid bilayer, 298
  membrane transport, 301f
  mitochondria, 299
  nucleus, 299
  organic molecules, 299
  plasma membrane, 300
  prokaryotic, 298
  ribosomes, 299
  role of testosterone, 300
  steroid hormones, 300
chemical and biochemical sensing, 491f, 492, 493
  electrochemical sensors, 491
  atomic force microscopy, 494
  biological recognition elements, 493
  biosensor performance, 494
  ELISA, 494
  immunoassay, 493
  low Reynolds number vortices, 495
  nanopores and nanopore membranes used as, 496
  nucleic acid biosensors, 493
  optical detection methods, 494
  optical sensors, 491
  surface plasmon resonance, 494
chemical equilibrium, 236
chemical potential, 236f
chromatography
  gas, 330
  liquid, 330
classification of partial differential equations, 128f
  characteristic curves, 129
  elliptic equations, 128
  hyperbolic equations, 128
  parabolic equations, 128
colloid science, 29
colloidal systems
  van der Waals forces in, 31
  DLVO theory, 31

continuity equation, 88
continuum approximation, 75
Couette flow, 45
Crick, Francis, 285

Debye–Hückel approximation, 256
Derjaguin approximation, 275
dielectric constant, 25
dielectrics, 24
diffusion coefficient, 43, 48f
  human serum albimun in serum, 293
  Stokes–Einstein equation, 48
dimensional analysis, 117f
  dimensionless parameters, 118
  drag coefficient, 118
  dynamic similarity, 117
  fully developed flow, 119
DNA
  base pair, 286
  hydrogen bond in, 286
  role in synthesizing proteins, 286
  transcription, 286
DNA transport, 476f
  and biochemical sensing, 478
  comparison with experiment, 483
  current calculation, 482
  direction and surface charge density, 477
  mathematical model, 479
  through α-hemolysin pores, 476
  translocation, 476, 481
drug delivery
  compartment models, 412
  pharmacokinetics, 412
dynamic viscosity, 16, 45f

electric field, 214, 214f
  Coulomb's law, 214
  due to a charged plate, 217
  due to a charged wire, 216
  Gauss's law, 218
electrical conductivity, role in Joule heating, 343
electrical conductors, 24
electrical double layer, 26f, 231, 251f
  anode, 28
  asymmetric multivalent mixtures, 259f
  Boltzmann distribution, 254
  cathode, 28
  on a cylinder, 265
  Debye length defined, 27
  Debye–Hückel picture, 251
  electroosmotic pump, 28
  excluded volume effect, 253
  Gouy–Chapman picture, 252
  Helmholtz picture, 251
  history, 20
  inner Helmholtz plane, 253
  ionic strength, 27, 72, 239, 256, 264, 268, 308, 324, 345
  Stern layer, 27, 251
  triple layer model, 252
electrical insulators, 24
electrical permittivity, 54, 214
  relative permittivity, 54
electrochemical potential, 243f
  Donnan equilibrium, 244
  Nernst equation, 244
electrochemistry, 230f
  ζ-potential, 251
  ζ-potential liquid side view, 261
  ζ-potential solid side view, 262
  acids and bases, 244
  buffer solution, 244
  charge polarization, 251
  covalent bond, 231
  Debye length, 254
  Derjaguin approximation, 276
  electrical double layer, 251
  equilibrium constant, 245
  hydration of ions, 25, 31, 232
  hydrophilic molecules, 232
  hydrophobic molecules, 232
  neutral pH, 246
  polar molecule, 232
  Stern layer, 27, 251
  surface element integration, 277
electrokinetic phenomena, 306f
  capillary electrophoresis, 331
  discovery of electro-osmosis, 19
  electro-osmosis, 306
  electro-osmotic flow, 306
  electro-osmotic mobility, 311
  electromigration, 319
  electrophoresis, 306, 331
  electrophoretic mobility, 333
  electrophoretic velocity, 333
  gel electrophoresis, 331
  Joule heating, 342
  relationship between electromigration and electrophoresis, 331
  sedimentation potential, 306, 341
  streaming current, 340f, 347
  streaming potential, 306, 338f
electrokinetic remediation, 332
electrolyte solution, 14, 25f
  current and current density, 320f
  electrical conductivity, 267
  as electrical conductors, 24
  Lewis acid defined, 26
  Lewis base defined, 26
electro-osmotic flow, 307f
  in an annulus, 318f
  chromatography, 330
  current and current density, 316f
  dispersion in, 328f
  electro-osmotic pump, 3
  field flow fractionation, 330
  in nozzles and diffusers, 320f
  volume flow rate in, 9, 310
electrophoresis, 331f
  full numerical solution for the electrophoretic mobility, 336
  Henry's solution, 334

electrostatics, 213f
  Coulomb's law, 24
  current, 225
  current density, 225
  differential form of Gauss's law, 223
  dipole moment, 221
  electric dipole, 221
  electric field definition, 24
  electric potential, 220
  electrical conductance, 225
  electrical conductivity, 225
  electrical permittivity of a vacuum, 24
  electrical resistance, 225
  Gauss's law, 218
  image charge, 228
  induced dipole, 222
  Maxwell's equations, 226
  nonpolar molecules, 221
  Ohm's law, 225
  Poisson equation, 100
  Poisson's equation, 223
  polar molecules, 221
  surface charge density, 216
  volume charge density, 216
energy equation
  conduction heat transfer, 104
  derivation, 102f
equation of charge conservation, 101
Euler's equations, 16
experimental methods and Brownian motion, 133

Fick's law, 43
  history, 18
Fourier, Joseph, 17, 188
Fourier's law, 42, 103
  history, 17
Franklin, Rosalind, 285

Gauss's law, volume charge density in, 219
gels, 30
Gibbs function, 230, 236, 236f
  equilibrium constant and, 241
  role in equilibrium, 240
Grahame's equation, 261

Hamiltonian, 349, 419, 455
heat transfer, 180f
  forced convection, 22
  free convection, 22
  Graetz problem, 190f
  heat transfer coefficient, 186
  heat transfer resistance, 182
  in a channel, 182
  Joule heating, 183
  mean temperature, 186
  Nusselt number, 182, 189, 191, 206
  Peclet number, 189
  thermal entrance length, 184, 190
  thermally fully developed, 186
heat, mass, momentum, and electrical analogies, 123
hydration of ions, 25
  Albert Einstein's contribution, 234
hydrophilic surface, 51, 58, 108, 112
hydrophobic surface, 58, 108, 112, 148

ideal solution, 239
ill-conditioned matrix, 396
initial value problems
  Euler forward formula, 414
  Euler methods, 413
  modified Euler method, 414
  Taylor series method, 413
interpolation
  cubic splines, 367
  difference table, 361
  divided difference table, 363
  Lagrangian polynomial interpolation, 361
  linear interpolation, 360
  Newton interpolation formulas, 363
  truncation error, 364
ion channels, 7, 34, 301
  gating of, 302
  transmembrane potential, 302
ions, 14
  valence, 26

kinematics, 77
  Eulerian description of fluid motion, 78
  Lagrangian description of fluid motion, 78
  material derivative, 79
  pathlines, 80
  streaklines, 81
  stream function, 80
  streamlines, 79

Lab-on-a-chip, 4
law of mass action, 412
lubrication, 160f
  lubrication approximation, 162
  Reynolds equation, 164
lyophilic colloids, 30
lyophobic colloids, 30

MacLaurin series, 352
mass flow rate, 60
mass spectrometry, 34
mass transfer
  chemical reaction, 22, 26
  convection, 22
  diffusion, 22
  electrical migration, 99
  entrance length, 188
  mass averaged velocity, 98
  mass density, 94
  mass fraction, 95
  molar concentration, 94
  molar mass, 94

mass transfer (*cont.*)
  molarity, 94
  mole fraction, 95
  Nernst–Planck equation, 99
  species velocity, 98
  in a thin film, 192f
matched asymptotic expansions, 313, 344
micro total analysis systems, 4
molecular bonding, 233f
  atoms, 233, 449
  covalent bond, 233
  hydrogen bond, 232
  ionic bond, 233
  valence, 233
  van der Waals force, 233
molecular dynamics, 447f
  and electro-osmotic flow, 465
  Ewald sum, 460
  Lennard Jones potential, 453
  minimum image convention, 460
  periodic boundary conditions, 457
  Poiseuille flow, 468
  postprocessing, 465, 473
  for protein structure and conformation, 283
molecular simulations, 447f
  ab initio molecular dynamics, 450, 453, 471
  Born–Oppenheimer approximation, 450
  empirical potentials, 450, 452
  ensemble, 451
  Schröedinger equation, 449

nanopore membranes, 4, 33, 140
  for diabetes treatment, 499
  for nanofiltration, 499
  for reverse osmosis, 499
Navier–Stokes equations, 88f
  history, 16
  indicial notation, 90
  stress tensor, 90
  surface and body stresses, 83f
  vorticity, 86
Newtonian fluid, 16, 87
no-slip condition, history, 17, 107, 148
Non-Newtonian fluids, 126f
nucleic acids, 285
  chromosomes, 285
  DNA, 285
  genes, 285
  RNA, 285
numerical analysis, 348f
  Adams–Moulton methods, 418
  central difference approximation, 374
  curve-fitting, 370
  definite integral, 377
  derivative boundary conditions, 407
  Gauss Elimination, 386
  Gauss-Seidel Iteration, 391
  ill conditioned matrix, 396
  ill-conditioning and stability of linear systems of equations, 396
  indefinite integral, 377
  initial value problems, 411
  Jacobi iteration, 391
  Michaelis-Menten problem, 425
  numerical differentiation, 373
  Romberg integration, 384
  round-off errors, 350
  Runge-Kutta methods, 415
  sets of linear equations, 386
  Simpson's rules for integration, 379
  singular perturbation problem, 401
  solution of linear boundary value problems, 399
  solution of nonlinear ordinary differential equations, 403
  stability of initial value solvers, 424
  stiff initial value problems, 424
  successive overrelaxation, 392
  symplectic integration methods, 420
  Thomas algorithm, 396
  trapezoidal rule integration, 377
  tri-diagonal linear equations, 386
  tri-diagonal systems of equations, 393
  truncation error, 350
  validation, 435
  verification, 435
numerical differentiation
  backward difference approximation, 374
  forward difference approximation, 374
  central difference approximation, 374
  sloping difference formula, 375
numerical solution of partial differential equations, 430f
  elliptic equations, 431
  parabolic equations, 430
numerical solution of the boundary layer equations, 405
numerical solution of the Poisson-Nernst-Planck equations, 428f

Ohm's law, 101
osmotic pressure, 280

parallel plate viscometer, 45
pH, 26
pipe and channel flow, 141f
  entrance length, 142
  fully developed regime, 143
point-of-care systems, 492
Poiseuille flow, 143f
  history, 17
  solution derived, 144
  volume flow rate in, 10, 144
  volume flowrate in, 144
Poiseuille flow analogy, 224
Poisson equation, 100
Poisson–Boltzmann distribution, 255
polymers, 249
  biopolymers, 30, 250
  dimer, 250
  monomer, 249
  PDMS, 250
  PMMA, 250

tetramers, 250
trimers, 250
polysacchrides, 286
glucose, 287
porous media, 162
protein synthesis, 288
gene expression, 288
gene role in, 288
proteins, 287f
amino acids role in structure, 290
antibodies, 288, 295
antigen, 288, 295
apolipoproteinE, 294
binding site, 295
bioinformatics, 292
C-terminus, 290
conformation, 290
denatured, 290
enzymes, 288
function, 288
human insulin, 294
human serum albumin, 290, 292, 484
igG, 294
insulin sequencing, 291
interferon, 294
ion channels, 292
ligand, 295
lysozyme, 294
movement proteins, 289
N-terminus, 290
nutrient proteins, 289
peptide bond, 289
polypeptide backbone, 289
primary structure, 289
protein binding, 295
quartenary structure, 292
regulatory proteins, 289
secondary structure, 291
structural proteins, 289
structure, 289
tertiary structure, 291
transport proteins, 289
use of molecular dynamics simulations, 292

Redlich-Kwong equation of state, 353
Reuss, F. F., 19
RNA, 286

semipermeable membranes, 271f
concentration polarization, 275
current–voltage relationship, 273
induced charge electrical double layers, 275
limiting current, 274
silica surface, 246f
equilibrium constants, 247
isoelectric point, 247
Nernstian–Stern layer model, 249
polyethelene glycol treatment, 247
potential determining ions, 247
surface charge density, 248
silicon, 246
singular perturbation problem, 314
ssDNA, 285
Stokes flow past a sphere, 169f
drag coefficient, 172
history, 16
pressure, 171
Stokes–Einstein equation, 50, 171
Stokes' drag law, sedimentation, 172
streaming potential, history, 19
Sturm–Liouville problem, 191
suction flows, 155f
developing, 158
fully developed, 155
surface and body forces, 83f
surface charge density
liquid side view, 260f
related to potential, 43*f*
solid side view, 248f, 262
surface tension, 55, 55f
capillary number, 58
hydrophilic surface, 58
hydrophobic surface, 58
surfactants, 59
Weber number, 58
Young's law, 58

Taylor, G. I., 194
Taylor–Aris dispersion, 195f
dispersion coefficient, 197, 198
Taylor series, 351f, 413f
temperature and concentration boundary layers, 205f
thermal conductivity, 42, 52, 182f
thermal sciences, 22
thermodynamics, 40f, 62f
closed system, 63
equilibrium, 63
first law of thermodynamics, 65
and head loss, 69
and losses in pipe and channel flows, 68
open system, 63
perfect gas, 62
second law of thermodynamics, 66
thermodynamic property, 64
transport properties, 18, 41f

unsteady mass transport in a membrane, 201f

van der Waals forces, 31, 453, 456
Verlet, Loup, 422

Water, 231f, 452f
extended simple point charge model, 232
simple point charge model, 232
structure of, 231
well-posed problems, 130f

zerofinding, 353f
bisection method, 354
Newton's method, 354
secant method, 354